ANNUAL REVIEW OF BIOPHYSICS AND BIOMOLECULAR STRUCTURE

ANNUAL REVIEW OF BIOPHYSICS AND BIOMOLECULAR STRUCTURE

VOLUME 25, 1996

ROBERT M. STROUD, *Editor*
University of California, San Francisco

WAYNE L. HUBBELL, *Associate Editor*
University of California, Los Angeles

WILMA K. OLSON, *Associate Editor*
Rutgers, The State University of New Jersey

MICHAEL P. SHEETZ, *Associate Editor*
Duke University, Durham

http://annurev.org science@annurev.org 415-493-4400

ANNUAL REVIEWS INC. 4139 EL CAMINO WAY P.O. BOX 10139 PALO ALTO, CALIFORNIA 94303-0139

ANNUAL REVIEWS INC.
Palo Alto, California, USA

International Standard Serial Number: 1056-8700
International Standard Book Number: 0-8243-1825-0
Library of Congress Catalog Card Number: 79-188446

⊗ The paper used in this publication meets the minimum requirements of American National Standard for Information Sciences—Permanence of Paper for Printed Library Materials. ANSI Z39.48-1984.

TYPESET BY MARYLAND COMPOSITION CO., INC.
PRINTED AND BOUND IN THE UNITED STATES OF AMERICA

PREFACE

In 105 AD, Ts'ai Lun, an official of the Chinese Imperial Court, presented the Emperor Ho Ti with the first samples of sheets of pressed pulp: paper. Without pretension, his discovery is documented in the official history of the Han Dynasty. There is nothing more permanent nor more compelling than a printed page. Ts'ai Lun's invention is one of the longest-lived inventions still in current use: it continues archive and document the latest discoveries. But now, Ts'ai Lun's invention too turns a page. Paper, the essential precursor to printing, has been the preeminent means of communicating and archiving science and the history of ideas on our planet. Ts'ai Lun's discovery was recognized (remarkable in that period)—he became rich, received aristocratic title, and eventually, through his involvement in palace intrigue, took his own life.

Like its wise inventor, Ts'ai Lun's invention presages the superseding of its own self. After a full 1890 years, the printed page is rising beyond itself and seeding new networks at an explosive rate. Electronic communication and electronic publishing will soon replace even the largest current libraries with a handful of disks occupying the space of only a few printed volumes. The compaction ratio of disk storage volume to book volume, ever decreasing, is now one small 100-Megabyte disk to 100 books. A lifetime of a person's reading is held in the volume previously occupied by a comprehensive dictionary.

Electronic communication transforms the way we think about and do science. Another step in efficiency, another leap in horizons. The way we research each other's contributions and the format of the way citations are referenced in citation indices both need to be thought about and revised. Long-range collaborations thankfully become short-range. The *Annual Reviews* series will shortly be available in fully "electronic" form also. As we approach the end of the millennium, we hope that soon we will no longer need to recycle paper! Trees will grow again. And Ts'ai Lun's paper will probably stay with us through the 2000-year mark, as we will all still communicate, scratch out ideas, write poetry, paint, and write music on paper.

We are indebted to our author colleagues for intellectual focus and effort, and especially to Amanda Suver, Bonnie Meyers, and Naomi Lubick for their untiring efforts in the editorial office.

ROBERT M. STROUD
EDITOR

Annual Review of Biophysics and Biomolecular Structure
Volume 25 (1996)

CONTENTS

SOME RELATED ARTICLES IN OTHER *ANNUAL REVIEWS*

From the *Annual Review of Biochemistry,* Volume 65, 1996:

Structural Basis of Lectin-Carbohydrate Recognition, WI Weis, K Drickamer

Connexins, Connexons, and Intercellular Communication, DA Goodenough, JA Goliger, DL Paul

Relationships Between DNA Repair and Transcription, EC Friedberg

DNA Topoisomerases, JC Wang

Biochemistry and Structural Biology of Transcription Factor IID (TFIID), SK Burley, RG Roeder

Hematopoietic Receptor Complexes, JA Wells, AM de Vos

Mechanisms of Helicase-Catalyzed DNA Unwinding, TM Lohman, KP Bjornson

Protein Prenylation: Molecular Mechanisms and Functional Consequences, FL Zhang, PJ Casey

Structure and Function of the 20S and 26S Proteasomes, O Coux, K Tanaka, AL Goldberg

Electron Transfer in Proteins, HB Gray, JR Winkler

From the *Annual Review of Cell and Developmental Biology,* Volume 11, 1995:

Integrins: Emerging Paradigms of Signal Transduction, MH Ginsberg, MA Schwartz, MD Schaller

Biological Atomic Force Microscopy: From Microns to Nanometers and Beyond, Z Shao, J Yang, AP Somlyo

Protein Import into the Nucleus: An Integrated View, GR Hicks, NV Raikhel

The Nucleolus, PJ Shaw, EG Jordan

Heat Shock Transcription Factors: Structure and Regulation, C Wu

Control of Actin Assembly at Filament Ends, DA Schafer, JA Cooper

Receptor-Mediated Protein Sorting to the Vacuole in Yeast: Roles for Protein Kinase, Lipid Kinase, and GTP-Binding Proteins, JH Stack, B Horazdovsky, SD Emr

COPs Regulating Membrane Traffic, TE Kreis, M Lowe, R Pepperkok

Unconventional Myosins, MS Mooseker, RE Cheney

How MHC Class II Molecules Acquire Peptide Cargo: Biosynthesis and Trafficking Through the Endocytic Pathway, PR Wolf, HL Ploegh

From the *Annual Review of Genetics,* Volume 29, 1995:

Yeast Transcriptional Regulatory Mechanisms, K Struhl

Light-Harvesting Complexes in Oxygenic Photosynthesis: Diversity, Control, and Evolution, AR Grossman, D Bhaya, KE Apt, DM Kehoe

Trinucleotide Repeat Expansion and Human Disease, CT Ashley Jr, ST Warren

Membrane Protein Assembly: Genetic, Evolutionary, and Medical Perspectives, C Manoil, B Traxler

The Genetics of Proteasomes and Antigen Processing, JJ Monaco, D Nandi

Cystic Fibrosis: Genotypic and Phenotypic Variations, J Zielenski, LC Tsui

Homologous Recombination Proteins in Prokayotes and Eukaryotes, RD Camerini-Otero, P Hsieh

From the *Annual Review of Microbiology,* Volume 50, 1996:

What Size Should a Bacterium Be? A Question of Scale, AL Koch

Mechanisms of Adhesion by Oral Bacteria, PE Kolenbrander

The F0F1-Type ATP Synthases of Bacteria: Structure and Function of the F0 Complex, K Altendorf, G Deckers-Hebestreit

Osmoadaptation by Rhizosphere Bacteria, Karen J Miller, Janet M Wood

From the *Annual Review of Pharmacology and Toxicology,* Volume 35, 1995:

The Pharmacology of the Gastric Acid Pump: The H^+, K^+ ATPase, G Sachs, JM Shin, C Briving, B Wallmark, S Hersey

An Evaluation of the Role of Calcium in Cell Injury, AW Harman, MJ Maxwell

Molecular Strategies for Therapy of Cystic Fibrosis, JA Wagner, AC Chao, P Gardner

Nitric Oxide in the Nervous System, J Zhang, SH Snyder

The ARYL Hydrocarbon Receptor Complex, O Hankinson

Pharmacology of Cannabinoid Receptors, AC Howlett

P_2-Purinergic Receptors: Subtype-Associated Signaling Responses and Structure, TK Harden, JL Boyer, RA Nicholas

From the *Annual Review of Physical Chemistry,* Volume 46, 1995:

Collective Variable Description of Native Protein Dynamics, S Hayward, N Go

From the *Annual Review of Physiology,* Volume 58, 1996:

ANNUAL REVIEWS INC. is a nonprofit scientific publisher established to promote the advancement of the sciences. Beginning in 1932 with the *Annual Review of Biochemistry,* the Company has pursued as its principal function the publication of high-quality, reasonably priced *Annual Review* volumes. The volumes are organized by Editors and Editorial Committees who invite qualified authors to contribute critical articles reviewing significant developments within each major discipline. The Editor-in-Chief invites those interested in serving as future Editorial Committee members to communicate directly with him. Annual Reviews Inc. is administered by a Board of Directors, whose members serve without compensation.

1996 Board of Directors, Annual Reviews Inc.

Management of Annual Reviews, Inc.

Annu. Rev. Biophys. Biomol. Struct. 1996. 25:1–28

CIRCULAR OLIGONUCLEOTIDES: New Concepts in Oligonucleotide Design

Eric T. Kool

Department of Chemistry, University of Rochester, Rochester, New York 14627

KEY WORDS: triple helix, antisense, circular DNA, topologic modification, rolling circle, polymerase

ABSTRACT

Recent progress in the synthesis and properties of circular oligonucleotides as ligands for DNA and RNA and as templates for polymerase enzymes is described. Small synthetic circular DNAs, RNAs, and chimeric analogues ranging from 28 to 74 nucleotides in size have been synthesized with the use of a nonenzymatic ligation strategy. Some of these were designed to undergo triplex formation with single-stranded DNA and RNA targets, and many bind with affinities and sequence selectivities considerably greater than those seen for linear oligonucleotides. Design strategies and modes of binding are discussed in the light of possible use of such molecules as hybridization probes, molecular diagnostics, and sequence-specific inhibitors of gene expression. Small circular oligonucleotides have also been shown to act as unusually efficient templates for DNA and RNA polymerases, which produce long, repeating copies of the circular sequence by a rolling circle process.

CONTENTS

1056-8700/96/0610-0001$08.00

PERSPECTIVES AND OVERVIEW

DNA Oligonucleotides as Useful Biological Tools and Potential Therapeutics

The ready availability of synthetic oligonucleotides has led to the rapid development of molecular biological and medical diagnostic technologies for studying and identifying specific genetic sequences. The polymerase chain amplification reaction is just one example of this technologic advancement (46). Oligonucleotides are also used as nucleic acid hybridization agents and, when combined with labels or other identifying groups, can be used in the location of specific genes (68).

Currently, DNA oligonucleotides and analogues are also being studied vigorously worldwide as a new class of potential therapeutic agent (11, 73). Dozens of research groups have been constructing DNA and related analogues to inhibit expression of specific genes. The idea of targeting a specific disease-related genetic sequence is appealing. The implication is that a disease can be attacked at its molecular origins.

Further, a general approach to treatment of many diseases is possible, as most diseases have a genetic component.

Possible Limitations of Standard DNA Oligonucleotides

DNA is built from nucleotides joined by phosphodiester bonds. One chief problem that arises from this structure is that nuclease enzymes that degrade DNA exist in biological media. In human serum, a short linear oligodeoxynucleotide has a half-life of only approximately 30 min (13, 63, 86). This degradation necessarily limits the amount of active oligonucleotide that can exert its intended effect. Another potential limitation arises from the fact that the DNA backbone is charged and, therefore, may pass through cell membranes with less efficiency than might be desired (11, 73). Finally, standard DNA oligonucleotides may not bind complementary sequences tightly enough to have a biological effect, particularly if blocking of the progression of polymerase enzymes or ribosomes is desired (73).

In response to these perceived limitations, many research groups have focused on ways to modify the DNA backbone in an effort to slow degradation, increase cellular uptake, or increase binding affinity. Some workers have pursued the construction of noncharged (backbone-modified) DNA analogues (15, 47, 71), as a possible solution to some of these problems. Although degradation generally is not a concern for such analogues, cellular uptake is not necessarily enhanced [indeed, it may even be slowed (48, 87)]. In addition, although one or two of these analogues bind more tightly than does natural DNA under conditions of low ionic strength (22, 74), most lose their advantage in the presence of multivalent cations, such as Mg^{2+} and spermine, at levels found in the body.

Recent Advances Solve Some of These Problems

Recent studies have shown that DNA phosphodiester oligonucleotides can be modified to combat some of the above-mentioned problems. DNA oligonucleotides do, in fact, enter cells at a significant rate (87), and several approaches aimed at increasing uptake and aiding delivery are showing promise (1). In addition, the degradation of phosphodiester oligonucleotides can be slowed considerably by blocking one or both ends of the chain, because the primary degrading enzymes present in the body are of the exonuclease type (26, 63). To this end, modified DNA oligonucleotides that evade degradation by formation of a tight hairpin at the 3′-end have been reported (35). Finally, several research groups have demonstrated that by using a triple helical, rather than a simple double helical, structure, DNA oligonucleotides can be designed that bind their complementary target with much greater affinity and sequence selectivity than do the older, "standard" Watson-Crick DNAs (19, 27, 36, 52, 53). These recent advances, which may be useful in the improvement of diagnostic and therapeutic agents, are the focus of this report.

BEYOND WATSON AND CRICK: MULTIPLE STRANDS AND SHAPES

Higher Order Helices

The discovery that nucleic acids can form not only double helices (84) but also triple (24) and even quadruple (69) helical structures has opened the door to a variety of new ways to target single- and double-stranded DNAs and RNAs. Because this report addresses the properties of triple helical DNAs, a short review of their structure is appropriate. Figure 1 shows the common base triads found in triplex DNAs. There are two structural classes of triplexes: In the parallel motif (24, 45), a third strand (which is T,C-rich) binds a purine-rich (i.e. A,G-rich) site in Watson-Crick duplex DNA and lies parallel to the purine strand with which it makes contact. In the antiparallel motif (5, 21, 43), a third strand is G,A-rich or G,T-rich and binds a purine site in duplex DNA by lying antiparallel to the purine strand with which it makes contact. The hydrogen bonds between the third strand and the purine strand of the duplex are termed Hoogsteen bonds, and the third strand is sometimes called the Hoogsteen strand (Figure 1). The parallel motif requires protonation at N-3 of cytosine for greatest stability and, therefore, is usually stabilized by pH values less than neutral (20, 38); the antiparallel motif is largely insensitive to pH. In general, triplex DNAs maintain the right-handed turn of duplex DNA without distorting it greatly (2, 54).

Topologic Modification of DNA

Early on, the triplex structure was viewed as a way to target a site in duplex DNA, by simple construction of a third strand complementary to a target site in the duplex (45). Although this approach remains attractive for targeting duplex DNA, we can now devise ways to use modified DNAs to form a triplex structure when the target is single-stranded (see below). If only one strand of the triplex is considered the target, then the other two strands together can be considered a ligand for binding it (19, 27, 36, 48, 50, 52, 53). Topologic modification, such as linking the two strands by a loop (27, 66) or doubly linking two

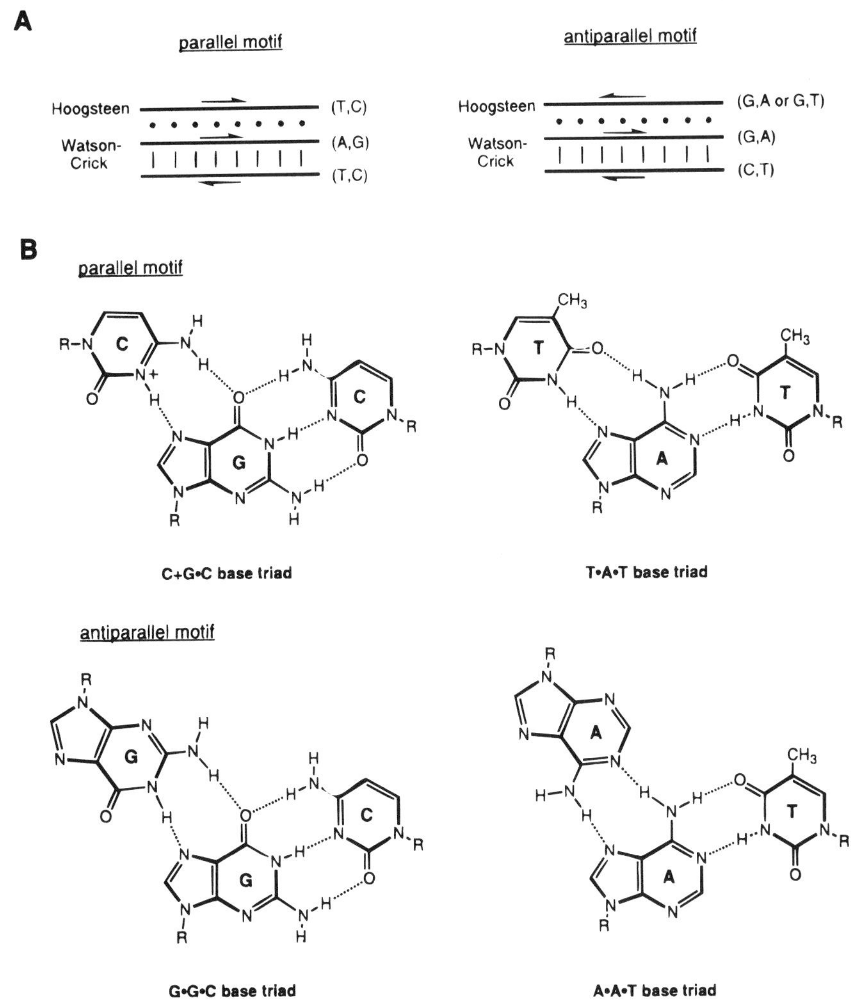

Figure 1 (A) Strand orientation and composition for the parallel and antiparallel motifs of triplex DNA. (B) Structures of the common base triads for both types of triplexes.

binding domains in a circular oligonucleotide (36, 52, 53), can result in DNA ligands that bind single-stranded targets with very high levels of efficiency.

Topologic modification is attractive because it is not limited to phosphodiester backbones. Most strategies outlined below, therefore, probably could be used to improve binding properties for many different backbone-modified DNAs, as well as for those composed of the natural linkage.

CIRCULAR OLIGONUCLEOTIDES

Self-Paired ("Dumbbell") Circular Oligonucleotides: Decoys for Proteins

Circular DNAs (or RNAs) that are internally self-paired in the Watson-Crick fashion were prevalent in the early literature of small circular oligonucleotides (3, 10, 17, 23, 41, 55, 85) (Figure 2). Such molecules essentially are duplex DNAs that are capped by nucleotide loops [or, more recently, nonnucleotide loops (55)] on either end and often have high values of melting transition temperature (T_m). This strong secondary structure makes this kind of circle simpler to construct than those that lack a strong internal structure, because a linear circle precursor folds itself into the nicked dumbbell shape in solution. Enzymatic or chemical ligation (3) yields the covalently closed molecule.

Small DNA dumbbells (~20–56 nucleotides in length) were first constructed to study the structure and energetics of DNA helices and hairpin loops (3, 17, 23, 85). Recently, it was proposed that certain of these molecules might serve as regulators of gene expression by acting as decoys for specific transcription factors (6a, 9a, 10). If the duplex portion of a dumbbell corresponds to the recognition sequence for a protein that regulates a specific gene, then the decoy, by being present at higher levels of concentration, could bind the protein and inhibit its action on the natural site. Because the molecule is circular and mostly in duplex form, it would be expected to be relatively resistant to degradation by nuclease enzymes in biological media (see below).

In addition to DNA dumbbells, self-paired circular RNA dumbbells also have been receiving attention recently (41). Such proteins as *tat* and *rev,* from the human immunodeficiency virus type 1, are known to be necessary for viral replication. As part of this process, these

A

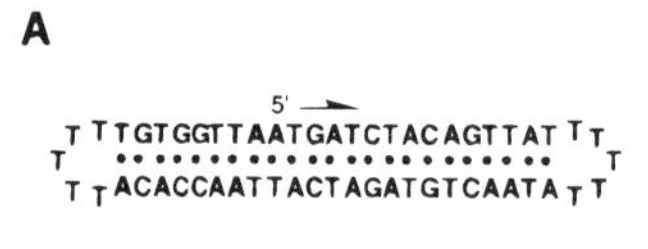

B

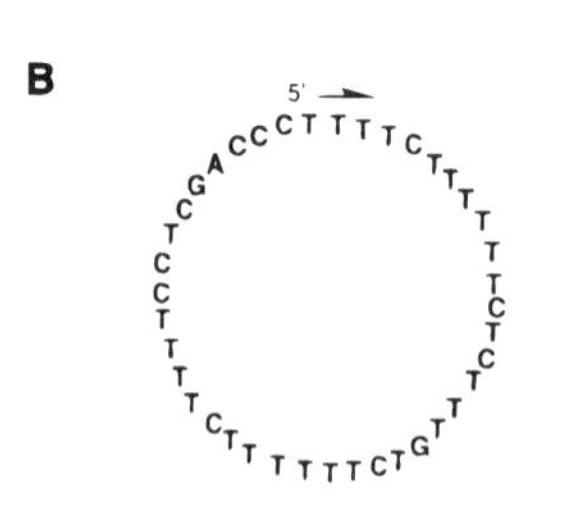

Figure 2 Examples of two classes of synthetic circular DNAs designed for biological targets. (A) DNA dumbbell (decoy DNA) containing a binding site for the Hepatocyte Nuclear Factor I (10). (B) Open circular DNA designed to form a triplex with a site in the human c-*myc* gene promoter (62).

proteins bind to specific double-stranded elements in the RNA viral sequence (the TAR element and *rev* response element, respectively). As with the DNA decoy strategy, dumbbell RNA molecules might bind and sequester such control proteins (41), thereby inhibiting viral replication.

Unpaired Circular Oligonucleotides

Until recently, unpaired circular DNAs (Figure 2) had been found rarely in the literature, probably because synthetic methods for preparing them were not well developed. The earliest studies focused on the synthesis of very small DNA circles, 2–14 nucleotides in length (7, 16). The first methods for cyclization relied on chance for closure of the bond; for the most part, therefore, larger circles were inaccessible. It was not until later that the use of separate DNA templates for closing circles was first demonstrated (36, 52) (see below). Template-directed circle closures for both DNA and RNA are now increasingly common in the literature (9, 18, 49).

Unpaired circular DNAs that do not have strong internal structure are free to form complexes with other nucleic acid sequences. We first described the strategy of using unpaired circular oligonucleotides as triplex-forming ligands for DNA in 1991 (36, 52). Since then, their binding affinities (53) and the effects of ionic strength and pH have been characterized (20). Their sequence selectivity of binding (36) and the kinetics of association and dissociation (78) also have been studied. We have examined many circle sizes (ranging from 26 to 74 nucleotides) and binding site sizes (from 4 to 32 nucleotides) (36, 62). In addition, much effort has been directed toward optimizing loop structures (6, 77) and lengths (64). Details of these studies are given below.

Synthetic Methods for Preparing Circular Oligonucleotides

Methods for the synthesis of circular DNAs and RNAs have been advanced as the need for these structures has increased. Several published studies have focused on nontemplated (random) cyclization of short synthetic oligonucleotides (7, 16). Both solid-phase and solution-phase methods have been investigated. In practice, cyclization is difficult to achieve entropically, as yields drop off rapidly with increasing length. To date, no one has successfully cyclized oligomers longer than approximately 14 nucleotides with the use of random approaches.

To overcome the entropic barrier to cyclization, we developed a simple, templated-directed strategy for the nonenzymatic construction of circular oligonucleotides from linear precursors that are phospho-

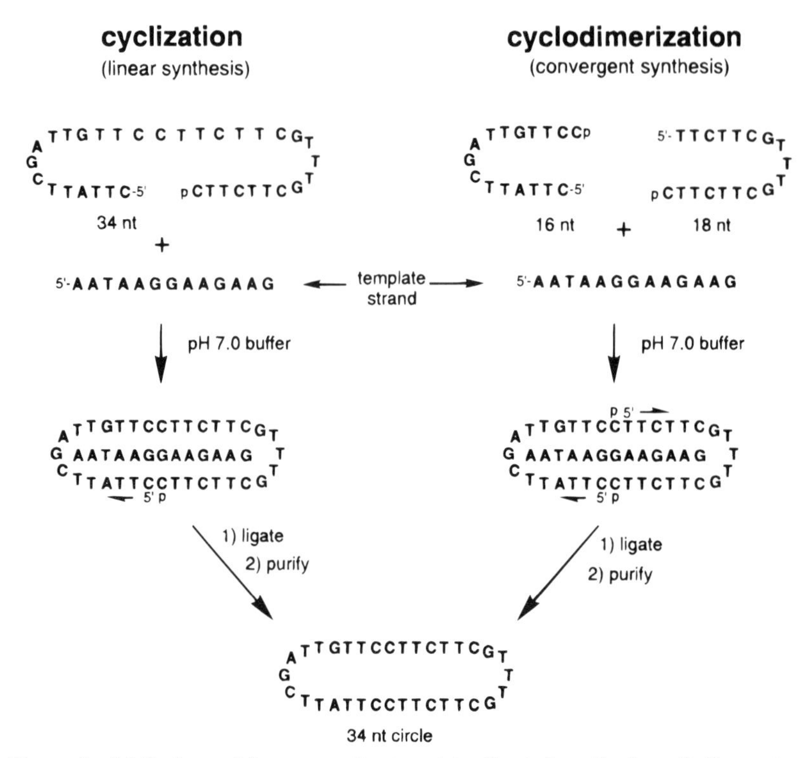

Figure 3 Methods used for preparative template-directed cyclization of oligonucleotides in aqueous solution. Conversion from linear to circular forms often is greater than 50–60% and is sometimes greater than 90%. The precursor strands contain a phosphate group at one end and a normal hydroxyl group at the other. Standard cyclization involves use of a full-length precursor along with a template to align the ends. Cyclodimerization uses two half-length oligonucleotides that are joined twice in one step to form a circle.

rylated on one end (Figure 3). Reagents, such as BrCN/imidazole (32), or a water-soluble carbodiimide, such as 1-Ethyl-3-(3-dimethylamino-propyl)-carbodiimide (3), can activate a phosphate for attack by a hydroxyl group. Addition of these reagents to an aqueous solution of a linear phosphorylated precursor of approximately 30 nucleotides, however, does not yield a cyclic product. Formation of very large macrocycles can be difficult (31), because the ends must find each other for reaction to occur. We solved this entropic problem with the use of a short oligonucleotide as a template that aligns the ends directly adjacent to each other (Figure 3).

A cyclization reaction takes a few hours to complete and can be carried out with the use of crude unpurified oligomers for circle precursor and template. Conversions from linear to circular forms are on the order of 50–95% (53, 62), depending on sequence, and the circular product is separated easily from the mixture by either preparative denaturing gel electrophoresis or ion exchange high-performance liquid chromatography.

Recently, a convergent approach has been developed in which two half-length oligomers are ligated twice in one step to yield a circular product (Figure 3) (62). This approach can be used to increase overall yields in the synthesis to be competitive with yields for synthesis of linear DNAs of the same size.

Characterization of a circular product requires different methods than those used for a linear sequence. For example, the use of standard sequencing would require that the circle first be nicked at a specific site. One common characterization of circles involves the comparison of relative electrophoretic mobility in gels of different percentages of polyacrylamide. For example, a 34-nucleotide circular sequence travels at 0.9 times the rate of a linear 34mer in a 20% denaturing polyacrylamide gel electrophoresis (53).

Perhaps the most reliable confirmation of circularity comes from exposure to endonuclease enzymes or chemical cleaving agents (53, 79). With a circular oligonucleotide, the initial cleavage by a single-strand–specific nuclease (such as S1) creates a distinctive product: instead of producing two fragments, as occurs with a linear sequence, it produces one linear product, with more rapid mobility than the circular structure. Another test for circularity of synthetic oligonucleotides is exposure to exonuclease-type activities, because circles are, by definition, completely resistant. For example, circles are uncleaved by T4 DNA polymerase, which carries a 3′ exonuclease activity, under conditions in which a linear 5′-phosphate precursor is cleaved completely to mononucleotides (53). We also have radiolabeled a linear precursor on the 5′ end with ^{32}P-phosphate and then cyclized it. The product then is resistant to removal of the label by calf alkaline phosphatase, whereas the uncyclized precursor has its label removed completely (53).

TRIPLEX FORMATION WITH SINGLE-STRANDED TARGETS

The Advantages of Connecting Two Binding Domains

In a DNA triple helix, the three strands are in contact with each other through specific hydrogen bonds. The central strand of the three is nearly always purine rich, because purines have two H-bonding faces.

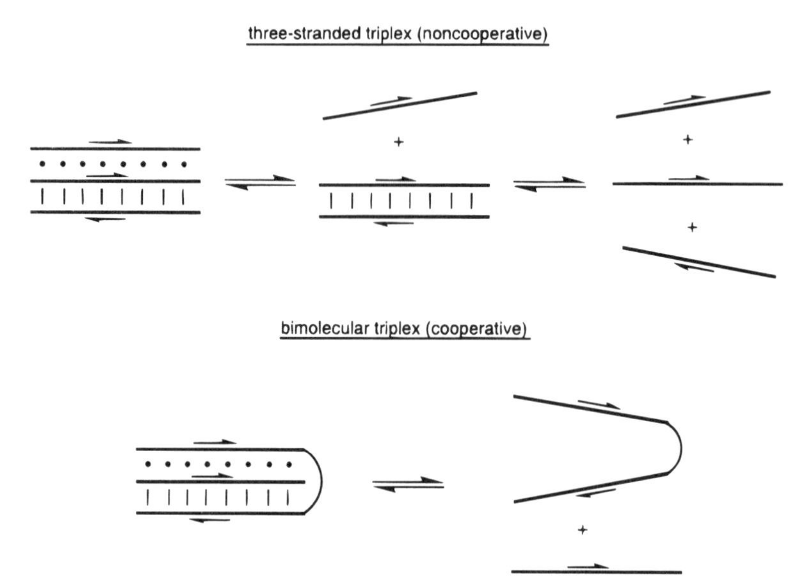

Figure 4 Termolecular and bimolecular association of DNA strands to form triple helices. Linking of two of the strands leads to greater cooperativity and tighter binding, largely because of the entropic benefit.

The other two strands bind the central strand but do not make direct H-bonded contact with each other (2, 54). As mentioned above, one can envision the formation of a triplex with a single-stranded target if two appropriate binding domains are used as a ligand. Although two physically separate DNA strands can be used, this approach has entropy working against it: The formation of one complex from three separate strands means net loss of translational and rotational entropy for two molecular entities (Figure 4). In practice, two separate strands usually bind to a target with no greater affinity than that of a simple Watson-Crick complement (36, 53), because the dissociation of the two binding interactions are independent events. Thus, such a triple-helical complex is noncooperative.

The independence of these two binding interactions can be overcome by linking two strands together (Figures 4 and 5). The ligand is then unimolecular, and the act of binding is bimolecular, as for Watson-Crick duplex formation. Such two-domain ligands often act cooperatively in their binding; thus, the two binding interactions can reinforce each other (19, 27, 36, 52, 53). This preorganization of the ligand (12) can result in higher levels of binding affinity and sequence selectivity, as well.

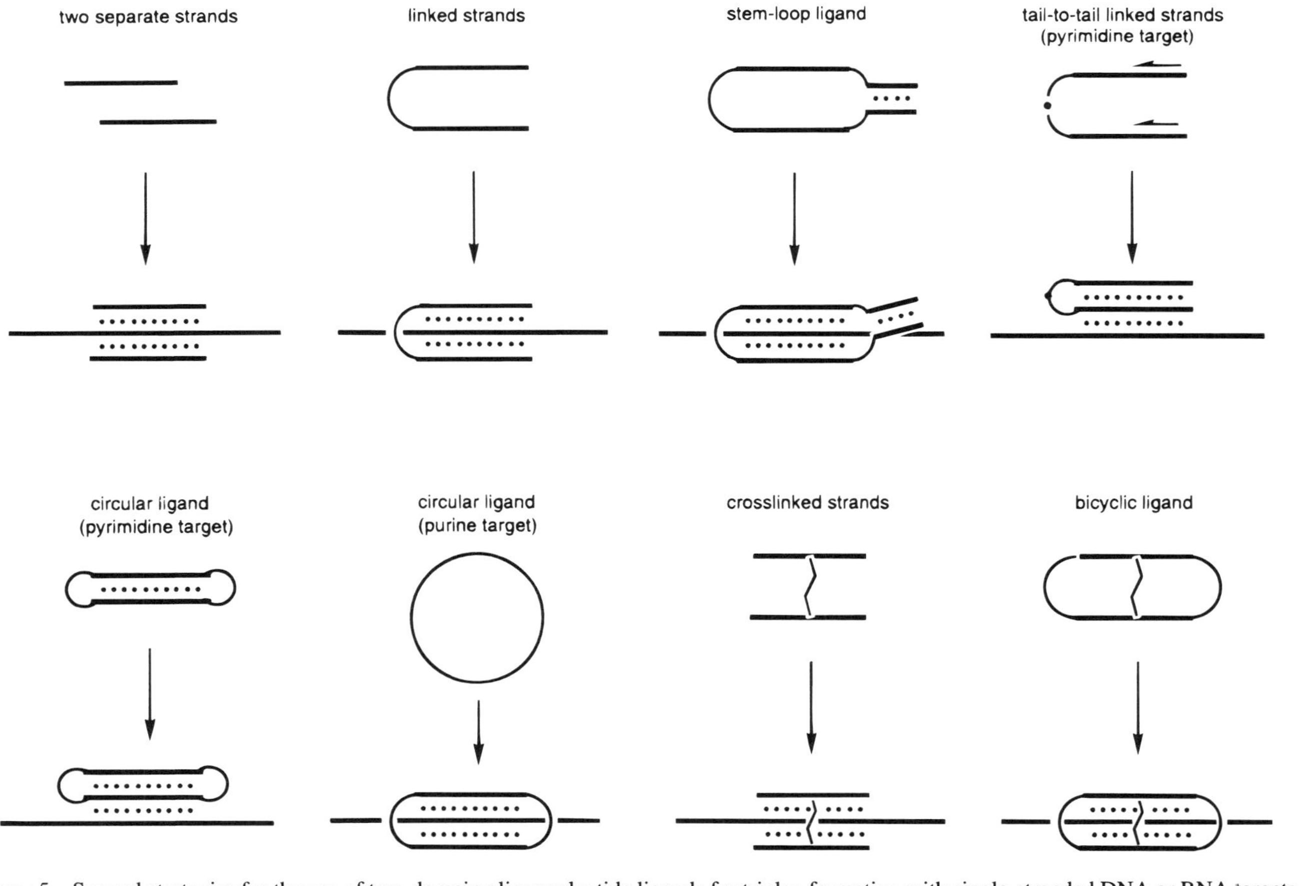

Figure 5 Several strategies for the use of two-domain oligonucleotide ligands for triplex formation with single-stranded DNA or RNA targets.

Several Structural Possibilities Exist

Since the first reports of the strategy of triplex formation on a single-stranded target in 1991 (19, 27, 36, 52, 53), numerous studies on the properties of such complexes and on several different molecular strategies have been published (4, 28–30, 33, 34, 70, 75, 78– 81). Figure 5 illustrates some of the ways in which two phosphodiester DNA sequences can be linked to form a two-domain ligand for a single-stranded target. Two binding domains can be joined by a nucleotide loop (19, 36, 52) or by nonnucleotide linkers of appropriate length (27, 64). The two domains can be linked in standard 5′ to 3′ orientation or in a head-to-head or tail-to-tail orientation (34). In addition, they can be linked across the center by a disulfide bond (8). All these approaches result in greater binding affinity than does the standard Watson-Crick approach. The highest levels of affinity arise when the two binding domains are linked at both ends (Figure 5), such as for a stem-loop oligonucleotide, in which a nucleotide loop bridges one end and Watson-Crick stem closes the other (19), and for a fully circular oligonucleotide, in which both ends of the domains are linked by nucleotide or nonnucleotide loops (36, 52, 53) (see below).

CIRCULAR OLIGONUCLEOTIDES AS TRIPLEX-FORMING LIGANDS FOR DNA AND RNA

High Binding Affinity

A simple comparison shows that circular oligonucleotides bind a complementary strand of DNA not only more tightly than does a standard Watson-Crick strand but also with greater affinity than does a singly linked, triplex-forming ligand (53). Figure 6 illustrates this effect. The complexes are formed at pH 7.0 in a buffer that contains 100 mM Na^+ and 10 mM Mg^{2+}, an ionic strength near physiologic conditions. Binding affinities are measured through thermal melting experiments, in which the complex is slowly heated while absorbance of the DNA (260 nm) is monitored. The T_m and free energies derived by computer fitting of the melting data allow for comparison of thermal and thermodynamic stabilities of the complexes.

In the experiment shown in Figure 6, the target strand is a purine-rich DNA dodecamer. A standard pyrimidine-rich Watson-Crick complement binds with a T_m of 44°C, whereas a ligand constructed from two pyrimidine binding domains linked by a loop has a greater T_m of 55°C. The closed circular version of this same sequence, however, binds with a T_m that is 18°C greater than the Watson-Crick complement and

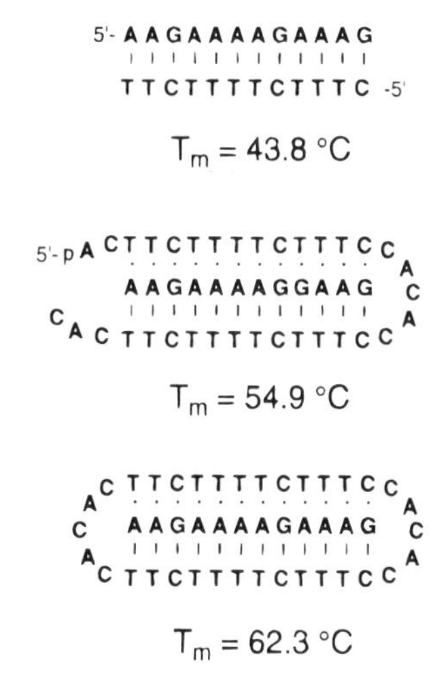

Figure 6 Comparison of binding properties of a standard Watson-Crick complementary oligonucleotide, a singly linked triplex-forming oligonucleotide, and a doubly linked circular triplex-forming compound, all targeted to the same sequence at pH 7.0 (100 mM Na^+, 10 mM Mg^{2+}).

a free energy that is 6 kcal/mol more favorable. This finding corresponds to a difference in association constant of 4 orders of magnitude (the K_d for the circle is ~10^{-11} M). This kind of binding advantage has held true for circles of different sequences and sizes (53, 62).

High Sequence Selectivity

The use of DNA ligands that contain two binding domains not only increases affinity but also can result in greater sequence selectivity. This finding was first observed with circular DNAs in 1991 (36) and has since been characterized by its relative association and dissociation rates for binding (78). Sequence selectivity is, of course, important in practical application, because the human genome contains three billion base pairs of sequence and, therefore, offers many closely related sequences as potential targets (11, 73).

Experiments that show that circular oligonucleotides have increased selectivity relative to linear strands are illustrated in Figure 7. One centrally located nucleotide in a target 12mer DNA was varied to test the effect of a mismatch. We define selectivity as the difference in binding affinity for the complementary target relative to mismatched ones. The results show that, although a linear oligonucleotide has selectivity against mismatches of approximately 4 kcal/mol, a circular ligand has much greater selectivity (among the same targets) of 6–8 kcal/mol range of free energies (36, 78). Recent studies indicate that the increase in selectivity arises from both the circular preorganized structure and the protonation of cytosines in the Hoogsteen strand (78).

complex (X = A,T,G,C)	X	$-\Delta G°_{25}$ (complex)	selectivity	half-life for dissociation
	A	11.9 kcal•mol^{-1}	--	85 sec
3'-TTCTTTC TTTT C	G	10.0	1.9 kcal•mol^{-1}	2.9 sec
5'-AAGAAXGAAAAG	T	9.0	2.9	0.5 sec
	C	8.9	3.0	0.7 sec
A C TTCTTTC TTTT CC A	A	18.1 kcal•mol^{-1}	--	27 days
C AAGAAXGAAAAG C	G	11.5	6.6 kcal•mol^{-1}	27 sec
A C TTCTTTC TTTT CC A	T	10.6	7.5	12 sec
	C	11.1	7.0	41 sec

Figure 7 Measured sequence selectivities for a standard Watson-Crick DNA oligonucleotide, as compared with that for a circle binding the same series of target strands. Also, calculated half-lives for dissociation of the complexes at pH 7.0, 37°C (78).

Circular oligonucleotides clearly can have not only a high level of binding affinity but also a high level of sequence selectivity. Both properties arise from the same modification; many strategies have been described for increasing the binding affinity of oligonucleotides (73, 74), but few have been demonstrated to increase selectivity simultaneously. Increasing the level of binding affinity alone will result in tighter binding of undesired genetic targets.

Kinetics of Binding

High-binding affinity in a bimolecular complex can arise from a rapid rate of association, a slow rate of dissociation, or some combination of the two. Bimolecular association rates between short DNAs that form a duplex have been measured in several studies (72); interestingly, the rates apparently do not depend on the sequence or length of the DNA or on whether there is a mismatch in the sequence. Rate constants for duplex association are commonly in the range 10^6 to 10^7 $M^{-1}sec^{-1}$ (72). Because mismatches decrease the equilibrium binding constant, this finding implies that mismatches cause increased dissociation rates rather than affect association.

With the use of stopped-flow methods, we have studied the association rates for circular DNAs and RNAs binding complementary and mismatched nucleic acid strands (78). Triplex formation among these species occurs with rate constants that are virtually the same as those measured for simple duplex formation; our measured triplex rate constants are in the range of 10^6 to 10^7 $M^{-1}sec^{-1}$. This finding is unusual for triplex formation: Third-strand binding to a preformed duplex is relatively slow, with rate constants 2–4 orders of magnitude less than the above-mentioned rates (42, 58).

We also found that, as for duplex DNAs, the presence of mismatches in the target strand did not affect the association rate significantly (78). Dissociation rates (derived from binding constants and association rates) were calculated for these complexes (Figure 7), and mismatches caused a large increase in dissociation rate. For example, a DNA circle complexed to a complement has a half-life for dissociation (37°C) of almost 1 month; with a single mismatch in the target, however, this half-life decreases to 12–41 s. Circular DNAs and RNAs, therefore, bind with very long half-lives to their designed targets. Because of their high level of selectivity, however, they dissociate from mismatched targets relatively rapidly.

Effects of RNA vs DNA Backbone

We have recently constructed circular oligonucleotides that consist of an RNA backbone (79). Interestingly, circular RNA oligonucleotides have some very different binding properties than do DNA circles of the same sequence (see Figure 8). Circular DNAs can form tight triplexes (as discussed above) with DNA target strands; with RNA target strands, however, a DNA circle is bound only by the Watson-Crick domain and not by the Hoogsteen side. This finding is consistent with studies that show dramatic differences in stabilities of triplexes, depending on whether they are constructed from DNA or RNA strands (or combinations thereof) (57).

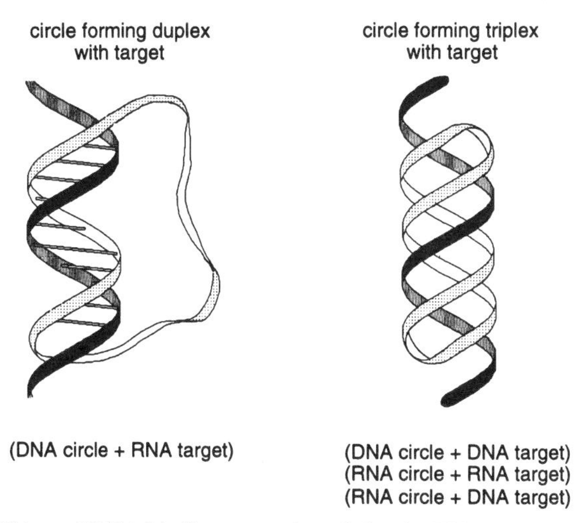

Figure 8 DNA- and RNA-binding properties of circular RNA and DNA oligonucleotides that have the same sequence. Three cases form strong triplexes, whereas circular DNAs form weaker, duplex-type complexes with RNA targets (80).

Circular triplex-forming RNAs, however, can bind both DNA and RNA targets with similar, high levels of affinity (Figure 8). We have shown recently that chimeric circular oligonucleotides that contain part DNA and part RNA structure can also form tight triplexes with RNA targets (81). Moreover, the methyl group of thymine (present in DNA and absent in RNA) is responsible for a good deal of stabilization in the triplex (82), whereas the 2′-hydroxyl of RNA has much to do with many of the differences in binding preference for the two backbones.

One interesting aspect of the differences in circle complexes with DNA and RNA is the high level of selectivity of a DNA circle for a DNA target over an RNA target (82). This feature leads to the possibility of choosing which type of polynucleotide to bind even if both are present. A normal Watson-Crick DNA strand shows very little selectivity over the two types of target.

Further Preorganization of DNA Structure: Bicyclic Oligonucleotides

The linking of two separate binding domains leads to a DNA ligand that binds more cooperatively and, thus, more tightly. Further, this preorganization of an oligonucleotide ligand also can lead to a higher level of sequence selectivity. Because we have shown that two such linkages between the opposing binding domains (i.e. a circular ligand) can lead to better advantages than one, it would seem that further preorganization (rigidification) might improve DNA-binding properties even more.

To this end, we examined molecular models of triple helical complexes to see whether linking of the two pyrimidine strands in a parallel-motif triplex might be possible. The models indicated that a short linking chain might bridge from one pyrimidine strand to the other without disrupting the binding of a central purine-rich target strand. One possible linkage might be a disulfide bond formed between two C-5-thiopropyne–modified uracil bases (Figure 9) (8). Such a cross-link could be used to join two binding domains near their center (or at other positions), rather than at the end, where most loops are positioned. The formation of a cross-link in a circular oligonucleotide would lead to formation of a bicyclic oligonucleotide. Interestingly, analogous disulfide–cross-linked bicyclic peptide antibiotics that bind to specific sequences of duplex DNA are known (76).

Recently, we synthesized a novel thiopropyne-modified uracil nucleoside and have further incorporated it into DNA strands intact (8). This cross-linking base can pair normally and without destabilization with adenine in DNA. In addition, two such groups in opposite pyrimidine

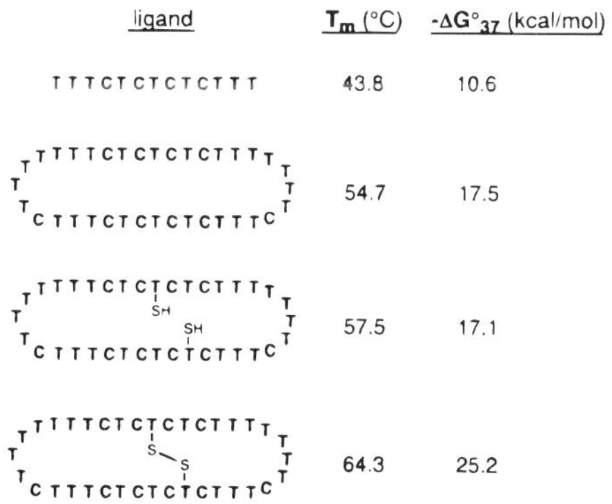

ligand	T_m (°C)	$-\Delta G°_{37}$ (kcal/mol)
TTTCTCTCTCTTT	43.8	10.6
TTTTTCTCTCTCTTTTT / T T T T / CTTTCTCTCTCTTTC (circle)	54.7	17.5
TTTTTCTCTCTCTTTTT / T SH SH T T T / CTTTCTCTCTCTTTC (circle)	57.5	17.1
TTTTTCTCTCTCTTTTT / T S—S T T T / CTTTCTCTCTCTTTC (bicycle)	64.3	25.2

Figure 9 Binding properties of a bicylic disulfide-linked oligonucleotide (pH 7.0) (8) to the oligopurine target d(AAAGAGAGAGAAA), as compared with unlinked and unmodified circles and a standard Watson-Crick oligonucleotide.

strands in a triplex can be cross-linked efficiently in aerobic conditions. This linking leads to tighter binding than can be achieved simply by linking two domains by a nucleotide loop between them. Finally, we constructed a bicyclic oligonucleotide that has such a disulfide bridge (Figure 9) (8). This compound binds its target with affinity considerably greater than that of an unmodified circular DNA. In addition, as was hoped, the added preorganization improves selectivity, thereby giving measured sequence selectivities of 10–12 kcal/mol against a single mismatch (8). The level of selectivity of this bicyclic compound is among the highest ever measured for a DNA-binding ligand.

Expanding the Range of Possible Target Sequences

The triplex-forming oligonucleotides described to this point were all designed to bind a purine-rich strand of DNA or RNA. That is a general limitation of all known triplexes: Hoogsteen bonds (involving natural DNA bases) can be formed only when the acceptor strand is purine rich. Third-strand binding to duplex DNA occurs when the target contains all or mostly purines. Similarly, until recently, two-domain oligonucleotide ligands could bind only to a purine-rich single strand (19, 27, 36, 52, 53). We recently showed, however, that a two-domain oligonucleotide that can target pyrimidine-rich strands can be constructed by using the antiparallel triplex motif, thus complementing the previous strategy (80).

Figure 10 illustrates this newer approach. In the antiparallel motif, the third (Hoogsteen) strand and the central strand are purine rich, and the far strand is pyrimidine rich. If we view the last strand as the target,

target strand complex	T_m (°C)	$-\Delta G°_{37}$ (kcal)
3'-GAGGAGGGAGGA 5'-CTCCTCCCTCCT	56.9	14.2
ACGTGGTGGGTGGT C ACGAGGAGGGAGGA-5' 5'-CTCCTCCCTCCT	68.4	16.9
ACGTGGTGTGTGGT C x ACGAGGAGGGAGGA-5' 5'-CTCCTCCCTCCT	61.9	12.4
ACGTGGTGGGTGGTCA C C ACGAGGAGGGAGGACA 5'-CTCCTCCCTCCT	71.0	17.3

Figure 10 The binding of single-stranded pyrimidine sequences by singly and doubly linked purine-rich oligonucleotide ligands.

we then can use two purine-rich strands as the ligand. In analogy to our previous experiments, it seemed that connecting the two binding domains with one or two loops might lead to further enhancement of binding properties.

Indeed, this strategy is successful: A singly looped, purine-rich oligonucleotide binds a 12mer pyrimidine complement (pH 7.0, 10 mM Mg^{2+}, 100 mM Na^{+}) with a T_m 12°C higher than does a standard Watson-Crick oligomer (80) (Figure 10). The addition of a second loop (thus making a purine-rich circular DNA ligand) adds another 3°C to the T_m of the complex. Interestingly, the binding of RNA works as well as DNA in this new strategy. The level of sensitivity to backbone differences apparently is lower in this case than in those examined previously. The addition of this second strategy effectively doubles the number of biological targets that can be bound by triplex formation.

In three other recent reports, related tactics for the binding of pyrimidine targets have been explored (34, 56, 67, 75). In one of these reports, a somewhat different approach was taken: The parallel motif was used to accomplish the binding of pyrimidines by making two-domain phosphodiester ligands that have an unnatural connection between the domains (34). If a pyrimidine-rich domain is linked to a purine-rich domain by a tail-to-tail (3′-3′) connection, it reverses the directionality of the strand midway. Such a two-domain ligand can also bind a complementary pyrimidine target strand with enhanced affinity, especially when the target is RNA (see Figure 5).

Multisite Binding by Conformational Switching

There has been some interest recently in the use of combinations of oligonucleotides to inhibit gene expression (39, 44). The binding of more than one site can sometimes lead to synergistic effects in inhibition. In addition, when a viral sequence is targeted, the virus could escape inhibition by mutating the target site for oligonucleotide binding. If more than one site is bound simultaneously, however, such an escape by mutation becomes more difficult (39).

With these ideas in mind, new strategies were developed for multisite DNA recognition. In one approach, Rubin et al (61) designed a circular oligonucleotide molecule to bind more than one sequence by switching conformation. This strategy involves the alternating use of binding domains and loop domains in circular triplex-forming oligonucleotides (Figure 11). By using one or the other of a set of binding domains, such a ligand can bind tightly and with a high level of selectivity to multiple sequences of DNA. An early report described a cyclic 36mer that can bind two different 9-base sequences (61). A later study showed that the complexity of this approach could be expanded; in that case, a cyclic 35mer was shown to bind six different 8-base sequences of DNA by switching conformations (Figure 11) (60).

Binding of Duplex DNA

Studies in which a fluorescent-labeled, short DNA duplex was used showed that a circular triplex-forming DNA complementary to one of the two strands could bind it by displacing the other strand from the duplex (51). Interestingly, this exchange of strands occurs much more rapidly than does simple dissociation of the duplex; thus, the circle acts as a catalyst for unwinding the DNA. Kinetics measurements indicated that the mechanism involves an intermediate step in which the circle is bound to the duplex.

The binding of duplex DNA is interesting because transcription of specific disease-related genes could be inhibited (73) and because many medical diagnostic methods involve the identification of specific sequences of duplex. More recent studies with relatively long duplex DNAs (both synthetic and plasmid derived) indicate that circular triplex-forming oligonucleotides can, in fact, bind to complementary sites in a duplex at neutral pH (K Ryan and ET Kool, in preparation). Preliminary results have shown that at least two different types of complexes are formed; studies are under way to characterize the binding energetics, kinetics, and structure.

Resistance to Degradation in Biological Media

As mentioned previously, blocking the ends of a phosphodiester oligonucleotide can slow its degradation in some biological media, because the primary degrading activity is an exonuclease enzyme that degrades

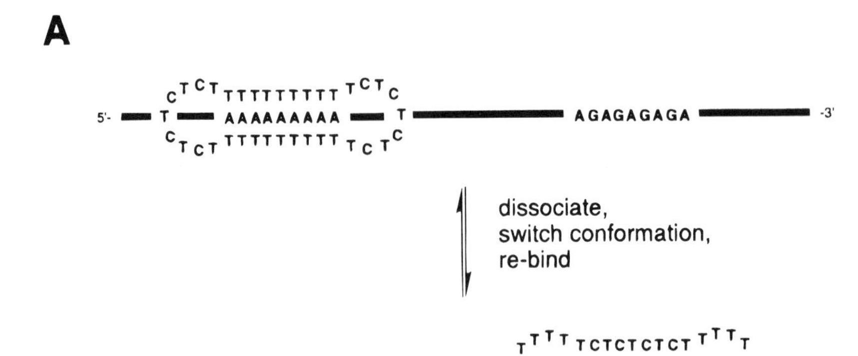

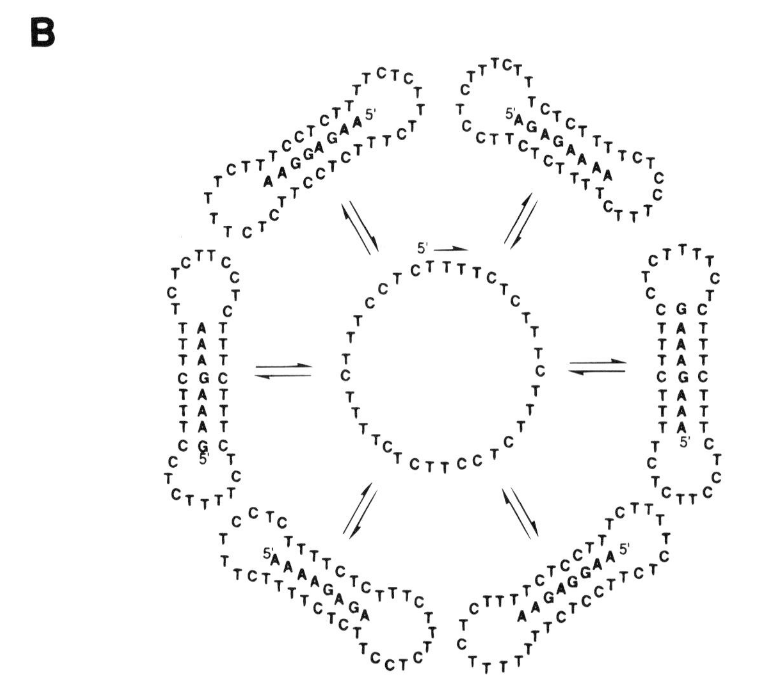

Figure 11 Molecular strategies for multisite recognition of DNA. A single circular DNA molecule can be designed to switch conformations and bind more than one sequence. (A) A cyclic 36mer that binds two 9mer sequences of DNA (61). (B) A cyclic 35mer that binds six different 8mer sites by conformational switching (60).

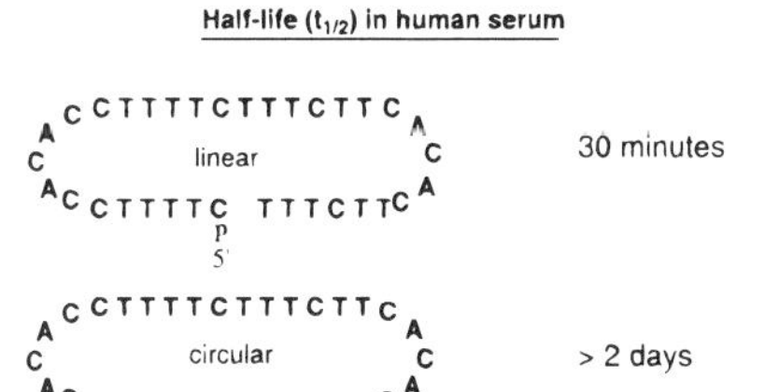

Figure 12 The relative stability of linear and circular DNA oligonucleotides in undiluted human serum at 37°C, as measured by analytic gel electrophoresis (63). Half-life is given for the first cleavage event.

DNA mostly from the 3′ end (26). Circular oligonucleotides have no ends; by definition, therefore, they almost are not substrates for exonuclease degradation. A simple experiment (63) illustrates this effect (Figure 12).

A linear phosphodiester oligonucleotide, 34 bases long, was incubated in fresh undiluted human serum at 37°C. Its structure was monitored by analytic gel electrophoresis; for comparison, the same oligonucleotide sequence was also tested in circular form. This sequence is highly pyrimidine rich and is unlikely to form any sort of duplex-type structure with itself.

The results showed convincingly that the linear oligonucleotide was unstable, whereas the circular version was stable. In fact, although linear DNAs commonly are degraded in minutes in human serum, this circular DNA was completely stable for at least 2 days (63). It remains to be seen whether this advantage in stability also is operative inside cells. One study carried out in mice showed that a circular DNA was apparently susceptible to endonuclease cleavage (67a). Interestingly, this immunity of circular structure to degradation apparently does not extend to RNA: We have found in preliminary experiments, again in human serum, that circular and linear RNAs are degraded considerably more rapidly than even linear DNAs are (S Wang and ET Kool, unpublished data).

BIOLOGICAL EFFECTS OF TRIPLEX FORMATION

Inhibition of DNA and Protein Synthesis

Because the strategies described in this report have been developed only within the past 5 years, few reports of biological activity for such ligands exist currently. One study described the properties of a two-domain ligand linked by nucleotides and carrying an intercalator (28).

The ligand was tested for inhibition of DNA synthesis by DNA polymerase I, and its binding site was situated downstream from the initiation site for the enzyme. The results showed that the triplex-forming oligonucleotide caused the polymerase progression to stall, whereas a Watson-Crick complement to the same site had less effect. Similar results were seen recently with triplex-forming, purine-rich ligands targeted to a pyrimidine site in DNA (67).

A second case involved the use of two physically separate, purine-rich oligonucleotide strands to bind a pyrimidine target in messenger RNA near the translation start codon (56). These compounds were constructed with the methylphosphonate backbone modification. The data from in vitro translation studies showed that the triplex-forming oligomers inhibited translation efficiently, whereas a standard Watson-Crick complementary phosphodiester oligomer had little or no effect.

Circular DNAs in Cell Culture

We have reported preliminary findings that involve the use of circular DNA oligonucleotides targeted to the *bcr/abl* mutation junction in cell culture (59). This mutation occurs in chronic myeloid leukemia (CML). With collaborators, we examined effects of linear phosphodiester DNAs, linear phosphorothioate DNAs, circular antisense compounds, and circular scrambled controls on the growth of the K562 and BV173 CML cell lines. All compounds were added only once at the beginning of the experiment, in concentrations of 1–32 μM and with no added enhancers of cellular uptake. Results showed that the circular antisense compounds inhibit cell growth at a 50% level at 8 μM, whereas circular controls and both forms of linear compounds have no effect on growth, even at 32 μM, under these conditions. The circular antisense compounds in the study were targeted to RNA and, thus, were not expected to bind tightly. Nonetheless, an advantageous effect, which may arise in part from added resistance to degradation, has been noted. Additional studies currently are under way.

ROLLING CIRCLE RNA/DNA SYNTHESIS: CIRCULAR OLIGONUCLEOTIDES AS TEMPLATES FOR POLYMERASES

The studies described above establish that circular DNAs can be useful as ligands for binding specific sequences of DNA and RNA and for proteins. These compounds therefore serve as efficient recognition elements. DNA in nature, however, is more than just a recognition element; it also encodes information for replication. Interestingly, recent reports have shown that small circular oligonucleotides can also encode information, by serving as templates for polymerase enzymes (14, 25). These findings may result in useful new methods for producing specific DNA and RNA sequences.

Our laboratory began testing synthetic circular DNAs as templates for polymerases in 1992. The first studies (37; D Liu, SL Daubendiek, MA Zillmann, and ET Kool, submitted) were carried out with the use of circles that were 34 nucleotides in length and were originally designed as DNA-binding ligands. We found that, by adding a polymerase (the Klenow enzyme), a primer, and the four deoxynucleotide triphosphates to a circle in buffer solution, long DNAs were being synthesized (Figure 13). We subsequently showed that these products were long multimeric repeats of the circle-encoded sequence and termed the process "rolling circle DNA synthesis." Recently, Fire & Xu (25) confirmed this finding. They showed that circles 34–52 nucleotides in size can be used as templates for *Escherichia coli* DNA polymerase I. We have now established that T4 DNA polymerase also carries out this reaction efficiently and that circles as small as 26 nucleotides can behave as templates (D Liu, SL Daubendiek, MA Zillmann, ET Kool, submitted).

Perhaps the most surprising part of this finding is the fact that the circles serve as efficient enzyme substrates despite their small size and curvature. Models of a 28mer DNA circle that take into account the extended nature of a single-stranded oligonucleotide indicate that it has a diameter of approximately 45Å, which is considerably smaller than most polymerase enzymes themselves. The polymerase is apparently

Rolling circle DNA synthesis

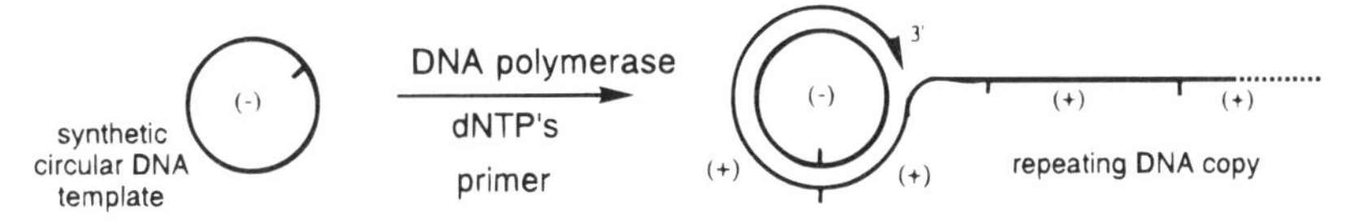

Rolling circle RNA synthesis

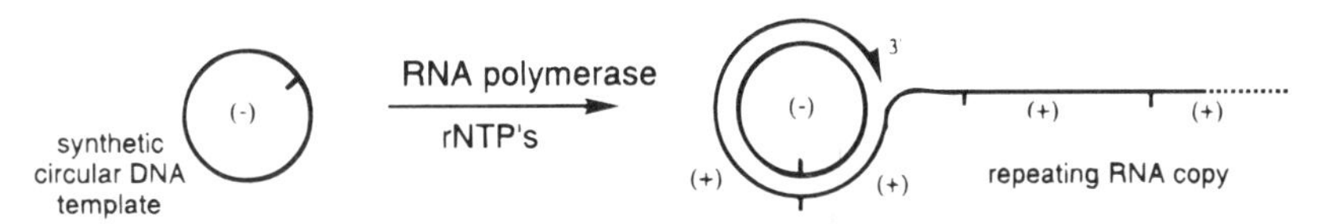

Figure 13 The synthesis of long DNA or RNA multimer strands with the use of small synthetic circular DNAs as catalytic templates.

not affected adversely by the tendency toward curvature in these substrates; indeed, the curvature may aid the process by helping unwind the DNA strand that is being synthesized from the template.

Just as interesting is the related finding that similar DNA circles can also behave as templates for RNA synthesis by the enzyme T7 RNA polymerase (Figure 13) (14). Although the circles contain no conserved RNA promoter sequences, at least some circles can serve as highly efficient templates. The products are, as with DNA synthesis, long (~9,000 nucleotides) repeating multimers of RNA, encoded by the circle sequence. In analogy to the above reaction, we use the term ''rolling circle RNA synthesis'' (14). It remains to be shown how general this rolling synthesis of RNA is, i.e. will there be restrictions on the operative sequences and structures? On the whole, it appears to be a novel way to generate specific RNA sequences in amplified amounts.

CONCLUSIONS

These studies establish that topologic modification of oligonucleotides can lead to large and advantageous differences in their properties. By careful design of oligonucleotide sequences, it is possible to design linked and circular DNA- and RNA-binding ligands that have very high levels of affinity, sequence selectivity, and resistance to degradation. These results can be acheived by modification of the topology of the DNA rather than the chemical makeup of its backbone. There is no particular reason, however, why such topologic modifications cannot also be carried out with backbone-modified DNAs, if desired. In addition, the topologic change in DNA from linear to circular can result in small synthetic molecules that behave as efficient templates for polymerase enzymes. Some of these strategies for structural and functional modification likely will find their way into oligonucleotide analogues that act as diagnostic tools or as catalysts for nucleic acid synthesis or have useful biological activity. These possibilities await future studies.

ACKNOWLEDGMENTS

I thank my coworkers, whose names are listed in the references, for their effort and enthusiasm, the National Institutes of Health (GM46625), and the Army Research Office for support. I also acknowledge a Beckman Foundation Young Investigator Award, an Office of Naval Research Young Investigator Award, a Dreyfus Foundation Teacher-Scholar Award, an Alfred P. Sloan Foundation Fellowship, and an American Cyanamid Faculty Award.

Literature Cited

1. Akhtar S, ed. 1995. *Delivery Systems for Antisense Oligonucleotide Therapeutics*. Boca Raton, FL: CRC
2. Arnott S, Selsing E. 1974. Structures for the polynucleotide complexes poly(dA)poly(dT) and poly(dT)(dA)poly(dT). *J. Mol. Biol.* 88:509–21
3. Ashley GA, Kushlan DM. 1991. Chemical synthesis of oligodeoxynucleotide dumbbells. *Biochemistry* 30: 2927–33
4. Bandaru R, Hashimoto H, Switzer C. 1995. An inverted motif for oligonucleotide triplexes: adenosine-pseudouridine-adenosine. *J. Org. Chem.* 60: 786–88
5. Beal BA, Dervan PB. 1991. Second structural motif for recognition of DNA by oligonucleotide-directed triple-helix formation. *Science* 251: 1360–63
6. Booher MA, Wang S, Kool ET. 1994. Base pairing and steric interactions between pyrimidine strand bridging loops and the purine strand in DNA Pyr Pur Pyr triple helices. *Biochemistry* 33:4645–51
6a. Brelinska A, Shivdasani RA, Zhang L, Nabel GJ. 1990. Regulation of gene expression with double-stranded phosphorothioate oligonucleotides. *Science* 250:997–1000
7. Capobianco ML, Carcuro A, Tondelli L, Garbesi A, Bonora GM. 1990. One pot solution synthesis of cyclic oligodeoxyribonucleotides. *Nucleic Acids Res.* 18:2661–69
8. Chaudhuri NC, Kool ET. 1995. Very high affinity DNA recognition by crosslinked and bicyclic oligonucleotides. *J. Am. Chem. Soc.* 117: 10434–42
9. Chen CY, Sarnow P. 1995. Initiation of protein synthesis by the eukaryotic translational apparatus on circular RNAs. *Science* 268:415–17
9a. Chu BCF, Orgel LE. 1992. Crosslinking transcription factors to their recognition sequences with Pt^{II} complexes. *Nucleic Acids Res.* 20:2497–2502
10. Clusel C, Ugarte E, Enjolras N, Vasseur M, Blumenfeld M. 1993. Ex vivo regulation of specific gene expression by nanomolar concentration of double-stranded dumbbell oligonucleotides. *Nucleic Acids Res.* 21:3405–11
11. Cohen JS, ed. 1989. *Oligodeoxynucleotides: Antisense Inhibitors of Gene Expression*. Boca Raton, FL: CRC
12. Cram DJ. 1987. Molecular cells, their guests, portals, and behavior. *Chemtech* 17:120–25
13. Dagle JM, Weeks DL, Walder JA. 1991. Pathways of degradation and mechanism of action of antisense oligonucleotides in *Xenopus laevis* oocytes. *Antisense Res. Dev.* 1:11–20
14. Daubendiek SL, Ryan K, Kool ET. 1995. Rolling circle RNA synthesis: circular oligonucleotides as efficient substrates for T7 RNA polymerase. *J. Am. Chem. Soc.* 117:7818–19
15. De Mesmaeker A, Waldner A, Lebreton J, Hoffmann P, Fritsch V, et al. 1994. Amides as a new type of backbone modification in oligonucleotides. *Angew. Chem. Intl. Ed. Engl.* 33: 26–29
16. DeNapoli L, Messere A, Montesarchio D, Piccialli G, Santacroce C. 1993. PEG-supported synthesis of cyclic oligodeoxyribonucleotides. *Nucleosides Nucleotides* 12:21–30
17. Doktycz MJ, Goldstein RF, Paner TM, Gallo FJ, Benight AS. 1992. Studies of DNA dumbbells. I. Melting curves of 17 DNA dumbbells with different duplex stem sequences linked by T4 end loops: evaluation of the nearest-neighbor stacking interactions in DNA. *Biopolymers* 32:849–64
18. Dolinnaya NG, Blumenfeld M, Merenkova IM, Oretskaya TS, Krynetskaya NF, et al. 1993. Oligonucleotide circularization by template-directed chemical ligation. *Nucleic Acids Res.* 21:5403–7
19. D'Souza DJ, Kool ET. 1992. Strong binding of single-stranded DNA by stem-loop oligonucleotides. *J. Biomol. Struct. Dyn.* 10:141–52
20. D'Souza DJ, Kool ET. 1994. Solvent, pH, and ionic effects on the binding of single-stranded DNA by circular oligodeoxynucleotides. *Bioorg. Med. Chem. Lett.* 4:965–70
21. Durland RH, Kessler DJ, Gunnell S, Duvic M, Pettit BM, Hogan ME. 1991.

Binding of triple helix forming oligonucleotides to sites in gene promoters. *Biochemistry* 30:9246–55
22. Egholm M, Buchardt O, Christensen L, Behrens C, Freier SM, et al. 1993. PNA hybridizes to complementary oligonucleotides obeying the Watson-Crick hydrogen bonding rules. *Nature* 365:566–68
23. Erie DA, Jones RA, Olson WK, Sinha NK, Breslauer KJ. 1989. Melting behavior of a covalently closed, single-stranded, circular DNA. *Biochemistry* 28:268–73
24. Felsenfeld G, Davies DR, Rich A. 1957. Formation of a three-stranded polynucleotide molecule. *J. Am. Chem. Soc.* 79:2023–24
25. Fire A, Xu SQ. 1995. Rolling replication of short DNA circles. *Proc. Natl. Acad. Sci. USA* 92:4641–45
26. Gamper HB, Reed MW, Cox T, Virosco JS, Adams AD, et al. 1993. Facile preparation of nuclease resistant 3′ modified oligodeoxynucleotides. *Nucleic Acids Res.* 21:145–50
27. Giovannangeli C, Montenay-Garestier T, Rougée M, Chassignol M, Thuong NT, Hélène C. 1991. Single-stranded DNA as a target for triple-helix formation. *J. Am. Chem. Soc.* 113:7775–76
28. Giovannangeli C, Thuong NT, Hélène C. 1993. Oligonucleotide clamps arrest DNA synthesis on a single-stranded DNA target. *Proc. Natl. Acad. Sci. USA* 90:10013–17
29. Gryaznov SM, Lloyd DH. 1993. Modulation of oligonucleotide duplex and triplex stability via hydrophobic interactions. *Nucleic Acids Res.* 21: 5909–15
30. Hudson RHE, Damha MJ. 1993. Association of branched nucleic acids. *Nucleic Acids Symp. Ser.* 29:97–99
31. Illuminati G, Mandolini L, Masci B. 1977. Ring-closure reactions. 9. A comparison with related cyclization series. *J. Am. Chem. Soc.* 99:6308–12
32. Kanaya E, Yanagawa H. 1986. Template-directed polymerization of oligoadenylates using cyanogen bromide. *Biochemistry* 25:7423–30
33. Kandimalla ER, Agrawal S. 1994. Single strand-targeted triplex formation: stability, specificity, and RNAse H activation properties. *Gene* 149:115–21
34. Kandimalla ER, Agrawal S, Vekataraman G, Sasisekharan V. 1995. Single strand targeted triplex formation: parallel-DNA hairpin duplexes for targeting pyrimidine strands. *J. Am. Chem. Soc.* 117:6416–17
35. Khan IM, Coulson JM. 1993. A novel method to stabilise antisense oligonucleotides against exonuclease degradation. *Nucleic Acids Res.* 21: 2957–58
36. Kool ET. 1991. Molecular recognition by circular oligonucleotides. Increasing the selectivity of DNA binding. *J. Am. Chem. Soc.* 113:6265–66
37. Kool ET. 1993. *US Patent.* Applied
38. Lipsett MN. 1964. Complex formation between polycytidylic acid and guanine oligonucleotides. *J. Biol. Chem.* 239:1256–60
39. Lisziewicz J, Sun D, Klotman M, Agrawal S, Zamecnik P, Gallo R. 1992. Long-term treatment of human immunodeficiency virus-infected cells with antisense oligonucleotide phosphorothioates. *Proc. Natl. Acad. Sci. USA* 89:11209–13
40. Deleted in proof
41. Ma MYX, Reid LS, Climie SC, Lin WC, Kuperman R, et al. 1993. Design and synthesis of RNA miniduplexes via a synthetic linker approach. *Biochemistry* 32:1751–58
42. Maher LJ, Dervan PB, Wold BJ. 1990. Kinetic analysis of oligodeoxyribonucleotide-directed triple-helix formation on DNA. *Biochemistry* 29:8820–26
43. Marck C, Thiele D. 1978. Poly(dG) - poly(dC) at neutral and alkaline pH: the formation of triple stranded poly(dG)poly(dG)poly(dC). *Nucleic Acids Res.* 5:1017– 28
44. Morgan R, Edge M, Colman A. 1993. A more efficient and specific strategy in the ablation of mRNA in *Xenopus laevis* using mixtures of antisense oligos. *Nucleic Acids Res.* 21:4615–20
45. Moser HE, Dervan PB. 1987. Sequence-specific cleavage of double helical DNA by triple helix formation. *Science* 238:645–50
46. Mullis KB, Faloona FA. 1987. Specific synthesis of DNA in vitro via a polymerase-catalyzed chain reaction. *Methods Enzymol.* 155:335–50
47. Nielsen PE, Egholm M, Berg RH, Buchardt O. 1991. Sequence selective recognition of DNA by strand displacement with a thymine-substituted polyamide. *Science* 254:1497–1500
48. Nielsen PE, Egholm M, Buchardt O. 1994. Peptide nucleic acid (PNA). A DNA mimic with a peptide backbone. *Bioconjugate Chem.* 5:3–7
49. Nilsson M, Malmgren H, Samiotaki M, Kwiatkowski M, Chowdhary BC, Landegren U. 1994. Padlock probes: circularizing oligonucleotides for lo-

calized DNA detection. *Science* 265: 2085–88
50. Noll DM. O'Rear JL, Cushman CD, Miller PS. 1994. Interaction of oligodeoxyribonucleotides through formation of chimeric duplex/triplex complexes. *Nucleosides Nucleotides* 13: 997–1005
51. Perkins TA, Goodman JL, Kool ET. 1993. Accelerated displacement of duplex DNA strands by a synthetic circular oligodeoxynucleotide. *J. Chem. Soc. Chem. Commun.* 215–17
52. Prakash G, Kool ET. 1991. Molecular recognition by circular oligonucleotides: strong binding of single-stranded DNA and RNA. *J. Chem. Soc. Chem. Commun.* 1161– 63
53. Prakash G, Kool ET. 1992. Structural effects in the recognition of DNA by circular oligonucleotides. *J. Am. Chem. Soc.* 114:3523–27
54. Rajagopal P, Feigon J. 1989. Triple-strand formation in the homopurine: homopyrimidine DNA oligonucleotides $d(G\text{-}A)_4$ and $d(T\text{-}C)_4$. *Nature* 339:637–40
55. Rentzeperis D, Ho J, Marky LA. 1993. Contribution of loops and nicks to the formation of DNA dumbbells: melting behavior and ligand binding. *Biochemistry* 32:2564–72
56. Reynolds MA, Arnold LJ, Almazan MT, Beck TA, Hogrefe RI, et al. 1994. Triple-strand-forming methylphosphonate oligodeoxynucleotides targeted to mRNA efficiently block protein synthesis. *Proc. Natl. Acad. Sci. USA* 91:12433–37
57. Roberts RW, Crothers DM. 1992. Stability and properties of double and triple helices: dramatic effects of RNA or DNA backbone composition. *Science* 258:1463–66
58. Rougée M, Faucon B, Barcelo F, Giovannangeli C, Garestier T, Hélène C. 1992. Kinetics and thermodynamics of triple-helix formation: effects of ionic strength and mismatches. *Biochemistry* 31:9269–78
59. Rowley PT, Thomas MA, Kosciolek BA, Kool ET. 1993. Circular antisense oligonucleotides inhibit proliferation of chronic myeloid leukemia cells in a sequence-specific manner. *Blood* 82: 330 (Abstr.)
60. Rubin E, Kool ET. 1994. Strong, selective binding of six different DNA sequences by a single conformation-switching DNA macrocycle. *Angew. Chem.* 106:1057–59; *Angew. Chem. Int. Ed. Engl.* 33:1004–7
61. Rubin E, McKee TL, Kool ET. 1993. Binding of two different DNA sequences by conformational switching. *J. Am. Chem. Soc.* 115:360–61
62. Rubin E, Rumney S, Kool ET. 1995. Convergent oligodeoxynucleotide synthesis. A cyclodimerization approach to construction of circular oligodeoxynucleotides. *Nucleic Acids Res.* 23:3547–53
63. Rumney S IV, Kool ET. 1992. DNA recognition by hybrid oligoether-oligodeoxynucleotide macrocycles. *Angew. Chem.* 104:1686–1689; *Angew. Chem. Int. Ed. Engl.* 31:1617–19
64. Rumney S IV, Kool ET. 1995. Structural optimization of non-nucleotide loop replacements for duplex and triplex DNAs. *J. Am. Chem. Soc.* 117: 5635–46
65. Deleted in proof
66. Salunkhe M, Wu T, Letsinger RL. 1992. Control of folding and binding of oligonucleotides by use of a nonnucleotide linker. *J. Am. Chem. Soc.* 114: 8768–72
67. Samadashwily GM, Dayn A, Mirkin SM. 1993. Suicidal nucleotide sequences for DNA polymerization. *EMBO J.* 12:4975–83
67a. Sands H, Gorey-Feret J, Ho SP, Bao Y, Cocuzza AJ, et al. 1995. Biodistribution and metabolism of internally 3H-labeled oligonucleotides. *Mol. Pharmacol.* 47:636–46
68. Strobel SA, Doucette-Stamm LA, Riba L, Housman DE, Dervan PB. 1991. Site-specific cleavage of human chromosome 4 by triple-helix formation. *Science* 254:1639–42
69. Sundquist WI, Klug A. 1989. Telomeric DNA dimerizes by formation of guanine tetrads between hairpin loops. *Nature* 342:825–29
70. Trapane TL, Christopherson MS, Roby CD, Ts'o POP, Wang D. 1994. DNA triple helices with C-nucleosides (deoxypseudouridine) in the second strand. *J. Am. Chem. Soc.* 116: 8412–13
71. T'so POP, Miller PS, Aurelian L, Murakami A, Agris C, et al. 1987. An approach to chemotherapy based on base sequence information and nucleic acid chemistry: MATAGEN. *Ann. NY Acad. Sci.* 507:220–41
72. Turner DH, Sugimoto N, Freier SM. 1990. In *Nucleic Acids* (subvolume C), ed. W Saenger, pp. 201–27. Berlin: Springer-Verlag
73. Uhlmann E, Peyman A. 1990. Antisense oligonucleotides: a new therapeutic principle. *Chem. Rev.* 90:543
74. Varma RS. 1993. Synthesis of oligo-

nucleotide analogues with modified backbones. *SYNLETT* 621–37
75. Vo T, Wang S, Kool ET. 1995. Binding of pyrimidine sequences by triple helix formation. structural optimization of binding. *Nucleic Acids Res.* 23: 2937–44
76. Wang AHJ, Ughetto G, Quigley GJ, Hakoshima T, van der Marel GA, et al. 1984. The molecular structure of a DNA-triostin A complex. *Science* 225:1115–21
77. Wang S, Booher MA, Kool ET. 1994. Stabilities of nucleotide loops bridging the pyrimidine strands in DNA pyr - pur pyr triple helices: special stability of the CTTTG loop. *Biochemistry* 33: 4639–44
78. Wang S, Friedman AM, Kool ET. 1995. Origins of high sequence selectivity: a stopped-flow kinetics study of hybridization by duplex- and triplex-forming oligonucleotides. *Biochemistry*. 34:9774–84
79. Wang S, Kool ET. 1994. Circular RNA oligonucleotides: synthesis, nucleic acid binding properties, and a comparison with circular DNAs. *Nucleic Acids Res.* 22:2326–33
80. Wang S, Kool ET. 1994. Recognition of single-stranded nucleic acids by triplex formation: the binding of pyrimidine sequences. *J. Am. Chem. Soc.* 116:8857–58
81. Wang S, Kool ET. 1995. Relative stabilities of triple helices containing DNA, RNA, and 2′-O-Methyl-RNA: chimeric circular oligonucleotides as probes. *Nucleic Acids Res.* 23: 1157–64
82. Wang S, Kool ET. 1995. Origins of large differences in stability between DNA and RNA helices: C-5 methyl and 2′-hydroxyl effects. *Biochemistry* 34:4125–32
83. Deleted in proof
84. Watson JD, Crick FHC. 1953. Molecular structure of nucleic acids. *Nature* 171:737–38
85. Wemmer DE, Benight AS. 1985. Preparation and melting of single strand circular DNA loops. *Nucleic Acids Res.* 13:8611–21
86. Wickstrom E. 1986. Oligodeoxynucleotide stability in subcellular extracts and culture media. *J. Biochem. Biophys. Methods* 13:97–102
87. Zhao Q, Matson S, Herrera CJ, Fisher E, Yu H, Krieg AM. 1993. Comparison of cellular binding and uptake of antisense phosphodiester, phosphorothioate, and mixed phosphorothioate and methylphosphonate oligonucleotides. *Antisense Res. Dev.* 3:53–66

Annu. Rev. Biophys. Biomol. Struct. 1996. 25:29–53

THE DYNAMICS OF WATER–PROTEIN INTERACTIONS

Robert G. Bryant

Department of Chemistry, University of Virginia, Charlottesville, Virginia 22901

KEY WORDS: NMR, NMR dispersion, relaxation, magnetic resonance imaging, water binding, diffusion, water dynamics

ABSTRACT

The magnetic field and temperature dependence of the water proton nuclear spin-lattice relaxation rate requires that the motion timescale for water molecules in contact with proteins is close to that for pure water at room temperature. Nevertheless, there are a few water molecules, which may be detected by high-resolution, cross-relaxation spectroscopy, that must have relatively long protein-bound lifetimes and that carry the bulk of the relaxation coupling between the protein and the water. The water–protein magnetic coupling affects the interpretation of water relaxation rates in heterogeneous protein systems, such as tissues, and provides new ways to extract useful information about the immobilized components through the effects on the water NMR spectrum. The discussion shows that the conclusions concerning the rapid water molecule motions at the interface are not in conflict with the observations of many water oxygen atom positions in protein crystal structures.

CONTENTS

1056-8700/96/0610-0029$08.00

PERSPECTIVE

Water, the ubiquitous biochemical solvent, is a particularly complex liquid. Although enormous progress has been made, the effort to understand water itself remains no less significant than the struggle to understand how water interacts energetically and dynamically with solutes, particularly such macromolecular solutes as proteins, saccharides, and nucleic acids (40). With rare exceptions, the issue is not solvation in the sense of molecular contact. Water molecules will populate, at least transiently, the spaces immediately adjacent to solutes, macromolecular or not. Rather, the issues are details of solvent organization and the lifetimes of the aggregate structures or individual water molecules in the regions of solute contact or interface.

The concept of a hydration shell, which has long been prevalent, was perhaps made most vivid by the iceberg structures suggested by Frank & Evans (29) and was applied rapidly to proteins (73). Diffraction experiments support this concept, which was developed originally on the basis of less direct evidence (30, 36, 105). This suggestion, however, which is clearly important in discussions of the energetic landscape for a folded protein or polynucleotide, suppresses a crucial aspect of the hydration problem, namely the dynamics of the interactions. A sound understanding of the dynamics is needed to understand the kinetics of molecular rearrangements, the displacement of water in binding interactions, and the dielectric properties of the interfacial region.

The water molecule dynamics may span the range of time scales from roughly 0.1 ps, which is associated with the fast and subtle reorganization that is related to a perturbation of the electric environment, to much slower motions, which are associated with proton exchange. At neutral pH, the lifetime of a proton on a water molecule is on the order of 0.4 ms (88, 93). Nuclear magnetic resonance and magnetic relaxation spectroscopy provide several tools for examining events that occur on these different time scales. This review is concerned primarily with the application of these tools to understand the structure and dynamics of water in the immediate vicinity of proteins and other macromolecules. The earlier contributions of magnetic resonance to this problem were somewhat controversial. A wide variety of experimental results, however, now suggest convergence to a common picture that accounts for the present data and appears to interface well with data from complementary approaches, such as diffraction methods. The review focuses on proton spectroscopy and relaxation because of limited space and the central role of proton spectroscopy in structural determinations, medical imaging, and in vivo spectroscopy. The review shows that three types of interactions contribute to the water proton magnetic relaxation response: proton exchange, water molecule exchange with a few specific protein sites, and transient collisions between the protein and the water.

SPECTROSCOPY AND RELAXATION

The structural information in NMR spectroscopy is carried predominantly by the chemical shift, the scalar coupling constants, and the cross-relaxation rates, all of which depend on the local environment of the nucleus observed. Even in complex heterogeneous environments, however, the NMR spectrum of water usually consists of a single resonance line, so that structural information must be derived from the relaxation or cross-relaxation effects.

There are many measurable magnetic relaxation times, but the simplest is the longitudinal or spin-lattice relaxation rate, $1/T_1$, which describes the rate at which the nuclear spin system comes to equilibrium with the other degrees of freedom in the system or achieves the population of magnetic energy levels in the magnetic field determined by the temperature and the Boltzmann distribution law. The spin-lattice relaxation requires that the spin system exchange quanta of energy with the nonspin parts of the heat capacity, usually called the lattice even in liquids. The size of these quanta is determined by the strength of the magnetic field imposed on the sample. Because spontaneous emission is very improbable at the relatively low frequencies of NMR experiments, the energy exchanges are stimulated and require the spin system to couple to fluctuating fields in the sample. The relaxation occurs when a component of the fluctuation spectrum matches the nuclear Larmor frequency and stimulates a spin flip. Components at twice the Larmor frequency are also effective (1). The relaxation times for protons are on the order of 3 s in pure water. The relaxation measurement, therefore, provides information about the dynamics at the Larmor frequencies only with the aid of quantum statistical theories that relate a model for the molecular motions to the intensity of the magnetic fluctuations at the required transition frequencies.

The usual approach is to develop a model for the fluctuations in the time domain by constructing a correlation function that depends on the details of the motions involved, such as translation, rotation, and restricted diffusion. The simplest case is random rotational diffusion, for which the correlation function decays exponentially. The relaxation rate is proportional to the strength of the magnetic dipole–dipole coupling between the two protons in the water molecule multiplied by the components of the Fourier transform of the time correlation function at the Larmor frequency and at twice the Larmor frequency. This particular mechanism yields the familiar relaxation equation

$$\frac{1}{T_1} = \frac{3}{10}\frac{\hbar^2\gamma^4}{\mathbf{r}^6}\left(\frac{\tau_c}{1 + (\omega\tau_c)^2} + \frac{4\tau_c}{1 + (2\omega\tau_c)^2}\right) \qquad 1.$$

where ω is the Larmor frequency, γ is the magnetogyric ratio of the spin, and τ_c is the correlation time for the reorientation of the intermoment vector, $\mathbf{r}$, which connects the two protons in the water molecule (1). It is clear by inspection that although the rate, $1/T_1$, may be near unity, the dynamical information contained is at the Larmor frequency; at 500 MHz, the Larmor frequency corresponds to fluctuations in the range of tenths of nanoseconds. Slower motions may be sensed by using lower field strengths or by measuring other relaxation times that depend on different fluctuation frequencies. When nuclear spin relaxation rates are used to build a better understanding of molecular dynamics, the issue becomes how to construct an accurate and complete model for the correlation functions (89, 92).

The intramolecular dipole–dipole relaxation that involves identical spins, as shown in Equation 1, accounts for approximately half the relaxation in pure water; the rest is caused by intermolecular proton–proton coupling that is modulated by relative translational motions (35, 48, 86). In protein solutions and gels, intermolecular contributions and cross-relaxation effects may become critical (25, 61). The basic magnetic dipole–dipole coupling is the same in these complex systems. The intermoment distances may be time dependent, however, and the spins may have different spectroscopic properties, such as chemical shift or linewidth, that make them experimentally distinguishable. Cross-relaxation effects may be key to understanding the observed relaxation rates and may also be used to demonstrate molecular proximity because of the dependence of the relaxation efficiency on $\mathbf{r}^{-6}$, as in Equation 1. Although space does not permit review of the major relaxation equations that have been developed, a qualitative picture is sufficient to gain a considerable understanding of the magnetic resonance results in protein systems.

PROTEIN SOLUTIONS

Early observations on the NMR spectra of water in protein solutions were remarkable, because the effects on the water resonance were found to be small (91). In part, this observation resulted from the very high concentration of water relative to the protein, so that a significant dilution of the spectroscopic effects was found.

At the IBM Watson Laboratories, Redfield et al (104) developed instrumentation that provided a major advance in this field. Koenig, Brown, and collaborators (75) applied this instrumentation to proteins. Koenig & Schillinger (83) first observed the magnetic field dependence of the water proton spin-lattice relaxation rate and found a dispersion that is a nearly Lorentzian decrease in the relaxation rate with increasing field strength. The inflection point was identified with the protein rotational correlation time and was subsequently shown to depend on the protein size or molecular weight appropriately (45).

The interpretation of the relaxation dispersion profile in protein solutions has been somewhat controversial. An early model discussed by Koenig & Schillinger (83) was a chemical exchange model in which the relaxation was assumed to be the weighted average of relaxation rates of water molecules free in solution and those that were presumed to be bound to the protein and rotated with the rotational correlation time of the protein. Even if one neglected intermolecular effects, this picture led to the conclusion that the magnitude of the relaxation effects was so small that only a few molecules could be bound and move as slowly as the protein; otherwise, the relaxation rates at low magnetic field strengths would be much larger. Eisenstadt (27) discussed such models very carefully. The problem with this interpretation was that the protein is obviously in contact with hundreds of water molecules. How could only a few of these molecules contribute to the water relaxation rate significantly? Consequently, the interpretation of these data evolved through a series of attempts to involve many more water molecules. Included were ideas of local anisotropic motion near the protein and specialized hydrodynamic coupling between the protein and the water (19, 20, 42–44, 74, 82); none of these attempts was completely satisfactory.

A second possible contribution to the magnetic relaxation coupling between the protein and the water is proton exchange between protein ionizable groups and the water molecule (49–52, 83). Such an exchange mixes the spin-lattice relaxation rates, as shown in Equation 2.

$$\frac{1}{T_1} = \frac{1}{T_{1\text{free}}} + \sum_i \frac{P_{\text{Hi}}}{T_{1,\text{i}} + \tau_{\text{Hi}}}, \qquad 2.$$

where the sum runs over *i* protein–proton exchange sites occupied with probability P_{Hi} each characterized by a relaxation time, $T_{1,i}$, and an exchange lifetime, τ_{Hi}. This contribution was generally recognized to be small because of the relatively large values of the exchange time relative to the $T_{1,i}$. Further, in protein solutions of dimethyl sulfoxide in D_2O, the relaxation dispersion could be observed in methyl protons for which the proton exchange is unimportant on the time scale of these experiments (13). Thus, the relaxation coupling between the water and the protein protons may involve a proton transfer to produce the relaxation dispersion, but the transfer is not necessary.

What we have called the whole molecule exchange mechanism (12) was revitalized by the observations of the Wüthrich laboratory that one could detect specific nuclear Overhauser effects between water protons and specific protons in bovine pancreatic trypsin inhibitor (97–100). Since these early observations, similar results have been reported for many proteins by a variety of laboratories (18, 28, 34, 38, 39, 85, 103, 122). The essence of the results is that a few molecules are bound for a sufficiently long time to have an effective magnetization transfer between the water and protein protons by an intermolecular cross-relaxation path. These sites may then be identified; in bovine pancreatic trypsin inhibitor, for example, four sites were identified. The remainder of the water associated with the protein showed no significant cross-relaxation with the protein residues. It was argued that for the vast majority of water–protein contacts, the weakness of the intermolecular proton–proton dipole–dipole coupling between the water and protein protons resulted from very short-lived interactions, 300 ps or less. The consequence of this short lifetime was cross-relaxation rates too slow to make significant contributions to the observations.

These results support the idea that there are a few specific water molecule binding sites of sufficient lifetime to have effective magnetic interactions between the protein protons and the water protons. Because the observable relaxation rates are slow, on the order of 1 s, there is plenty of time for exchange between these bound or buried sites and the bulk water that provides the requisite coupling. The exchange of buried solvent molecules also accounts for the observations of relaxation dispersions in nonaqueous solvent components for which proton exchange is not possible.

An important detail of the protein solution data is that the magnetic field dependence of the spin-lattice relaxation rate of the water protons is not described accurately by the Lorentzian shape implied by Equation 1. Specifically, the curve is somewhat broader and a little asymmetric, with a tail at high magnetic field strengths (45). This problem has been handled by assuming a broadening function or distribution of effective correlation times implicit in a function such as the Cole-Cole dispersion expressions. Such approaches may provide a parameterization of the data but not an understanding of the underlying mechanisms. Although a distribution of correlation times is reasonable, the distribution almost certainly is narrow for the water molecules bound to the protein. For example, free rotation of a bound water molecule about a single hydrogen bond changes the correlation time by a factor of only three compared with the rigid case (109,116). Equation 2, however, suggests an heuristically helpful approach. We rewrite this equation to include other contributions:

$$\frac{1}{T_1} = \frac{1}{T_{1\mathrm{free}}} + \sum_s \frac{P_s}{T_{1,s} + \tau_s} + \sum_i \frac{P_{Hi}}{T_{1,i} + \tau_{Hi}} + \sum_f \frac{P_f}{T_{1,s} + \tau_f} \qquad 3.$$

The first sum is over the s buried molecule binding sites that dominate the relaxation dispersion profile. Limits placed on $\tau_{s,}$, which are discussed later, make its contribution to the denominator of the first sum negligible. The relatively slow proton exchange times, however, make the contributions from the second sum in Equation 3 small. The field dependence is contained in the $T_{1,i}$, and the result of the slow exchange or large τ_{Hi} in Equation 3 is not only to make the contributions smaller but also to shift the inflection for each contribution to higher field strengths and contribute slightly to the high field tail of the dispersion. The last sum includes contributions from molecules that stick to the protein for times shorter than the rotational correlation time of the protein. The effective correlation time for each is given as $\tau_{cf} = (1/\tau_{\mathrm{rotation}} + 1/\tau_f)^{-1}$, where the sum runs over the distribution of exchange times, τ_f. To assume a Lorentzian form for each $T_{1,s}$, the distribution of exchange times must be centered at very short values, i.e. on the order of the translational diffusion correlation times for the bulk water. Only the long time tail of this distribution contributes significantly to the relaxation dispersion (128). The relative contributions are summarized schematically in Figure 1. Kimmich and coworkers (64, 68) have presented a more elegant and theoretically satisfying approach. They treat the rapid motion of water at the protein surface in terms of surface translational diffusion at the level of the correlation function. The essence of the dynamical picture is the same, i.e. the water at the protein surface moves very rapidly compared with the protein.

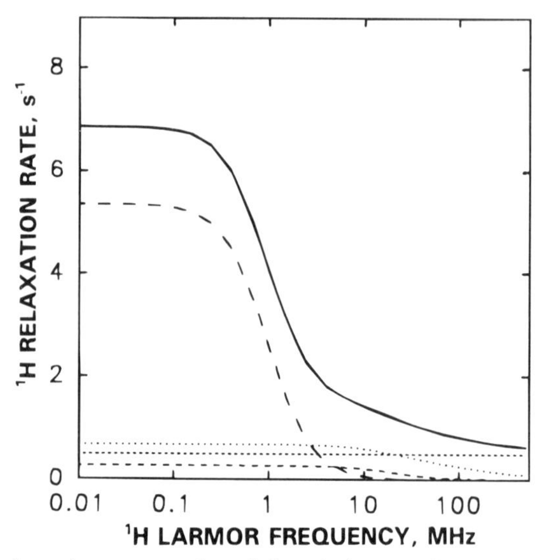

Figure 1 A schematic representation of the relative contributions to the total water proton spin-lattice relaxation rate for a 15% bovine serum albumin solution at 298 and neutral pH. In order from the top: total relaxation rate (*solid line*), buried water molecule contribution (*dashed line*), rapidly exchanging surface water contribution (*dotted line*), bulk water rate (*dashed line*), protein–proton exchange contribution (*lowest dashed line*), the amplitude of which is a function of pH.

TRANSLATIONAL DIFFUSION

Characterization of water molecule translational motion near a protein provides a fundamental characterization of the water dynamics and permits evaluation of the intermolecular proton–proton dipole–dipole coupling between water and protein protons. As noted above, this coupling is apparently weak enough that intermolecular cross-relaxation effects are not generally observed between surface protein protons and water protons. The correlation times for the motion, therefore, must be much shorter than the rotational correlation times of the protein. Direct characterization of the water translational motion supports this picture.

The most direct way to measure translational motion of water is to measure the attenuation of proton spin echoes in a magnetic field gradient (41, 112). The experiment works by phase encoding nuclear spins during a preparation period with a magnetic field gradient such that their Larmor frequency, hence phase accumulation, depends linearly on position in the sample. After a delay, a refocusing pulse is applied that produces a spin echo. If the molecule moves to a different position, hence a different Larmor frequency, the result of motion is loss of intensity in the refocused spin echo that may be related quantitatively to the translational diffusion constant and the magnetic field gradient strength.

The difficulty in applying this experiment to a protein solution is that the experiment will return the average diffusion constant of water in the solution, which is little changed from pure water because of the relatively low concentration of protein compared with water. Kimmich and coworkers (66, 84, 107) overcame this problem, however, by controlling the sample composition and by studying frozen protein solutions. It has long been known that a significant water proton NMR signal, which permits characterization of the aqueous layers immediately adjacent to the protein, persists in frozen protein solutions well below the freezing point of water (87). Kimmich and coworkers have applied the field gradient methods to frozen protein solutions and have reported that the water translational diffusion constant is reduced from the bulk water values by a factor less than ten.

Polnaszek & Bryant (101, 102) took a completely different approach to characterization of the water motion at the protein surface. They used a nitroxide spin label to place a large electron spin magnetic moment on the protein surface. The magnetic field from the unpaired electron is approximately 1000 times larger than those from the protons, so that it is possible to isolate the paramagnetic contribution to the water proton relaxation easily. In this case, the relaxation mechanism is an electron magnetic dipole-water proton-dipole interaction. Because the electron spin relaxation time of the nitroxide radical is long, the correlation time for the intermolecular coupling is that for the relative translational motion of the proton-electron pair. Because the protein moves slowly compared with the water, the effective correlation time for the coupling is the translational correlation time for the water near the nitroxide on the protein surface. Measurement of the water proton relaxation rate over a wide range of magnetic field strengths permits extraction of the translational diffusion constant for the water near the nitroxide. The relaxation is dominated by the effects within approximately 10 Å of the surface, and the diffusion constants found are within a factor of five of the values for bulk water, which is in excellent agreement with the pulsed field gradient results.

In summary, three independent magnetic resonance evaluations of the water dynamics at the protein surface demonstrate that motions of water essentially in contact with the protein are very rapid and nearly as fast as those in the bulk water. These assessments are the translational diffusion constant measured from the field dependence of the paramagnetic contribution to water relaxation in protein solutions, the pulsed field gradient measurements of the water translational diffusion constant in frozen solutions, and the absence of the water-surface residue cross-relaxation or Overhauser effects. Additional evidence for the rapid motions of water adjacent to the protein derives from magnetic resonance experiments conducted on systems in which the protein is immobilized rotationally and a bulk water phase may not exist.

ROTATIONALLY IMMOBILIZED PROTEINS

Rotationally immobilized protein systems include dry proteins, cross-linked systems, crystals, gels, and more complex molecular aggregates, such as those found in tissues. Rotationally immobilized protein systems provide excellent models for characterizing water in more complex heterogeneous systems in which the considerations are similar (106, 108, 117, 118).

Dry Systems

In a dry system, there is not enough water present to create a bulk solution phase; all the water in the system is adsorbed. Study of these systems provides characterization of the surface interactions by default. Like most adsorbed liquid systems, water adsorbed on proteins presents a relatively narrow resonance line, the properties of which depend on the level of hydration. The key feature of the magnetic relaxation of water protons in hydrated protein systems is the magnetic coupling between the liquid water spins and the protein protons that makes the spin-lattice relaxation nonexponential and interpretation, in terms of Equation 1, inappropriate. Although nonexponential relaxation was well known (14, 56, 58–60) Edzes & Samulski (25, 26) first demonstrated the relaxation coupling directly and presented a discussion of the relaxation in terms of the coupled relaxation in collagen systems. Additional contributions were made by the laboratories of Fung (31, 32), Bryant (16, 53, 57, 81, 110), Andrew (4–10, 33), Koenig (75–80), Schleich (17, 108, 123, 124), Yeung (2, 125, 126), Henkelman (47, 94), and others (3, 11, 113). Although a quantitative discussion of the relaxation coupling is not possible here, Figure 2 illustrates the dynamic implications of the measurements.

Data set A results from a measurement of the dry protein proton, T_1, as a function of temperature at a resonance frequency of 57.5 MHz (110). The response is nearly classic, in that one observes a minimum in T_1 with decreasing temperature, as implied by Equation 1. The relaxation is actually more complex than is suggested by Equation 1, however, because the relaxation time at the minimum is too large. One

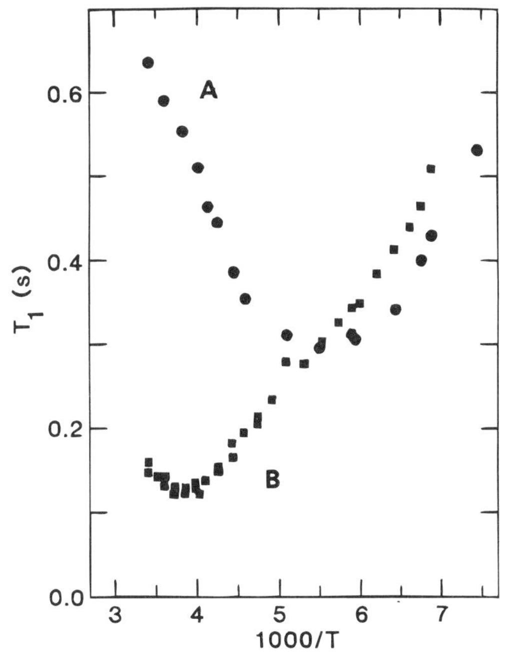

Figure 2 Proton spin-lattice relaxation time vs reciprocal temperature for lysozyme at 57.5 MHz. (*A*) A dry lysozyme; (*B*) 20.6 g water per 100 g lysozyme (110).

would usually expect this situation to correspond to a solid, with T_1 on the right side of the graph. However, although the protein is immobilized rotationally, the rapidly rotating methyl groups in the protein act as relaxation sinks and relax the whole protein population, which is coupled to the sinks by relatively efficient intramolecular spin diffusion in the solid. At room temperature, therefore, the solid protein falls on the high temperature side of the T_1 minimum usually associated with liquids, and the minimum occurs when the methyl rotation rate approximately matches the Larmor frequency. At room temperature, the methyl rotation is so fast that the mechanism is inefficient and the relaxation times are long, approximately 0.5 s.

The addition of water generates data set B in Figure 2. As noted above, the water protons are coupled strongly to the protein–proton relaxation. At room temperature, the adsorbed water, which rotates more slowly than do the methyl groups, relaxes the protein protons, and the detected relaxation rates increase because of the addition of many water proton relaxation sinks. A quantitative analysis of these data, which considers the cross-relaxation problem completely, is discussed in detail elsewhere (110). The result, however, is that the correlation times for reorientation of the adsorbed water molecules are in the range several hundreds of picoseconds, even at coverages that are relatively low, e.g. 0.2 g water/g protein.

The motions of the water molecules slow down with decreasing temperature and water content. As the water molecule motion slows below the Larmor frequency at low temperature, the water protons no longer relax efficiently, and the hydrated system relaxation rates cross those for the dry protein. At a low enough temperature, therefore, the water protons simply add to the relaxation load of the more rapidly rotating methyl protons, and the hydrated system relaxes more slowly than the dry system does. These results on the hydrated protein systems are entirely consistent with the solution-phase results, which also indicate that the surface mobility of most water protons is sufficiently high that little contribution is made to the protein–proton relaxation or cross-relaxation rates.

Protein Gels and Tissues

The protein gel is a dynamically heterogeneous system that, at one level, may be divided into two components. One component behaves magnetically like a solid (the rotationally immobilized protein). This component is in intimate contact with the second component, a liquid (the water). The cross-linked protein system or protein gels are valuable as physical models of more complex systems, such as tissues in which molecular and dynamic diversity makes quantitative interpretation using a detailed models intractable. The possibility for high-resolution protein–proton experiments is lost as soon as the decreased rotational mobility of the protein broadens the resonance lines. As in the case of a frozen protein solution, a narrow water resonance remains, in both simple gels and tissues, that is the primary observable for in vivo diagnostic imaging and spectroscopy.

The effects of magnetic cross-relaxation or relaxation coupling between the protein and the water are profound in the cross-linked systems, as shown in Figure 3 (90). The lower relaxation dispersion profile shows that the typical solution phase has nearly Lorentzian dispersion, as discussed above. Cross-linking the protein, while the other variables in the sample remain constant, yields the upper relaxation dispersion profile, which is characteristic of many dynamically heterogeneous systems, particularly tissues (15). The spin-lattice relaxation remains non-exponential, so that interpretations in terms of Equation 1 are inappropriate. The shape of the relaxation dispersion profile has been observed in dry and hydrated protein powders and gels or in cross-linked protein

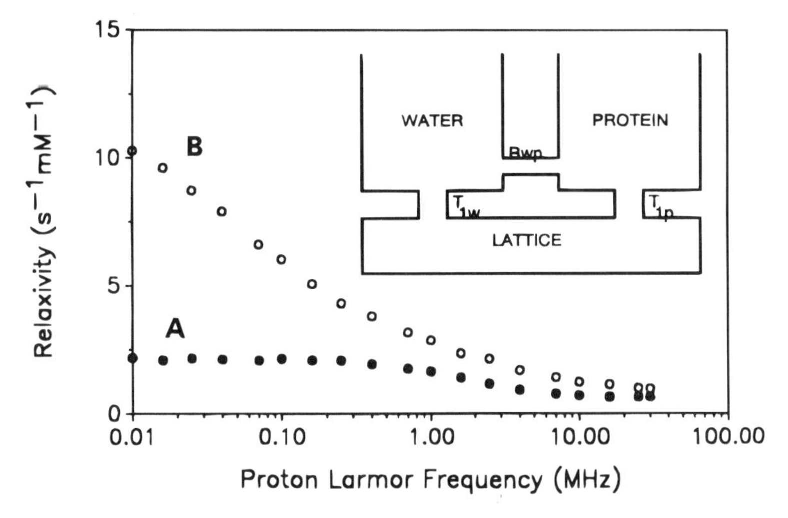

Figure 3 Proton spin-lattice relaxation rates expressed as rates per millimole of protein for a 1.8 mM bovine serum albumin solution (*filled circles*) and a cross-linked sample of the same composition at 298 K (adapted from Reference 14). (*inset*) The predominant relaxation pathways for the water and protein spin populations that are coupled.

systems by detecting the slowly relaxing component of the biexponential longitudinal magnetization decay. The rate may be described by

$$\frac{1}{T_1} = A\omega^{-B}, \qquad 4.$$

where A is a constant, ω is the Larmor frequency, and B is generally between 0.5 and 0.75. Kimmich and coworkers (63, 64, 67–70, 95, 96, 108, 111, 114, 115) have discussed this dependence in terms of a defect diffusion model originally applied to engineering polymer systems; other models, however, may provide a similar magnetic field dependence. Lester & Bryant (90) have shown that the observable relaxation profiles in hydrated protein systems were dominated by the cross-relaxation between the water and the protein, with the water essentially reporting a scaled version of the protein–proton relaxation that serves as a relaxation sink for the water spins. This cross-relaxation mechanism accounts for the main features of water–proton relaxation in tissues (13).

Because the rotationally immobilized components of a heterogeneous system have very different NMR spectra, cross-relaxation can be exploited in ways similar to, but fundamentally different from, those in the solution case. The rotational immobilization makes the protein–proton spin system behave magnetically like a solid. The transverse relaxation time is short, typically the order of 10 μs, which yields a line width between 30 and 40 kHz. The T_1 is long, however, which makes the spin system susceptible to saturation effects. Unlike the solution case, the protein spins are well coupled to each other by the dipole–dipole interaction, which makes intramolecular spin diffusion efficient (33). The broad line is homogeneous, so that an attempt to saturate the broad line at one frequency affects the whole spectrum (17, 37, 47, 119, 121, 125).

These characteristics of the protein spin system permit a class of experiments that exploit the relaxation coupling to the liquid. The simplest experiment is summarized in Figure 4. A low-level preparation pulse is applied to the solid spin system at a frequency offset, Δ, from the water resonance that partially saturates the protein–proton spins. The cross-relaxation between the protein and the water protons causes

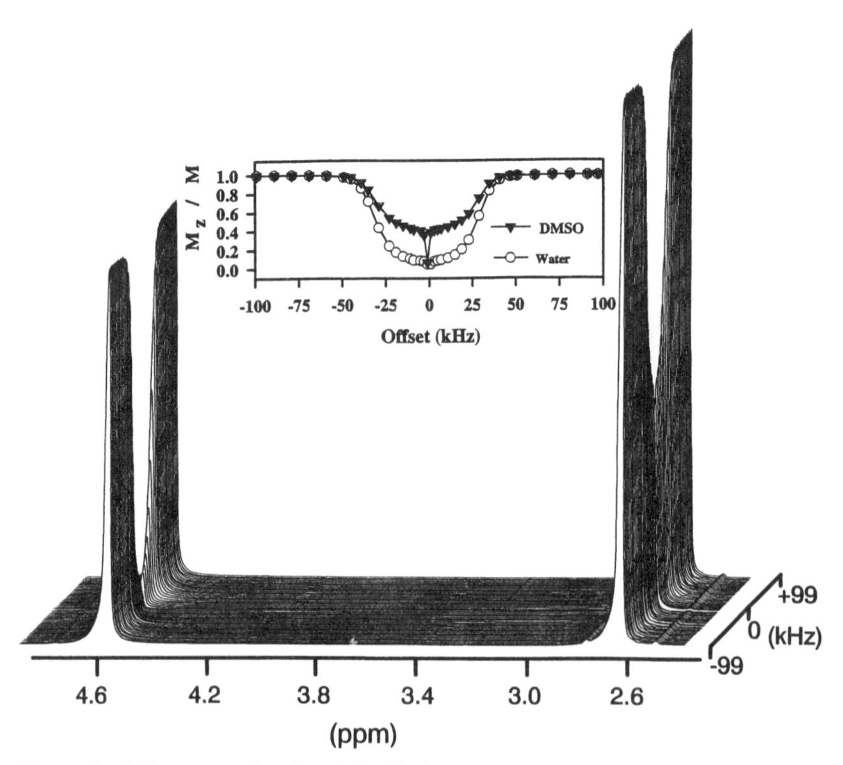

Figure 4 A Z spectrum for dimethylsulfoxide and water in a 6.8% cross-linked bovine serum albumin sample obtained at 500 MHz. (Adapted from Reference 54.)

the water proton magnetization to decrease as well. A pulse on-resonance with the water signal at the conclusion of the preparation pulse reads the effect on the water or other solvent resonance. A plot of the water signal intensity as a function of the preparation pulse offset yields a representation of the rotationally immobilized protein–proton spectrum. This spectrum is sometimes called a Z-spectrum because of its origin in population or Z-magnetization transfer (37). This basic experiment has many applications.

The acquisition of the protein–proton spectrum by detecting water proton signals is convenient experimentally, but an imaging sequence may be substituted equally well for the detection portion of the experiment. Using this relaxation coupling, a water proton–based image may be made to report intensities that are proportional to spin characteristics of the rotationally immobilized or solid components in the sample. The technique has obvious applications in medical diagnostic imaging, where some pathologies may involve changes in the more solid components of the tissue (119). Several process control applications are also apparent in the food industry (24, 120). Of importance to this review is the mechanism for the magnetization transfer between the rotationally immobilized protein protons and the water proton system.

The Z-spectrum may be recorded efficiently for solvent systems that have no exchangeable protons, such as dimethylsulfoxide, so that the magnetization transfer is not predominantly carried by proton exchange (37, 90). The intensity of the Z-spectrum is not a strong function of pH in protein systems, which implies that the proton exchange mechanisms may make a significant, but not dominant, contribution to the total magnetization transfer rates. The mechanisms that permit efficient communication between the protein protons and the water protons when the protein is not rotating freely are the same as in the protein solution case: proton transfer, whole water molecule exchange, and transient dipolar couplings modulated by the translational motion of the water molecule as it moves by the protein (12). What is fundamentally different from the solution case is the efficient dipolar coupling among protons in the immobilized protein system that makes spin diffusion efficient and makes the whole protein–proton population respond collectively.

The translational motion of the water molecule is not coupled significantly to rotational motion of the protein; thus, the translational motion of the water near the protein should be approximately the same whether the protein rotates or not. Because the translational correlation times for the water molecules are very short, the proton relaxation contributions from proton dipole–dipole interactions that are modulated by

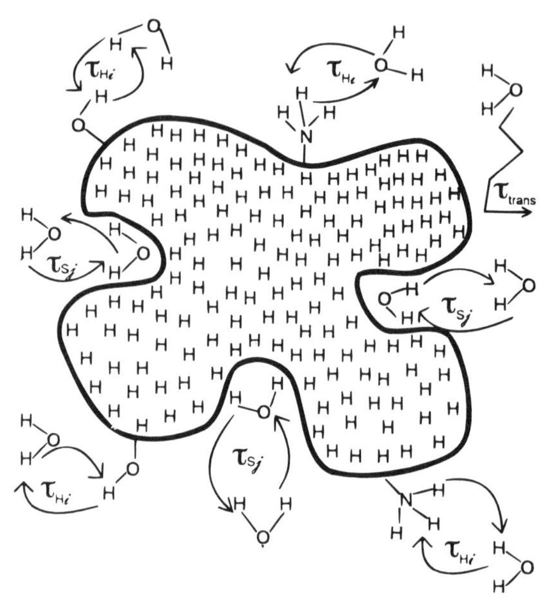

Figure 5 A schematic representation of the magnetically important interactions of water with a protein characterized by correlation times for buried water molecule exchange, τ_{Sj} , for proton exchange with ionizable groups, τ_{Hi}, and for translation near the surface, τ_{trans}.

translational diffusion near the protein will remain small compared with contributions from proton and whole molecule exchange paths.

Figure 5 represents a summary of these mechanisms. In the protein gel or cross-linked case, the dominant contribution is the whole molecule exchange path, which is consistent quantitatively with the few water molecule binding sites that may be identified for small proteins from the solution-phase two-dimensional nuclear Overhauser (2D NOE) experiments discussed above. The magnitude of the magnetization transfer rate constant has been measured and analyzed in detail for cross-linked protein systems by using a second order kinetic approach (54, 55, 127). In the rotationally immobilized system, the magnetization transfer rate for a water molecule bound inside the protein so that it does not rotate is well approximated as $1/T_{2S}$, i.e. the transverse relaxation rate for the solid spin system (4, 10, 33, 55). That is, when bound to the few specific sites in the protein, the water molecule protons behave like the protein protons and come to thermal equilibrium with the protein–proton population in a time the order of the T_2 of the solid protein protons, which is the order of 10 μs. The measured magnetization transfer rates may be computed from this limiting value for serum albumin, with the assumption that there are on the order of ten specific water molecule binding sites per protein molecule, which is consistent with the original solution-phase data of Schillinger & Koenig (83), as well as subsequent measurements.

This mechanistic picture is consistent with the response of nonaqueous cosolvents in cross-linked protein systems. Z-spectrum experiments conducted on mixed solvent gel systems have demonstrated clearly significant cross-relaxation effects that are detected easily in nonaqueous solvent components, as shown in Figure 4. Further, it is easy to distinguish between competitive and noncompetitive sites for different solvents. For the systems studied to date, the organic solvents do not alter the water proton cross-relaxation efficiency, but there is a significant difference in the temperature dependence of the water Z-spectrum compared with nonaqueous components. The water cross-relaxation efficiency increases with increasing temperature; the opposite is true, however, for cosolvents like methanol (54). These effects are understood easily in terms of the two cross-relaxation pathways available to water, namely whole molecule and proton exchange processes. The latter increases in importance with increasing temperature because of the temperature dependence of the proton exchange rate constant. There are no exchangeable protons on the methyl groups of the organic solvents. Therefore, the relaxation efficiency is proportional to the number of molecules bound to the protein, which apparently decreases with increasing temperature for the molecules studied to date, namely methanol, acetonitrile, dimethylsulfoxide, acetone, and dimethylformamide.

BURIED WATER MOLECULE LIFETIMES

The feature of these relaxation mechanisms that remains unresolved quantitatively is the exchange rate of the buried water molecules. These water molecules do not necessarily all have the same exchange time, because they are generally in structurally different environments within the protein. The Wüthrich group (98) used a chemical shift reagent to move the water resonance line from its usual position and noted that the chemical shift of the observable buried water molecule resonances followed the chemical shift of the bulk water resonance. This finding permits the conclusion that the chemical exchange rate must be fast compared with the chemical shift difference created, or on the order of 0.1 ms. An exchange lifetime this fast or faster is more than sufficient to account for the observations reviewed here. It is very likely, however, that the exchange lifetimes are considerably shorter. For example, the recent reports by Denisov & Halle (21–23) suggest significantly shorter lifetimes on the basis of ^{17}O relaxation dispersion data, which also verify the importance of the whole molecule exchange pathway.

Experiments that demonstrate individual water molecule lifetimes have not been reported. Important constraints can be deduced, however, on the basis of the magnitude of the total relaxation rate at low field strengths in protein solutions. For the solution-phase magnetic relaxation dispersion, the low-field portion of the data set provides a measure of the total coupling between the water and the protein. In the spirit of the original Schillinger & Koenig discussion (83), we may estimate the minimum value of the bound water molecule relaxation rate and then use this value in combination with the measurements to compute the maximum number of water molecules that may be bound to the protein for a time as long as or longer than the rotational correlation time for the protein or other macromolecule. The essence of the assumptions that lead to the bound molecule relaxation rates is that only water proton–water proton intramolecular relaxation is considered for the bound sites in which the water molecules are assumed to rotate with the correlation time of the protein. Thus, water proton–protein–proton intermolecular relaxation is neglected and leads to an underestimate of the bound relaxation rate. Consequently, when this number is combined with the experimental relaxation rate, the number of bound molecules is overestimated. Nevertheless, when this strategy is followed for bovine pancreatic trypsin inhibitor, the number of bound molecules deduced is 2 or 3, far less than the number of molecules identified in the crystal structures, and even less than the number of water molecule binding sites identified from the 2D NOE experiments. In addition, this estimate places an upper bound on the number of tightly held water molecules. Therefore, the assumptions of the model are inadequate, and, on average, the molecules cannot be bound for lifetimes as long as the rotational correlation time of the protein (128). Nevertheless, some of the molecules must be bound long enough to produce a significant Overhauser effect, which requires lifetimes on the order of a few nanoseconds.

Although this approach does not permit precise measurements of the water molecule lifetimes, it does provide significant constraints on the nature of the lifetime distribution that are also consistent with the ^{17}O data (21–23). Even for the molecules identified by NMR as relatively long lived, the average lifetime must be less than the rotational correlation time of the protein under the conditions of the magnetic relaxation dispersion experiment, or approximately 20 ns. Similar results have been obtained for the DNA dodecamer, $d(CGCGAATTCGCG)_2$ (129).

RELATION TO DIFFRACTION RESULTS

It is not surprising that there are specific sites on or in the protein for water molecule binding, given the number of crystallographic reports that identify water molecule oxygen atom positions. Of the water molecule positions that appear in crystal structures, the NMR data demonstrate that very few are sufficiently long lived to contribute significantly to the NMR cross-relaxation rates either in the solution or in the rotationally immobilized case.

The NMR and diffraction data are not in conflict. The diffraction experiment consists of the superposition of a large number of scattering events, each of which occurs on a time scale that is very short compared with the molecular reorientations that may be present in the system. The result is a spread in the data that corresponds to the superposition of structures that represents essentially a population average. For water, these data indicate that the potential surfaces have local minima into which the water molecules fall reproducibly. Therefore, these structures contribute significantly to the overall energy minimization, because the water molecule electric dipole and the hydrogen bonding capabilities may affect the protein structure markedly.

None of these observations conflict with the NMR spectroscopic or relaxation data reviewed here. The NMR data require that the lifetimes of practically all these water molecule structures be on the order of tens of nanoseconds or less. In energetic terms, these results simply mean either that the potential minima are not very deep or that the activation barriers for their displacement are small. This conclusion is not surprising. Consider, for example, the energies associated with alkali metal ion hydration (62). In the gas phase, the enthalpy change for adding successive water molecules to the ion may be very large; however, the water molecule lifetimes in the first coordination sphere of these ions in aqueous solutions may also be in the range of tens of picoseconds.

In summary, although the water near a protein surface may be ordered by a variety of interactions, such as hydrogen bonds and ion-dipole effects, with the exception of a very few molecules, the residence time at any one position is very short, on the order of tens of picoseconds.

Acknowledgments

I am grateful to the many scientists, friends, and collaborators who have participated in the water struggle and discussed a number of ideas over the years. In particular, I thank Drs. Scott D. Kennedy, Cathy C. Lester, Scott D. Swanson, Jonathan Grad, Cynthia L. Jackson, Denise

Hinton, Dawei Zhou, Huiming Zhang, William Shirley, Edward Hsi, Seymour H. Koenig, Rodney D. Brown, Bing M. Fung, Hong Yeung, R. Mark Henkelman, Jack Freed, Jean-Pierre Korb, Wilmer Miller, Rufus Lumry, and Thomas Schleich. I also thank Mindy Whaley for technical assistance. This work was supported by National Institutes of Health, GM-34541, The National Science Foundation, The University of Minnesota, The University of Rochester, and The University of Virginia.

Literature Cited

1. Abragam A. 1961. Liquids and gases. In *Principles of Nuclear Magnetism*, 8:264–353. Oxford, UK: Clarendon. 599 pp.
2. Adler RS, Yeung HN. 1993. Transient decay of longitudinal magnetization in heterogeneous spin systems under selective saturation. III. Solution by projection operators. *J. Magn. Reson. Ser. A* 104:321–30
3. Akasaka K. 1981. Longitudinal relaxation of proton under cross saturation and spin diffusion. *J. Magn. Reson.* 45:337–43
4. Andrew ER, Bone DN, Bryant DJ, Cashell EM, Gaspar R, Meng QA. 1982. Proton relaxation studies of dynamics of proteins in the solid state. *Pure Appl. Chem.* 54:585–94
5. Andrew ER, Bryant DJ, Cashell EM. 1980. Proton magnetic relaxation of proteins in the solid state: molecular dynamics of ribonuclease. *Chem. Phys. Lett.* 69:551–54
6. Andrew ER, Bryant DJ, Cashell EM, Meng QA. 1981. A proton NMR study of relaxation and dynamics in polycrystalline insulin. *FEBS Lett.* 126:208–10
7. Andrew ER, Bryant DJ, Cashell, EM Meng QA. 1982. Solid state dynamics of proteins by nuclear magnetic relaxation. *Phys. Lett.* 88A:487–90
8. Andrew ER, Bryant DJ, Tizvi TZ. 1983. The role of water in the dynamics and proton relaxation of solid proteins. *Chem. Phys. Lett.* 95:463–66
9. Andrew ER, Gaspar R Jr, Vennart W. 1978. Proton magnetic relaxation in solid poly(L-alanine), poly(L-Lucine), poly(L-valine), and polyglycine. *Biopolymers* 17:1913–25
10. Andrew ER, Green TJ, Hoch MJR. 1978. Solid-state proton relaxation of biomolecular components. *J. Magn. Reson.* 29:331–39
11. Aso M, Kakishita M. 1985. Evaluation of the cross relaxation rate between water protons and macromolecule protons by time dependent NOE. *Magn. Reson. Med.* 5:33–39
12. Bryant RG. 1995. Magnetization transfer, cross relaxation in tissue. In *Encyclopedia of Magnetic Resonance*, ed. IR Young. New York: Elsevier. In press
13. Bryant RG, Jarvis M. 1984. Nuclear magnetic relaxation dispersion in protein solutions: a test of proton-exchange coupling. *J. Phys. Chem.* 88: 1323–24
14. Bryant RG, Jentoft JE. 1974. NMR relaxation in lysozyme crystals. *J. Am. Chem. Soc.* 96:297–99
15. Bryant RG, Mendelson D, Lester CC. 1991. The magnetic field dependence of protein proton spin relaxation in tissues. *Magn. Reson. Med.* 21: 117–26
16. Bryant RG, Shirley WM. 1980. Dynamical deductions from NMR relaxation measurements at the water-protein interface. *Biophys. J.* 32:3–16
17. Caines GH, Schleich T, Rydzewski JM. 1991. Incorporation of magnetization transfer into the formalism for rotating-frame spin-lattice proton NMR relaxation in the presence of

an off-resonance irradiation field. *J. Magn. Reson.* 95:558–66
18. Clore GM, Bax A, Omichinski JG, Gronenborn AM. 1994. Localization of bound water in the solution structure of a complex of the erythroid transcription factor GATA-1 with DNA. *Structure* 2:89–94
19. Conti S. 1986. Proton magnetic relaxation dispersion in aqueous biopoymer systems I. Fibrinogen solutions. *Mol. Phys.* 59:449–82
20. Conti S. 1986. Proton magnetic relaxation dispersion in aqueous biopolymer systems. II. Fibrin gels. *Mol. Phys.* 59:483–505
21. Denisov VP, Halle B. 1994. Dynamics of internal and external hydration of globular proteins. *J. Am. Chem. Soc.* 116:10324–25
22. Denisov VP, Halle B. 1995. Hydrogen exchange and protein hydration: the deuteron spin relaxation dispersions of bovine pancreatic trypsin inhibitor and ubiquitin. *J. Biol. Chem.* 245:698–709
23. Denisov VP, Halle B. 1995. Protein hydration dynamics in aqueous solution: a comparison of bovine pancreatic trypsin inhibitor and ubiquitin by oxygen-17 spin relaxation dispersion. *J. Mol. Biol.* 245:682–97
24. Eads TM, Axelson DE. 1995. Nuclear cross relaxation spectroscopy and single point imaging measurements of solids and solidity in foods. In *Magnetic Resonance in Food Science,* ed. PS Belton, I Delgadillo, AM Gil, GA Webb, pp. 230–42. London: R. Soc. Chem. 292 pp.
25. Edzes HT, Samulski ET. 1977. Cross relaxation and spin diffusion in the proton NMR of hydrated collagen. *Nature* 265:521–23
26. Edzes HT, Samulski ET. 1978. Measurement of cross relaxation effects in ^{1}H NMR spin-lattice relaxation of water in biological systems: hydrated collagen and muscle. *J. Magn. Reson.* 31:207–29
27. Eisenstadt M. 1985. NMR relaxation of protein and water protons in diamagnetic hemoglobin solutions. *Biochemistry* 24:3407–21
28. Ernst JA, Clubb RT, Zhou H-X, Gronenborn AM, Clore GM. 1995. Demonstration of positionally disordered water within a protein hydrophobic cavity by NMR. *Science* 267: 1813–17
29. Frank HS, Evans MW. 1945. Free volume and entropy in condensed systems. III. Entropy in binary liquid mixtures: partial molal entropy in dilute solutions: structure and thermodynamics in aqueous electrolytes. *J. Chem. Phys.* 13:507–32
30. Frey M. 1993. Water structure of crystallized proteins: high resolution studies. In *Water and Biological Macromolecules*, ed. E Westhof, pp. 98–147. Boca Raton, Fla: CRC Press. 466 pp.
31. Fung BM, McGaughy TW. 1980. Cross relaxation in hydrated collagen. *J. Magn. Reson.* 39:423–20
32. Fung BM, McGaughy TW. 1979. Study of spin-lattice and spin-spin relaxation times of ^{1}H, ^{2}H, and ^{17}O in muscle water. *Biophys. J.* 28: 293–304
33. Gaspar R, Andrew ER, Bryant DJ, Cashell EM. 1982. Dipolar relaxation and slow molecular motions in solid proteins. *Chem. Phys. Lett.* 86: 327–30
34. Gerothanassis IP. 1994. Multinuclear and multidimensional NMR methodology for studying individual water molecules bound to peptides and proteins in solution: principles and applications. *Prog. Nucl. Magn. Reson. Spectrosc.* 26:171–237
35. Glasel JA, 1972. Nuclear magnetic resonance studies on water and ice. In *Water: A Comprehensive Treatise,* ed. F Franks, 1:215–54. New York: Plenum
36. Goodfellow JM, Thanki N, Thornton JM. 1993. Hydration of amino acids in protein crystals. In *Water and Biological Macromolecules*, ed. E Westhof, pp. 63–97. Boca Raton, Fla: CRC Press. 466 pp.
37. Grad J, Bryant RG. 1990. Nuclear magnetic cross-relaxation spectroscopy. *J. Magn. Reson.* 90:1–8
38. Grazesiek S, Bax A. 1993. The importance of not saturating water in protein NMR. Applications to sensitivity enhancement and NOE measurements. *J. Am. Chem. Soc.* 115: 12593–94
39. Grazesiek S, Bax A, Nicholson LK, Yamazaki T, Wingfield P, et al. 1994. NMR evidence for the displacement of conserved interior water molecule in HIV protease by non-peptide cyclic urea based inhibitor. *J. Am. Chem. Soc.* 116:1581–82

40. Gregory RB, ed. 1995. *Protein Solvent Interactions*. New York: Dekker
41. Hahn E. 1950. Spin echos. *Phys. Rev.* 80:580–94
42. Halle B, Andersson T, Forsen S, Lindman B. 1981. Protein hydration from water oxygen-17 magnetic relaxation. *J. Am. Chem. Soc.* 103: 500–8
43. Halle B, Piculell L. 1986. Water spin relaxation in colloidal systems. III. Interpretation of the low-frequency dispersion. *J. Chem. Soc. Faraday Trans. 1* 82:415–29
44. Halle B, Wennerstrom H. 1981. Interpretation of magnetic resonance data from water nuclei in heterogeneous systems. *J. Chem. Phys.* 75: 1928–43
45. Hallenga K, Koenig SH. 1976. Protein rotational relaxation as studied by solvent ^{1}H and ^{2}H magnetic relaxation. *Biochemistry* 15:4255–64
46. Deleted in proof
47. Henkelman RM, Huang X, Xiang QS, Stanisz GJ, Swanson SD, Bronskill MJ. 1993. Quantitative interpretation of magnetization transfer. *Magn. Reson. Med.* 29:759–66
48. Hertz HG. 1967. Microdynamic behavior of liquids studied by NMR relaxation times. *Prog. Nucl. Magn. Reson. Spectrosc.* 3:159–230
49. Hills BP. 1992. The proton cross-relaxation model of water relaxation in biopolymer systems. II. The sol and gel states of gelatine. *Mol. Phys.* 76: 509–23
50. Hills BP. 1992. The proton exchange cross-relaxation model of water relaxation in biopolymer systems. *Mol. Phys.* 76:489–508
51. Hills BP, Takacs SF, Belton PS. 1989. The effects of proteins on the proton NMR transverse relaxation times of water. I. Native bovine serum albumin. *Mol. Phys.* 67: 903–18
52. Hills BP, Takacs SF, Belton PS. 1989. The effects of proteins on the proton NMR transverse relaxation time of water. II. Protein aggregation. *Mol. Phys.* 67:919–37
53. Hilton BD, Hsi E, Bryant RG. 1977. ^{1}H nuclear magnetic resonance relaxation of water on lysozyme powders. *J. Am. Chem. Soc.* 99:8483–90
54. Hinton DP, Bryant RG. 1995. ^{1}H magnetic cross-relaxation between multiple solvent components and rotationally immobilized protein. *Magn. Reson. Med.* In press
55. Hinton DP, Bryant RG. 1995. Measurement of Protein Preferential Solvation by Z-spectroscopy. *J. Phys. Chem.* 98:7939–41. Correction, 98: 12458
56. Hsi E, Bryant RG. 1975. NMR relaxation in frozen lysozyme solutions. *J. Am. Chem. Soc.* 97:3220–21
57. Hsi E, Bryant RG. 1977. Nuclear magnetic resonance relaxation in cross-linked lysozyme crystals: an isotope dilution experiment. *Arch. Biochem. Biophys.* 183:588–91
58. Hsi E, Bryant RG. 1977. Nuclear magnetic resonance relaxation studies of carbonic anhydrase derivatives in frozen solutions. *J. Phys. Chem.* 81:462–65
59. Hsi E, Jentoft JE, Bryant RG. 1976. Nuclear magnetic resonance relaxation in lysozyme crystals. *J. Phys. Chem.* 80:422–16
60. Hsi E, Mason R, Bryant RG. 1976. Magnetic resonance studies of alphachymotrypsin crystals. *J. Phys. Chem.* 80:2592–97
61. Kalk A, Berendsen HJC. 1976. Proton magnetic relaxation and spin diffusion in proteins. *J. Magn. Reson.* 24:343–66
62. Kebarle P. 1977. Ion thermochemistry and solvation from gas phase ion equilibria. *Annu. Rev. Phys. Chem.* 28:445–76
63. Kimmich R. 1985. Characteristic power laws and dimensionality of dynamic processes in condensed polymer systems as seen by NMR techniques. *Helv. Phys. Acta* 58:102– 20
64. Kimmich R. 1990. Dynamic processes in aqueous protein systems. Molecular theory and NMR relaxation. *Makromol. Chem.* 34:237–48
65. Kimmich R, Doster W. 1976. Monte Carlo study of defect fluctuations and reptations in polymer melts. *J. Polym. Sci.: Polym. Phys. Ed.* 14: 1671–82
66. Kimmich R, Gneiting T, Kotitschke K, Schnur G. 1990. Fluctuations, exchange processes, and water diffusion in aqueous protein systems: a study of BSA by diverse NMR techniques. *Biophys. J.* 58:1183–97
67. Deleted in proof
68. Kimmich R, Nusser W, Gneiting T. 1990. Molecular theory for NMR relaxation in protein solutions and tis-

sue. Surface diffusion and free volume analogy. *Colloids Surf.* 45: 283–302

69. Kimmich R, Schnur G, Kopf M. 1988. The tube concept of macromolecular liquids in the light of NMR experiments. *Prog. Nucl. Magn. Reson. Spectrosc.* 20:385–421
70. Kimmich R, Winter F. 1985. Double-diffusive fluctuations and the $\{n\}^{3/4}$ law of spin-lattice relaxation in biopolymers. *Prog. Colloid Polym. Sci.* 71:66–70
71. Kimmich R, Winter F, Nusser W, Spohn KH. 1986. Interactions and fluctuations deduced from proton field-cycling relaxation spectroscopy of polypeptides, DNA, muscles and algae. *J. Magn. Reson.* 68:263–82
72. Deleted in proof
73. Klotz IM. 1958. Protein hydration and behavior. *Science* 128:815–22
74. Koenig SH. 1980. The dynamics of water-protein interactions. Results from measurements of nuclear magnetic relaxation dispersion. In *Water in Polymers,* ed. SP Rowland, 127: 157–76. *ACS Symp. Ser.* Washington, DC: Am. Chem. Soc.
75. Koenig SH, Brown RD III. 1984. Determinants of proton relaxation rates in tissue. *Magn. Reson. Med.* 1: 437–49
76. Koenig SH, Brown RD III. 1991. Field cycling relaxometry of protein solutions and tissue: implications for MRI. *Prog. Nucl. Magn. Reson. Spectrosc.* 22:487–567
77. Koenig SH, Brown RD III. 1993. A molecular theory of relaxation and magnetization transfer applied to cross-linked BSA. A model for tissue. *Magn. Reson. Med.* 30:685–95
78. Koenig SH, Brown RD III, Pande A, Ugolini R. 1993. Rotational inhibition and magnetization transfer in α-crystallin solutions. *J. Magn. Reson. Ser. B* 101:172–77
79. Koenig SH, Brown RD III, Ugolini R. 1993. A unified view of relaxation in protein solutions and tissue including hydration and magnetization transfer. *Magn. Reson. Med.* 19: 77–83
80. Koenig SH, Brown RD III, Ugolini R. 1993. Magnetization transfer in cross-linked BSA at 200 MHz. A model for tissue. *Magn. Reson. Med.* 29:311–16
81. Koenig SH, Bryant RG, Hallenga K, Jacobs GS. 1978. Magnetic cross-relaxation among protons in protein solution. *Biochemistry* 17:4348–58
82. Koenig SH, Hallenga K, Shporer M. 1975. Protein-water interactions studied by solvent ^{1}H, ^{2}H, and ^{17}O magnetic relaxation. *Proc. Natl. Acad. Sci. USA* 72:2667–71
83. Koenig SH, Schillinger WE. 1969. Nuclear magnetic relaxation dispersion in protein solutions. *J. Biol. Chem.* 244:3283–89
84. Kotitschke K, Kimmich R, Rommel E, Parak F. 1990. NMR study of diffusion in protein hydration shells. *Prog. Colloid Polym. Sci.* 83:211–15
85. Kriwacki RW, Hill RB, Flanagan JM, Caradonna JP, Prestegard JH. 1993. New NMR methods for the characterization of bound waters on macromolecules. *J. Am. Chem. Soc.* 115:8907–11
86. Krynicki K. 1966. Proton spin-lattice relaxation in pure water between 0°C and 100°C. *Physica* 32:167–80
87. Kuntz ID Jr, Brassfield TS, Law GD, Purcell GV. 1969. Hydration of macromolecules. *Science* 163:1329–31
88. Lamb WJ, Brown DR, Jonas J. 1981. Temperature and density dependence of the proton lifetime in liquid water. *J. Phys. Chem.* 85:3883–87
89. Lenk R. 1986. *Fluctuations, Diffusion and Spin Relaxation.* Amsterdam: Elsevier
90. Lester CC, Bryant RG. 1991. Water proton nuclear magnetic relaxation in heterogeneous systems: hydrated lysozyme results. *Magn. Reson. Med.* 22:143–53
91. Lumry R, Matsumiya H, Bovey FA, Kowalsky A. 1961. The study of the structure and denaturation of heme proteins by nuclear magnetic relaxation. *J. Phys. Chem.* 65:837–43
92. McConnell J. 1987. *The Theory of Nuclear Magnetic Relaxation in Liquids.* Cambridge/New York: Cambridge Univ. Press
93. Meiboom S. 1961. Nuclear magnetic resonance study of the proton transfer in water. *J. Chem. Phys. 34*:375–88
94. Morrison C, Henkelman RM. 1995. A model for magnetization transfer in tissues. *Magn. Reson. Med.* 33: 475–82
95. Nusser W, Kimmich R. 1990. Protein backbone fluctuations and NMR field cycling relaxation spectroscopy. *J. Phys. Chem.* 94:5637–39

96. Nusser W, Kimmich R, Winter F. 1988. Solid state NMR study of protein polypeptide backbone fluctuations interpreted by multiple trapping of dilating defects. *J. Phys. Chem.* 92:6806–14
97. Otting G, Liepinsh E, Farmer BT II, Wuthrich K. 1991. Protein hydration studied with homonuclear 3D ^{1}H NMR experiments. *J. Biol. Nucl. Magn. Reson.* 1:209–15
98. Otting G, Liepinsh E, Wuthrich K. 1991. Proton exchange with internal water molecules in the protein BPTI in aqueous solution. *J. Am. Chem. Soc.* 113:4363–69
99. Otting G, Liepinsh E, Wuthrich K. 1991. Protein hydration in aqueous solution. *Science* 254:974–80
100. Otting G, Wuthrich K. 1989. Studies of protein hydration in aqueous solution by direct NMR observation of individual protein-bound water molecules. *J. Am. Chem. Soc.* 111: 1871–75
101. Polnaszek CF, Bryant RG. 1984. Nitroxide radical induced solvent proton relaxation: measurement of localized translational diffusion. *J. Chem. Phys.* 81:4038–45
102. Polnaszek CF, Bryant RG. 1984. Self diffusion of water at the protein surface: a measurement. *J. Am. Chem. Soc.* 106:428–29
103. Qi PX, Urbauer JL, Fuentes EJ, Leopold MF, Wand AJ. 1994. Structural water in oxidized and reduced horse heart cytochrome c. *Struct. Biol.* 1: 378–82
104. Redfield AG, Fite W, Bleich HE. 1968. Precision high speed current regulators for occasionally switched inductive loads. *Rev. Sci. Instrum.* 39:710–15
105. Saenger W. 1987. Structure and dynamics of water surrounding biomolecules. *Annu. Rev. Biophys. Biophys. Chem.* 16:93–114
106. Schauer G, Kimmich R, Nusser W. 1988. Deuteron field cycling relaxation spectroscopy and translational water diffusion in protein hydration shells. *Biophys. J.* 53:397– 404
107. Deleted in proof
108. Schleich T, Caines CH, Rydzewski JM. 1992. Off-resonance rotating frame spin-lattice relaxation: theory and in vivo MRS and MRI applications. In *Biological Magnetic Resonance*, ed. LJ Berliner, J Reuben, 11: 55–134. New York: Plenum
109. Shimizu H. 1962. Effect of molecular shape on nuclear magnetic relaxation. *J. Chem. Phys.* 37:765–78
110. Shirley WM, Bryant RG. 1982. Proton nuclear spin relaxation and molecular dynamics in the lysozyme-water system. *J. Am. Chem. Soc.* 104: 2910–18
111. Spohn KH, Kimmich R. 1983. Characterization of the mobility of various chemical groups in the purple membrane of halobacterium halobium by ^{13}C, ^{31}P and ^{2}H solid state NMR. *Biophys. Biochem. Res. Commun.* 114: 713–20
112. Stejskal EO, Tanner JE. 1965. Spin diffusion measurements: spin echos in the presence of a time-dependent field gradient. *J. Chem. Phys.* 42: 288–92
113. Stoesz JD, Redfield AG, Malinowski D. 1978., Cross relaxation and spin diffusion effects on the ^{1}H NMR of biopolymers in water. *FEBS Lett.* 91: 320–24
114. Winter F, Kimmich R. 1982. Spin lattice relaxation of dipole nuclei ($I = 1/2$) coupled to quadrupole nuclei ($S = 1$). *Mol. Phys.* 45:33–49
115. Winter F, Kimmich R. 1985. ^{14}N–^{1}H and ^{2}H–^{1}H cross relaxation in hydrated proteins. *Biophys. J.* 48: 331–35
116. Woessner DE. 1962. Nuclear spin relaxation in ellipsoids undergoing rotational brownian motion. *J. Chem. Phys.* 37:647–54
117. Woessner DE. 1977. Nuclear magnetic relaxation and structure in aqueous heterogeneous systems. *Mol. Phys.* 34:899–920
118. Woessner DE. 1980. An NMR investigation into the range of the surface effects on the rotation of water molecules. *J. Magn. Reson.* 39:297–308
119. Wolff SD, Balaban RS. 1989. Magnetization transfer contrast (MTC) and tissue water proton relaxation in vivo. *Magn. Reson. Med.* 10:135–44
120. Wu JY, Bryant RG, Eads TM. 1992. Detection of solid-like domains in starch by cross-relaxation NMR spectroscopy. *J. Agric. Food Chem.* 40:449–55
121. Wu X. 1991. Lineshape of magnetization transfer via cross relaxation. *J. Magn. Reson.* 94:186–90
122. Xu RX, Meadows RP, Fesik SW.

1993. Heteronuclear 3D NMR studies of water bound to an FK506 binding protein/immunosuppressant complex. *Biochemistry* 32:2473–80

123. Yang H, Schleich T. 1994. Modified Jeneer solid echo pulse sequences for the measurement of proton dipolar spin-lattice relaxation time (T_{1D}) of tissue solid-like macromolecular components. *J. Magn. Reson. Ser. B* 105:205– 10
124. Yang H, Schleich T. 1994. T_1 discrimination contributions to proton magnetization transfer in heterogeneous biological systems. *Magn. Reson. Med.* 32:16– 22
125. Yeung HN, Adler RS, Swanson SD. 1994. Transient decay of longitudinal magnetization in heterogeneous spin systems under selective saturation. IV. Reformulation of the spin-bath-model equations by the Redfield-Provotorov theory. *J. Magn. Reson. Ser. A* 106:37–45
126. Yeung HN, Swanson SD. 1992. Transfer decay of longitudinal magnetization in heterogeneous systems under selective saturation. *J. Magn. Reson.* 99:466–79
127. Zhou D, Bryant RG. 1994. Magnetization transfer, cross-relaxation, and chemical exchange in rotationally immobilized protein gels. *Magn. Reson. Med.* 32:725–32
128. Zhou D, Bryant RG. 1995. ^{1}H magnetic relaxation dispersion in protein solutions. unpublished
129. Zhou D, Bryant RG. 1995. Water molecule binding and lifetimes on the DNA, d(CGCGAATTCGCG)$_2$. *J. Biomolec. NMR*. Submitted

Annu. Rev. Biophys. Biomol. Struct. 1996. 25:55–78

USING SELF-ASSEMBLED MONOLAYERS TO UNDERSTAND THE INTERACTIONS OF MAN-MADE SURFACES WITH PROTEINS AND CELLS

Milan Mrksich and George M. Whitesides

Department of Chemistry, Harvard University, Cambridge, Massachusetts 02138

KEY WORDS: biocompatibility, biomaterials, biosurfaces

ABSTRACT

Self-assembled monolayers (SAMs) formed on the adsorption of long-chain alkanethiols to the surface of gold or alkylsilanes to hydroxylated surfaces are well-ordered organic surfaces that permit control over the properties of the interface at the molecular scale. The abililty to present molecules, peptides, and proteins at the interface make SAMs especially useful for fundamental studies of protein adsorption and cell adhesion. Microcontact printing is a simple technique that can pattern the formation of SAMs in the plane of the monolayer with dimensions on the micron scale. The convenience and broad application offered by SAMs and microcontact printing make this combination of techniques useful for studying a variety of fundamental phenomena in biointerfacial science.

CONTENTS

1056-8700/96/0610-0055$08.00

PERSPECTIVES AND OVERVIEW

Man-made surfaces in contact with biological environments are important in biology, biotechnology, and medicine. These surfaces occur in tools and reagents for studies in molecular and cell biology; in substrates for enzyme-linked immunosorbent assay (ELISA), cell culture, and tissue engineering; in materials for contact lenses, dental prostheses, and devices for drug delivery; in coatings for catheters, indwelling sensors, and implant devices; and in materials for chromatography and storage of proteins (50, 65). The first event that usually occurs on contact of the synthetic material with a medium that contains dissolved protein is the adsorption of protein to the surface: other responses, such as the attachment of cells, are secondary and depend on the nature of the adsorbed layer of protein. Because of its central importance, the adsorption of protein to man-made surfaces has been studied extensively. Although much has been learned, there are still no mechanistic models that rationalize (or predict) the interaction of a protein with a surface in molecular detail. A broad goal of research in this area is to understand the interactions of proteins with surfaces at the level of detail that is now common for characterization of the interactions of proteins with water, ligands, and other proteins in solution.

Model systems designed to elucidate these mechanisms must have three components: (*a*) a protein with known high-resolution structure and properties (e.g. stability, conformational dynamics, and tendency to aggregate); (*b*) a structurally well-defined surface with properties that can be tailored and controlled simply and allows the complex functionality relevant to biochemistry to be introduced at the surface; and (*c*) one or several analytical techniques that can measure adsorption of protein in situ and in real time. Numerous proteins suitable for these types of studies are now available. Analytical methodologies appropriate for these studies (e.g. surface plasmon resonance spectroscopy and fluorescence spectroscopy) are becoming available. However, X-ray and electron diffraction studies of adsorbed crystalline monolayers are still the only techniques that can provide information at the molecular scale, and these techniques are applicable only in special cases. The absence of methods to prepare well-defined surfaces has been a problem; for those surfaces that are well defined, such as metals, metal oxides, and crystals, surface properties cannot be controlled precisely.

Self-assembled monolayers (SAMs)—particularly those formed by the adsorption of long-chain alkanethiols on gold—are a recently developed class of organic surfaces that are well suited for studying interactions of surfaces with proteins and cells. The ability to control the composition and properties of SAMs precisely through synthesis, combined with the simple methods that can pattern their functional groups in the plane of the monolayer, makes this class of surfaces the best now available for fundamental mechanistic studies of protein adsorption and cell adhesion. Here we review the use of two classes of SAMs, alkanethiolates on gold and alkylsiloxanes on hydroxylated surfaces, in the study of processes that occur at the interface between a man-made surface and a biological medium. We do not discuss much of the excellent work in which other classes of surfaces, such as Langmuir-Blodgett films, lipid bilayers, and polymers, have been used (for reviews, see 44, 60, 61, 75).

BACKGROUND

Man-Made Surfaces That Contact Biological Media

Many early studies of the interactions of artificial surfaces with biological media were motivated by problems associated with the formation of thrombus on foreign surfaces that contact blood—a process that is now relatively well understood and involves the adsorption of proteins intimately (73). For the past 3 decades, researchers have sought to understand the interactions of man-made materials with proteins (and processes dependent on protein adsorption), with applications in biocompatible materials as a central motivation. Although many materials have been identified that are compatible with tissue to varying degrees [e.g. titanium (implants), polymethylmethacrylate (contact lenses), ceramics (dental prosthesis), polyurethanes (artificial heart), pyrolytic carbon (heart valves)] there has been much less progress in the elucidation of the mechanisms by which these materials function.

Adsorption of Protein

A survey of the extensive literature on the adsorption of proteins to man-made surfaces is not practical in this review (for reviews, see 44, 61, 63, 75). The adsorption processes are complicated. Even in the simplest case, where a single, well-defined protein adsorbs to a uniform, well-defined surface, a substantial range of processes is usually involved (Figure 1). After an initial adsorption of a protein to a surface, the protein can (Figure 1*a*) dissociate from the surface and return to solution; (*d*) change orientation; (*d*) change conformation but retain biological activity; (*f*) denature and lose activity; or (*g*) exchange with other proteins in solution. These processes are complicated further by a range of conformationally altered and/or denatured states accessible to the adsorbed protein and by the many different microenvironments at the surface created by heterogeneities in the surface and the presence and conformations of other proteins. Lateral protein–protein interactions may dominate the protein–surface interactions. Because many of these processes are essentially irreversible, models must emphasize the kinetic aspects of protein adsorption (59).

ORIENTATION AND CONFORMATION OF ADSORBED PROTEIN Most studies have generated empirical models by analyzing the amount and rates of

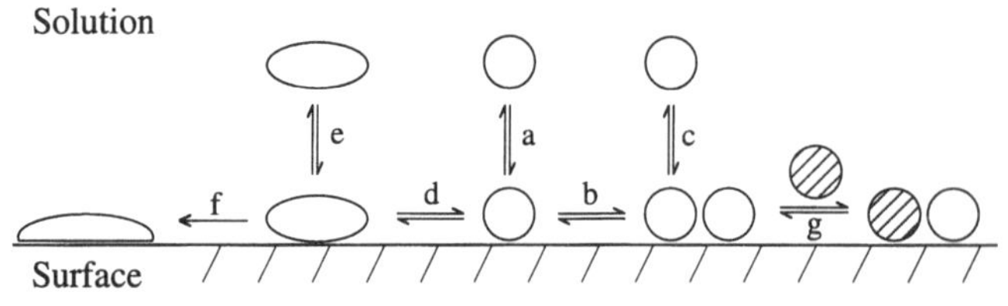

Figure 1 The complexities associated with studies of protein adsorption. Several equilibria must be considered on adsorption of a protein to a surface (*a*); lateral mobility of the adsorbed protein (*b*); dissociation of a protein adjacent to another protein (*c*); reversible denaturation and changes in conformation of the protein (*d*); dissociation of the altered protein (*e*); denaturation of the protein that results in irreversible adsorption (*f*); and exchange of the protein with a protein from solution (*g*). This scheme is not complete but is complicated further by the many different conformations and environments available to an adsorbed protein.

protein adsorption; few have investigated adsorption at the molecular level. The information required for a detailed molecular description of these processes, including information concerning the orientation and conformation of adsorbed proteins as a function of time and conditions, has been difficult to obtain. Few analytical methods exist—and none with the power to reveal structure comparable to X-ray crystallography or multidimensional NMR spectroscopy—that can characterize the conformation and orientation of proteins adsorbed to surfaces. Infrared spectroscopy has been used to assess the degree of denaturation of fibronectin adsorbed to SAMs that present different functional groups (7). Lee & Belfort (37) correlated the activity of adsorbed RNase A to a model that involved conversion between two orientations of the protein at the surface. Darst et al (14) probed the orientation and conformation of myoglobin adsorbed to a polydimethylsiloxane surface by using a panel of five monoclonal antibodies that had known epitopes on the protein. This latter method and related footprinting methods that use proteases (87) or selective chemical reagents for modification of residues of proteins (69) are perhaps the most general methods for the direct characterization of unlabeled proteins adsorbed to surfaces. Kornberg and coworkers (13) have used electron diffraction to determine the structure of two-dimensional crystals of the protein streptavidin adsorbed to a biotinylated lipid layer. Rennie and coworkers (18) have used neutron reflection to determine the structure of β-casein adsorbed to hydrophobic alkylsiloxane monolayers. An emerging technique based on X-ray standing waves also provides direct structural information with near-atomic resolution for cases in which the layer of protein is ordered (6).

MEASURING PROTEIN ADSORPTION The most useful of the many analytical techniques that have been used to measure protein adsorption are those that are compatible with a variety of surfaces and that provide measurements in situ and in real time (59). Methods that measure the dielectric properties of an interface [surface plasmon resonance (SPR) spectroscopy (41), waveguide interferometry (62), and ellipsometry (45)] and those that measure changes in the resonance frequency of a piezoelectric material [quartz crystal microbalance (77), surface acoustic wave (77), and acoustic plate mode (12) devices] are particularly well suited. The primary drawback of these methods is that they measure bulk properties of the interface and provide little or no detail about atomic-level interactions. Surface plasmon resonance spectroscopy is particularly well suited for use in conjunction with SAMs, because both techniques use thin films of gold as substrates: A commercial SPR

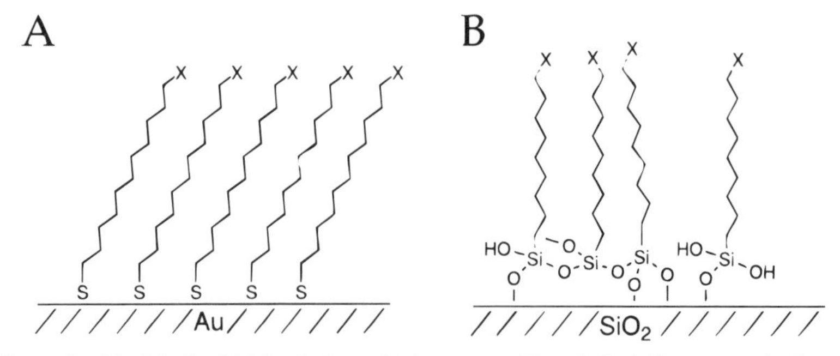

Figure 2 Models for SAMs of alkanethiolates on gold and alkylsiloxanes on hydroxylated surfaces. (*A*) The thiol groups coordinate to the hollow threefold sites of the gold (111) surface, and the alkyl chains pack in a quasi-crystalline array. (*B*) The conformations of alkylsilanes and the details of their bonding to surface hydroxyl groups are less clear; a mixture of possible conformations and geometries is probably involved. The surface properties of both SAMs are controlled by controlling the terminal function group X.

instrument has demonstrated many of the characteristics required for efficient studies of adsorption of proteins to functionalized SAMs on gold (53).

Self-Assembled Monolayers of Alkanethiolates on Gold

Self-assembled monolayers of alkanethiolates on gold form on the adsorption of a long-chain alkanethiol [$X(CH_2)_nSH$, $n = 11 - 18$) from solution (or vapor) to a gold surface (Equation 1). The structure of these SAMs is now well established (16, 79):

$$RSH + Au(0)_n \rightarrow RS^-Au(I) \;\; Au(0)_n + \tfrac{1}{2}H_2(?).$$

The sulfur atoms coordinate to the gold atoms of the surface, and the trans-extended alkyl chains are tilted approximately 30 degrees from the normal to the surface (Figure 2). The properties of the interface depend on the terminal functional group X of the precursor alkanethiol; even structurally complex groups can be introduced onto the surface through straightforward synthesis. The surface chemistry of SAMs can be controlled further by forming so-called mixed SAMs from solutions of two or more alkanethiols. Self-assembled monolayers are stable in air or in contact with water or ethanol for periods of several months; they desorb at temperatures greater than 70°C or when irradiated with UV light in the presence of oxygen. They have been used in cell culture for periods of days. Self-assembled monolayers that are supported on gold 5–10 nm in thickness (on glass slides) are transparent, and those supported on gold of thickness greater than 100 nm are opaque and reflective (15); even the thin films of gold are electrically conductive.

Self-assembled Monolayers of Alkylsiloxanes

Alkylsiloxanes are obtained by reaction of a hydroxylated surface (usually the native oxide of silicon or glass) with a solution of alkyltrichlorosilane (or alkyltriethoxysilane) (48, 56, 72). The reactive siloxane groups condense with water and with hydroxyl groups of the surface and neighboring siloxanes to form a cross-linked network; the bonding arrangement is not well defined but depends on the conditions used to form the SAM (Figure 2). These SAMs are significantly more stable thermally than alkanethiolates on gold and do not require evaporation of a layer of metal for preparation of substrates. The siloxane monolayers are limited, however, in the range of functional groups that can be displayed at the surface by the reactivity of the alkyltrichlorosilane groups of the precursors and by the technical difficulty of introducing functional groups once the monolayer has formed.

The Physical-Organic Chemistry of Self-Assembled Monolayers

An important goal in interfacial science is to understand the relationship between the microscopic structure of a surface and its macroscopic properties; this relationship is particularly relevant in studies of protein adsorption where hydrophobic forces are dominant (71). Self-assembled monolayers on gold—and, to a lesser extent, alkylsiloxanes—offer the level of structural control required for detailed studies of adsorption processes. Studies of the influence of a terminal functional group X of a SAM on the wettability of the surface reveal that the hydrophobicity of the surfaces can be controlled precisely (2). Self-assembled monolayers that present polar functional groups (e.g. carboxylic acid and hydroxyl) are wetted by water. Those that present nonpolar, organic groups (e.g. trifluoromethyl and methyl) are autophobic and emerge dry from water. Monolayers that present fluorinated groups are more water repellent than Teflon.

INTERACTIONS OF PROTEINS WITH SELF-ASSEMBLED MONOLAYERS

Adsorption of Proteins to Self-Assembled Monolayers

The adsorption of several model proteins to SAMs that present different functional groups (e.g. alkyl, perfluoroalkyl, amide, ester, alcohol, nitrile, carboxylic acid, phosphonic, boric acids, amines, and heterocycles) correlates approximately with the hydrophobicity of the surfaces (43, 57); the degree of denaturation, as inferred by the density of a layer of adsorbed protein, increases with the hydrophobicity of the surface and decreases with the concentration of protein in the contacting solution. Adsorption on hydrophobic surfaces often is irreversible kinetically, but the protein adlayer can be removed with detergents or replaced by other proteins in solution. Although SAMs that present ionic groups have been used extensively to control protein adsorption, less is known about the relationships between the properties of charged surfaces and the structure and properties of the layer of protein. Vroman (73) has used ellipsometry and immunologic identification of adsorbed proteins to characterize the exchange of plasma proteins at hydrophilic glass surfaces.

Surfaces That Resist the Adsorption of Proteins

Much effort has been directed toward the identification of biologically "inert" materials, i.e. materials that resist the adsorption of protein. The most successful method to confer this resistance to the adsorption of protein has been to coat the surface with poly(ethylene glycol) (PEG) (20, 22); a variety of methods, including adsorption, covalent immobilization, and radiation cross-linking, have been used to modify surfaces with PEG (22). Polymers that comprise carbohydrate units also passivate surfaces, but these materials are less stable and less effective than PEG (41, 76). A widely used strategy is to preadsorb a protein—usually bovine serum albumin—that resists adsorption of other proteins. This strategy suffers from problems associated with denaturation of the blocking protein over time (3) or exchange of this protein with others in solution. A further limitation of this strategy is the inability to present other groups (e.g. ligands, antibodies) at the surface in controlled environments.

Self-assembled monolayers that are prepared from alkanethiols terminated in short oligomers of the ethylene glycol group [$HS(CH_2)_{11}(OCH_2CH_2)_nOH$: $n = 2 - 7$] resist the adsorption of several model proteins, as measured by both ex situ ellipsometry and in situ SPR spectroscopy (53, 58). Even SAMs that contain as much as 50% methyl-terminated alkanethiolates, if mixed with oligo(ethylene glycol)-terminated alkanethiolates, resist the adsorption of protein. Self-assembled monolayers that present oligo(ethylene glycol) groups are useful as controls in studies of the adsorption of proteins to surfaces. The ability to prepare SAMs that present derivatives of these and other groups will be useful for investigating the mechanisms by which these surfaces resist adsorption.

De Gennes, Andrade, and coworkers (29, 30) have proposed that surfaces modified with long PEG chains resist the adsorption of protein by "steric stabilization." In aqueous solution, the PEG chains are solvated and disordered. Adsorption of protein to the surface causes the glycol chains to compress, with concomitant desolvation. Both the energetic penalty of transferring water to the bulk and the entropic penalty incurred on compression of the layer serve to resist protein adsorption. It is not clear that this analysis applies to thin, dense films of oligo(ethylene glycol) groups, as De Gennes and Andrade predicted that surfaces comprising densely packed, nearly crystalline chains of PEG might not resist the adsorption of protein. It is remarkable that SAMs presenting densely packed tri(ethylene glycol) groups resist the adsorption of protein. These layers are almost certainly different in comformational flexibility and solvation than are long, dilute PEG chains. We presume there is sufficient free volume in the glycol layer of the SAMs to allow solvation by water. An understanding of the properties of these SAMs may permit the design of new classes of inert surfaces.

Immobilization of Proteins to Self-Assembled Monolayers

The immobilization of proteins to substrates is important in many areas, ranging from ELISA and cell culture to biosensors; consequently, many strategies have been developed to confine proteins to surfaces (51, 63). Methods that rely on noncovalent association of proteins with surfaces—with the use of both hydrophobic and electrostatic interactions—are the most common and experimentally simplest, but they are also the least well controlled.

Methods that rely on covalent coupling of proteins to surfaces are inherently more controlled and give layers of protein that cannot dissociate from the surface or exchange with other proteins in solution. A variety of surface chemistries have been used; the most successful have been based on the formation of amide and disulfide bonds (26, 74, 78). The selectivity and rapid reaction of thiols with α-haloacetyl groups constitutes a particularly attractive protocol (38). The use of well-defined surfaces and proteins that have only a small number of reactive groups permits a high degree of control over the attached protein. For example, genetic engineering was used to construct a mutant of cytochrome c that had only a single cysteine group; immobilization of this protein to a SAM terminated in thiol groups gave a uniformly oriented layer of protein (26).

A common problem that limits the use of immobilized proteins is denaturation of the protein, with concomitant loss in activity. Methods to increase the lifetimes of these proteins have involved coupling to "inert" materials (76). A commercial SPR instrument that quantitates protein–protein interactions uses a gel layer of carboxylated dextran to stabilize immobilized proteins (41). Self-assembled monolayers that are terminated in oligo(ethylene glycol) groups may have broad usefulness as inert supports, because a variety of reactive groups can be incorporated in SAMs in controlled environments.

Biospecific Adsorption of Proteins to Self-Assembled Monolayers

The design of surfaces to which analytes bind specifically is important for biosensors and other technologies, e.g. affinity chromatography, cell culture, coatings for implants, and artificial organs. These surfaces must possess specificity for a particular protein and simultaneously resist the nonspecific adsorption of other proteins.

Immobilization schemes based on the biotin–streptavidin interaction have been investigated widely. Spinke et al (68) studied the recognition of streptavidin by SAMs that present biotin ligands. The effectively irreversible complexation in this system is useful for many applications, such as immobilization of proteins or nucleic acids, but is not relevant to the weak, reversible recognition that is more common in biology. In an early model system, Mosbach and coworkers (45) used ellipsometry to characterize the reversible binding of lactate dehydrogenase to SAMs that present analogues of nicotinamide adenine dinucleotide (NAD). Several groups have studied the recognition of immobilized antigen by antibodies, because of the availability of antibodies to a variety of antigens and the high specificity displayed by antibodies (35).

Self-assembled monolayers that present oligo(ethylene glycol) groups were used as supports to which ligands for proteins were attached (Figure 3). With the use of SPR spectroscopy, Sigal et al (64) measured the binding of a His-tagged T-cell receptor to a SAM presenting a Ni(II) complex. Likewise, carbonic anhydrase bound to SAMs that presented a benzenesulfonamide group (Figure 4) (52). In both cases, the amount of protein that bound increased with the density of ligand on the SAM. Both SAMs also resisted the nonspecific adsorption of other proteins. The SAM terminated in benzenesulfonamide groups could be used to measure the concentration of carbonic anhydrase (CA) in a complex mixture that contains several other proteins (52). The effectiveness of the oligo(ethylene glycol) groups to resist nonspecific

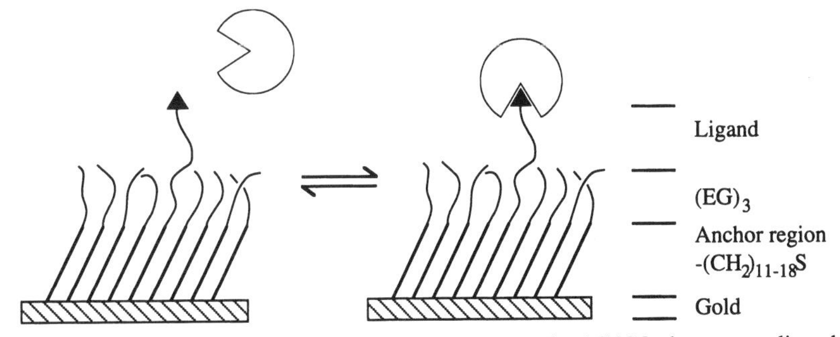

Figure 3 Design of SAMs for biospecific adsorption. Mixed SAMs that present ligands and oligo(ethylene glycol) groups permit control over the density of adsorbed protein. The glycol layer is effective at preventing nonspecific adsorption of protein.

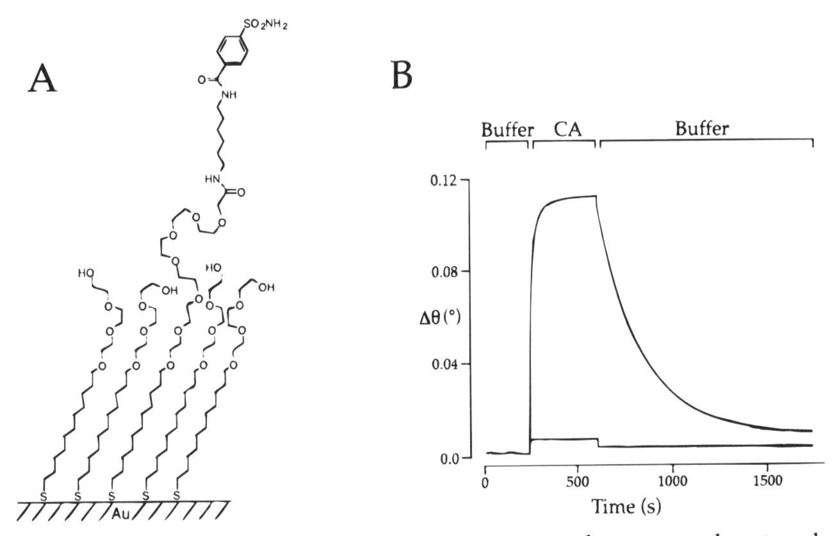

Figure 4 Surface plasmon resonance spectroscopy was used to measure the rate and quantity of binding of CA to a SAM terminated in EG_3 groups and benzenesulfonamide groups (*A*). The change in resonance angle ($\Delta\theta$) of light reflected from the SAM/gold is plotted against time; the time over which the solution of CA (5 μM) was allowed to flow through the cell is indicated at the top of the plot (*B*). (*upper curve*) Binding (and dissociation) of CA to a SAM containing approximately 5% of the ligand-terminated alkanethiolate. Carbonic anhydrase did not adsorb to a SAM that presented only ethylene glycol groups (*lower curve*). A response caused by the change in index of refraction of the CA-containing solution was observed on introduction of protein into the flow cell (evident in *lower curve*). The difference between the measured response and this background signal represents binding of the CA to the SAM.

adsorption, combined with the ability of SAMs to present a range of groups in controlled environments, makes this system well suited for other studies of biospecific adsorption and for applications dependent on specific adsorption.

Attachment of Cells to Self-Assembled Monolayers

The attachment and spreading of anchorage-dependent cells to surfaces are mediated by proteins of the extracellular matrix, e.g. fibronectin, laminin, and collagen. A common strategy for controlling the attachment of cells to a surface therefore relies on controlling the adsorption of matrix proteins to the surface. Both hydrophobic (42, 46), and ionic (34) SAMs have been used as substrates for cell culture. A significant problem with these preparations is the lack of control over the adsorption process. It generally has been assumed that the density of matrix protein is the parameter that influences the behavior of attached cells. Studies of the differentiation response of fibroblasts and neuroblastoma cells on siloxane SAMs terminated in different groups that had been coated with fibronectin suggested that cell behavior depended on the conformation of fibronectin and not on the density of protein (7, 40). The role of protein adsorption in most instances remains poorly understood.

Cell attachment to and spreading on fibronectin involve binding of integrin receptors of the cell to the tripeptide RGD of the matrix. Massia & Hubbell (47) demonstrated that a siloxane SAM that presents the RGD peptide supported the attachment and spreading of fibroblast cells. These synthetic culture substrates have advantages over the traditional matrix-coated substrates of increased reproducibility in culture and utility for fundamental studies of cell-matrix interactions. Corresponding work has focused on the development of substrates for serum-free cell culture by immobilizing essential growth factors at the surface of the substrate (86).

CONTROL OVER SPATIAL ADSORPTION OF PROTEIN

Patterning Self-Assembled Monolayers

MICROCONTACT PRINTING Microcontact printing (μCP) (36, 54, 81) provides a new and convenient method for patterning SAMs of alkanethiolates on gold with features of sizes ranging down to 1 μm (Figure

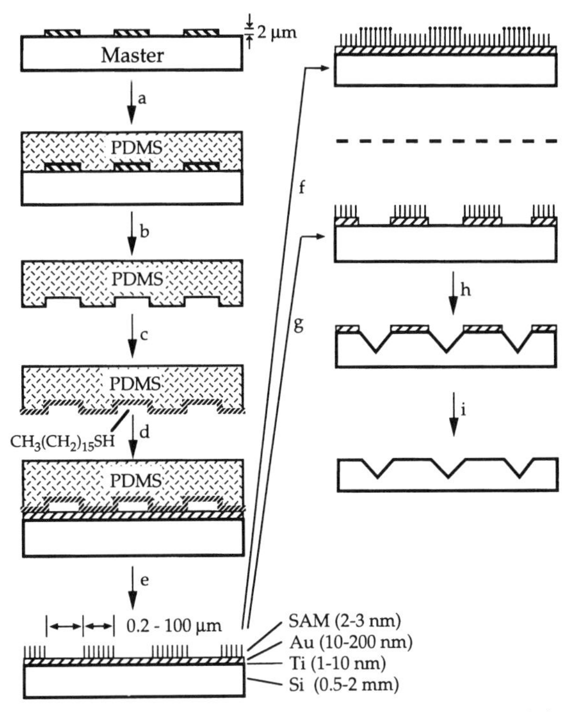

Figure 5 Microcontact printing starts with a master template containing a pattern of relief (*a*); this master can be fabricated by photolithography or by other methods. A PDMS stamp cast from this master (*b*) is inked with a solution of alkanethiol in ethanol (*c*) and used to transfer the alkanethiol to surface of gold (*d*); a SAM is formed only at those regions where the stamp contacts the surface (*e*). The bare regions of gold can then be derivatized with a different SAM by rinsing with a solution of a second alkanethiol (*f*). The initial patterned SAM can also be used to protect the underlying gold from dissolution in a corrosive etchant (*g*). Anisotropic etching of the exposed silicon gives contoured surfaces (*h*). The gold mask can be removed by washing with *aqua regiai*; the resulting silicon substrates are useful as new masters from which stamps can be cast or as substrates for a variety of applications.

5); features as small as 200 nm have been formed with the use of this technique (85). Microcontact printing starts with an appropriate relief structure from which an elastomeric stamp is cast; this "master" template usually is generated photolithographically, but any substrate that has an appropriate pattern of relief can be used. The polydimethylsiloxane (PDMS) stamp is "inked" with a solution of alkanethiol in ethanol, dried, and manually brought into contact with a surface of gold. The alkanethiol is transferred to the surface only at those regions where the stamp contacts the surface. This process produces a pattern of SAM that is defined by the pattern of the stamp. Conformal contact between the elastomeric stamp and surface allow surfaces that are rough (at the scale of 100 nm) to be patterned over areas several square centimeters in size with edge resolution of the features better than 50 nm. Multiple stamps can be cast from a single master, and each stamp can be used hundreds of times. Microcontact printing also was used to pattern siloxanes on the surfaces of SiO_2 and glass (84) and to pattern SAMs on nonplanar and contoured surfaces (27). Because μCP relies on molecular self-assembly and does not require stringent control over the laboratory environment, it can produce μm-scale patterns conveniently and at low cost relative to methods that use photolithography.

PHOTOLITHOGRAPHY Photolithographic methods illuminate a surface with UV light through a mask (5, 17, 34). In the common "lift-off" method (34), a silicon oxide substrate is coated with a thin layer of photoresist. The resist is exposed to UV light through a mask, and the exposed regions are subsequently removed in a developing bath; this process creates a pattern of silicon dioxide that can be derivatized with an alkylsiloxane SAM. The remaining regions of photoresist are then removed, and a different SAM is formed on the complementary regions. Other variants of photolithography create patterns by using UV light to damage, or modify, a SAM. Wrighton and coworkers (19) prepared SAMs of alkanethiolates terminated in an aryl azide group; near-UV irradiation of the SAM through a lithographic mask and a thin film of an amine resulted in the attachment of the amine in the exposed regions. Hickman et al (24) irradiated thiol-terminated siloxanes through a mask in the presence of oxygen to form sulfonate groups. These methods are less well controlled and less general than such methods as μCP, which pattern the adsorption of preformed components. Photolithographic methods can produce patterns that have features down to 1 μm conveniently. Capital costs for the equipment and controlled environment facilities, however, make this technique expensive and inconvenient for the biological researcher.

FABRICATION OF CONTOURED SURFACES Both μCP (32) and photolithography (10) have been used to pattern silicon substrates with a layer of resist that protects the substrate from dissolution in a chemical etchant (Figure 4). Chemical etching of these patterned substrates produces contoured features whose shapes depend on the orientation of the silicon and the time of etching; anisotropic etching of a silicon $\langle 100 \rangle$ surface produces controlled V-shaped grooves. The properties of these etched substrates can be tailored either by forming an alkylsiloxane SAM or by evaporating a layer of gold and forming a SAM of alkanethiolates. Alternatively, the topographic pattern can be transferred to other substrates (e.g. prepolymers) with the use of a PDMS stamp cast from the etched master.

Patterning Adsorption of Protein on Self-Assembled Monolayers

Patterned SAMs on gold have been used extensively to control the adsorption of protein to surfaces (Figure 6). This method relies on the ability of a SAM terminated in oligo(ethylene glycol) groups to resist the adsorption of protein. Microcontact printing was used to pattern a SAM into regions terminated in methyl groups and oligo(ethylene glycol) groups (42, 54). Immersion of these SAMs in aqueous solutions that contain proteins resulted in the adsorption of a monolayer of protein only on the methyl-terminated regions; this pattern of protein could be imaged by scanning electron microscopy (43). Bhatia et al (4) patterned a siloxane film terminated in thiol groups by irradiation with UV light through a mask. The fluorescent protein phycoerythrin was immobilized to the thiol groups in regions that were protected from the UV light with the mask; photo-induced oxidation of the thiol groups in regions of the surface that were irradiated presumably gave negatively charged sulfonate groups, which resisted the adsorption of protein.

Patterned Attachment of Cells on Self-Assembled Monolayers

The same methods used to pattern the adsorption of proteins to surfaces have been used to direct the attachment of cells to surfaces (5, 23, 34, 66, 67). In an early example, Kleinfeld et al (34) used photolithography to pattern siloxane SAMs into regions terminated in methyl and amino groups. When plated in the presence of serum, neural cells attached and spread selectively on the amino-terminated regions; in the absence of serum, the cells attached to all regions. Others have also found that amino-terminated siloxanes are excellent substrates for culture of neural cells (24). We presume that proteins of the serum that do not support cell attachment adsorbed to the hydrophobic areas. Regions terminated in perfluoro groups have also been used to resist the attachment of cells (3, 67, 70).

Self-assembled monolayers patterned into regions terminated in methyl and oligo(ethylene glycol) groups permit spatial control over the attachment of cells. Microcontact printing was used to pattern SAMs into adhesive lines ranging from 10 to 100 μm in width; after coating these substrates with fibronectin, endothelial cells were confined to

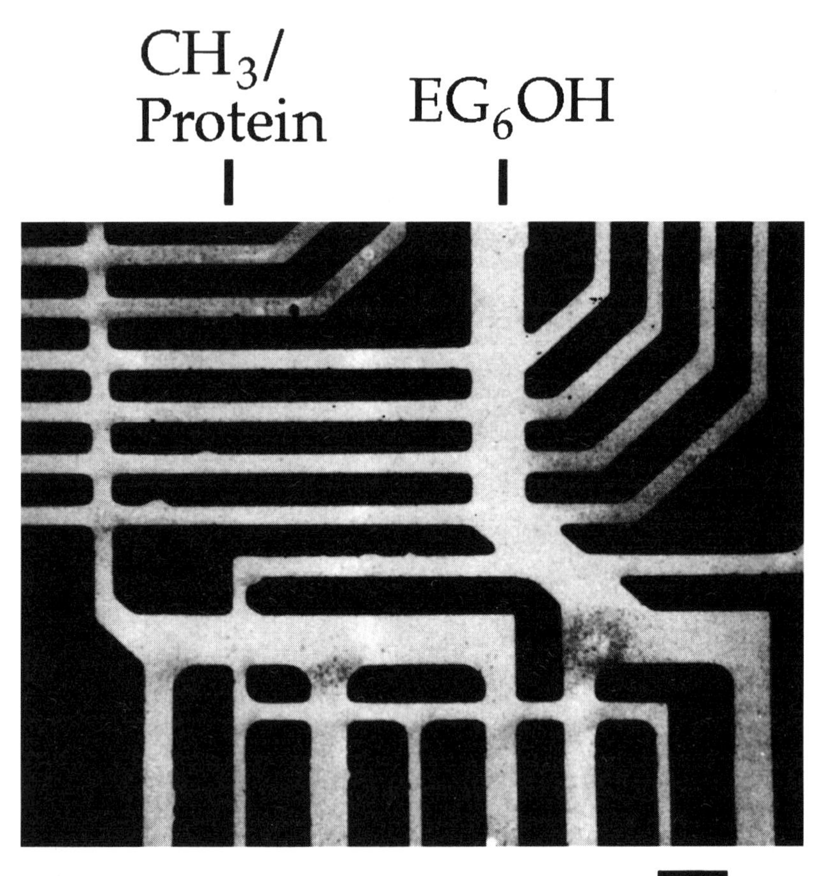

Figure 6 Scanning electron micrograph of fibrinogen adsorbed on a patterned SAM. A patterned hexadecanethiolate SAM on gold was formed by μCP, and the remainder of the surface was derivatized by immersion in a solution containing a hexa(ethylene glycol)-terminated alkanethiol [$HS(CH_2)_{11}(OCH_2CH_2)_6OH$]. The patterned substrate was immersed in a solution of fibrinogen (1 mg/mL) in phosphate-buffered saline for 2 h, removed from solution, rinsed with water, and dried. Fibrinogen adsorbed only to the methyl-terminated regions of the SAM, as illustrated by the dark regions in the micrograph: Secondary electron emission from the underlying gold is attenuated by the protein adlayer.

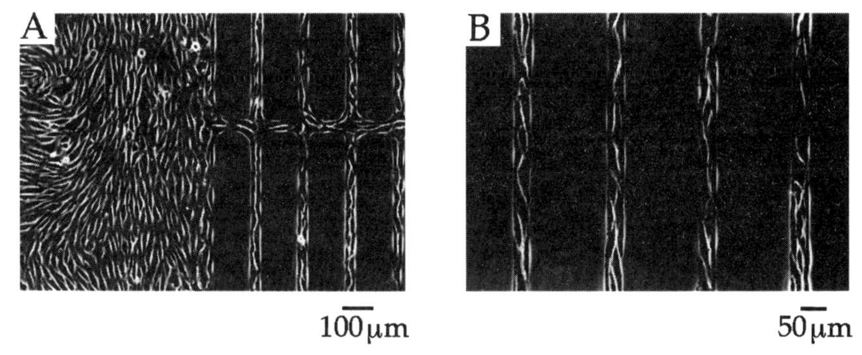

Figure 7 Control over the attachment of bovine capillary endothelial cells to planar substrates that were patterned into regions terminated in methyl groups and tri(ethylene glycol) groups using μCP. The substrates were coated with fibronectin before cell attachment; fibronectin adsorbed only to the regions of methyl-terminated SAM. (*A*) An optical micrograph showing attachment of endothelial cells to a nonpatterned region (*left*) and to lines 30 μm in width. (*B*) A view at higher magnification of cells attached to the lines.

grow on these lines (Figure 7). This technique was also used to pattern SAMs into adhesive regions approximately 20 × 50 μm in size that were surrounded by EG_6-terminated SAM (66). After the hydrophobic regions were coated with laminin, hepatocytes attached to the rectangular islands and conformed to the shape of the underlying pattern. The size of the islands controlled DNA synthesis, cell growth, and protein secretion of the attached cells. The ability to pattern the attachment of individual cells may be useful for single cell manipulation, toxicology and drug screening.

Attachment of Cells on Contoured Surfaces

Several groups have used surfaces contoured into grooves and ridges, which were fabricated with the use of photo- or electron-beam lithography, to study their effects on the behavior and growth of attached cells (9, 10, 25, 49). Chou et al (9) found that human fibroblasts adherent to surfaces contoured into V-shaped grooves had increased levels of fibronectin synthesis and secretion. Surfaces with arrays of grooves of varying dimensions controlled the alignment and orientation of attached mammalian cells (10, 49). Surfaces with arrays of ridges directed the motility and induced differentiation of the fungus *Uromyces* (25).

A simple technique based on μCP and micromolding (33) was

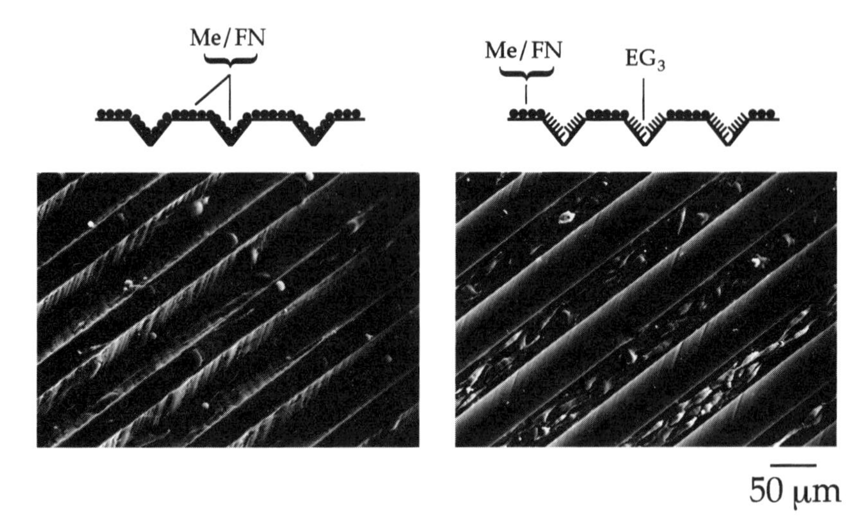

Figure 8 Control over the attachment of endothelial cells to contoured surfaces using SAMs. The substrates are films of polyurethane (supported on glass slides) that were coated with gold and modified with SAMs of alkanethiolates terminated in methyl groups and tri(ethylene glycol) groups; the substrates were coated with fibronectin before cell attachment. (*left*) Cells attached to both the ridges and grooves of substrates that present fibronectin at all regions. (*right*) Cells attached only to the ridges when the grooves were modified with a SAM presenting tri(ethylene glycol) groups.

used to fabricate contoured surfaces of optically transparent films of polyurethane on glass coverslips. After evaporating a thin film of gold onto these substrates, the ridges were derivatized with a SAM by stamping with a flat PDMS stamp; a different SAM was formed in the grooves by immersing the substrate in a solution of alkanethiol. By modifying the ridges with a SAM of hexadecanethiolate, and the grooves with a SAM terminated in oligo(ethylene glycol) groups, endothelial cells were confined to attach and spread only on the ridges (Figure 8). By the reverse process, cells were confined to attach and grow in the grooves.

APPLICATIONS OF SURFACES BASED ON SELF-ASSEMBLED MONOLAYERS IN BIOCHEMISTRY

Drug Design and Screening

Several combinatorial strategies for drug design screen mixtures of potential ligands to identify those with a desired property—usually the ability to bind a protein. Methods that use immobilized libraries have the primary advantage that the identity of each ligand is defined by its location in the matrix; amplification and sequencing of selected ligands are not required. Fodor and colleagues (28) used photolithography and solid-phase synthesis to pattern SAMs of siloxanes into arrays of hundreds of different peptides. Confocal microscopy was used to assay the in situ binding of a monoclonal antibody to each of these peptides in a single experiment. This technology also was used to create an array containing 256 octanucleotides that served as a hybridization probe for sequencing DNA. The demonstration of libraries that contain non-natural biopolymers (8), and the potential to screen for reactions at surfaces (78), widens the scope of this technology. Scanning probe microscopies may be useful for screening libraries, because these techniques can assay functionalized surfaces at the sub-micron scale rapidly (39, 80).

Biosensors

Self-assembled monolayers are finding increasing use to tailor the molecular recognition properties of surfaces used in biosensing. The surface of a TiO_2–SiO_2 waveguide was modified with an alkylsiloxane monolayer terminated in amino groups to which the antibody anti-HBsAg was conjugated; this difference interferometer measured the binding of hepatitis B down to a concentration of 2×10^{-13} M in undiluted serum (62). Self-assembled monolayers have been used in similar ways to control the properties of sensors on the basis of the quartz crystal microbalance (77), acoustic plate modes (31), and surface plasmon resonance (52, 53, 64). Nonspecific adsorption of protein is a common problem with these devices. A commercial technology uses gel layers of dextran to control unwanted adsorption (41), although most applications use bovine serum albumin–coated surfaces. Recent work with ligands immobilized on SAMs terminated in oligo(ethylene glycol) groups provides a route to biospecific surfaces with a high control over the properties of the surfaces (52, 64).

Electrochemical Methods

Self-assembled monolayers can either mediate or inhibit the transfer of electrons from the underlying gold to electrolytes in solution. The properties of SAMs terminated in electroactive groups (e.g. ferrocene or quinone groups) could be switched reversibly by adjusting the potential at the underlying gold (1). Several groups have studied the transfer of electrons from gold to electroactive proteins immobilized to SAMs (11, 82). There is now commercial technology for analysis of biological analytes on the basis of electrochemiluminescence from tris-bipyridine ruthenium(II) tags (21): This technology is well suited for SAMs.

PROSPECTS FOR SELF-ASSEMBLED MONOLAYERS IN BIOLOGY

Self-Assembled Monolayers as Model Surfaces

Self-assembled monolayers of alkanethiolates on gold provide the best system available with which to understand interactions of proteins and cells with man-made surfaces. The ease with which complex and delicate groups of the sorts relevant to biochemistry can be presented in controlled environments, combined with simple methods that can pattern the formation of SAMs in the plane of the monolayer, make these surfaces well suited for studies of fundamental aspects of biointerfacial science. Other advantages with this system include the optical transparency of SAMs when supported on thin films of gold, the electrical conductivity of the underlying gold, the compatibility of these substrates with a range of analytical methodologies, the stability of these substrates during storage and in contact with biological media, and the range of surfaces, including curved and nonplanar substrates, that can be used.

Self-Assembled Monolayers in Cell Biology

Designed substrates permit strategies for the noninvasive control over the activity of attached cells. Langer and coworkers (83) cultured endothelial cells on optically transparent films of electrically conducting polypyrrole. Cells attached and spread normally on fibronectin-coated polypyrrole in the oxidized state; on application of a reducing potential, however, the extension of cells and the synthesis of DNA were both inhibited. Okano et al (55) grafted thermoresponsive gels of poly(*N*-isopropylacrylamide on cell culture substrates. Hepatocytes attached to the substrates normally at 37°C; when the culture was chilled to 4°C, the cells detached from the substrate. The features of SAMs described in this review make them well-suited model surfaces for studies in biology that require substrates with tailored properties.

Literature Cited

1. Abbott N, Whitesides GM. 1994. Potential-dependent wetting of aqueous solutions on self-assembled monolayers formed from 15-(ferrocenylcarbonyl)pentadecanethiol on gold. *Langmuir* 10:1493–97
2. Bain CD, Whitesides GM. 1989. Modeling organic surfaces with self-assembled monolayers. *Adv. Mat. Angew. Chem.* 101:522–28
3. Bekos EJ, Ranieri JP, Aebischer P, Gardella JA, Bright FV. 1995. Structural changes of bovine serum albumin upon adsorption to modified fluoropolymer substrates used for neural cell attachment studies. *Langmuir* 11: 984–89
4. Bhatia SK, Hickman JJ, Ligler FS. 1992. New approach to producing patterned biomolecular assemblies. *J. Am. Chem. Soc.* 114:4432–33
5. Britland S, Clark P, Connolly P, Moores G. 1992. Micropatterned substratum adhesiveness: a model for morphogenetic cues controlling cell behavior. *Exp. Cell Res.* 198:124–29
6. Caffrey M, Wang J. 1995. Membrane-structure studies using x-ray standing waves. *Annu. Rev. Biophys. Biomol. Struct.* 24:351–78
7. Cheng S-S, Chittur KK, Sukenik CN, Culp L, Lewandowska K. 1994. The conformation of fibronectin on self-assembled monolayers with different surface composition: an FTIR/ATR study. *J. Colloid Interface Sci.* 162: 135–43
8. Cho CY, Moran EJ, Cherry S, Stephens J, Fodor SPA, et al. 1993. An unnatural biopolymer. *Science* 261: 1303–5
9. Chou L, Firth JD, Uitto V-J, Brunette DM. 1995. Substratum surface topography alters cell shape and regulates fibronectin mRNA level, mRNA stability, secretion and assembly in human fibroblasts. *J. Cell. Sci.* 108: 1563–73
10. Clark P, Connolly P, Curtis ASG, Dow JAT, Wilkinson CDW. 1990. Topographical control of cell behavior: II. multiple grooved substrata. *Development* 108:635–44
11. Collinson M, Bowden EF, Tarlov. 1992. Voltammetry of covalently immobilized cytochrome c on self-assembled monolayer electrodes. *Langmuir* 8:1247–50
12. Dahint R, Grunze M, Josse F, Renken J. 1994. Acoustic plate mode sensor for immunochemical reactions. *Anal. Chem.* 66:2888–92
13. Darst SA, Ahlers M, Meller PH, Kubalek EW, Blankenburg R, et al. 1991. Two-dimensional crystals of streptavidin on biotinylated lipid layers and their interactions with biotinylated macromolecules. *Biophys. J.* 59: 387–96
14. Darst SA, Robertson CR, Berzofsky JA. 1988. Adsorption of the protein antigen myoglobin affects the binding of conformation-specific monoclonal antibodies. *Biophys. J.* 53:533–39
15. DiMilla PA, Folkers JP, Biebuyck HA, Harter R, Lopez GP, Whitesides GM. 1994. Wetting and protein adsorption of self-assembled monolayers of alkanethiolates supported on transparent films of gold. *J. Am. Chem. Soc.* 116:2225–26
16. Dubois LH, Nuzzo RG. 1992. Synthesis, structure, and properties of model organic surfaces. *Annu. Rev. Phys. Chem.* 43:437–63
17. Dulcey CS, Georger JH, Krauthamer V, Stenger DA, Fare TL, Calvert JM. 1991. Deep UV photochemistry of chemisorbed monolayers: patterned coplanar molecular assemblies. *Science* 252:551–54
18. Fragneto G, Thomas RK, Rennie AR, Penfold J. 1995. Neutron reflection study of bovine β-casein adsorbed to OTS self-assembled monolayers. *Science* 267:657–60
19. Frisbie CD, Wollman EW, Wrighton MS. 1995. High lateral resolution imaging by secondary ion mass spectrometry of photopatterned self-assembled monolayers containing aryl azide. *Langmuir* 11:2563–71
20. Gombotz WR, Guanghui W, Horbett TA, Hoffman AS. 1991. Protein adsorption to poly(ethylene oxide) surfaces. *J. Biomed. Mater. Res.* 15: 1547–62
21. Gudibands SR, Kenten JH, Link J, Friedman K, Massey RJ. 1992. *Mol. Cell Probes* 6:495–503
22. Harris JM, ed. 1992. *Poly(ethylene glycol) Chemistry: Biotechnical and Biomedical Applications*. New York: Plenum
23. Healy KE, Lom B, Hockberger PE. 1994. Spatial distribution of mammalian cells dictated by material surface chemistry. *Biotechnol. Bioeng.* 43: 792–800
24. Hickman JJ, Bhatia SK, Quong JN, Shoen P, Stenger DA, et al. 1994. Rational pattern design for in vitro cellu-

lar networks using surface photochemistry. *J. Vac. Sci. Technol.* 12:607–16
25. Hoch HC, Staples RC, Whitehead B, Comeau J, Wolf ED. 1987. Signaling for growth orientation and cell differentiation by surface topography in *Uromyces*. *Science* 235:1659–62
26. Hong H-G, Jiang M, Sligar SG, Bohn PW. 1994. Cysteine-specific surface tethering of genetically engineered cytochromes for fabrication of metalloprotein nanostructures. *Langmuir* 10: 153–58
27. Jackman RJ, Wilbur JL, Whitesides GM. 1995. Fabrication of submicron features on curved substrates by microcontact printing. *Science* 269: 664–66
28. Jacobs JW, Fodor SPA. 1994. Combinatorial chemistry—applications of light-directed chemical synthesis. *Trends Biotechnol.* 12:19–26
29. Jeon SI, Andrade JD. 1991. Protein-surface interactions in the presence of polyethylene oxide: effect of protein size. *J. Colloid Interface Sci.* 142: 159–66
30. Jeon SI, Lee JH, Andrade JD, De Gennes PG. 1991. Protein-surface interactions in the presence of polyethylene oxide: simplified theory. *J. Colloid Interface Sci.* 142:149–58
31. Kepley LJ, Crooks RM, Ricco AJ. 1992. Selective surface acoustic wave-based organophosphonate chemical sensor employing a self-assembled composite monolayer: a new paradigm for sensor design. *Anal. Chem.* 64: 3191–93
32. Kim E, Kumar A, Whitesides GM. 1995. Combining patterned self-assembled monolayers of alkanethiolates on gold with anisotropic etching of silicon to generate controlled surface morphologies. *J. Electrochem. Soc.* 142:628–33
33. Kim E, Xia Y, Whitesides GM. 1995. Making polymeric microstructures: capillary micromolding. *Nature* 376: 581–84
34. Kleinfeld D, Kahler KH, Hockberger PE. 1988. Controlled outgrowth of dissociated neurons on patterned substrates. *J. Neurosci.* 8:4098–120
35. Kooyman RPH, van den Heuvel DJ, Drijhout JW, Welling GW. 1994. The use of self-assembled receptor layers in immunosensors. *Thin Solid Films* 244:913–16
36. Kumar A, Biebuyck HA, Whitesides GM. 1994. Patterning self-assembled monolayers: applications in material science. *Langmuir* 10:1498–511
37. Lee C-S, Belfort G. 1989. Changing activity of ribonuclease A during adsorption: a molecular explanation. *Proc. Natl. Acad. Sci. USA* 86: 8392–96
38. Lee YW, Reed-Mundell J, Sukenik CN, Zull JE. 1993. Electrophilic siloxane-based self-assembled monolayers for thiol-mediated anchoring of peptides and proteins. *Langmuir* 9: 3009–14
39. Leggett GJ, Roberts CJ, Williams PM, Davies MC, Jackson DE, Tendler SJB. 1993. Approaches to the immobilization of proteins at surfaces for analysis by scanning tunneling microscopy. *Langmuir* 9:2356–62
40. Lewandowska K, Pergament E, Sukenik CN, Culp LA. 1992. Cell-type-specific adhesion mechanisms mediated by fibronectin adsorbed to chemically derivatized substrata. *J. Biomed. Mater. Res.* 26:1343–63
41. Liedberg B, Lundstrom I, Stenberg E. 1993. Principles of biosensing with an extended coupling matrix and surface plasmon resonance. *Sens. Actuators B* 11:63–72
42. Lopez GP, Albers MW, Schreiber SL, Carroll R, Peralta E, Whitesides GM. 1993. Convenient methods for patterning the adhesion of mammalian cells to surfaces using self-assembled monolayers of alkanethiolates on gold. *J. Am. Chem. Soc.* 115:5877–78
43. Lopez GP, Biebuyck HA, Harter R, Kumar A, Whitesides GM. 1993. Fabrication and imaging of two-dimensional patterns of proteins adsorbed on self-assembled monolayers by scanning electron microscopy. *J. Am. Chem. Soc.* 115:10774–81
44. Lundstrom I, Ivarsson B, Jonsson U, Elwing H. 1987. Protein adsorption and interaction at solid surfaces. In *Polymer Surfaces and Interfaces,* ed. WJ Feast, HS Munro, 1:201–30. New York: Wiley & Sons
45. Mandenius CF, Welin S, Danielsson B, Lundstrom I, Mosbach K. 1984. The interaction of proteins and cells with affinity ligands covalently coupled to silicon surfaces as monitored by ellipsometry. *Anal. Biochem.* 137: 106–14
46. Margel S, Vogler EA, Firment L, Watt T, Haynie S, Sogah DY. 1993. Peptide, protein, and cellular interactions with self-assembled monolayer model surfaces. *J. Biomed. Mater. Res.* 27: 1463–76
47. Massia SP, Hubbell JA. 1990. Covalent surface immobilization of arg-gly-

asp- and tyr-ile-gly-ser-arg-containing peptides to obtain well-defined cell-adhesive substrates. *Anal. Biochem* 187:292–301
48. McGovern ME, Kallury KMR, Thompson M. 1994. Role of solvent on the silanization of glass with octadecyltrichlorosilane. *Langmuir* 10: 3607–14
49. Meyle J, Gultig K, Brich M, Hammerle H, Nisch W. 1994. Contact guidance of fibroblasts on biomaterial surfaces. *J. Mater. Sci. Mater. Med.* 5: 463–66
50. Mikos AG, Murphy RM, Bernstein H, Peppas NA, eds. 1994. *Biomaterials for Drug and Cell Delivery*. Pittsburgh: PA. 331 pp.
51. Mosbach K. 1987–1988. Immobilized enzymes and cells. *Methods Enzymol.* Vol. 135–37
52. Mrksich M, Grunwell JR, Whitesides GM. 1995. Bio-specific adsorption of carbonic anhydrase to self-assembled monolayers of alkanethiolates on gold that present benzenesulfonamide groups. *J. Am. Chem. Soc.* In press
53. Mrksich M, Sigal GB, Whitesides GM. 1995. Surface plasmon resonance permits in situ measurement of protein adsorption on self-assembled monolayers of alkanethiolates on gold. *Langmuir*. 11:4383–85
54. Mrksich M, Whitesides GM. 1995. Patterning self-assembled monolayers using microcontact printing: a new technology for biosensors? *Trends Biotechnol.* 13:228–35
55. Okano T, Yamada N, Okuhara M, Sakai H, Sakurai Y. 1995. Mechanism of cell detachment from temperature-modulated, hydrophilic-hydrophobic polymer surfaces. *Biomaterials* 16: 297–303
56. Parikh AN, Allara DL, Azouz IB, Rondelez F. 1994. An intrinsic relationship between molecular structure in self-assembled n-alkylsiloxane monolayers and deposition temperature. *J. Phys. Chem.* 98:7577–90
57. Prime KL, Whitesides GM. 1991. Self-assembled organic monolayers: model systems for studying adsorption of proteins at surfaces. *Science* 252: 1164–67
58. Prime KL, Whitesides GM. 1993. Adsorption of proteins onto surfaces containing end-attached oligo(ethylene oxide): a model system using self-assembled monolayers. *J. Am. Chem. Soc.* 115:10714–21
59. Ramsden JJ. 1993. Experimental methods for investigating protein adsorption kinetics at surfaces. *Q. Rev. Biophys.* 27:41–105
60. Ramsden JJ. 1995. Puzzles and paradoxes in protein adsorption. *Chem. Soc. Rev.* 73–78
61. Sadana A. 1992. Protein adsorption and inactivation on surfaces. Influence of heterogeneities. *Chem. Rev.* 92: 1799–818
62. Schlatter D, Barner R, Fattinger Ch, Huber W, Hubscher J, et al. 1993. The difference interferometer: application as a direct affinity sensor. *Biosens. Bioelectron.* 8:109–16
63. Scouten WH, Luong JHT, Brown RS. 1995. Enzyme or protein immobilization techniques for applications in biosensor design. *Trends Biotechnol.* 13: 178–85
64. Sigal GB, Bamdad C, Barberis A, Strominger J, Whitesides GM. 1995. A self-assembled monolayer for the binding and study of histidine-tagged proteins by surface plasmon resonance. *Anal. Chem.* In press
65. Silver FH, Doillon C. 1989. *Biocompatibility: Interactions of Biological and Implantable Materials*. New York: VCH Publ. 310 pp.
66. Singhvi R, Kumar A, Lopez GP, Stephanopoulos GN, Wang DIC, et al. 1994. Engineering cell shape and function. *Science* 264:696–98
67. Spargo BJ, Testoff MA, Nielsen TB, Stenger DA, Hickman JJ, Rudolph AA. 1994. Spatially controlled adhesion, spreading, and differentiation of endothelial cells on self-assembled molecular monolayers. *Proc. Natl. Acad. Sci. USA* 91:11070–74
68. Spinke J, Liley M, Schmitt F-J, Guder H-J, Angermaier L, Knoll W. 1993. Molecular recognition at self-assembled monolayers: optimization of surface functionalization. *J. Chem. Phys.* 99:7012–19
69. Stark GR. 1970. Recent developments in chemical modification and sequential degradation of proteins. *Adv. Protein Chem.* 24:261–308
70. Stenger DA, Georger JH, Dulcey CS, Hickman JJ, Rudolph AS, et al. 1992. Coplanar molecular assemblies of amino- and perfluorinated alkylsilanes: characterization and geometric definition of mammalian cell adhesion and growth. *J. Am. Chem. Soc.* 114: 8435–42
71. Tilton RD, Robertson CR, Gast AP. 1991. Manipulation of hydrophobic interactions in protein adsorption. *Langmuir* 7:2710–18
72. Ulman A. 1991. *An Introduction to Ul-*

trathin Organic Films: From Langmuir-Blodgett to Self-Assembly. London: Academic. 442 pp.
73. Vroman L, Adams AL. 1986. Adsorption of proteins out of plasma and solutions in narrow space. *J. Colloid Interface Sci.* 111:391–402
74. Wagner P, Kernen P, Hegner M, Ungewickell E, Semenze G. 1994. Covalent anchoring of proteins onto gold-directed NHS-terminated self-assembled monolayers in aqueous buffers: SMF images of clathrin cages and triskelia. *FEBS Lett.* 356:267–71
75. Wahlgren M, Arnebrant T. 1991. Protein adsorption to solid surfaces. *Trends Biotechnol.* 9:201–8
76. Wang P, Hill TG, Wartchow CH, Huston ME, Oehler LM, et al. 1992. New carbohydrate-based materials for the stabilization of proteins. *J. Am. Chem. Soc.* 114:378–80
77. Ward MD, Buttry DA. 1990. In situ interfacial mass detection with piezoelectric transducers. *Science* 249: 1000–7
78. Whitesell JK, Chang HK, Whitesell CS. 1994. Enzymatic grooming of organic thin films. *Angew. Chem. Int. Ed. Engl.* 33:871–73
79. Whitesides GM, Gorman CB. 1995. Self-assembled monolayers: models for organic surface chemistry. In *Handbook of Surface Imaging and Visualization,* AT Hubbarded, ed. Boca Raton, FL: CRC Press, pp. 713–32
80. Wilbur JL, Biebuyck HA, MacDonald JC, Whitesides GM. 1995. Scanning force microscopies can image patterned self-assembled monolayers. *Langmuir* 11:825– 31
81. Wilbur JL, Kumar A, Biebuyck HA, Kim E, Whitesides GM. 1995. Microcontact printing of self-assembled monolayers: applications in microfabrication. *Nanotechnology*. In press
82. Willner I, Katz E, Riklin Z, Kasher R. 1992. Mediated electron transfer in glutathione reductase organized in self-assembled monolayers on Au electrodes. *J. Am. Chem. Soc.* 114: 10965–66
83. Wong JY, Langer R, Ingber DE. 1994. Electrically conducting polymers can noninvasively control the shape and growth of mammalian cells. *Proc. Natl. Acad. Sci. USA* 91:3201–4
84. Xia Y, Mrksich M, Kim E, Whitesides GM. 1995. Microcontact printing of siloxane monolayers on the surface of silicon dioxide, and its application in microfabrication. *J. Am. Chem. Soc.* 117:9576–78
85. Xia Y, Whitesides GM. 1995. Use of controlled reactive spreading of liquid alkanethiol on the surface of gold to modify the size of features produced by microcontact printing. *J. Am. Chem. Soc.* 117:3274–75
86. Zheng J, Ito Y, Imanishi Y. 1994. Cell growth on immobilized cell-growth factor. *Biomaterials* 15:963–68
87. Zhong M, Lin L, Kallenbach NR. 1995. A method for probing the topography and interactions of proteins: footprinting of myoglobin. *Proc. Natl. Acad. Sci. USA* 92:2111–15

Annu. Rev. Biophys. Biomol. Struct. 1996. 25:79–112

ANTIGEN-MEDIATED IgE RECEPTOR AGGREGATION AND SIGNALING: A Window on Cell Surface Structure and Dynamics

David Holowka and Barbara Baird

Department of Chemistry, Cornell University, Ithaca, New York 14853-1301

KEY WORDS: mast cells, cell surface receptors, ligand binding, membrane protein mobility, membrane domains, detergent insolubility, fluorescence

ABSTRACT

The high-affinity receptor for immunoglobulin E, FcϵRI, serves as an archtype for multisubunit immunoreceptors that mediate cell activation in response to foreign antigens. Antigen-mediated aggregation of this receptor at the surface of mast cells and basophils initiates a biochemical cascade that uses nonreceptor tyrosine kinases as key participants in the earliest steps of this signal transduction process. Cross-linking of FcϵRI with ligands of well-defined structure and valency has revealed detailed information about the fundamental requirements for functionally active receptor aggregates. Cross-linking–dependent changes in the interaction of these receptors with other cellular components have been characterized with biochemical and biophysical methods to develop a more complete view of signal initiation. Recent evidence suggests that this process involves the interaction of aggregated FcϵRI with specialized plasma membrane domains that may localize important signaling molecules in the vicinity of aggregated receptors. Although these various studies were aimed toward understanding the operation of one cell surface receptor, they provide new insights into plasma membrane structure and dynamics that are generally relevant to the function of most nucleated mammalian cells.

1056-8700/96/0610-0079$08.00

CONTENTS

INTRODUCTION AND SCOPE

Cell surface receptors in the immune system have evolved to respond to foreign substances for host defense in animals. A particularly challenging aspect of this task is to deliver a consistent and highly regulated response to an enormous number of different chemical structures. These foreign substances (antigens) have literally millions of different structures that must be recognized with strong discrimination. In addition, their physical state, whether free in solution or bound to cells, provides an additional diversity factor that the immune system must accommodate. The biosynthetic process by which T and B lymphocytes generate the necessarily vast repertoire of recognition structures is now well understood. In this process, the production of variable region domains of immunoglobulin (Ig) heavy and light chains and of T-cell receptor (TCR) α and β subunits involves genetic recombination mechanisms that are, in several respects, unique to these cells and these genes. Several recent reviews have addressed the current understanding of the generation of diversity (118) and the structural details of Ig-specific (99) and TCR-specific (59) antigen-binding sites.

This enormous diversity of chemical structures has resulted in antigen receptors that have complementary diversity in their binding sites, whereas other structural parts remain constant. Unlike some other families of receptors, such as the heterotrimeric G protein receptors (135), these immunoreceptors have not evolved to respond to their specific ligands by transmittal of induced conformational changes. Rather, the three types of immunoreceptors become activated when they are aggregated on the cell surface by ligands that bind two or more receptors simultaneously. Thus, B lymphocytes respond to multivalent soluble antigens or to bivalent anti-Ig antibodies that bind the surface Ig of the B-cell–receptor (BCR) complex. B lymphocytes do not respond to monovalent analogues of these same antigens or to monovalent Fab fragments of the same anti-Ig antibodies (20, 30). T-cell receptors recognize small peptides that are bound in the clefts of major histocompatibility complex proteins when these peptides are presented on the surfaces of other cells (59). T-cell receptors also can mediate responses initiated by aggregation through anti-TCR antibodies in certain cases (76).

A third type of immunoreceptor that mediates responses to foreign antigens is the Fc receptor for soluble Ig. A specific Fc receptor binds to the constant Fc segment of a particular Ig class (or subclass); the variable domains in each of the two Ig Fab segments provide specific antigen recognition. A particularly well-characterized representative of the Fc receptors is FcϵRI, the high-affinity receptor for IgE (Figure 1) that is found on mast cells and basophils and mediates the immediate hypersensitivity responses of these cells (132). Common structural features within different types of multisubunit antigen-binding immunoreceptors have been reviewed previously (54, 145).

Among the immunoreceptors, the necessity of cross-linking for functional activation has been most firmly established for FcϵRI, and Metzger has eloquently described this subject (see 87 for review). In this review, we focus on biophysical investigations of this receptor system, which can serve as a model for the larger families of immunoreceptors. Since its development in 1973, the rat basophilic leukemia (RBL) cell line (34), which is derived from mucosal mast cells (130), often is used as an experimentally attractive subject for biochemical and biophysical investigations of FcϵRI. These cells are grown in large quantities in cell culture, and a functional subline, RBL-2H3, typically expresses $2–3 \times 10^5$ FcϵRI per cell on the cell surface (7). Unlike native mast cells or basophils, whose FcϵRI are variably occupied with physiologically derived IgE, the RBL cells can be sensitized to a controlled extent with exogenous IgE of selected specificity that may be labeled with a fluorescent or other probe. Immunoglobulin E binds monomerically

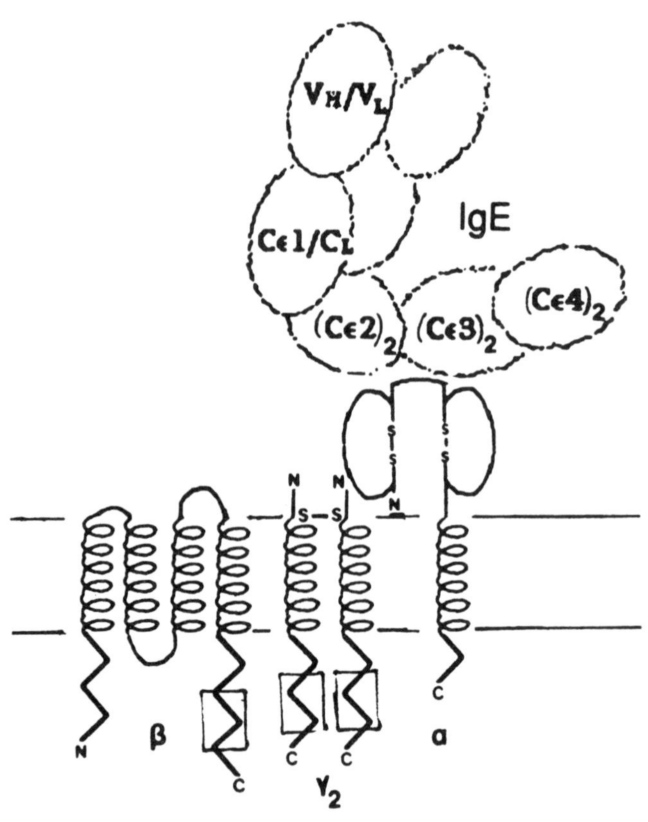

Figure 1 Polypeptide composition of the IgE-FcεRI complex. FcεRI subunit stoiciometries and structural features predicted from DNA sequence (18) are indicated: extracellular Ig-like α subunit domains are represented by the disulfide-containing loops (~110 residues each); α, β, and γ transmembrane regions are indicated by the helical segments (22–25 residues each); Y-X-X-L-$X_{6\text{-}7}$-Y-X-X-L sequences in the β and γ cytoplasmic segments are indicated by the boxes. The conformation and orientation of FcεRI-bound IgE was determined by resonance energy transfer measurements (6). Adapted from 54.

and tightly to FcεRI (K_a ~10^{10} M^{-1}) and dissociates very slowly (57, 68); for many experimental time scales, therefore, the IgE-FcεRI complex can be considered irreversible.

In the studies summarized here, the binding and functional properties of well-defined ligands have been examined to elucidate the features of FcεRI aggregation that determine the character of the biological response. Structural consequences of FcεRI cross-linking on the cell surface also have been investigated extensively. In addition, new insights into the dynamic behavior of cell surface proteins, including apparent associations with membrane domains, have been derived from these studies. Finally, we consider the relationship of these physical changes caused by cross-linking ligands to the consequent signaling events and discuss the implications of these results for an emerging view of a compartmentalized, but dynamic, plasma membrane.

LIGAND BINDING AND CROSS-LINKING OF IGE-FCεRI

Critical Features of Cross-Linking for FcεRI-Mediated Signaling

Because immunoreceptor aggregation is essential for the initiation of signal transduction, as established for FcεRI (87), a substantial effort to elucidate the critical details of ligand binding that facilitate this process has been undertaken. Early studies indicated that small, symmetric, bivalent ligands are sufficient to trigger the degranulation of mast cells and basophils in vivo, as measured by a local skin reaction called passive cutaneous anaphylaxis (22, 98). In vitro studies with rabbit basophils and bis(benzylpenicilloyl)-ligands, which contain spacers of different lengths, confirmed that bridging of cell surface IgE-FcεRI by simple bivalent ligands can trigger cellular degranulation effectively, as measured by the release of histamine (133). Because IgE has a binding site for hapten at the end of each of its two Fab segments, however, bivalent ligands can make a large variety of complexes, including different lengths of linearly cross-linked IgE-FcεRI and cyclized complexes. There also can be monomeric complexes in which the bivalent ligand binds simultaneously to both sites of IgE, but these complexes require ligands of a certain minimal length. The closest approach of these binding sites for IgE has been measured with fluorescence resonance energy transfer as > 80 Å (6).

The complexities of bivalent ligands that cross-link bivalent receptors, such as IgE-FcεRI, were circumvented in the studies of Metzger and colleagues (45, 129), who chemically cross-linked monomers of IgE as ligands for FcεRI to investigate the minimal aggregate size sufficient to trigger a degranulation response. In an initial study, Segal et al (129) used IgE covalent oligomers, which were cross-linked with 3,3′-dimethylsuberimidate and column fractionated, to show that dimers are as good as larger aggregates in stimulating peritoneal mast cells. In a subsequent study, however, Fewtrell & Metzger (45) showed that highly purified covalent dimers stimulated the RBL cells poorly, whereas covalent trimers were effective, and larger oligomers even more potent. Because these same covalent dimers effectively stimulated degranulation in peritoneal mast cells or caused the skin reactions, the authors suggested that their ability to stimulate degranulation of the RBL cells was limited by the responsiveness of those cells and not by an inherent inability of FcϵRI dimers to mediate a functional response (45).

Further analysis showed that the dimeric IgE could bind bivalently, which indicated that FcϵRI cross-linked in this manner were much less effective per receptor than were FcϵRI cross-linked with trimers and higher oligomers (45). In these experiments, significant degranulation was triggered by substantially fewer than 10^3 trimers or larger oligomers bound per cell, which corresponds to a small fraction of the $> 10^5$ FcϵRI per RBL cell. Similarly, small numbers of aggregates were estimated to be sufficient for threshold responses in basophils from bivalent ligand studies without direct binding data (24, 33). It appears, therefore, that the number of FcϵRI per cell exceeds the number necessary for a functional response by 1–2 orders of magnitude, which enables mast cells and basophils in vivo to respond to a wide variety of antigens that are recognized by the spectrum of IgE molecules with different binding site specificities. Although biologically interesting, this situation poses a real challenge to the biophysicist who is attempting to discern mechanism: The physical changes caused by the cross-linked IgE-FcϵRI are usually measured as the average over the whole population; however, these properties must be related to the delivery of the transmembrane signals that, at any moment, may be occurring with only a small fraction of receptors.

Orientational Constraints as Possible Limitations to Receptor-Mediated Signaling

The observation that covalent IgE dimers are much less effective, per cross-linked FcϵRI, than are trimeric IgE in stimulating RBL cells (45) provoked a number of investigations into the molecular basis. As described below, dimeric IgE causes significantly less reduction in lateral mobility than does trimeric IgE, as measured by fluorescence photobleaching recovery (FPR) (85). Further, dimers of IgE-FcϵRI made with an anti–IgE-Fc monoclonal antibody (mAb) were not immobilized and did not stimulate degranulation (86). These results suggested that dimeric IgE-FcϵRI do not interact effectively with other cellular components that restrict the lateral mobility of trimeric and larger complexes. In contrast to these results, studies with six different anti-FcϵRI mAbs from two different laboratories showed that FcϵRI cross-linked with these mAbs stimulate degranulation effectively (8, 97). Binding studies with monovalent Fab fragments indicated that all but one of these mAbs recognize a single epitope per FcϵRI, and all compete with IgE for binding to the FcϵRI α subunit (8, 97). On the basis of the binding studies, Ortega et al (96) compared the degranulation dose-response curves for three different anti-FcϵRI mAbs with cross-linking curves predicted from binding studies and concluded that these anti-FcϵRI have differential potency per unit cross-link. They suggested that different orientations of the cross-linked FcϵRI may affect the initiation of the functional response significantly.

This suggestion raises the interesting possibility that covalent IgE dimers, and at least some of the anti-IgE mAbs, might hold their bound FcϵRI in sterically restricted orientations that are less favorable for delivering a signal to the RBL cells. In recent studies, Benhamou et al (12) showed that, although oligomeric IgE bind to Fcγ receptors on RBL cells, these FcγR do not stimulate degranulation in response to oligomeric IgE or oligomeric IgG. In contrast, other mast cells and basophils possess FcγRIII that can mediate degranulation responses (29, 137). It remains possible, therefore, that the responses observed with dimeric IgE on peritoneal mast cells (44) or basophils (61) involve these FcγR in addition to FcϵRI. If FcϵRI aggregated by dimeric IgE do not signal because of steric constraints, then trimeric IgE might be a more potent activator simply because these longer oligomers allow additional spacing between FcϵRI bound to the first and third of the connected IgE. Consistent with this hypothesis, Kane et al (63) found that bivalent antigens constructed of linear avidin polymers, 2–6 avidins in length with a hapten located at each end, are more effective at triggering degranulation in RBL cells than is a single avidin with the same hapten bivalency. As described below, studies with a bivalent ligand that forms cyclic complexes of IgE-FcϵRI on RBL cells also suggest that orientational constraints on cross-linked FcϵRI can restrict their functional effectiveness greatly.

The Importance of Ongoing Cross-Linking in the Functional Response to FcϵRI

Robust degranulation of human basophils (61) and RBL cells (45) that is stimulated by covalent oligomers of IgE suggests that static cross-links, as expected from the very slow dissociation of monomerically bound IgE (57, 68), are not obstacles to sustained signaling. A more recent study (65) showed further that covalent oligomers stimulate tyrosine phosphorylation of the β and γ subunits of FcϵRI on RBL cells under several different experimental conditions. The addition of monovalent IgE in large excess after the initiation of signaling by covalent oligomers causes only a small reduction in the ongoing response, which indicates that ongoing formation of new cross-links with FcϵRI is not necessary to sustain the response. This finding appears to contrast with other receptor systems, such as ligand-gated ion channels or ligand-mediated activation of heterotrimeric G proteins, in which ligand binding commonly leads to a transient state of activation, followed by a desensitization process (16, 52). Two different interpretations of the FcϵRI results are possible: First, cross-links made by covalent oligomers are active for an extended period of time, i.e they are not desensitized readily. Second, these cross-links deliver a sustained signal that is no longer dependent on the state of the receptors.

Well-established experiments have provided arguments against the latter possibility for the general case of cross-linked FcϵRI. These experments have shown that when multivalent antigen is used to aggregate IgE-FcϵRI complexes, the cellular response is rapidly halted with the addition of a large excess of monovalent hapten. This result, which was observed initially in degranulation studies on human basophils (134), also has been obtained with RBL cells for degranulation (44) and for the earliest events of transmembrane signaling, including tyrosine phosphorylation (100), phosphatidyl inositol hydrolysis (75), and Ca^{2+} mobilization (75, 144). The simple explanation for these results, i.e. excess monovalent ligand eliminates FcϵRI aggregation by completely reversing the cross-linking reaction, is not sufficient. Seagrave et al (127) used flow cytometry to monitor the dissociation of a fluorescent phycobili protein that was multiply conjugated with dinitrophenyl (DNP) from anti-DNP IgE-sensitized RBL cells and showed that addition of a monovalent DNP-ligand caused a large fraction of the bound antigen to dissociate but on a much slower timescale than the degranulation stopped. In our laboratory, we made similar observations with an ^{125}I-labeled multivalent DNP antigen and found conditions under which more than 90% of the bound antigen remained associated for many tens of minutes, whereas degranulation halted within 1 min after adding the monovalent DNP-ligand (38, 53). Under the conditions used, monovalent ligands dissociate in less than 1 min, which indicates that most of the cell-associated antigen observed in that experiment was bound multivalently.

Two interpretations of the monovalent ligand effect are consistent with these results. First, the population of cross-linked IgE-FcϵRI that is responsible for the functional response could represent a very small subset that is readily disaggregated by the addition of monovalent ligand. Second, activation could be maintained by ongoing formation of transiently active FcϵRI. In this case, halting of the functional response is caused by prevention of continual formation of cross-links between new pairs of FcεRI. The first interpretation is more consistent with the results obtained with nondissociable covalent oligomers of IgE, as monomeric IgE does not stop degranulation under conditions in which ongoing formation of new cross-links is prevented (65). The second interpretation, however, is supported further by experiments with a symmetric bivalent DNP ligand, $(DCT)_2$-cys. These experiments are summarized below. Structural differences between oligomeric IgE–cross-linked FcεRI and antigen–cross-linked IgE-FcεRI could provide some explanation for the different results obtained in these two cases if, for example, the rates of FcεRI desensitization varied.

Bivalent Ligand Binding to Investigate Requirements for Competent Signaling

Dembo & Goldstein (31) developed equilibrium theory for the binding of bivalent ligands to cell-bound IgE, in which the degranulation dose-response curve was predicted to be symmetric about the concentration of maximal cross-link formation. This formation occurs at the concentration of free ligand equal to $1/2K_I$, where K_I is the intrinsic equilibrium constant for the ligand binding monovalently to IgE (see, for example, Figure 2). Results from early functional studies with a bivalent

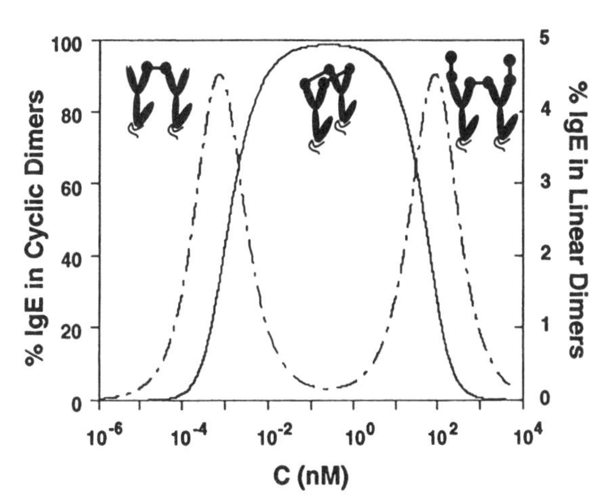

Figure 2 Equilibrium distributions of IgE-FcεRI cross-linked into cyclic (*solid line*; *left axis*) and linear dimers (*dashed line*; *right axis*) by $(DCT)_2$-cys as a function of free bivalent ligand concentration. The curves, predicted from Dembo-Goldstein theory (31) and experimentally determined binding parameters, are symmetric about $1/2K_I$, = 0.25 nM. The predominant dimeric species in three concentration regimes are indicated. Adapted from 104.

benzylpenicilloyl ligand (32), and later with a bivalent DNP ligand (74), on human basophils were consistent with the predictions of this model, but no direct binding data were available for these cells.

Erickson et al (40) developed a method for monitoring the binding of DNP ligands to anti-DNP IgE. This method is based on the observation that the fluorescence of fluorescein-5′-isothiocyanate, which is covalently conjugated to anti-DNP IgE, is quenched significantly when DNP ligands occupy the IgE binding sites. In collaboration with Goldstein (see 41 for review), these authors developed this fluorescence method further to be used with intact cells and subnanomolar concentrations of labeled IgE to provide high signal-over-background binding data with seconds time resolution (39, 41). These data can be analyzed with appropriate theoretical models to yield kinetic and equilibrium binding parameters with considerable rigor. A systematic approach to investigating complex binding events can take advantage of monovalent analogues of the bivalent IgE and bivalent ligands. In addition, IgE can be evaluated both in solution and bound to cells.

This method was used to examine the binding of DCT, the monovalent analogue of the bivalent DNP ligand, $(DCT)_2$-cys (N, N′-bis[ε-[(2,4-DNP)amino]caproyl]-L-tyrosyl]-L-cystine), and affinity and rate constants were readily determined. Detailed analysis of the binding kinetics for DCT and IgE-FcεRI on cells confirmed quantitatively the theoretical prediction that increasing the density of receptors confined to a surface causes decreases in the rate constants (per receptor) for both forward binding because of competition effects (39) and dissociation because of rebinding effects (49) (the equilibrium constant remains the same; see 48 for further discussion).

Because of cross-linking, the binding of the bivalent ligand, $(DCT)_2$-cys, to bivalent IgE at the cell surface is a complex process, but elucidation has been aided by comparison to simpler situations. $K_I = 2 \times 10^9 M^{-1}$ for $(DCT)_2$-cys binding monovalently has been established from a variety of experiments (104). The dissociation of $(DCT)_2$-cys from anti-DNP IgE in solution (103) in the presence of excess DCT is very similar to that for dissociation from IgE on the cell surface (42) and can be described as a two-step process with kinetic coefficients (rate constants including statistical factors) that differ from each other by approximately 1 order of magnitude ($3 \times 10^{-2} s^{-1}$; $3 \times 10^{-3} s^{-1}$). The binding results also suggested that cyclic dimers, which contain two $(DCT)_2$-cys and two IgE, are a highly stable species. For Fab fragments of this same anti-DNP IgE, a single exponential expression provides a good fit to the dissociation data with $(DCT)_2$-cys under cross-linking conditions. The value of the rate constant for this dissociation process, which includes both breaking cross-links and dissociation of the monovalently bound bivalent ligand, is similar to the smaller dissociation rate constant determined for the case with bivalent IgE ($4 \times 10^{-3}s^{-1}$; 136). This comparison indicates that the faster dissociation process seen with bivalent IgE is the opening of cyclic complexes (103). The bivalent $F(ab')_2$ fragment of this IgE also yields a single exponential dissociation curve similar to the Fab fragment, even though it forms the same dimeric IgE complexes as intact IgE, as revealed by gel permeation chromatography (136). These results indicate that the Fc segments of intact IgE increase the dissociation rate of the cyclic complexes, which implies that juxtaposition of these segments increases the steric constraints significantly. Similar conclusions were reached from measurements of energy transfer in a related situation (6).

$(DCT)_2$-cys has proved to be a useful ligand for comparing binding and functional properties. This bivalent ligand cross-links IgE-FcϵRI on RBL cells readily, yet it stimulates degranulation and Ca^{2+} responses poorly compared with such multivalent ligands as DNP-conjugated to bovine serum albumin (62). This finding is somewhat surprising, considering the potential for the formation of long linear chains of IgE-receptor complexes with this high-affinity bivalent ligand. In addition, the previous studies have demonstrated good stimulation of degranulation with human (32) and rabbit (133) basophils with the use of small bivalent ligands. These apparent differences in response may reflect the more discriminating requirements for cross-linking of RBL cells described above for covalent oligomers (45). Substantial evidence now indicates that $(DCT)_2$-cys forms highly stable cyclic dimers with IgE. These dimers prevent the formation of trimers and larger complexes effectively, although a relatively small amount of linear dimers can form at some concentrations of this ligand (Figure 2; 104). This conclusion is supported by direct comparison of the distributions of cross-linked aggregates, as predicted by extension of the Dembo-Goldstein theory, and cellular changes, including lateral diffusion (with FPR), degranulation, and Ca^{2+} responses measured at the same concentrations. The FPR measurements can differentiate between IgE-FcϵRI dimers, which show a small reduction in lateral mobility compared with monomers and higher aggregates that cause substantial immobilization.

Degranulation responses of anti-DNP–sensitized RBL cells after cross-linking with $(DCT)_2$-cys are not always detectable. When they occur, however, the maximal response is observed at concentrations of $(DCT)_2$-cys that are 1–2 orders of magnitude greater than the concentration for maximal cross-linking ($1/2K_I$), where the cyclic dimers are overwhelmingly predominant (Figure 2; 38, 62, 104). Similarly, small Ca^{2+} responses to $(DCT)_2$-cys in individual adherent RBL cells were observed to be maximal at a concentration 100 times greater than $1/2K_I$ (112). More substantial and dependable degranulation responses can be obtained when an independently binding anti-IgE mAb, B1E3, is combined with $(DCT)_2$-cys. Binding and FPR studies indicated that B1E3 cross-links IgE-FcϵRI on the cell surface into aggregates no larger than dimers (64, 104). As expected from these results, this mAb stimulates little or no Ca^{2+}mobilization or degranulation by itself (104). It cross-links together the IgE-FcϵRI bound to $(DCT)_2$-cys, however, and robust responses are observed at concentrations 10–100 times greater than $1/2K_I$ (109). This higher concentration corresponds to the predicted maximum for linear dimers of IgE-FcϵRI cross-linked by $(DCT)_2$-cys, and this finding is discussed further below. It is striking that the maximal response does not occur at $\langle 1/2K_I \rangle$, which indicates strongly that IgE-FcϵRI forced into cyclic dimers by $(DCT)_2$-cys cannot stimulate Ca^{2+} or degranulation responses, even when these complexes are joined together in larger aggregates.

Schweitzer-Stenner and colleagues (123–125) also have found strong evidence for the formation of cyclic complexes in their solution studies with bivalent DNP-ligands of lengths similar to $(DCT)_2$-cys (but different chemical structures) and a monoclonal anti-DNP IgE with lower affinity for DNP ligands. In this work, they developed a novel method for analysis of nonequilibrium binding data to determine kinetic constants for the cross-linking reaction that is especially useful for ligands with low affinity (124, 125). The propensity for cyclic complex formation with these bivalent ligands and IgE, together with their incompetence in cellular signaling, suggests that structural constraints may prevent the proper interaction of the FcϵRI with appropriate signaling molecules (104). Indications of structural rigidity in membrane-bound IgE-FcϵRI complexes formed with small bivalent ligands come from studies on the segmental flexibility of IgE-FcϵRI (55) and from measurements of energy transfer between FcϵRI-bound IgE in these complexes (6). Linear dimers formed with the same ligands probably would have greater orientational freedom than cyclic dimers, which might explain the functional responses seen at the higher concentration of $(DCT)_2$-cys. Initial results obtained with a bispecific IgE that binds DNP in only one of its Fab segments are consistent with this interpretation. This IgE can form linear, but not cyclic, dimers with $(DCT)_2$-cys, and the dose-response curve for degranulation is shifted toward lower concentrations of this ligand expected for maximal cross-linking (K

Subramanian, C Hine, D Holowka, and B Baird, manuscript in preparation).

As described above, in the dose-response curve for $(DCT)_2$-cys and anti-DNP IgE on RBL cells, maximal degranulation and Ca^{2+} responses are observed at concentrations that are 10–100 times greater than the maximum predicted for the predominating cyclic dimers, and this higher concentration corresponds to a maxima in the linear dimers (Figure 2). The theoretical curve for the equilibrium distribution of cross-linked aggregates is symmetric, and no response is detected at a $(DCT)_2$-cys concentration 10–100 times lower than $\langle 1/2K_I \rangle$ (104). Other factors appear to affect the functional outcome of cross-linking as well. One significant difference between the linear cross-linked complexes at their two predicted maxima in the dose-response curve is likely to be the lifetime of each complex; during this time, interactions that lead to the initiation of signaling and possibly desensitization occur. At the high concentration maximum [~10^{-7} M $(DCT)_2$-cys], most bivalent ligands are in solution; all IgE binding sites are filled, but only a small fraction ($< 5\%$) are predicted to be involved in linear aggregates (Figure 2; 104). On the basis of recent assignments of dissociation rate constants (136), the half-life of a linearly cross-linked complex in this situation is approximately 200 s at 25°C (i.e. $0.69/k_{-2}$), as cross-links broken by dissociation of one end of the bivalent ligand are not likely to reform in the presence of a large excess of bivalent ligand in solution. This half-life is longer than the time required to achieve a maximal Ca^{2+} response in the presence of B1E3 (~100 s; R Posner, D Holowka and B Baird, unpublished results), so the duration of a cross-link probably would not limit its effectiveness.

At the low concentration regime for maximal linear dimer formation, most IgE binding sites are free, and there is virtually no $(DCT)_2$-cys in solution. Cross-links that are broken, therefore, are likely to reform between the same pair of IgE-FcϵRI if they do not move too far apart before rebinding can occur. Fluorescence photobleaching recovery (104) and phosphorescence anisotropy (90, 91) measurements show that IgE-FcϵRI in the presence of B1E3 and $(DCT)_2$-cys are laterally and rotationally immobile. Even in the absence of B1E3, $(DCT)_2$-cys on cells are observed to undergo a time-dependent loss of dissociability in the presence of excess IgE in solution (42), which indicates further that cellular interactions can retard the ability of aggregated receptors to dissociate. Continual involvement of the same FcϵRI might make them more likely to be shunted into a desensitized state that could be mediated by phosphorylation or some other mechanism (see section on Membrane Structure and IgE-FcϵRI Signaling). Evidence for antigen-mediated desensitization in RBL cells has been described (144). A recent study by Schweitzer-Stenner et al (126) is also relevant to these issues. This study indicates that there is a critical lower threshold for the lifetime of cross-links made with anti-FcϵRI mAb that is required for productive signaling. Interestingly, even in this situation, where aggregates are limited to dimeric FcϵRI complexes, too much cross-linking leads to inactivation (126).

It is attractive to suggest that the dynamic formation of cross-links between new pairs of IgE-receptor complexes at the high concentration regime for $(DCT)_2$-cys provides a more favorable situation for ongoing signal transduction than the continual reformation of cross-links between the same pairs of complexes at the low concentration regime. As discussed above, this model is not consistent with the ability of virtually nondissociable covalent IgE oligomers to deliver a sustained signaling response. Similarly, we found recently that streptavidin–cross-linked, biotinylated IgE-FcϵRI complexes stimulate a strong Ca^{2+} response that is not halted by the addition of a large excess of biotin, which prevents the ongoing formation of new cross-links (L Pierini, D Holowka and B Baird, unpublished results). Dynamic cross-linking of FcϵRI may be required for multivalent ligands that cross-link reversibly through IgE binding sites, but other factors must be found to explain the striking differences observed with the irreversible cross-linkers.

PHYSICAL CONSEQUENCES OF FcϵRI AGGREGATION

IgE-FcϵRI as a Prototype for Studies of Lateral Diffusion of Cell Surface Proteins

In 1976, FcϵRI was one of the first specific receptors to be analyzed by the newly developed technique of FPR (119). Since that initial work, more than ten studies from at least five different laboratories have measured lateral diffusion coefficients (D_L) for the unaggregated IgE-FcϵRI complex, and most of these studies have indicated average values in the range of $2–4 \times 10^{-10}$ cm^2/s. The percentage of these monomeric FcϵRI found to be laterally mobile (%R) varied from 60 to 100% (45, 63, 78, 85, 104, 138). These values for D_L and %R are considered "typical" for cell surface proteins, although D_L and %R can range widely for different proteins on various cell types (see 35 for a more comprehensive review). The D_L for IgE-FcϵRI complexes and several other membrane proteins on cells is at least 1 order of magnitude less than the D_L on membrane vesicles derived from chemically treated cells (commonly called "blebs"; 143). Bleb vesicles derived from RBL cells have a lower level of density of FcϵRI than the native cells have, and the vesicular FcϵRI exhibit diffusion properties that approximate those expected in simple lipid bilayers on the basis of theoretical predictions (115). In contrast, IgE-FcϵRI on cells that have become swollen after suspension in hypertonic solutions or after permeabilization by streptolysin O or digitonin exhibit values of D_L that are only two- to fivefold greater than on native cells (43). For the permeabilized cells, values of $\%R$ are in the same range as those on native cells (i.e. ~60–80%), whereas those on blebs and cells in hypertonic solution usually are 100% (139). For all the swollen cells, the microfilament-associated cytoskeleton and cytosolic granules are observed to be retracted from the cell periphery, such that there is a region of clear cytoplasm that separates the plasma membrane from the nuclear-cytoskeletal complex (43). These results indicate that some restriction on the lateral diffusion of these receptors is imposed by components of the plasma membrane other than the native cytoskeletal connections to the nucleus and other internal structures. This result is not entirely surprising, as membrane proteins occupy a large fraction of the plasma membrane surface (113), and Saxton (116, 117) has theorized that collisions with mobile, as well as immobile, proteins will restrict the lateral diffusion of a tracer protein.

In the case of FcϵRI, deletions of individual cytoplasmic segments of each of the α, β, and γ subunits do not affect the lateral diffusion of monomeric IgE-receptors complexes (78). This finding contrasts with changes that have been observed for certain other cell surface proteins, such as major histocompatibility complex class II molecules after deletions in their cytoplasmic segments (142). The binding of IgE on the extracellular side of FcϵRI has only small effects on the value of D_L (two- to fourfold increase) as determined by back-diffusion after electric field–induced redistribution (83). In more recent studies, we found that the attachment of a low density lipoprotein (LDL) particle to the IgE-FcϵRI complex causes a large reduction in the value of $\%R$ (from ~80% to $< 20\%$). This reduction correlates with observations that these DiI-labeled LDL–IgE-FcϵRI complexes exhibit a very limited range of lateral diffusion in single particle tracking experiments (JP Slattery, D Holowka, WW Webb, and B Baird, manuscript in preparation). These results suggest that the large mass of LDL (~2×10^6 daltons) on the extracellular surface, about tenfold greater than IgE, provides a substantial drag on FcϵRI as it moves, possibly because of barriers of the glycocalyx and/or the extracellular matrix that these larger complexes encounter. Further investigations of lateral mobility with this versatile receptor are likely to reveal a more detailed understanding of the cellular factors that restrict the long-range mobility of membrane proteins.

Changes in Lateral Mobility Caused by FcεRI Aggregation

Small aggregates of IgE-FcεRI made with stable trimers and higher oligomers of IgE or formed with anti-IgE mAb lose most of the lateral mobility exhibited by monomeric IgE-FcεRI (85, 86). In these situations, the value of $\%R$ is typically reduced from 70–100% for monomers to less than 20% for the aggregates, whereas IgE-FcεRI that remain mobile usually have D_L values that are similar to the monomeric complexes. In the case of trimers and tetramers, the increased size of the complexes formed cannot account for this large reduction in lateral mobility; the established hydrodynamic theory predicts only a logarithmic dependence on intramembrane molecular volume (115). The immobilization process occurs rapidly on binding of the cross-linking ligand, does not depend on cellular signaling that is inhibited at 4°C or by NaN3/deoxyglucose, and can be readily reversed if monovalent hapten is added to antigen-aggregated IgE-FcεRI complexes within a short time (minutes) after antigen is bound (86). As indicated above, aggregates too small to trigger significant degranulation exhibited little or no immobilization, and these observations suggested that cross-linking–dependent immobilization of FcεRI represents interactions with other cellular components that can precede and may be important for signal transduction (85, 86).

Support for this interpretation comes from studies with FcεRI in which the cytoplasmic segments of each of the three different subunits were deleted by molecular genetic techniques, and the lateral mobilities of these mutants in the presence of monomeric or oligomeric IgE were investigated (78). All the mutant FcεRI bound to monomeric IgE exhibited lateral mobility similar to the wild-type receptors, but one of two mutants defective in signaling (the mutant with a C-terminal–deleted β subunit) showed a significant reduction in the aggregation-dependent immobilization. The molecular basis for this effect is not yet known, but recent studies described below suggest that interactions of the β subunit with the tyrosine kinase $p53/56^{lyn}$ may contribute to the aggregation-dependent immobilization observed. Studies in which sensitive kinase assays have been used have shown co-immunoprecipitation with FcεRI of small amounts of members of the src tyrosine kinase family, particularly $p53/56^{lyn}$, and enhanced kinase activity on FcεRI aggregation has been detected in these immunoprecipitates (14, 36, 56). A recent chemical cross-linking study (147) demonstrated enhanced co-purification of p53/56^{lyn} such that, in the absence of IgE-FcϵRI aggregation, approximately 2–4% of the receptors have p53/56^{lyn} associated, whereas after aggregation with multivalent antigen, 10–13% have p53/56^{lyn} associated. Under these latter conditions, there is also substantially increased FcϵRI-associated tyrosine kinase activity (147).

These biochemical results are very provocative, as they point to aggregation-dependent interactions with important signaling molecules; they may also be relevant to the biophysical observations of receptor immobilization. The enhanced interaction of FcϵRI with p53/56^{lyn}, which is anchored to the membrane by an N-terminal fatty acid modification (108), is not likely by itself to account for the large reduction in lateral mobility observed on cross-linking. The recruitment of other proteins, however, through interactions between their SH2 domains and phosphotyrosine residues on FcϵRI or its associated p53/56^{lyn} could contribute to this process. Tyrosine kinase activation occurs efficiently at 4°C on aggregation of BCR (23) and FcϵRI (46), so the aggregation-dependent immobilization observed under these conditions could be at least partially dependent on tyrosine kinase activation. The suggestion that phosphotyrosine-mediated interactions can lead to large "signaling particles" in the case of certain growth factor receptors (140) and immunoreceptors (101, 145) has not yet been demonstrated unequivocally. If this process does occur with a large enough fraction of aggregated FcϵRI, however, it could contribute to the loss of lateral mobility that is observed.

A different, but possibly related, explanation for aggregation-dependent FcϵRI immobilization arises from two different studies that indicate that specialized membrane domains interact with aggregated FcϵRI. In one study, Thomas et al (139) found that large-scale aggregation of IgE-FcϵRI at the cell surface caused by anti-IgE is accompanied by co-redistribution of membrane domains that are labeled with certain fluorescent lipid analogues, including the carbocyanine derivative $DiIC_{16}$. Aggregation of a GD_{1b} ganglioside with a specific mAb also causes co-redistribution of these labeled domains (139). In a separate study, this ganglioside was shown to be a marker for biochemically isolated membrane domains that are resistant to solubilization by Triton X-100 and contains abundant p53/56^{lyn} that appears to be highly active (46). Related membrane domains, known as caveolae, have been hypothesized to play important roles in cellular signaling and membrane trafficking (2, 72, 101a). Under mild conditions of cell lysis, the aggregated IgE-FcϵRI are preferentially co-isolated with the Triton X-100 resistant domains on RBL-2H3 cells (46; KA Field, D Holowka, and B Baird, unpublished results). As discussed below, recent results suggest that this interaction may be relevant to the earliest known biochemical events in FcϵRI-mediated signaling: tyrosine phosphorylation of the β and/or γ subunits of FcϵRI by p53/56^{lyn} (101).

Although possibly necessary, IgE-FcϵRI immobilization appears to be insufficient for cellular activation, as indicated by the above-described cyclic complexes that are formed with the bivalent ligand $(DCT)_2$-cys in the presence of the anti-IgE mAb B1E3 (104). At nanomolar concentrations of $(DCT)_2$-cys, a small reduction in the FPR value of *%R* is observed, which is consistent with the formation of dimers, and B1E3 added alone causes a similarly small change. However, the combination of these bivalent ligands causes reduction in *%R* to 20% or less, which is indicative of highly restricted lateral mobility (104). As discussed in the section, Bivalent Ligand Binding to Investigate Requirements for Competent Signaling, little or no degranulation is observed in this concentration regime. We recently found that nanomolar $(DCT)_2$-cys causes efficient tyrosine phosphorylation of the γ subunit of FcϵRI but no detectable tyrosine phosphorylation of the β subunit. In this situation, the tyrosine kinase p72^{syk} is not activated, and no stimulated tyrosine phosphorylation of whole cell lysates is detected (N Troy, D Holowka, and B Baird, manuscript in preparation). This finding indicates that γ phosphorylation (probably mediated by p53/56^{lyn}) can occur normally in the sterically constrained cyclic dimers formed under these conditions but that the tyrosine kinase cascade that ordinarily is initiated by p72^{syk} activation (see below) is prevented. The correlation between γ phosphorylation and the loss of lateral mobility under these conditions is consistent with the hypothesis that the interaction of aggregated FcϵRI with the p53/56^{lyn}-containing domains mediates both processes.

Cross-Linking–Dependent Resistance to FcϵRI Solubilization by Mild Detergents

A biochemical characteristic of aggregated immunoreceptors that may be related to their loss of lateral mobility is their induced resistance to solubilization by nonionic detergents, such as Triton X-100, or zwitterionic detergents, such as CHAPS. This phenomenon, commonly referred to as "detergent insolubility," was first described for BCR aggregated by anti-Ig antibodies on B lymphocytes (19) and has since been documented for FcϵRI (3, 110) and TCR (47, 80). In all these cases, most monomeric, unaggregated receptors are readily solubilized by Triton X-100 or NP-40, but at least some fraction of the receptors become resistant to solubilization after aggregation at the cell surface or after internalization into endosomes (110). The extent of cross-linking needed to detect significant detergent insolubility varies with the ligands, receptors, and cells examined. In the case of TCR, an anti-receptor mAb that makes dimeric complexes is sufficient to cause the insolubility of 30–40% of those receptors on T-cell lines (80; N Marano, D Holowka, and B Baird, unpublished results). More extensive cross-linking of TCR is necessary, however, for peripheral blood T lymphocytes (47). For FcϵRI on RBL cells, dimeric complexes, as well as monomeric receptors, are solubilized, even in the presence of a chemical cross-linker that enhances detergent insolubility of trimeric oligomers or multivalent antigen (110).

Mao et al (77) showed that, in the absence of a stabilizing chemical cross-linker, detergent insolubility of FcϵRI caused by multivalent antigen is transient and appears to be lost as cross-links between IgE-receptor complexes dissociate in the presence of Triton X-100. This finding may explain why some ligands that have weaker binding constants, such as the anti-IgE mAb A2, do not cause significant detergent insolubility, even though they can cause extensive immobilization of FcϵRI on the cell surface (86, 110). A strong argument for a structural relationship between lateral immobilization at the cell surface and detergent insolubility is the sensitivity of both to specific deletion of the C-terminal segment of the β subunit of FcϵRI (77, 78). As indicated above, the importance of this cytoplasmic segment in FcϵRI-mediated signaling suggests that these processes also have some relationship to signal transduction. (See section on Membrane Structure and IgE-FcϵRI Signaling for further discussion.)

Apgar (4) showed that cross-linking–dependent detergent insolubility of FcϵRI is not a consequence of a variety of early signaling events, including protein kinase C activation or inositol phosphate hydrolysis. With polyclonal anti-IgE–mediated detergent insolubility, tyrosine kinase inhibitors have no detectable effect on this process (D Holowka, unpublished results). These results indicate that detergent insolubility caused by cross-linking proceeds independently of signaling. Oliver and colleagues (95) have suggested that this process may limit the extent of signal transduction by shunting functionally active receptors into an inactive compartment that prevents them from participating in the signaling process. The dose-dependence for detergent insolubility commonly is shifted to higher concentrations of cross-linking ligand compared with functional responses, such as Ca^{2+} mobilization or degranulation, and the functional response often declines at concentrations of ligand at which detergent insolubility is maximal (69, 110, 128). In support of this effect as a mechanism for desensitization, Seagrave & Oliver (128) observed that a large excess of monovalent hapten added after a large dose of multivalent antigen substantially reduced the amount of detergent insolubility (from 26 to 6%) but enhanced the degranulation response.

Although Ca^{2+} and degranulation responses generally decline at doses of cross-linker that cause maximal detergent insolubility, the antigen-stimulated increase in cellular tyrosine phosphorylation appears to follow more closely the dose-response curves for detergent insolubility and for antigen binding (25, 69; K Xu, N Troy, D Holowka, and B Baird, unpublished results). This finding suggests that these processes are related more directly to the number of FcϵRI aggregated than are more downstream signaling events, which are maximal at submaximal amounts of tyrosine kinase activation. The implication is that the interaction of the aggregated receptors with cellular components that cause detergent insolubility does not prevent the receptors from participating in the stimulation of cellular tyrosine phosphorylation. Rather, these interactions can prevent the activation of this phosphorylation cascade from being translated into an effective signal for degranulation. A new model of signaling that incorporates these observations is described in the next section.

The structural basis for detergent insolubility of aggregated immunoreceptors has not yet been determined, but this goal seems to be important for understanding the molecular mechanisms by which functional responses are regulated. Woda & Wootin (146) showed that anti–Ig-stimulated detergent insolubility could be observed in plasma membrane preparations from B lymphocytes, and Apgar (3) made similar observations for antigen-dependent insolubility of FcϵRI on plasma membrane fragments made by nitrogen cavitation of RBL cells. In contrast, FcϵRI on plasma membrane blebs are solubilized efficiently before and after cross-linking (3), which indicates that a critical structural feature of the plasma membranes for detergent insolubility of receptors is absent in these vesicles. A role for the microfilament cytoskeleton in this phenomenon has been suggested (95). However, the actin polymerization inhibitor cytochalasin D has shown little (128) or no (3) effect on cross-linking–dependent FcϵRI insolubility in studies published to date, despite the observations that actin depolymerization conditions reduced the amount of detergent insolubility caused by anti-IgE (110). We recently observed that cytochalasin D can substantially reduce the detergent insolubility of FcϵRI in RBL cells caused by more limited cross-linking (D Holowka, unpublished results), which leaves open the possibility for cytoskeletal involvement, as discussed below.

Neither our laboratory (89) nor others (79) have been able to find significant amounts of new proteins selectively associated with aggregated, but not monomeric, FcϵRI under conditions in which a large fraction of the detergent-insoluble, chemical cross-link–stabilized receptors could be analyzed. This situation suggests that the interactions responsible for detergent insolubility could be caused by cellular components that do not appear on sodium dodecyl sulfate–polyacrylamide gels. One such candidate, the agorin family of proteins, is a major component of detergent-insoluble plasma membrane preparations from several different hematopoetic cells, including P815 mouse mastocytoma (5) and RBL (3) cells. Agorins are thought to play a structural role in the plasma membrane skeleton but have been difficult to study because of their unusual physical properties (5). In addition, certain lipids that play structural roles in the membrane domains, discussed above, may be involved. We recently detected modulations of FcϵRI signaling and detergent insolubility that result from alteration in the lipid composition of RBL cells. In these studies, treatment of RBL cells with cholesterol-depleting sphingomyelin liposomes caused a large reduction in cross-linking–dependent detergent insolubility of FcϵRI and an enhancement in stimulated degranulation (25; D Holowka, unpublished results). These results indicate that the lipid composition of the cells strongly influences these properties and suggest that normal association of aggregated receptors with specific, detergent-resistant membrane domains may play a significant role in both stimulating and desensitizing aspects of signaling.

Studies on the Rotational Motion of FcϵRI Before and After Aggregation

In contrast to FPR measurements of lateral diffusion of cell surface proteins over micron dimensions on a time scale of seconds to minutes, the rotational motion of these membrane proteins is measured on the timescale of 1–400 μs. On average, in the time it takes a monomeric IgE-FcϵRI to move a detectable lateral distance in FPR, i.e. approximately 0.5 μm, it has undergone many thousands of rotations (91). Clearly, the environmental influences on rotational motion are likely to be much different from those for lateral diffusion. In particular, rotational motion is likely to be more sensitive to the size of the rotating complex and to the membrane composition in the immediate vicinity of the receptor. In an ideal fluid membrane bilayer, the rotational diffusion is inversely proportional to membrane viscosity and the surface area of the transmembrane region of the protein (115). In the case of rhodopsin, which spans the membrane seven times in the disc outer segment membrane (28), or the structurally related protein bacteriorhodopsin in reconstituted fluid bilayers (27), the measured rotational correlation times (~15–20 μs) do not differ appreciably from those predicted by theory. FcϵRI in RBL cell membranes have the same number of predicted membrane-spanning segments as these other proteins but exhibit rotational correlation times that are typically at least several-fold larger (91, 102, 107, 149), which suggests that they experience significant retarding effects. These measurements have been made in several different laboratories with the use of two different methods. The more common method is phosphorescence anisotropy decay, in which a probe, such as erythrosin, is covalently attached to a protein ligand, such as IgE. Some of these studies have used labeled anti-FcϵRI mAb and their Fab fragments that compete for IgE binding (102, 107), and the results for unaggregated FcϵRI generally have been similar to those obtained with labeled IgE (91, 149).

Chang et al (26) found recently that mutants of FcϵRI that are missing C-terminal cytoplasmic tails of the β or γ subunits have significantly faster rates of rotation than the monomeric wild-type FcϵRI or FcϵRI with mutations in the N-terminal cytoplasmic segment of β or in the C-terminal cytoplasmic segment of α. This difference correlates with the functional inactivity of the β and γ C-terminal mutants and suggests that a physical connection to some functionally critical membrane component has been lost. As discussed above, it now appears that at least a small fraction of the receptors are pre-associated with the tyrosine kinase p53/56^{lyn} (36, 147). It seems unlikely, however, that even a stoichiometric association of this fatty acid–anchored membrane protein could account for the differences in rotation correlation times observed. One candidate for the component that loses interaction with the β and γ C-terminal mutants is a protein complex that can be chemically cross-linked to IgE-FcϵRI selectively to increase the apparent molecular size by almost twofold, as determined by gel permeation chromatography of cross-linked, solubilized receptor complexes (K Field, D Holowka, and B Baird, unpublished results).

Unlike lateral diffusion, the rotational diffusion of IgE-FcϵRI is affected substantially by dimer formation. At low concentrations the anti-IgE mAbs, B1E3 and A2, individually cross-link IgE-FcϵRI into dimers on the cell surface, which causes significant increases in the initial phosphorescence anisotropy values compared with monomers and little or no anisotropy decay during the measureable time course of 300–400

μs (91). A similar loss of rotational motion was observed for all the FcϵRI mutants when aggregated with B1E3, which indicates that the causative interactions are not dependent on signaling or on any individual cytoplasmic segment. On bleb vesicles, monomeric IgE-FcϵRI exhibit the same rotational correlation time as on intact cells. Unlike on cells, however, the dimers show substantial anisotropy decay with a rotational correlation time that is about twice the value for monomers (91). Larger IgE-FcϵRI aggregates on blebs and cells caused by multivalent antigen or polyclonal anti-IgE show little or no anisotropy decay (91), as expected from the correspondingly large increases in transmembrane areas (115). These results indicate that dimerized IgE-FcϵRI on the cell surface undergo a substantial change in their immediate environment that greatly restricts their rotational motion.

The phosphorescence anisotropy studies with erythrosin-labeled IgE-FcϵRI provide the view that a large fraction of the monomeric FcϵRI rotate within a fluid membrane environment, subject to some constraints (26). Some significant fraction of monomeric FcϵRI probably does not rotate detectably on the time scale of the anisotropy experiments, which may correspond to the small population of laterally immobile receptors observed with FPR (91). The loss of anisotropy decay that results from formation of IgE-FcϵRI dimers may reflect nonspecific entanglement of tethered receptors in some larger or less mobile structures, or possibly more specific interactions with a less mobile structure. In either case, FcϵRI lose significant rotational freedom on the cell surface after dimers or higher aggregate states are formed.

MEMBRANE STRUCTURE AND IGE-FCϵRI SIGNALING

Biochemical Consequences of FcϵRI-Mediated Activation of Tyrosine Phosphorylation

As indicated in the previous section, biochemical studies during the past 5 years have shown that aggregation of FcϵRI, TCR, and BCR activate a tyrosine kinase cascade. Tyrosine phosphorylation of the tandem sequences Y-X-X-L-X_7-Y-X-X-L (X is an unspecified amino acid; 109) on the γ subunit of FcϵRI (see Figure 1) by p53/56^{lyn} results in the recruitment of p72^{syk} to the receptor aggregate through interactions with the tandem SH2 domains of this kinase (60, 66). The net result of this process is activation of p72^{syk} and initiation of the cascade of cellular phosphorylation that results in a large increase in both tyrosine and serine/threonine phosphorylation of a number of different proteins (15, 100). Requirements for aggregate size, orientational freedom, and lifetime to initiate signaling have been discussed extensively above. Notably, there also appear to be minimal requirements for cellular integrity to enable the tyrosine kinase cascade. Pribluda & Metzger (105) found that aggregation-dependent tyrosine phosphorylation of the FcϵRI β and γ subunits can be obtained with broken cell preparations but that phosphorylation of other cellular substrates is largely absent. Among the tyrosine kinase substrates phosphorylated in intact cells is phospholipase C-γ1 (PLC-γ1) (109a). On activation, this substrate cleaves phosphatidyl inositol 1,4,5-trisphosphate to produce inositol 1,4,5-trisphosphate, which acts as a ligand to mediate the opening of Ca^{2+} channels (17), and 2,3-diacylglycerol, which activates several different isoforms of protein kinase C (67). The simultaneous activation of these two signaling pathways appears to be necessary and sufficient for the stimulation of cellular degranulation in mast cells (10).

Although the importance of tyrosine kinase activation in initiating such cellular responses as degranulation is now well established, the biochemical sequelae in this process and the participation of membrane structure are understood only partially. Pharmacologic studies are consistent with tyrosine kinase–dependent activation of PLC-γ1 as the earliest known downstream event in cellular activation (10), but the process appears to be complex. In addition, the involvement of several different tyrosine phosphatases (1), as well as other biochemical steps, have been suggested (148). Further, it is not yet clear whether the activation of PLC-γ1 is the essential step for the downstream signaling events. For example, a tyrosine kinase inhibitor from the tyrphostin family was found to inhibit FcϵRI-mediated degranulation and phosphatidyl inositol hydrolysis without affecting Ca^{2+} mobilization of RBL cells in suspension (71). Other studies have shown that FcϵRI-mediated Ca^{2+} mobilization in these cells is enhanced by pretreatment with cholera toxin, which is known to activate the heterotrimeric guanosine triphosphate–binding protein G_s (82, 93). We recently found that this protein is recruited to the Triton X-100–insoluble membrane domains in response to FcϵRI cross-linking (K Field, D Holowka, and B Baird, unpublished results). Other signaling molecules, including inositol 1,4,5-trisphosphate–gated Ca^{2+} channels, have been found to be associated with related domains in other cells (72).

As noted above, specialized regions of the plasma membrane have been implicated in several different transport and vesicle-trafficking processes, including uptake of extracellular proteins and cofactors and delivery of glycosylphosphatidyl inositol-linked proteins to the cell surface (2, 72, 101a). Other recent studies have indicated that receptor-activated Ca^{2+} influx (88) and exocytosis (122) occur at discrete regions of the plasma membrane in chromaffin cells and adrenal medullary cells, respectively. In addition, the complex of proteins known to participate in vesicle trafficking, as well as in regulated exocytosis, recently has been found to be enriched in membrane caveolae (120). The role played by detergent-resistant membrane domains (including caveolae) as specialized regions that mediate signaling and exocytosis in mammalian cells deserves careful consideration.

Possible Role for Triton X-100–Resistant Membrane Domains in FcεRI Function

We recently observed that aggregation of FcεRI on RBL cells is accompanied by recruitment of cellular $p53/56^{lyn}$ to Triton X-100–insoluble membrane domains that can be isolated on the basis of their low density by equilibrium sedimentation in sucrose gradients. A substantial fraction (~30%) of $p53/56^{lyn}$ co-isolates with the Triton X-100–insoluble domains from resting cells, and about twice as much $p53/56^{lyn}$ associates with these domains if the FcεRI has first been aggregated to cause cellular stimulation (46). Schroeder et al (121) recently suggested that the structural basis for Triton X-100–resistant domain formation is the co-aggregation of lipids and proteins that are anchored to the membrane by saturated hydrocarbon tails. Together with cholesterol, these components form bilayers with local order more similar to a gel phase than to fluid liquid crystalline membranes (121). Further, palmitoylation of certain src-family tyrosine kinases, including $p53/56^{lyn}$ and $p56^{lck}$ in T cells, has been shown to be important for the association of these proteins with Triton X-100–resistant membrane domains (108, 131). Thus, FcεRI-mediated stimulation of protein palmitoylation could be responsible for the recruitment of $p53/56^{lyn}$ and other proteins to these domains. The recruitment of $p53/56^{lyn}$ is accompanied by enhanced tyrosine phosphorylation of a unique set of membrane proteins that is associated with these domains. Further, stimulation results in an increase in tyrosine kinase activity, which probably is caused by the recruited $p53/56^{lyn}$ that copurifies with this domain (46). From the correlation of these events, it seems likely that the translocated $p53/56^{lyn}$ plays a significant role in mast cell responses stimulated by FcεRI.

Additional observations suggest that the interaction of aggregated IgE-FcεRI with these low-density, detergent-resistent membrane domains is important for signaling. Although aggregated FcεRI fractionates to a higher density in the sucrose gradients after extraction with 1% Triton X-100, homogenization, and ultracentrifugation, a substantial fraction of FcϵRI co-isolates with the detergent-resistant domains in a cross-linking–dependent manner when a lower Triton X-100 to cell ratio is used (46; K Field, D Holowka, and B Baird, unpublished results). These latter solubilization conditions also result in a marked enhancement of the antigen-dependent tyrosine kinase activity that co-immunoprecipitates with IgE-FcϵRI (106). Further, in vitro kinase assays of sucrose gradient fractions show that all the stimulated tyrosine phosphorylation of FcϵRI occurs on the domain-associated receptors (KA Field, D Holowka, and B Baird, unpublished results). As indicated above, $DiIC_{16}$-labeled domains that co-redistribute with aggregated IgE-FcϵRI on intact cells appear to be related to the Triton X-100–resistant domains because they share a common selective marker, the GD_{1b} ganglioside (46, 139). In contrast to the co-redistribution observed when IgE-FcϵRI are aggregated with soluble anti-IgE, these domains are excluded from the region of contact between RBL cells and cell-sized beads coated with IgE cross-linking ligands (L Pierini, D Holowka, and B Baird, unpublished results). Under these conditions, only a transient Ca^{2+} response is elicited by the beads. This response contrasts markedly to the sustained Ca^{2+} response observed when IgE-FcϵRI are aggregated by soluble cross-linking ligands, including anti-IgE (11, 87a).

These studies suggest a model for mast cell activation, in which aggregation of FcϵRI leads to recruitment of $p53/56^{lyn}$ into specialized membrane domains. This process is a link in the chain of FcϵRI-mediated signaling. The results described above further suggest that direct interaction of aggregated receptors with these domains is important for this activation process, possibly to facilitate the phosphorylation of FcϵRI γ subunit by active $p53/56^{lyn}$. Consequent activation of $p72^{syk}$ caused by association with tyrosine-phosphorylated FcϵRI would then initiate the cascade of tyrosine phosphorylation that is necessary for the activation of downstream signaling events. The proximity of these active kinases to membrane domains that contain proteins important in the downstream signaling events may be crucial for the functional outcomes, such as cellular degranulation. In addition, nucleation of preexisting membrane domains by FcϵRI co-aggregation or, possibly, de novo formation of such domains caused by the stimulation of protein palmitoylation or other enzyme activites may facilitate the coalescence of important signaling proteins that individually associate with these domains.

A Regulatory Role for the Cytoskeleton

How do the physical properties of the aggregated IgE-FcϵRI relate to their role in these processes? It is clear from the studies summarized above that even limited aggregation of FcεRI causes substantial restrictions on rotational and lateral diffusion. Although detectable only after substantial cross-linking, the interactions of aggregated FcεRI that lead to a loss of detergent solubility are likely to be related to those that lead to lateral immobilization. These interactions evidently do not prevent the FcεRI from activating the tyrosine kinase cascade; after some point in this process, however, they do limit FcεRI activation of downstream signaling events that lead to cellular degranulation. We recently found two different treatments that both reduce detergent insolubility of aggregated IgE-FcεRI and enhance antigen-mediated cellular degranulation. The first approach is pretreatment with cytochalasin D to prevent stimulated actin polymerization, and the second approach is pretreatment with sphingomyelin to reduce cellular cholesterol (25). Reduced insolubility (D Holowka, unpublished results) and enhanced secretion (95, 141) caused by cytochalasin D suggest that microfilaments normally regulate functional responses mediated by FcεRI, possibly by anchoring the Triton X-100–resistant membrane domains and thereby limiting their ability to redistribute with aggregated FcεRI. Similar effects of sphingomyelin treatment on FcεRI insolubility and secretion also may be related to its effects on microfilament–membrane domain interactions. We recently observed with confocal fluorescence microscopy that sphingomyelin treatment causes detachment of polymerized actin from these domains (D Holowka, unpublished results). Past studies implicated microfilaments in the regulation of the terminal steps of FcεRI-mediated cellular degranulation (92), in FcεRI-mediated membrane stiffening (73), and in the dramatic changes in cell surface topography that accompany these processes (95). Our recent results suggest that these different manifestations of FcεRI-cytoskeletal communication are mediated by microfilament interactions with the Triton X-100–resistant membrane domains. Such predictions can be tested directly with confocal fluorescence microscopy and other experiments.

CONCLUSIONS

It is becoming increasingly evident that functional responses to immunoreceptors depend on structural features of cells that are not readily preserved in subcellular preparations and usually only temporarily preserved in permeabilized cells (87, 105). At the same time, the existence of plasma membrane subregions, or domains, with distinct protein and lipid compositions is becoming more firmly established in the literature (2, 72, 101a). The recent data summarized above suggest a relationship between these two different sets of observations, namely, that immunoreceptors such as FcεRI function by communicating, possibly by direct interaction, with specialized domains that mediate some of the important cellular processes activated by these receptors. A corollary to this hypothesis is that connection of these domains with the microfilament cytoskeleton serves to regulate the activation of the more downstream pathways involved in the response.

This hypothesis is clearly preliminary in nature. As discussed above, however, it appears to account for several previously unexplained observations regarding the relationship between FcεRI-mediated signaling and physical changes that result from FcεRI aggregation. Testing the validity of this hypothesis will require full utilization of the most current technologies in cell biology, molecular genetics, and biophysics. The extent to which it is relevant for other immunoreceptors or other types of receptors is currently open to question. However, at least several aspects of growth factor–stimulated signaling also may be mediated by communication with cytoskeletally anchored membrane structures (70, 148), and several studies have suggested a role for specialized membrane domains in the function of β-adrenergic receptors (72). An increasing flow of information points to the complex and heterogeneous nature of plasma membrane structure as providing a basis for elaborate orchestratration of receptor-mediated signal transduction in hematopoetic and other mammalian cells.

ACKNOWLEDGMENTS

Many members of our laboratory and outside collaborators, whose names appear in the citations, contributed to our studies that are described in this review. We thank K Field for helpful comments on the manuscript. We gratefully acknowledge the support of National Institutes of Health research grants AI22449 and AI18306; National Institutes of Health training grants GM07273, GM08267, GM08384, and GM08210; and National Science Foundation grant GER9023463.

Literature Cited

1. Adamczewski M, Paolini R, Kinet JP. 1992. Evidence for two distinct phosphorylation pathways activated by high affinity immunoglobulin E receptors. *J. Biol. Chem.* 267: 18126–32
2. Anderson RGW. 1993. Caveolae: where incoming and outgoing messengers meet. *Proc. Natl. Acad. Sci. USA* 90:10909–13
3. Apgar JR. 1990. Antigen-induced cross-linking of the IgE receptor leads to an association with the deter-

gent-insoluble membrane skeleton of rat basophilic leukemia (RBL–2H3) cells. *J. Immunol.* 145:3814–22
4. Apgar JR. 1991. Association of the crosslinked IgE receptor with the membrane skeleton is independent of the known signalling mechanisms in rat basophilic leukemia cells. *Cell Regul.* 2:181–91
5. Apgar JR, Mescher MF. 1986. Agorins: major structural proteins of the plasma membrane skeleton of p815 tumor cells. *J. Cell Biol.* 103:351–60
6. Baird B, Zheng Y, Holowka D. 1993. Structural mapping of IgE-FcϵRII, an immunoreceptor complex. *Acc. Chem. Res.* 26:428–34
7. Barsumian EL, Isersky C, Petrino MKG, Siraganian RP. 1981. IgE-induced histamine release from rat basophilic leukemia cell lines: isolation of releasing and non-releasing clones. *Eur. J. Immunol.* 11: 317–23
8. Basciano LK, Berenstein EH, Kmak L, Siraganian RP. 1986. Monoclonal antibodies that inhibit IgE binding. *J. Biol. Chem.* 261:11823–31
9. Deleted in proof
10. Beaven MA, Metzger H. 1992. Signal transduction by Fc receptors: the FcϵRI case. *Immunol. Today* 14: 222–26
11. Beaven MA, Rogers J, Moore JP, Heketh TR, Smith GA, Metcalfe JC. 1984. The mechanism of the calcium signal and correlation with histamine release in 2H3 cells. *J. Biol. Chem.* 259:7129–36
12. Benhamou M, Berenstein EH, Jouvin MH, Siraganian RP. 1995. The receptor with high affinity for IgE on rat mast cells is a functional receptor for rat IgG2a. *Mol. Immunol.* 31: 1089–98
13. Benhamou M, Gutkind JS, Robbins KC, Siraganian RP. 1990. Tyrosine phosphorylation coupled to IgE receptor-mediated signal transduction and histamine release. *Proc. Natl. Acad. Sci. USA* 87:5327–30
14. Benhamou M, Ryba NJP, Kihara H, Nishikata H, Siraganian RP. 1993. Protein-tyrosine kinase $p72^{syk}$ in high affinity IgE receptor signaling. *J. Biol. Chem.* 268:23318–24
15. Benhamou M, Siraganian RP. 1992. Protein-tyrosine phosphorylation—an essential component of FcϵRI signaling *Immunol. Today* 13:195–97
16. Benovic J, Bouvier M, Caron M, Lefkowitz R. 1988. Regulation of adenylyl cyclase-coupled β adrenergic receptors. *Annu. Rev. Cell Biol.* 4: 405–28
17. Berridge MJ. 1993. Inositol phosphate and calcium signalling. *Nature* 361:315–25
18. Blank U, Ra C, Miller L, White K, Metzger H, Kinet J-P. 1989. Complete structure and expression in transfected cells of the high affinity IgE receptor. *Nature* 337:187–89
19. Braun J, Hochman PS, Unanue ER. 1982. Ligand-induced association of surface immunoglobulin with the detergent-insoluble cytoskeletal matrix of the B lymphocyte. *J. Immunol.* 128:1198–204
20. Braun J, Unanue ER. 1980. B lymphocyte biology studied with anti-Ig antibodies. *Immunol. Rev.* 52:3–28
21. Deleted in proof
22. Campbell DH, McCasland GE. 1944. In vitro anaphylatic response to polyhaptenic and monohaptenic simple antigens. *J. Immunol.* 49:315–25
23. Campbell MA, Sefton BM. 1990. Protein tyrosine phosphorylation is induced in murine B lymphocytes in response to stimulation with anti-immunoglobulin. *EMBO J.* 9:2125–31
24. Chabay R, DeLisi C, Hook WA, Siraganian RP. 1980. Receptor crosslinking and histamine release in basophils. *J. Biol. Chem.* 255:4628–35
25. Chang EY, Holowka D, Baird B. 1995. Alteration of lipid composition modulates FcϵRI signaling in RBL–2H3 cells. *Biochemistry* 34: 4373–84
26. Chang EY, Mao SY, Metzger H, Holowka D, Baird B. 1995. Effects of subunit mutation on the rotational dynamics of FcϵRI, the high affinity receptor for IgE, in transfected cells. *Biochemistry* 34:6093–99
27. Cherry RJ, Godfrey RE. 1981. Anisotropic rotation of bacteriorhodopsin in lipid membranes. *Biophys. J.* 36:257–76
28. Cone RA. 1972. Rotational diffusion of rhodopsin in the visual membrane. *Nat. New Biol.* 236:39–43
29. Daeron M, Bonnerot C, Latour S, Fridman WH. 1992. Murine recombinant FcγRIII, but not FcγRII, trigger serotonin release in rat basophilic leukemia cells. *J. Immunol.* 149: 1365–73
30. DeFranco AL, Gold MR, Jakway JP. 1987. B lymphocyte signal transduction in response to anti-immunoglobulin and bacterial lipopolysaccharide. *Immunol. Rev.* 95:161–76
31. Dembo M, Goldstein B. 1978. The-

ory of equilibrium binding of symmetric bivalent haptens to cell surface antibody: application to histamine resease from basophils. *J. Immunol.* 121:345–53

32. Dembo M, Goldstein B, Sobotka AK, Lichtenstein L. 1978. Histamine release due to bivalent penicilloyl haptens: control by the number of cross-linked IgE antibodies on the basophil plasma membrane. *J. Immunol.* 121: 354–58
33. Dembo M, Kagey-Sobotka A, Lichtenstein LM, Goldstein B. 1982. Kinetic Analysis of histamine release due to covalently linked IgE dimers. *Mol. Immunol.* 19:421–34
34. Eccleston E, Leonard BJ, Lowe JS, Welford HJ. 1973. Basophilic leukemia in the albino rat and a demonstration of the basopoetin. *Nat. New Biol.* 244:73–76
35. Edidin M. 1991. Translational diffusion of membrane proteins. In *The Structure of Biological Membranes*, ed. P Yeagle, pp. 539–72. Boca Raton, FL: CRC
36. Eiseman E, Bolen JB. 1992. Engagement of the high affinity IgE receptor activates src protein-related tyrosine kinases. *Nature* 355:78–80
37. Deleted in proof
38. Erickson JW. 1988. *Equilibrium and kinetic studies of a model ligand-receptor system: monovalent and bivalent ligand interactions with immunoglobulin E.* PhD thesis. Cornell Univ., Ithaca, NY
39. Erickson JW, Goldstein B, Holowka D, Baird B. 1987. The effect of receptor density on the forward rate constant for binding of ligands to cell surface receptors. *Biophys. J.* 52: 657–62
40. Erickson JW, Kane PM, Goldstein B, Holowka D, Baird B. 1986. Cross-linking of IgE-receptor complexes at the cell surface: a fluorescence method for studying the binding of monovalent and bivalent haptens to IgE. *Mol. Immunol.* 23:769–82
41. Erickson JW, Posner R, Goldstein B, Holowka D, Baird B. 1991. Analysis of ligand binding and cross-linking of receptors in solution and on cell surfaces. In *Biophysical and Biochemical Aspects of Fluorescence Spectroscopy*, ed. TG Dewey, pp. 169–95. New York: Plenum
42. Erickson JW, Posner R, Goldstein B, Holowka D, Baird B. 1991. Bivalent ligand dissociation kinetics from receptor-bound immunoglobulin E: evidence for a time-dependent increase in ligand rebinding at the cell surface. *Biochemistry* 30:2357– 63
43. Feder TJ, Chang EY, Holowka D, Webb WW. 1994. Disparate modulation of plasma membrane protein lateral mobility by various cell permeabilizing agents. *J. Cell. Physiol.* 158: 7–16
44. Fewtrell C. 1985. Activation and desensitization of receptors for IgE on tumor basophils. In *Calcium in Biological Systems*, ed. RP Ruben, GB Weiss, JW Putney Jr, pp. 129–36. New York: Plenum
45. Fewtrell C, Metzger H. 1980. Large oligomers of IgE are more effective than dimers in stimulating rat basophilic leukemia cells. *J. Immunol.* 125:701–10
46. Field KA, Holowka D, Baird B. 1995. FcϵRI-mediated recruitment of p53/56^{lyn} to detergent resistant membrane domains accompanies cellular signaling. *Proc. Natl. Acad. Sci. USA* 92:9201–5
47. Geppert TD, Lipsky PE. 1991. Association of various T cell-surface molecules with the cytoskeleton. *J. Immunol.* 146:3298–305
48. Goldstein B. 1989. Diffusion limited effects of receptor clustering. *Comments Theor. Biol.* 1:109–27
49. Goldstein B, Posner R, Torney D, Erickson J, Holowka D, Baird B. 1989. Competition between solution and cell surface receptors for ligand: the dissociation of hapten bound to surface antibody in the presence of solution antibody. *Biophys. J.* 56:955–66
50. Deleted in proof
51. Deleted in proof
52. Hess GP, Udgaonkar JB, Olbricht WL. 1987. Chemical kinetic measurements of transmembrane processes using rapid reaction techniques: acetylcholine receptor. *Annu. Rev. Biophys. Biophys. Chem.* 16:507
53. Holowka D, Baird B. 1990. Structure and function of the high affinity receptor for immunoglobulin E. In *Cellular and Molecular Mechanisms of Inflammation*, ed. C Cochran, M Giambrone, 1:173–95. San Diego: Academic
54. Holowka D, Baird B. 1992. Recent evidence for common signalling mechanisms for immunoreceptors responsive to foreign antigens. *Cell. Signal.* 4:339–49
55. Holowka D, Wensel T, Baird B. 1990. A nanosecond fluorescence depolarization study on the segmental

flexibility of receptor-bound immunoglobulin E. *Biochemistry* 29: 4607–12
56. Hutchcroft JE, Geahlen RL, Deanin GG, Oliver JM. 1992. FcεRI-mediated tyrosine phosphorylation and activation of the 72-kDa protein-tyrosine kinase PTK72, in RBL–2H3. *Proc. Natl. Acad. Sci. USA* 89: 9107–111
57. Isersky C, Rivera J, Mims S, Triche TJ. 1979. The fate of IgE bound to rat basophilic leukemia cells. *J. Immunol.* 122:1926–36
58. Deleted in proof
59. Jorgensen JL, Reay PA, Ehrich EW, Davis MM. 1992. Molecular components of T-cell recognition. *Annu. Rev. Immunol.* 10:835–73
60. Jouvin MH, Adamczewski M, Numerof R, Letourneur O, Valle A, Kinet JP. 1994. Differential control of the tyrosine kinases lyn and syk by the two signaling chains of the high affinity immunoglobulin E receptor. *J. Biol. Chem.* 269:5918–25
61. Kagey-Sobotka A, Dembo M, Goldstein B, Metzger H, Lichtenstein LM. 1981. Qualitative characteristics of histamine release from human basophils by covalently cross-linked IgE. *J. Immunol.* 127:2285–91
62. Kane P, Erickson J, Fewtrell C, Baird B, Holowka D. 1986. Crosslinking of IgE-receptor complexes at the cell surface: structural requirements of bivalent haptens for the triggering of mast cells and tumor basophils. *Mol. Immunol.* 23:783–90
63. Kane PM, Holowka D, Baird B. 1988. Cross-linking of IgE-receptor complexes by rigid bivalent antigens >200 Å in length triggers cellular degranulation. *J. Biol. Chem.* 107: 969–80
64. Keegan A, Fratazzi C, Shopes B, Baird B, Conrad D. 1991. Characterization of monoclonal rat anti-mouse IgE antibodies and their use to map the site on mouse IgE that interacts with FcεRI and FcγRII. *Mol. Immunol.* 28:1149–54
65. Kent UM, Mao SY, Wofsy C, Goldstein B, Ross S, Metzger H. 1994. Dynamics of signal transduction after aggregation of cell-surface receptors: studies on the type I receptor for IgE. *Proc. Natl Acad. Sci. USA* 91: 3087–91
66. Kihara H, Siraganian RP. 1994. Src homolology 2 domains of syk and lyn bind to tyrosine-phosphorylated subunits of the high affinity IgE receptor. *J. Biol. Chem.* 35:22427–32
67. Kikkawa U, Kishimoto A, Nishizuka Y. 1989. The protein kinase C family heterogeneity and its implications. *Annu. Rev. Biochem.* 58:31–44
68. Kulczycki A, Metzger H. 1974. The interaction of IgE with rat basophilic leukemia cells. II. Quantitative aspects of the binding reaction. *J. Exp. Med.* 140:1676–95
69. Labrecque G, Holowka D, Baird B. 1989. Antigen-triggered membrane potential changes in IgE-sensitized rat basophilic leukemia cells: evidence for a repolarization response that is important in the stimulation of cellular degranulation. *J. Immunol.* 142:236–43
70. Landreth GE, Williams LK, Rieser GD. 1985. Association of the epidermal growth factor receptor kinase with the detergent-insoluble cytoskeleton of A431 cells. *J. Cell Biol.* 101:1341–50
71. Liotta MA. 1993. *Studies of mechanisms of signal transduction by two receptors for foreign antigen, FcεRI and the T cell receptor*. PhD thesis. Cornell Univ., Ithaca, NY
72. Lisanti MP, Scherer PE, Tang ZL, Sargiacomo M. 1994. Caveolae, caveolin and caveolin-rich membrane domains: a signalling hypothesis. *Trends Cell Biol.* 4:231–35
73. Liu Z-Y, Young J-I, Elson EL. 1987. Rat basophilic leukemia cells stiffen when they secrete. *J. Cell Biol.* 105: 2933–43
74. MacGlashan DW, Dembo M, Goldstein B. 1985. Test of a theory relating to the crosslinking of IgE antibody on the surface of human basophils. *J. Immunol.* 135:4129–34
75. Maeyama K, Hohman RJ, Ali H, Cunha-Melo JR, Beaven MA. 1988. Assessment of IgE-receptor function through measurement of hydrolysis of membrane inositol phospholipids. *J. Immunol.* 140:3919–27
76. Manger B, Weiss A, Weyand C, Goronzy J, Stobo JD. 1985. T cell activation: differences in the signals required for IL-2 production by nonactivated and activated T cells. *J. Immunol.* 135:3669–73
77. Mao SY, Alber G, Rivera J, Kochan J, Metzger H. 1992. Interaction of aggregated native and mutant IgE receptors with the cellular skeleton. *Proc. Natl. Acad. Sci. USA* 89: 2222–26
78. Mao SY, Varin-Blank N, Edidin M,

Metzger H. 1991. Immobilization and internalization of mutated IgE receptors in transfected cells. *J. Immunol.* 146:958–66
79. Mao SY, Yamashita T, Metzger H. 1995. Chemical crosslinking of IgE-receptor complexes in RBL-2H3 cells. *Biochemistry* 34:1968–77
80. Marano N, Holowka D, Baird B. 1989. Bivalent binding of an anti-CD3 antibody to Jurkat cells induces association of the T cell receptor complex with the cytoskeleton. *J. Immunol.* 143:931–38
81. Deleted in proof
82. McCloskey MA. 1988. Cholera toxin potentiaes IgE-coupled inositol phospholipid hydrolysis and mediator secretion by RBL-2h3 cells. *Proc. Natl. Acad. Sci. USA* 85:7260–64
83. McCloskey MA, Liu ZY, Poo MM. 1984. Lateral electromigration and diffusion of Fcϵ receptors on rat basophilic leukemia cells: effects of IgE binding. *J. Cell Biol.* 99:778–87
84. Deleted in proof
85. Menon AK, Holowka D, Webb, WW, Baird B. 1986. Clustering, mobility and triggering activity of small oligomers of immunoglobulin E on rat basophilic leukemia cells. *J. Cell Biol.* 102:534–40
86. Menon AK, Holowka D, Webb WW, Baird B. 1986. Cross-linking of receptor-bound IgE to aggregates larger than dimers leads to rapid immobilization. *J. Cell Biol.* 102: 541–50
87. Metzger H. 1992. Transmembrane signaling: the joy of aggregation. *J. Immunol.* 149:1477–87
87a. Mohr FC, Fewtrell C. 1987. Depolarization of rat basophilic leukemia cells inhibits calcium uptake and exocytosis. *J. Cell Biol.* 104:783–92
88. Monck JR, Robinson IM, Escobar AL, Vergara JL, Fernandez JM. 1994. Pulsed Laser imaging of rapid Ca2+ gradients in excitable cells. *Biophys. J.* 67:505– 14
89. Monfalcone LM. 1989. *Investigation of the proteins associated with the immunoglobulin E-receptor complex on rat basophilic leukemia cells.* PhD thesis. Cornell Univ., Ithaca, NY
90. Myers JN. 1990. *Phosphorescence anisotropy studies of the rotational diffusion of small aggregates of immunoglobulin E-receptor complexes on rat basophilic leukemia cells,* pp. 122–37. PhD thesis. Cornell Univ., Ithaca, NY
91. Myers JN, Holowka D, Baird B. 1992. Rotational motion of monomeric and dimeric immunoglobulin E-receptor complexes. *Biochemistry* 31:567–75
92. NarasimhanV, Holowka D, Baird B. 1990. Microfilaments regulate the rate of exocytosis in rat basophilic leukemia cells. *Biochem. Biophys. Res. Comm.* 171:222– 29
93. Narasimhan V, Holowka D, Fewtrell C, Baird B. 1988. Cholera toxin increases the rate of antigen-stimulated calcium influx in rat basophilic leukemia cells. *J. Biol. Chem.* 263: 19626–32
94. Deleted in proof
95. Oliver JM, Seagrave JC, Stump RF, Pfeiffer JR, Deanin GG. 1988. Signal transduction and cellular response in RBL–2H3 mast cells. *Prog. Allergy* 42:195–245
96. Ortega E, Schweitzer-Stenner R, Pecht I. 1988. Possible orientational constraints determine secretory signals induced by aggregation of IgE receptors on mast cells. *EMBO J.* 7: 4101–09
97. Ortega E, Schweitzer-Stenner R, Pecht I. 1991. Kinetics of ligand binding to the type I Fcϵ receptor on mast cells. *Biochemistry* 30:3473–83
98. Ovary Z. 1961. Activite des substances a faible poids moleculaire dans les reaction antigene-anticorps in vivo et in vitro. *C. R. Acad. Sci. Paris* 253:582–83
99. Padlan EA. 1994. Anatomy of the antibody molecule. *Mol. Immunol.* 31: 169–217
100. Paolini R, Jouvin MH, Kinet JP. 1991. Phosphorylation and dephosphorylation of the high affinity receptor for immunoglobulin E immediately following receptor engagement and disengagement. *Nature* 353: 855–58
101. Paolini R, Numerof R, Kinet JP. 1992. Phosphorylation/dephosphorylation of high affinity IgE receptors: a mechanism for coupling/uncoupling a large signaling complex. *Proc. Natl. Acad. Sci. USA* 89:10733–37
101a. Parton RG, Simons K. 1995. Digging into caveolae. *Science* 269:1398–99
102. Pecht I, Ortega E, Jovin TM. 1991. Rotational dynamics of the Fcϵ receptor on mast cells monitored by specific monoclonal antibodies and IgE. *Biochemistry* 30:3450–08
103. Posner RG, Erickson JW, Holowka D, Baird B. 1991. Dissociation kinetics of bivalent ligand-immunoglobu-

lin E aggregates in solution. *Biochemistry* 30:2348–56
104. Posner RG, Subramanian K, Goldstein B, Thomas J, Feder T, et al. 1995. Simultaneous crosslinking by two non-triggering bivalent ligands causes synergistic signaling of IgE-FcεRI complexes. *J. Immunol.* 155: 3601–9
105. Pribluda VS, Metzger H. 1992. Transmembrane signaling by the high-affinity IgE receptor on membrane preparations. *Proc. Natl. Acad. Sci. USA* 89:11446–50
106. Pribluda VS, Pribluda C, Metzger H. 1994. Transphosphorylation as the mechanism by which the high-affinity receptor for IgE is phosphorylated upon aggregation. *Proc. Natl. Acad. Sci. USA* 91:11246–50
107. Rahman NA, Pecht I, Roess DA, Barisas BG. 1992. Rotational dynamics of type 1 FcεRI on individually-selected rat mast cells studied by polarized fluorescence depletion. *Biophys. J.* 61:334–46
108. Resh MD. 1994. Myristylation and palmitylation of src family members: the fats of the matter. *Cell* 76:411–13
109. Reth M. 1989. Antigen receptor tail clue. *Nature* 338:383–84
109a. Rhee SG, Choi KD. 1992. Regulation of inositol phospholipid-specific phospholipase C isozymes. *J. Biol. Chem.* 267§393–96
110. Robertson DR, Holowka D, Baird B. 1986. Crosslinking of immunoglobulin E-receptor complexes induces their interaction with the cytoskeleton of rat basophilic leukemia cells. *J. Immunol.* 136:4565–72
111. Deleted in proof
112. Ryan TA. 1989. *Signal transduction of immunoglobulin E receptor crosslinking*, pp. 107–49. PhD thesis. Cornell Univ., Ithaca, NY
113. Ryan T, Myers J, Holowka D, Baird B, Webb WW. 1988. Molecular crowding on the cell surface. *Science* 239:61–64
114. Deleted in proof
115. Saffman PG, Delbruck M. 1975. Brownian motion in biological membranes. *Proc. Natl. Acad. Sci. USA* 72:3111–13
116. Saxton MJ. 1987. Lateral diffusion in an archipelago: the effect of mobile obstacles. *Biophys. J.* 52:989–97
117. Saxton MJ. 1992. Lateral diffusion and aggregation: a Monte Carlo study. *Biophys. J.* 52:989–97
118. Schatz DG, Oettinger MA, Schlissel MS. V(D)J Recombination: molecular biology and regulation. 1992. *Annu. Rev. Immunol.* 10:359–83
119. Schlessinger J, Webb WW, Elson EL, Metzger H. 1976. Lateral motion and valence of Fc receptors on rat peritoneal mast cells. *Nature* 264: 550–52
120. Schnitzer JE, Liu J, Oh P. 1995. Endothelial caveolae have the molecular transport machinery for vesicle budding, docking and fusion including VAMP, NSF, SNARE, annexins and GTPases. *J. Biol. Chem.* 270: 14399–404
121. Schroeder R, London E, Brown D. 1994. Interactions between saturated acyl chains confer detergent resistance on lipids and glycosylphosphatidylinositol (GPI)-anchored proteins: GPI-anchored proteins in liposomes and cells show similar behavior. *Proc. Natl. Acad. Sci. USA* 91:12130–34
122. Schroeder TJ, Jankowski JA, Senyshyn J, Holz RW, Wightman RM. 1994. Zones of exocytotic release on bovine adrenal medullary cells in culture. *J. Biol. Chem.* 269:17215–20
123. Schweitzer-Stenner R, Licht A, Luscher I, Pecht I. 1987. Oligomerization and ring closure of immunoglobulin E class antibodies by divalent haptens. *Biochemistry* 26: 3602–12
124. Schweitzer-Stenner R, Luscher I, Pecht I. 1992. Dimerization kinetics of the IgE-class antibodies by divalent haptens I. The Fab-hapten interactions. *Biophys. J.* 63:551–62
125. Schweitzer-Stenner R, Luscher I, Pecht I. 1992. Dimerization kinetics of the IgE-class antibodies by divalent haptens II. The interaction between intact IgE and haptens. *Biophys. J.* 63:563–68
126. Schweitzer-Stenner R, Ortega E, Pecht I. 1994. Kinetics of FcεRI dimer formation by specific monoclonal antibodies on mast cells. *Biochemistry* 33:8813–25
127. Seagrave J, Deanin GG, Martin JC, David BH, Oliver JM. 1987. DNP-phycobiliproteins, fluorescent antigens to study dynamic properties of antigen-IgE-receptor complexes on RBL–2H3 rat mast cells. *Cytometry* 8:287–95
128. Seagrave J, Oliver JM. 1990. Antigen-dependent transition of IgE to a detergent-insoluble form is associated with reduced IgE receptor-dependent secretion from RBL-2H3

mast cells. *J. Cell. Physiol.* 144: 128–36
129. Segal D, Taurog J, Metzger H. 1977. Dimeric immunoglobulin E serves as a unit signal for mast cell degranulation. *Proc. Natl. Acad. Sci. USA* 74: 2993–97
130. Seldin DC, Adelman S, Austen KF, Stevens RL, Hein A, et al. 1985. Homology of the rat basophilic leukemia cell and the rat mucosal mast cell. *Proc. Natl. Acad. Sci. USA* 82: 3871–75
131. Shenoy-Scaria AM, Dietzen DJ, Kwong J, Link DC, Lublin DM. 1994. Cysteine3 of src family protein tyrosine kinase determines palmitoylation and localization in caveolae. *J. Cell Biol.* 126:353–63
132. Siraganian RP. 1988. Mast cells and basophils. In *Inflammation: Basic Principles and Clinical Correlates*, ed. J Gallin, J Goldstein, R Snyderman, p. 513. New York: Raven
133. Siraganian RP, Hook WA, Levine BB. 1975. Specific in vitro histamine release from basophils by bivalent haptens: evidence for activation by simple bridging of membrane bound antibody. *Immunochemistry* 12: 149–57
134. Sobotka AK, Dembo M, Goldstein B, Lichtenstein L. 1979. Antigen-specific desensitization of human basophils. *J. Immunol.* 122:511–17
135. Strader CD, Fong TM, Tota MR, Underwood D, Dixon RAF. 1994. Structure and function of G protein-coupled receptors. *Annu. Rev. Biochem.* 63:101–32
136. Subramanian K, Goldstein B, Holowka D, Baird B. The Fc segment of IgE influences the kinetics of dissociation of a symmetrical bivalent ligand from cyclic dimeric complexes. *Biochemistry*. Submitted
137. Takizawa F, Adamczewski M, Kinet JP. 1992. Identification of the low affinity receptor for immunoglobulin E on mouse mast cells and macrophages as FcγRII and FcγRIII. *J. Exp. Med.* 176:469–76
138. Thomas JL, Feder T, Webb WW. 1992. Effects of protein concentration on IgE receptor mobility in rat basophilic leukemia cell plasma membranes. *Biophys. J.* 61:1402–12
139. Thomas JL, Holowka D, Baird B, Webb WW. 1994. Large-scale co-aggregation of fluorescent lipid probes with cell surface proteins. *J. Cell Biol.* 125:795–802
140. Ullrich A, Schlessinger J. 1990. Signal transduction by receptors with tyrosine kinase activity. *Cell* 61: 203–12
141. Urata C, Siraganian RP. 1985. Pharmacologic modulation of the IgE or Ca2+ ionophore A23187 mediated Ca2+ influx, phospholipase activation, and histamine release in rat basophilic leukemia cells. *Int. Arc. Allery Appl. Immunol.* 78:92–100
142. Wade WF, Freed JH, Edidin M. 1989. Translational diffusion of class II MHC molecules is constrained by their cytoplasmic domains. *J. Cell Biol.* 109:3325– 31
143. Webb WW, Barak LS, Tank DW, Wu ES. 1981. Molecular mobility on the cell surface. *Biochem. Soc. Symp.* 46:191–205
144. Weetall M, Holowka D, Baird B. 1993. Heterologous desensitization of the high affinity receptor for IgE (FcεRI) on RBL cells. *J. Immunol.* 150:4072–83
145. Weiss A, Littman DR. 1994. Signal transduction by lymphocyte antigen receptors *Cell* 76:263–74
146. Woda BA, Woodin B. 1984. The interaction of lymphocyte membrane proteins with the lymphocyte cytoskeletal matrix. *J. Immunol.* 33: 2767–72
147. Yamashita T, Mao SY, Metzger H. 1994. Aggregation of the high-affinity IgE receptor and enhanced activity of p53/56^{lyn} protein-tyrosine kinase. *Proc. Natl. Acad. Sci. USA* 91: 11251–55
148. Yang LJ, Rhee SG, Williamson JR. 1994. Epidermal growth factor-induced activation and translocation of phospholipase C-γ1 to the cytoskeleton in rat hepatocytes. *J. Biol. Chem.* 269:7156–62
149. Zidovetzki R, Bartholdi M, Arndt-Jovin D, Jovin T. 1986. Rotational dynamics of the Fc receptor for immunoglobulin E on histamine-releasing rat basophilic leukemia cells. *Biochemistry* 25:4397–4401

Annu. Rev. Biophys. Biomol. Struct. 1996. 25:113–36

BRIDGING THE PROTEIN SEQUENCE–STRUCTURE GAP BY STRUCTURE PREDICTIONS

Burkhard Rost and Chris Sander

European Molecular Biology Laboratory, 69012 Heidelberg, Germany

KEY WORDS: multiple alignments, secondary structure, solvent accessibility, transmembrane helices, interresidue contacts, homology modeling, threading, knowledge-based mean-force potentials

ABSTRACT

The problem of accurately predicting protein three-dimensional structure from sequence has yet to be solved. Recently, several new and promising methods that work in one, two, or three dimensions have invigorated the field. Modeling by homology can yield fairly accurate three-dimensional structures for approximately 25% of the currently known protein sequences. Techniques for cooperatively fitting sequences into known three-dimensional folds, called threading methods, can increase this rate by detecting very remote homologies in favorable cases. Prediction of protein structure in two dimensions, i.e. prediction of interresidue contacts, is in its infancy. Prediction tools that work in one dimension are both mature and generally applicable; they predict secondary structure, residue solvent accessibility, and the location of transmembrane helices with reasonable accuracy. These and other prediction methods have gained immensely from the rapid increase of information in publicly accessible databases. Growing databases will lead to further improvements of prediction methods and, thus, to narrowing the gap between the number of known protein sequences and known protein structures.

CONTENTS

1056-8700/96/0610-0113$08.00

INTRODUCTION

Large-scale sequencing projects produce data of gene and, hence, protein sequences at a breathtaking pace. Although determination of protein three-dimensional structure by crystallography has become more efficient (51), the gap between the number of known sequences (45,000; 5, 7) and the number of known structures (3000; 8) is increasing rapidly. For many proteins, sequence determines structure uniquely, i.e. the entire information for the details of three-dimensional structure is contained in the sequence (4). In principle, therefore, protein structure could be predicted from physicochemical principles given only the sequence of amino acids. In practice, however, prediction from first principles, e.g. by molecular dynamics, is prevented by the high complexity of protein folding (with required computing time orders of magnitude too high) and by the inaccuracy of the experimental determination of basic parameters (93). Most protein structure prediction tools, therefore, are knowledge based, using a combination of statistical theory and empirical rules. Given a protein sequence of unknown structure (dubbed U), what can we uncover regarding the structure of U by using theoretical tools, or what can theory contribute to bridging the sequence-structure gap?

The most successful tool for predicting three-dimensional structure is homology modeling. An approximate three-dimensional model (which has a correct fold but inaccurate loop regions) can be constructed if U has significant similarity to a protein of known structure, evaluated in terms of pairwise sequence identity (i.e. by alignment) or sequence-structure fitness (i.e. threading). Homology modeling effectively raises the number of "known" three-dimensional structures from 3000 to approximately 10,000 (80). Threading methods may be used to make tentative predictions of three-dimensional structure for approximately

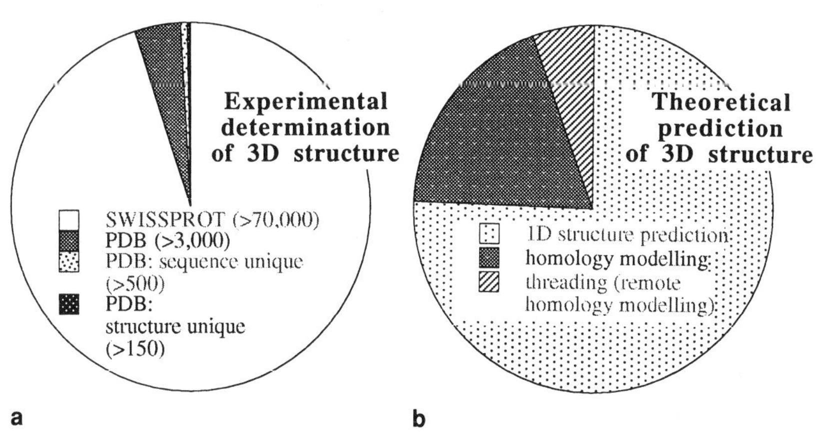

Figure 1 Bridging the sequence-structure gap by experiment and theory. The full-clock cycle corresponds to all protein sequences stored in the database SWISSPROT (release 31 with 44,000 sequences). (*a*) Fraction of proteins for which three-dimensional structure has been experimental determined (sequence unique: $< 25\%$ pairwise sequence identity; structure unique: unique overall fold-type, as defined by Reference 36). (*b*) Fraction of proteins for which three-dimensional structure can be predicted by homology modeling (estimated for threading). Note: unique three-dimensional structures cannot be predicted, yet.

an additional 3000 proteins. Consequently, theory-based tools already contribute significantly to bridging the sequence-structure gap (Figure 1). If U has no homologue of known three-dimensional structure, however, we are forced to resort to simplifications of the prediction problem. In the process, we can use the rich diversity of information in current databases. In this review, we focus on generic methods for prediction at three different levels of simplification (Figure 2), namely one, two, and three dimensions (Figure 3). We have included only methods that are available by automatic prediction services or programs and, thus, could be used to analyze large numbers of sequences, e.g. entire chromosomes (25, 42). The underlying question for every method is, What is the practical contribution of the method to the problems of protein structure prediction and analysis?

SEQUENCE ALIGNMENTS

At the level of protein molecules, selective pressure results from the need to maintain function, which in turn requires maintenance of the specific three-dimensional structure (21). This process is the basis for attempts to align protein sequences, i.e. to detect equivalent positions

```
    ....,....1....,....2....,....3
SeqKEGYLVKKSDGCKYGCLKLGENEGCDTECK
Sec    EE                 HHHHHHHH
Acc672402598503658398769497048506

    ....,....4....,....5....,....6....,
SeqAKNQGGSYGYCYAFACWCEGLPESTPTYPLPNKSC
Sec          EEEEE   EEEEE
Acc59825386051877124056168936468398688
```

1 D

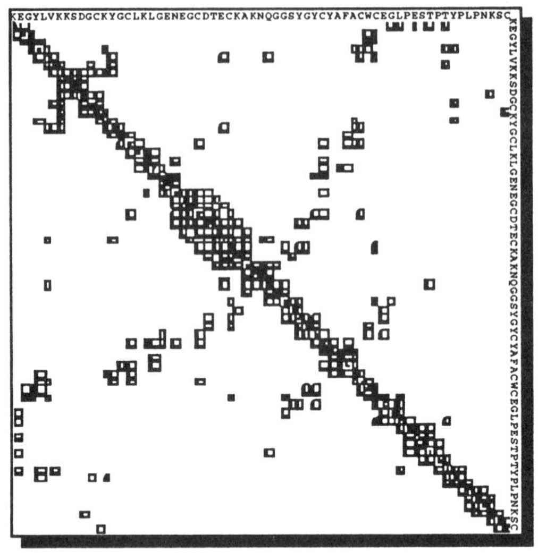

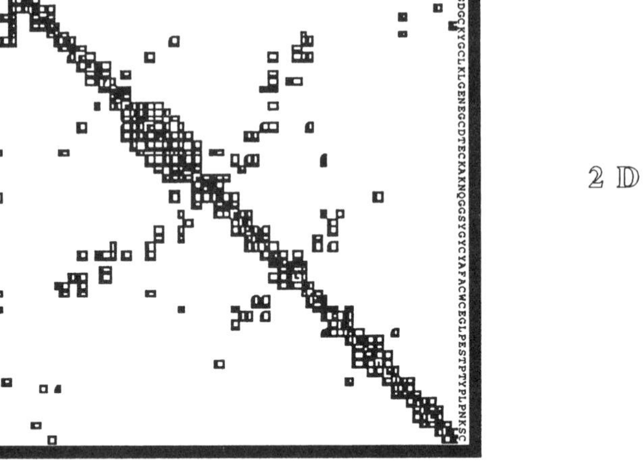

2 D

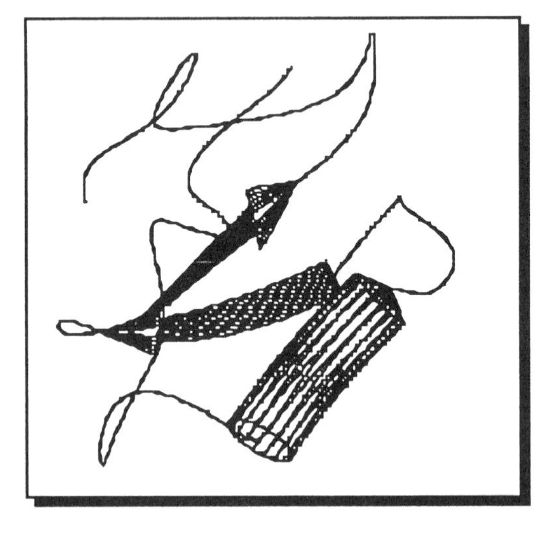

3 D

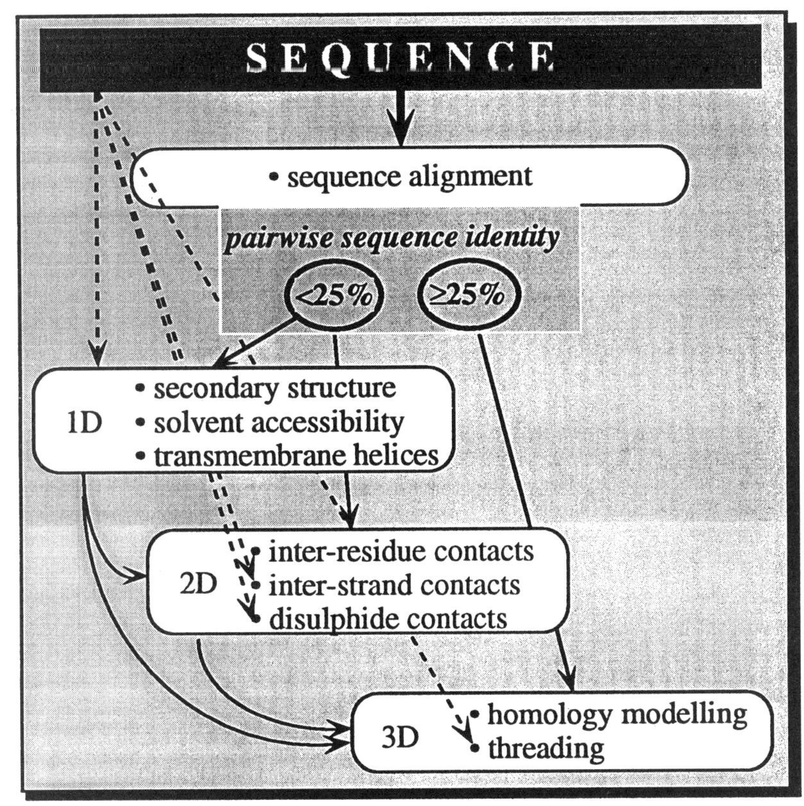

Figure 3 Summary of the tools available for sequence analysis reviewed here. The arrows indicate the input information used for a given method, e.g. secondary structure can be predicted from single sequences and alignments; the two-dimensional prediction can be used, in turn, for prediction of interstrand contacts and threading.

Figure 2 Representation of scorpion neurotoxin (PDB code 2sn3) in one, two, and three dimensions. Each of the representations gives rise to a different type of prediction. (*1D*) Seq, sequence in one-letter alphabet; Sec, secondary structure, with H for helix, E for strand, and blank for other; Acc, relative solvent accessibility (note: integer n codes for a relative accessibility of $n \times n\%$). (*2D*) Interresidue contact-map (sequence positions 1–65 plotted from left to right and from up to down); squares indicate that the respective residue pair is in contact. (*3D*) The trace of the protein chain in three dimensions is plotted schematically as a ribbon α-carbon trace. The two strands are indicated by arrows, the helix is marked by a cylinder. Graphs were generated with the use of WHAT IF, a molecular graphics package with modules for homology modeling, drug design, and protein structure analysis (96).

in strings of amino acid letters optimally. Accordingly, conservation and mutation patterns observed in alignments contain very specific information regarding three-dimensional structure. Surprisingly, much variation is tolerated without loss of structure: Two naturally evolved proteins with more than 25% identical residues (length > 80 residues) are very likely to be similar in three-dimensional structure (79). Even so, structure may be conserved in spite of much higher divergence (36). One naturally wonders how much data are required to detect structure-specific sequence motifs (67) and to align correctly even remote homologues (i.e. sequences with fewer than 25% pairwise identical residues)?

When the level of pairwise sequence identity is sufficient (say, > 40%), alignment procedures are (more or less) straightforward (24, 44, 79). With the use of fast alignment tools, one can scan entire databases that contain 100,000 sequences in minutes. Two fast sequence alignment programs are FASTA (65) and BLAST (3). For less similar protein sequences, however, alignments may fail (30, 94). The art of sequence alignment is to align related sequence segments accurately and to avoid aligning unrelated sequence stretches (20, 22, 31, 48, 52, 57, 75, 79, 92). Alignment techniques can be improved by incorporating information derived from three-dimensional structures (30). Profile-based multiple alignments appear to be sensitive and fast enough to scan entire databases if implemented on parallel machines (80).

One of the difficulties in comparing different alignment procedures is the lack of well-defined criteria for measuring the quality of an alignment. Very few papers have attempted to define such measures for the comparison of various methods (22, 30). The second problem for users is that most methods do not supply a cutoff criterion for distinguishing between homologous and nonhomologous sequences (i.e. false positive sequences). For some large sequence families, remote homologues can be aligned correctly (57, 92); for most cases, however, sequences with less than 25% sequence identity will be false positive, i.e. will have no structural or functional similarity to the guide sequence. A simple, length-dependent cutoff based on sequence identity is provided by MAXHOM, which is a profile-based, multiple-sequence alignment program that also runs in parallel complexes (79). This program, however, does not quantify the influence of (more subtle) similarities and of the occurrence of gaps.

EVALUATION OF PREDICTION METHODS

A systematic testing of performance is a precondition for any prediction to become reliably useful. For example, the history of secondary structure prediction has partly been a hunt for highest accuracy scores, with overly optimistic claims by predictors seeding the skepticism of potential users. In 1994, one major point about prediction methods became clear at the first international meeting for the evaluation of these methods in Asilomar, California (18): Exaggerated claims are more damaging than genuine errors. Even a prediction method of limited accuracy can be useful if the user knows what to expect. For the editors of scientific journals, this statement implies that a protein structure prediction method should be published only if it has been sufficiently cross-validated. This raises the difficult question of how to evaluate prediction methods.

When a data set is separated into a training set (used to derive the method) and a test set (or cross-validation set, used to evaluate performance), a proper evaluation (or cross-validation) of prediction methods needs to meet four requirements:

1. No significant pairwise sequence identity between training and test set. The proteins used for setting up a method (training set) and those used for evaluating it (test set) should have a pairwise sequence identity of less than 25% [length-dependent cutoff (79)], otherwise homology modeling could be applied that would be much more accurate than ab initio predictions (74, 76).
2. Comprehensive tests through using a large data set. All available unique proteins should be used for testing [currently > 400 (32)]. The reason for taking as many proteins as possible is simply that proteins vary considerably in structural complexity; certain features are easy to predict, others are harder (see Figure 5).
3. Avoid comparing apples with oranges. No matter which data sets are used for a particular evaluation, a standard set should be used for which results are also always reported (see Figure 4).
4. No optimization with respect to the test set. A seemingly trivial—and often violated—rule is that methods should never be optimized with respect to the data set chosen for final evaluation. In other words, the test set should never be used before the method is set up. (For example, using a cross-validation set to indicate when overtraining on the training data has occurred or to find out how many parameters should be used to describe the model is an implicit use of the cross-validation set in parameter optimization. The data reserved to test the method, therefore, should never be used in two ways.)

Most methods are evaluated in n-fold cross-validation experiments (splitting the data set into n different training and test sets). How many separations should be used, i.e. which value of n yields the best evaluation? A misunderstanding is often spread in the literature: the more separations (the larger n) the better. The exact value of n, however, is not important, provided that the test set is representative and comprehensive and that the cross-validation results are not misused to change parameters again. In other words, the choice of n is meaningless for the user.

PREDICTION IN ONE DIMENSION

Secondary Structure

The principal idea underlying most secondary structure prediction methods is the fact that segments of consecutive residues have preferences for certain secondary structure states (46). The prediction problem, therefore, becomes a pattern-classification problem tractable by computer algorithms. The goal is to predict whether the residue at the center of a segment of typically 13–21 adjacent residues is in a helix, a strand, or in no regular secondary structure. Many different algorithms have been applied to tackle this simplest version of the protein-structure prediction problem (70, 72). Until recently, however, performance accuracy seemed to have been limited to approximately 60% (percentage of residues correctly predicted in either α-helix, β-strand, or another conformation).

The use of evolutionary information in sequence has improved prediction accuracy significantly. The first method that reached a sustained level of a three-state prediction accuracy greater than 70% was the profile-based neural network program PHD, which uses multiple sequence alignments as input (70). By stepwise incorporation of more evolutionary information, prediction accuracy can be pushed to greater than 72% (72). A nearest-neighbor algorithm can be used to incorporate the same information with a similar performance (77) (Figure 4). A method that combines statistics and multiple alignment information (53) is clearly less accurate (Figure 4). Compared with methods that use single-sequence information only, methods that use the growing databases are 6–14 percentage points more accurate (Figure 4).

How good is a prediction accuracy of 72%? It is certainly reasonably good compared with the prediction of secondary structure by homology modeling (16, 74, 76). In addition, some residues within a structure are predicted at higher levels of accuracy than the mean value, i.e. prediction accuracy is 72% $\pm$ 9% (one standard deviation; Figure 5). Various applications of improved secondary-structure predictions prove

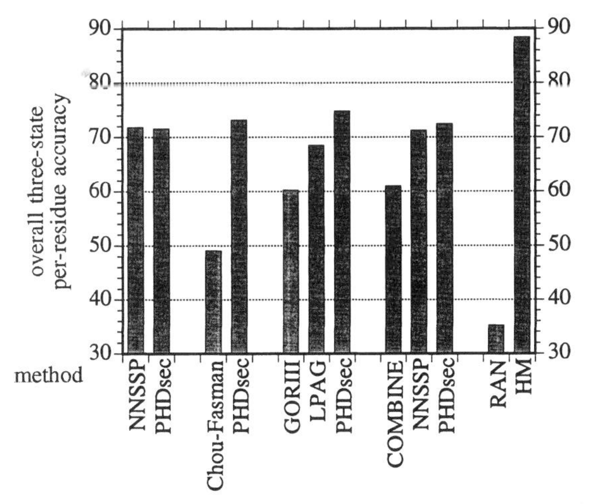

Figure 4 Accuracy of secondary structure prediction for various prediction methods. Abbreviations: RAN and HM, for comparison the results of the worst (random) and the best (homology modeling) possible predictions are given (74); Chou-Fasman, GORIII, and COMBINE, early prediction method based on single-sequence information (9, 15, 28) (these methods are still widely used by standard sequence analysis packages); LPAG, multiple alignment-based method using statistics (53); NNSSP, multiple alignment-based method using nearest neighbor algorithms (77); PHDsec, multiple alignment-based neural network prediction (72). The groups indicate identical test sets, e.g. GORIII is approximately eight percentage points less accurate than LPAG using the same algorithm but additional multiple alignments, and PHDsec is another six percentage points more accurate than LPAG by using neural networks instead of statistics.

that predictions are accurate enough to be of practical use [prediction-based threading, (40, 68); interstrand contact prediction, (39); chain tracing in X-ray crystallography; design of residue mutations]. One way to increase the 72% ± 9% accuracy level might be to predict secondary-structure content (proportion of residues in α-helix, β-strand, and other) and then use this initial classification to refine secondary structure prediction.

Proteins have been partitioned into various structural classes, e.g. on the basis of percentage of residues assigned to α-helix, β-strand, and other conformations (55). Such a coarse-grained classification, however, is not well defined (36). Consequently, given a protein sequence U, attempts to predict the secondary-structure content for U and then to use the result to predict the secondary structural class (i.e. all α, all β, or intermediates) is of limited practical use. Alignment-based predictions compare favorably with experimental means of determining the content in secondary structure. Surprisingly, PHD is, on average,

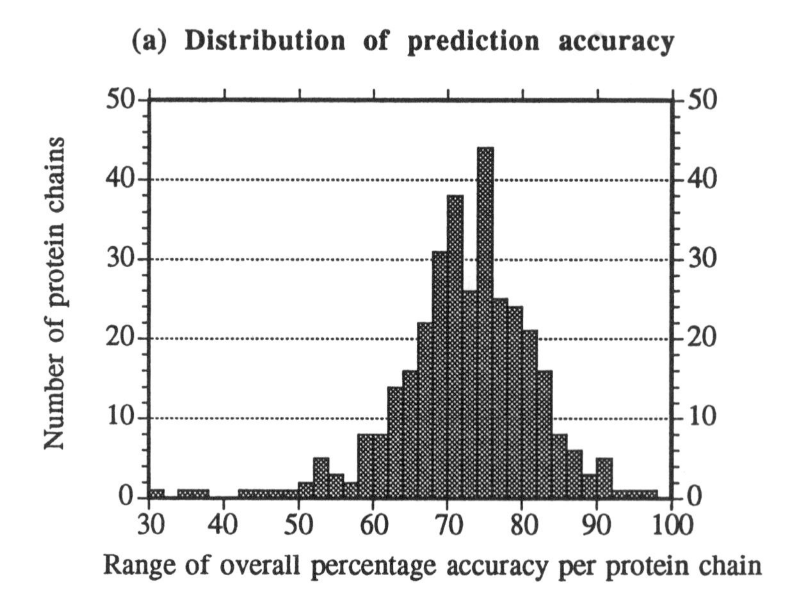
(a) Distribution of prediction accuracy
Number of protein chains
Range of overall percentage accuracy per protein chain
0
10
20
30
40
50
30
40
50
60
70
80
90
100

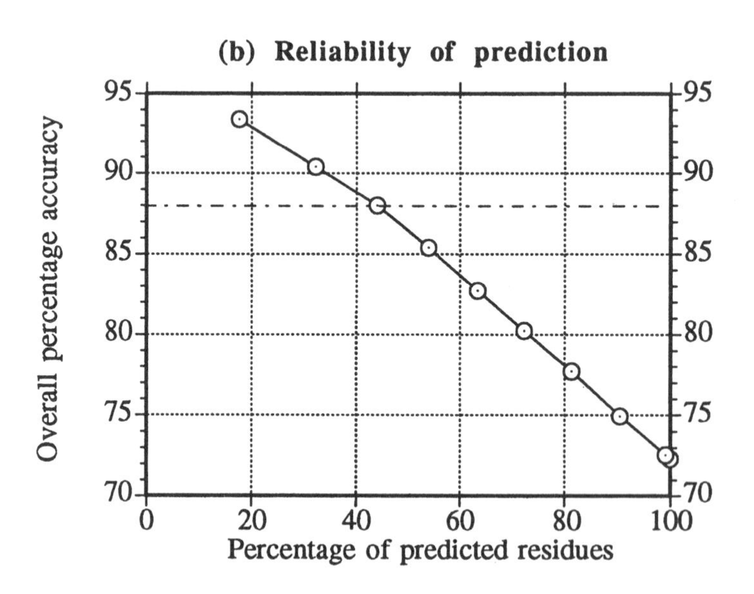
(b) Reliability of prediction
Overall percentage accuracy
Percentage of predicted residues
70
75
80
85
90
95
0
20
40
60
80
100

about as accurate as circular dichroism spectroscopy (70, 72). Of course, this finding does not imply that predictions can replace experiments. In particular, variation of secondary structure as a result of changes in environmental conditions (e.g. solvent) is generally accessible only experimentally.

One attempt to improve secondary structure predictions was to develop methods specifically for all-α helix proteins. Two points often have been confused in the literature. First, a two-state accuracy (helix, nonhelix) is not comparable to a three-state accuracy (helix, strand, other). For example, PHD of secondary structure (PHDsec) has an expected three-state accuracy of approximately 72% and an expected two-state accuracy of approximately 82% (71). Second, before a method specialized on all-α proteins can be applied to U, the structure type of U has to be predicted. Such a prediction has an expected accuracy of 70–80% (72). Even if the accuracy for determining whether U belongs to the all-α class reaches almost 100% (99), as recently claimed, specialized methods are still not very useful, as the improvement in accuracy by specializing on one class has been only marginal (71).

Solvent Accessibility

The principal goal is to predict the extent to which a residue embedded in a protein structure is accessible to solvent. Solvent accessibility can be described in several ways (73). The simplest is a two-state description distinguishing between residues that are buried (relative solvent accessibility $<16\%$) and exposed (relative solvent accessibility $\leq 16\%$). The classic method is to assign either of the two states, buried or exposed, according to residue hydrophobicity (for overview, see 73). A neural network prediction of accessibility, however, has been shown to be superior to simple hydrophobicity analyses (33).

Solvent accessibility at each position of the protein structure is conserved evolutionarily within sequence families (73). This fact has been used to develop methods for predicting accessibility using multiple

Figure 5 Secondary structure prediction accuracy for PHDsec evaluated on 337 protein families. (*a*) Prediction accuracy varies considerably between protein families. One standard deviation is nine percentage points, so prediction accuracy for most sequences is 63–81%, and the average accuracy is 72%. Because of this significant variation, prediction methods have to be evaluated on a sufficiently large set of unique proteins. (*b*) Residues with a higher reliability index are predicted with higher accuracy. For example, for 44% of all residues prediction accuracy is, on average, 88% (*dashed line*), i.e. comparable to homology modeling if it were applicable. In practice, attention should be focused on the most reliably predicted residues.

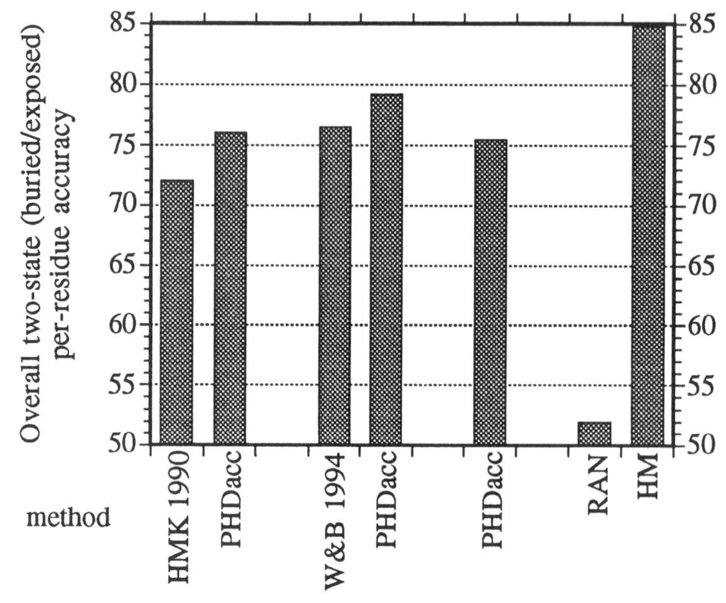

Figure 6 Two-state accuracy of predicting relative accessibility. Abbreviations: RAN and HM, for comparison the results of the worst (random) and the best (homology modeling) possible predictions are given (73); HMK 1990, neural network using single-sequence input (33); W&B 1994, multiple alignment–based prediction method using rather sophisticated expert rules and statistics (97); PHDacc, multiple alignment-based neural network prediction (73). The groups indicate identical test sets.

alignment information (6, 73, 97). Prediction accuracy is approximately 75% ± 7%, four percentage points higher than for methods not using alignment information (Figure 6). Predictions are accurate enough to be used as a seed for predicting secondary structure (6, 97) but not accurate enough to become useful as secondary structure predictions (68).

Transmembrane Helices

Even in the optimistic scenario that, in the near future, most protein structures will be either determined experimentally or predicted theoretically, one class of proteins will still represent a challenge for experimental determination of three-dimensional structure: transmembrane proteins. The major obstacle with these proteins is that they do not crystallize and are hardly tractable by NMR spectroscopy. For this class of proteins, therefore, structure prediction methods are needed even more than for globular water-soluble proteins. Fortunately, the prediction task is simplified by strong environmental constraints on transmembrane proteins: The lipid bilayer of the membrane reduces the degrees of freedom to such an extent that three-dimensional structure formation becomes almost a two-dimensional problem. Once the location of transmembrane segments is known for helical transmembrane proteins, three-dimensional structure can be predicted by exploring all possible conformations (91). Additionally, the prediction of the locations of these transmembrane helices is a much simpler problem than is the prediction of secondary structure for soluble proteins. Elaborated combinations of expert rules, hydrophobicity analyses, and statistics yield a two-state per-residue level of accuracy greater than 90% (43, 69, 83, 95).

Evolutionary information further improves prediction accuracy. For two methods, the use of multiple alignment information is reported to improve the level of accuracy of predicting transmembrane helices (66, 69). The best current prediction methods have a similar high level of accuracy of approximately 95%. As reliable data for the locations of transmembrane helices exist only for a few proteins, data used for deriving these methods originate predominantly from experiments in cell biology and gene-fusion techniques. Different authors often report different locations for transmembrane regions. Thus, the 95% level of accuracy is not verifiable. Despite this uncertainty in detail, the prediction of transmembrane helices is a valuable tool to scan entire chromosomes quickly (69). The classification into membrane/nonmembrane proteins has an expected error rate of less than 5%, i.e. approximately 5% of the proteins predicted to contain transmembrane regions will probably be false positive.

Cytoplasmic and extracellular regions have different amino acid compositions (61, 95). This difference allows for a successful prediction of the orientation of transmembrane helices with respect to the cell (pointing inside or outside the cell; 43, 83). Such predictions are estimated to be correct in more than 75% of all proteins (43). Going one step further, Taylor and colleagues (91) have correctly predicted the three-dimensional structure for the membrane-spanning regions of G-coupled receptors (seven helices) when starting from the known locations of the helices. For a successful automatic prediction of three-dimensional structure from sequence, the N- and C-terminal ends of transmembrane helices have to be predicted very accurately. It remains to be tested whether current prediction methods for the location of transmembrane helices are sufficiently accurate to predict three-dimensional structure of integral membrane proteins automatically.

PREDICTION IN TWO DIMENSIONS

Interresidue Contacts

Given all interresidue contacts or distances (see Figure 2), three-dimensional structure can be reconstructed by distance geometry (13, 63). Distance geometry is used for the determination of three-dimensional structures by NMR spectroscopy, which produces experimental data of distances between protons (13). Some fraction of interresidue contacts can be predicted. Helices and strands can be assigned on the basis of hydrogen-bonding patterns between residues (45). Thus, a successful prediction of secondary structure implies a successful prediction of some fraction of all the contacts. Contacts predicted from secondary structure assignment, however, are short ranged, i.e. between residues nearby in sequence. For a successful application of distance geometry, long-range contacts have to be predicted, i.e. contacts between residues far apart in the sequence. A few methods have been proposed for the prediction of long-range interresidue contacts. Two questions surround such methods: First, how accurate are these prediction methods on average. Second, are all important contacts predicted?

In sequence alignments, some pairs of positions appear to co-vary in a physicochemically plausible manner, i.e. a ''loss of function'' point mutation often is rescued by an additional mutation that compensates for the change (2). One hypothesis is that compensation would be most effective in maintaining a structural motif if the mutated residues were spatial neighbors. Attempts have been made to quantify such a hypothesis (62, 90) and to use it for contact predictions (29, 81). By applying a stringent significance cutoff in the prediction of contacts by correlated mutations, a small number of residue contacts can be predicted between 1.4 and 5.1 times better than random (29); further slight improvements are possible (D Thomas, unpublished data). These predictions are still not accurate enough to apply distance geometry to the results.

Analyzing correlated mutations is only one way to predict long-range interresidue contacts. Other methods use statistics (26), mean-force potentials (X Tamames and A Valencia, unpublished data), or neural networks (10). So far, none of the methods appears to find a path between the Scylla of missing too many true contacts and the Charybdis of predicting too many false contacts. Some of the methods, however, may provide sufficient information to distinguish between alternative models of three-dimensional structure (A Valencia, unpublished data). The ambitious goal to predict long-range interresidue contacts accurately enough will hopefully continue to attract intellectual resources.

Interstrand Contacts

One simplification of the problem to predict interresidue contacts focuses on predicting the contacts between residues in adjacent β-strands. Such an attempt is motivated by the hope that such interactions are more specific than sequence-distant (long-range) contacts in general and, hence, are easier to predict.

The only method published for predicting interstrand contacts is based on potentials of mean force (39) similar to those used in the evaluation of strand-strand threading (56). Propensities are compiled by database counts for 2 × 2 × 2 classes (parallel/antiparallel, H-bonded/non–H-bonded, N-/C-terminal). Each of the eight classes is divided further into five subclasses in the following way: Suppose the two strand residues at positions i and j are in close in space. Then, the following five residue pairs are counted in separate tables: $i/j-2$, $i/j-1$, i/j, $i/j+1$, $i/j+2$. Such pseudo-potentials identify the correct β-strand alignment in 35–45% of the cases.

Even if the locations of β-strands in the sequence are known exactly, the pseudo-potentials cannot predict the correct interstrand contacts in most cases (39). When using multiple alignment information, however, the signal-to-noise ratio increases such that interstrand contacts have been predicted correctly for most of the strands inspected in some test cases (39). For the purpose of reliable contact prediction, this result is inadequate, especially as the locations of the strands are not known precisely. The pseudo-potentials apparently can handle errors resulting from incorrect prediction of strands. Various test examples using predictions by PHDsec (72) as input to the β-strand pseudo-potentials indicate that the accuracy in predicting interstrand contacts drops (T Hubbard, unpublished data) but, in some cases, is still high enough to be useful for approximate modeling of three-dimensional structure (40).

Intercysteine Contacts

An extreme simplification of the contact prediction problem focuses on predicting contacts between cysteine residues (disulfide bridges). Previously, such contacts were obtained by experimental protein sequencing techniques. In the age of gene-sequencing projects, however, disulfide bridges are no longer part of the sequence information. Disulfide bond predictions are interesting for two reasons: First, disulfide bridges are crucial for structure formation of many proteins. Second, contacts between cysteines account for the most dominant signal in predicting interresidue contacts by mean-force potentials. The prediction of cysteine-bridges, therefore, is a subject of current interest.

One method for the prediction of disulfide-bonds uses a neural network to predict the bonding state of single cysteines (60), i.e. the goal is not to predict which cysteine pair is in contact but whether a cysteine residue is in contact to any other one. Strictly speaking, therefore, the method operates in one dimension. One result is that the cysteine bonding state appears to be influenced by the local sequence environment of up to 15 adjacent residues. Prediction accuracy in two states is claimed to be approximately 80%. The result, however, may be overly optimistic for two reasons. First, the test set was rather small (140 examples). Second, in the cross-validation experiments, training and testing examples were not separated on the basis of the level of pairwise sequence identity. Therefore, the question of how accurately intercysteine contacts can be predicted remains to be answered.

PREDICTION IN THREE DIMENSIONS

Homology Modeling

An analysis of the Protein Data Bank (PDB) of experimentally determined structures of protein reveals that all protein pairs with more than 30% pairwise sequence identity (for alignment length > 80; 79) have homologous three-dimensional structures, i.e. the essential fold of the two proteins is identical, but such details as additional loop regions may vary. Structure is more conserved than is sequence. This finding is the pillar for the success of homology modeling. The principal idea is to model the structure of U on the basis of the template of a sequence homologue of known structure. Consequently, the precondition for homology modeling is that a sequence homologue of known structure is found in PDB. Because homology modeling is currently the only theoretical means to predict three-dimensional structure successfully, this finding has two implications. First, homology modeling is applicable to "only" one quarter of the known protein sequences (see Figure 1). Second, as the template of a homologue is required, no unique three-dimensional structure can yet be predicted, i.e. no structure that has no similarity to any experimentally determined three-dimensional structure. If there is a protein with a sequence similar to U in PDB (say HU), is homology modeling straightforward?

The basic assumption of homology modeling is that U and HU have identical backbones. The task is to place the side chains of U into the backbone of HU correctly. For very high levels of sequence identity between U and HU (ideally differing by one residue only), side chains can be "grown" during molecular dynamics simulations (17, 47). For slightly lower levels (still of high-sequence similarity), side chains are built on the basis of similar environments in known structures (19, 23, 54, 58, 78, 89, 96). Rotamer libraries are used in the following way (19): 1. Rotamer distributions are extracted from a database of sequences that are not redundant. 2. Fragments of seven (helix, strand) or five residues (other) are compiled. 3. Fragments of the same length are shifted successively through the backbone of U. 4. For modeling the side chains of U, only those fragments from the rotamer library that have the same amino acid in the center as U, and for which the local backbone is similar to that around the evaluated position, are accepted. Over the whole range of sequence identity between U and HU for which homology modeling is applicable, the accuracy of the model drops with decreasing similarity. For levels of at least 60% sequence identity, the resulting models are quite accurate (19). (For even higher values, the models are as accurate as is experimental structure determination.) The limiting factor is the computation time required (34). How accurate is homology modeling for lower levels of sequence identity?

With decreasing sequence identity, the number of loops inserted grows. An accurate modeling of loop regions, however, implies solving the structure prediction problem. The problem is simplified in two ways. First, loop regions are often relatively short and can thus be simulated by molecular dynamics [note the central processing unit (CPU) time required for molecular dynamics simulations grows exponentially with the number of residues of the polypeptide to be modeled]. Second, the ends of the loop regions are fixed by the backbone of the template structure. Various methods are used to model loop regions. The best have the orientation of the loop regions correct in some cases (e.g. 1). With less than approximately 40% sequence identity, the accuracy of the sequence alignment used as the basis for homology modeling becomes an additional problem. Even down to levels of 25–30% sequence identity, however, homology modeling produces coarse-grained models for the overall fold of proteins of unknown structure.

Remote Homology Modeling (Threading)

As noted in the previous section, naturally evolved sequences with more than 30% pairwise sequence identity have homologous three-dimensional structures (79). Are all others nonhomologous? Not at all. In the current PDB database, there are thousands of pairs of structurally homologous pairs of proteins with less than 25% pairwise sequence identity (remote homologues) (36). If a correct alignment between U and a remote homologue RU (pairwise sequence identity to U $< 25\%$) is given, one could build the three-dimensional structure of U by homology modeling on the basis of the template of RU (remote homology modeling). A successful remote homology modeling must solve three different tasks: 1. RU has to be detected. 2. U and RU have to be aligned correctly. 3. The homology modeling procedure has to be tailored to the harder problem of extremely low sequence identity (with many loop regions to be modeled). Most methods developed so far have been addressed primarily to detect similar folds. The basic idea is to thread the sequence of U into the known structure of RU and to evaluate the fitness of sequence for structure by some kind of environment-based or knowledge-based potential (14, 86). Threading is, in some respects, a harder problem than is the prediction of three-dimensional structure (50, 86). Solving it, however, would enable the prediction of thousands of protein structures (see Figure 1). Can this hard nut be cracked?

The optimism generated by one of the first papers on threading published in the 1990s (11) has boosted attempts to develop threading methods (86). Most methods are based on pseudo-potentials and differ in the way such potentials are derived from PDB (98). One alternative is to use one-dimensional predictions for the threading procedure (68; G Barton, unpublished data; F Drabløs, unpublished data). The good news, after half a decade of intensive research by dozens of groups, is that all potentials capture different aspects, and it is likely that the correct remote homologue is found by at least one of these groups (82). The bad news is that no single method is accurate enough to identify the remote homologue correctly in most cases (82). Instead, evaluated on a larger test set, the correct remote homologue appears to be detected in approximately 30% of all cases (68). Unfortunately, this is only the first of the three tasks for successful remote homology modeling, the second (correct alignment of U and RU) is even harder. In many of the cases for which RU is identified correctly as a remote homologue of U, the alignment of U and RU is flawed in significant ways (unpublished data). This is fatal for the third step, the model-building procedure. Thus, is threading useful, at all?

Like all prediction methods, threading techniques are not error proof. One of the practical disadvantages of current tools is the lack of a successful measure for prediction reliability, such as that established for secondary structure prediction (see Figure 5). The conclusion seems to be that threading methods can be useful in the hands of rather skeptical expert users who can spot wrong hits and false alignments, even when the prediction method suggests a high confidence value for the error it generates. Three points may be added. First, threading techniques can clearly widen the range of successful sequence alignments (68). Second, some methods are accurate enough to be used in scanning entire chromosomes for remote homologues (12). Third, threading techniques may still become one of the most successful tools in structure prediction, but a lot of detailed work lies ahead.

ANALYSIS OF THREE-DIMENSIONAL STRUCTURES

A successful idea was to replace inductive force fields that capture the heuristics of physical principles by deductive, knowledge-based, mean-force potentials (e.g. 84). Such potentials, as well as more expert-knowledge–oriented approaches (49, 96), enable the detection of subtle stresses or possible errors in both experimentally determined three-dimensional structures and predicted models (85). Knowledge-based potentials of mean force appear to be valid even for proteins with properties not used for deriving the potentials [membrane proteins (85); coiled-coils, S O'Donoghue, unpublished data]. Because of this success, quality control tools that use these potentials are becoming a routine check applied to any experimentally determined structure or any structure predicted by homology modeling.

More and more frequently, a newly determined structure is identified to be remotely homologous to a known structure (38). Recently developed algorithms enable routine scans for possible remote homologues in PDB for any new structure (35, 41, 59, 64, 75, 88). Such searches are beginning to rival sequence database searches as a tool for discovering biologically interesting relationships (38). Similar techniques can often be exploited to determine domains in known structures (27, 37, 87).

CONCLUSION

Three-dimensional structure cannot yet be predicted reliably from sequence information alone. In other words, the only source for new, unique structures (structures for which no homologue exists in the database) are experiments. Given the amount of time needed to determine a protein structure experimentally, however, more non-unique structures can be predicted at atomic resolution by homology modeling in 1 month than have been determined by experiment during the past 3 decades. Unfortunately, such models typically have considerable coordinate errors in loop regions, and remote homology modeling (i.e. homology modeling for $< 25\%$ pairwise sequence identity) is not yet reliable. For a few cases, however, threading techniques already have resulted in accurate modeling of the overall fold (86).

The rich information contained in the growing sequence and structure databases has been used to improve the accuracy of predictions of some aspects of protein structure. Predictions of secondary structure, solvent accessibility, and transmembrane helices are becoming increasingly useful. This success is the result of both a better performance of multiple alignment–based methods and the ability to focus on more reliably predicted regions. Some methods have indicated that one-dimensional predictions can be useful as an intermediate step on the way to predicting three-dimensional structure (interstrand contacts, prediction-based threading). Another advantage of predictions in one dimension is that they are not very CPU-intensive, i.e. one-dimensional structure can be predicted for the protein sequence of, for example, entire yeast chromosomes overnight.

The prediction accuracy of chain-distant interresidue contacts is relatively limited so far. Analysis of correlated mutations can be used to distinguish between alternative models (e.g. for threading techniques). The prediction of interstrand contacts appears to be useful in some cases. An accurate method for the automatic prediction of contacts between residues not close in sequence remains to be developed.

Another encouraging development is the improvement of tools for the analysis of protein structures. Experimental inconsistencies can be spotted, and predicted models can be tested. The ease of scanning structure databases for remote homologues yields a rich amount of information with an effect on our understanding of protein structure and function.

ACKNOWLEDGMENT

We are grateful to Kimmen Sjölander (Santa Cruz, California) for her comprehensive help on improving the manuscript.

Literature Cited

1. Abagyan R, Totrov M. 1994. Biased probability Monte Carlo conformational searches and electrostatic calculations for peptides and proteins. *J. Mol. Biol.* 235:983–1002
2. Altschuh D, Vernet T, Moras D, Nagai K. 1988. Coordinated amino acid changes in homologous protein families. *Protein Eng.* 2:193–99
3. Altschul SF. 1993. A protein alignment scoring system sensitive at all evolutionary distances. *J. Mol. Evol.* 36:290–300
4. Anfinsen CB, Scheraga HA. 1975. Experimental and theoretical aspects of protein folding. *Adv. Protein Chem.* 29:205–300
5. Bairoch A, Boeckmann B. 1994. The SWISS-PROT protein sequence data bank: current status. *Nucleic Acids Res.* 22:3578–80
6. Benner SA, Badcoe I, Cohen MA, Gerloff DL. 1994. Bona fide prediction of aspects of protein conformation. *J. Mol. Biol.* 235:926–58
7. Benson D, Lipman DJ, Ostell J. 1993. GenBank. *Nucleic Acids Res.* 21: 963–75

8. Bernstein FC, Koetzle TF, Williams GJB, Meyer EF, Brice MD, et al. 1977. The protein data bank: a computer based archival file for macromolecular structures. *J. Mol. Biol.* 112:535–42
9. Biou V, Gibrat JF, Levin JM, Robson B, Garnier J. 1988. Secondary structure prediction: combination of three different methods. *Protein Eng.* 2: 185–91
10. Bohr H, Bohr J, Brunak S, Fredholm H, Lautrup B, et al. 1990. A novel approach to prediction of the 3-dimensional structures of protein backbones by neural networks. *FEBS Lett.* 261: 43–46
11. Bowie JU, Lüthy R, Eisenberg D. 1991. A method to identify protein sequences that fold into a known three-dimensional structure. *Science* 253: 164–69
12. Braxenthaler M, Sippl M. 1995. Screening genome sequences for known folds. In *Protein Structure by Distance Analysis*, ed. H Bohr, S Brunak. Boca Raton, FL: CRC
13. Brünger AT, Nilges M. 1993. Computational challenges for macromolecular structure determination by X-ray crystallography and solution NMR-spectroscopy. *Q. Rev. Biophys.* 26: 49–125
14. Bryant SH, Altschul SF. 1995. Statistics of sequence-structure threading. *Curr. Opin. Struct. Biol.* 5:236–44
15. Chou PY, Fasman GD. 1978. Prediction of the secondary structure of proteins from their amino acid sequence. *Adv. Enzymol.* 47:45–148
16. Colloc'h N, Etchebest C, Thoreau E, Henrissat B, Mornon J-P. 1993. Comparison of three algorithms for the assignment of secondary structure in proteins: the advantages of a consensus assignment. *Protein Eng.* 6: 377–82
17. Cornell WD, Howard AE, Kollman P. 1991. Molecular mechanical potential functions and their application to study molecular systems. *Curr. Opin. Struct. Biol.* 1:201–12
18. Defay T, Cohen FE. 1995. Evaluation of current techniques for ab initio protein structure prediction. *Proteins.* 23: 431–45
19. De Filippis V, Sander C, Vriend G. 1994. Predicting local structural changes that result from point mutations. *Protein Eng.* 7:1203–8
20. Deperieux E, Feytmans E. 1992. MATCH-BOX: a fundamentally new algorithm for the simultaneous alignment of several protein sequences. *Comput. Appl. Biosci.* 8:501–9
21. Doolittle RF. 1994. Convergent evolution: the need to be explicit. *Trends Biochem. Sci.* 19:15–8
22. Eddy SR. 1995. Multiple alignment using hidden Markov models. See Ref. 66a, pp. 114–20
23. Eisenmenger F, Argos P, Abagyan R. 1993. A method to configure protein side-chains from the main-chain trace in homology modelling. *J. Mol. Biol.* 231:849– 60
24. Flores TP, Orengo CA, Moss DS, Thornton JM. 1993. Comparison of conformational characteristics in structurally similar protein pairs. *Protein Sci.* 2:1811–26
25. Gaasterland T, Selkov E. 1995. Reconstruction of metabolic networks using incomplete information. See Ref. 66a, pp. 127–35
26. Galaktionov SG, Marshall GR. 1994. Properties of intraglobular contacts in proteins: an approach to prediction of tertiary structure. In *27th Hawaii Int. Conf. System Sciences, Wailea HI,* ed. L Hunter, pp. 326–35. Los Alamitos, CA: IEEE Comput. Soc.
27. Gerstein M, Sonnhammer ELL, Chothia C. 1994. Volume changes in protein evolution. *J. Mol. Biol.* 236: 1067–78
28. Gibrat J-F, Garnier J, Robson B. 1987. Further developments of protein secondary structure prediction using information theory. New parameters and consideration of residue pairs. *J. Mol. Biol.* 198:425–43
29. Goebel U, Sander C, Schneider R, Valencia A. 1994. Correlated mutations and residue contacts in proteins. *Proteins* 18:309–17
30. Henikoff S, Henikoff JG. 1993. Performance evaluation of amino acid substitution matrices. *Proteins* 17: 49–61
31. Henikoff S, Henikoff JG. 1994. Position-based sequence weights. *J. Mol. Biol.* 243:574–78
32. Hobohm U, Sander C. 1994. Enlarged representative set of protein structures. *Protein Sci.* 3:522–24
33. Holbrook SR, Muskal SM, Kim S-H. 1990. Predicting surface exposure of amino acids from protein sequence. *Protein Eng.* 3:659–65
34. Holm L, Rost B, Sander C, Schneider R, Vriend G. 1994. Data based modeling of proteins. In *Statistical Mechanics, Protein Structure, and Protein Substrate Interactions*, ed. S Doniach, pp. 277–96. New York: Plenum

35. Holm L, Sander C. 1993. Protein structure comparison by alignment of distance matrices. *J. Mol. Biol.* 233: 123–38
36. Holm L, Sander C. 1994. The FSSP database of structurally aligned protein fold families. *Nucleic Acids Res.* 22:3600–9
37. Holm L, Sander C. 1994. Parser for protein folding units. *Proteins* 19: 256–68
38. Holm L, Sander C. 1994. Searching protein structure databases has come of age. *Proteins* 19:165–73
39. Hubbard TJP. 1994. Use of β-strand interaction pseudo-potential in protein structure prediction and modelling. In *27th Hawaii Int. Conf. System Sciences, , Maui, HI,* ed. L Hunter, pp. 336–44. Los Alamitos, CA: IEEE Comput. Soc.
40. Hubbard TJP, Park J. 1995. Fold recognition and ab initio structure predictions using hidden Markov models and β-strand pair potentials. *Proteins.* In press
41. Johnson MS, Overington JP, Blundell TL. 1993. Alignment and searching for common protein folds using a data bank of structural templates. *J. Mol. Biol.* 231:735–52
42. Johnston M, Andrews S, Brinkman R, Cooper J, Ding H, et al. 1994. Complete nucleotide sequence of *Saccaromyces cerevisiae* chromosome VIII. *Science* 265:2077–82
43. Jones DT, Taylor WR, Thornton JM. 1992. A new approach to protein fold recognition. *Nature* 358:86–89
44. Jones DT, Taylor WR, Thornton JM. 1992. The rapid generation of mutation data matrices from protein sequences. *Comput. Appl. Biosci.* 8: 275–82
45. Kabsch W, Sander C. 1983. Dictionary of protein secondary structure: pattern recognition of hydrogen bonded and geometrical features. *Biopolymers* 22:2577–637
46. Kabsch W, Sander C. 1984. On the use of sequence homologies to predict protein structure: Identical pentapeptides can have completely different conformations. *Proc. Natl. Acad. Sci. USA* 81:1075–78
47. Karplus M, Petsko GA. 1990. Molecular dynamics simulations in biology. *Nature* 347:631–39
48. Krogh A, Brown M, Mian IS, Sjølander K, Haussler D. 1994. Hidden Markov models in computational biology: applications to protein modeling. *J. Mol. Biol.* 235:1501–31
49. Laskowski RA, Moss DS, Thornton JM. 1993. Main-chain bond lengths and bond angles in protein structures. *J. Mol. Biol.* 231:1049–67
50. Lathrop RH. 1994. The protein threading problem with sequence amino acid interaction preferences is NP-complete. *Protein Eng.* 7:1059–68
51. Lattman EE. 1994. Protein crystallography for all. *Proteins* 18:103–6
52. Lawrence CE, Altschul SF, Boguski MS, Liu JS, Neuwald AF, et al. 1993. Detecting subtle sequence signals: a Gibbs sampling strategy for multiple alignment. *Science* 262:208–14
53. Levin JM, Pascarella S, Argos P, Garnier J. 1993. Quantification of secondary structure prediction improvement using multiple alignments. *Protein Eng.* 6:849–54
54. Levitt M. 1992. Accurate modeling of protein conformation by automatic segment matching. *J. Mol. Biol.* 226: 507–33
55. Levitt M, Chothia C. 1976. Structural patterns in globular proteins. *Nature* 261:552–58
56. Lifson S, Sander C. 1980. Specific recognition in the tertiary structure of β-sheets in proteins. *J. Mol. Biol.* 139: 627–39
57. Livingstone CD, Barton GJ. 1993. Protein sequence alignments: a strategy for the hierarchical analysis of residue conservation. *Comput. Appl. Biosci.* 9:745–56
58. May ACW, Blundell TL. 1994. Automated comparative modelling of protein structures. *Curr. Opin. Biotechnol.* 5:355–60
59. Mitchell EM, Artymiuk PJ, Rice DW, Willett P. 1992. Use of techniques derived from graph theory to compare secondary structure motifs in proteins. *J. Mol. Biol.* 212:151–66
60. Muskal SM, Holbrook SR, Kim S-H. 1990. Prediction of the disulfide-bonding state of cysteine in proteins. *Protein Eng.* 3:667–72
61. Nakashima H, Nishikawa K. 1992. The amino acid composition is different between the cytoplasmic and extracellular sides in membrane proteins. *FEBS Lett.* 303:141–46
62. Neher E. 1994. How frequent are correlated changes in families of protein sequences? *Proc. Natl. Acad. Sci. USA* 91:98–102
63. Nilges M. 1993. A calculation strategy for the structure determination of symmetric dimers by 1H NMR. *Proteins* 17:297–309
64. Orengo CA, Brown NP, Taylor WT.

1992. Fast structure alignment for protein databank searching. *Proteins* 14: 139–67

65. Pearson WR, Lipman DJ. 1988. Improved tools for biological sequence comparison. *Proc. Natl. Acad. Sci. USA* 85:2444–48
66. Persson B, Argos P. 1994. Prediction of transmembrane segments in proteins utilising multiple sequence alignments. *J. Mol. Biol.* 237:182–92

66a. Rawlings C, Clark D, Altman R, Hunter L, Lengauer T, et al, eds. 1995. *3rd Int. Conf. on Intelligent Systems for Molecular Biology (ISMB), Cambridge, England.* Menlo Park, CA: AAAI

67. Rooman M, Wodak SJ. 1988. Identification of predictive sequence motifs limited by protein structure data base size. *Nature* 335:45–49
68. Rost B. 1995. TOPITS: Threading one-dimensional predictions into three-dimensional structures. See Ref. 66a, pp. 314–21
69. Rost B, Casadio R, Fariselli P, Sander C. 1995. Prediction of helical transmembrane segments at 95% accuracy. *Protein Sci.* 4:521–33
70. Rost B, Sander C. 1993. Prediction of protein secondary structure at better than 70% accuracy. *J. Mol. Biol.* 232: 584–99
71. Rost B, Sander C. 1993. Secondary structure prediction of all-helical proteins in two states. *Protein Eng.* 6: 831–36
72. Rost B, Sander C. 1994. Combining evolutionary information and neural networks to predict protein secondary structure. *Proteins* 19:55–72
73. Rost B, Sander C. 1994. Conservation and prediction of solvent accessibility in protein families. *Proteins* 20: 216–26
74. Rost B, Sander C, Schneider R. 1994. Redefining the goals of protein secondary structure prediction. *J. Mol. Biol.* 235:13–26
75. Russell RB, Barton GJ. 1992. Multiple protein sequence alignment from tertiary structure comparison: assignment of global and residue confidence levels. *Proteins* 14:309–23
76. Russell RB, Barton GJ. 1993. The limits of protein secondary structure prediction accuracy from multiple sequence alignment. *J. Mol. Biol.* 234: 951–57
77. Salamov AA, Solovyev VV. 1995. Prediction of protein secondary structure by combining nearest-neighbor algorithms and multiple sequence alignments. *J. Mol. Biol.* 247:11–15
78. Sali A, Blundell T. 1994. Comparative protein modelling by satisfaction of spatial restraints. In *Protein Structure by Distance Analysis*, ed. H Bohr, S Brunak, pp. 64–87. Amsterdam/Oxford/Washington: IOS Press
79. Sander C, Schneider R. 1991. Database of homology-derived structures and the structurally meaning of sequence alignment. *Proteins* 9:56–68
80. Sander C, Schneider R. 1994. The HSSP database of protein structure-sequence alignments. *Nucleic Acids Res.* 22:3597–99
81. Shindyalov IN, Kolchanov NA, Sander C. 1994. Can three-dimensional contacts in protein structures be predicted by analysis of correlated mutations? *Protein Eng.* 7:349–58
82. Shortle D. 1995. Protein fold recognition. *Nature Struct. Biol.* 2:91–92
83. Sipos L, von Heijne G. 1993. Predicting the topology of eukaryotic membrane proteins. *Eur. J. Biochem.* 213: 1333–40
84. Sippl MJ. 1993. Boltzmann's principle, knowledge based mean fields and protein folding. An approach to the computational determination of protein structures. *J. Comput. Aided Mol. Design* 7:473–501
85. Sippl MJ. 1993. Recognition of errors in three-dimensional structures of proteins. *Proteins* 17:355–62
86. Sippl MJ. 1995. Knowledge-based potentials for proteins. *Curr. Opin. Struct. Biol.* 5:229–35
87. Sternberg MJE, Hegyi H, Islam SA, Luo J, Russell RB. 1995. Towards an intelligent system for the automatic assignment of domains in globular proteins. See Ref. 66a, pp. 376–83
88. Subbiah S, Laurents DV, Levitt M. 1993. Structural similarity of DNA-binding domains of bacterophage repressors and the globin core. *Curr. Biol.* 3:141–48
89. Summers NL, Karplus M. 1990. Modeling of globular proteins. *J. Mol. Biol.* 216:991–1016
90. Taylor WR, Hatrick K. 1994. Compensating changes in protein multiple sequence alignments. *Protein Eng.* 7: 341–48
91. Taylor WR, Jones DT, Green NM. 1994. A method for α-helical integral membrane protein fold prediction. *Proteins* 18:281–94
92. Thompson JD, Higgins DG, Gibson TJ. 1994. Improved sensitivity of profile searches through the use of se-

quence weights and gab excision. *Comput. Appl. Biosci.* 10:19–29
93. van Gunsteren WF. 1993. Molecular dynamics studies of proteins. *Curr. Opin. Struct. Biol.* 3:167–74
94. Vingron M, Waterman MS. 1994. Sequence alignment and penalty choice. *J. Mol. Biol.* 235:1–12
95. von Heijne G. 1992. Membrane protein structure prediction. *J. Mol. Biol.* 225:487–94
96. Vriend G, Sander C. 1993. Quality of protein models: directional atomic contact analysis. *J. Appl. Crystallogr.* 26:47–60
97. Wako H, Blundell TL. 1994. Use of amino acid environment-dependent substitution tables and conformational propensities in structure prediction from aligned sequences of homologous proteins I. Solvent accessibility classes. *J. Mol. Biol.* 238:682–92
98. Wodak SJ, Rooman MJ. 1993. Generating and testing protein folds. *Curr. Opin. Struct. Biol.* 3:247–59
99. Zhu Z-Y. 1995. A new approach to the evaluation of protein secondary structure predictions at the level of the elements of secondary structure. *Protein Eng.* 8:103–8

Annu. Rev. Biophys. Biomol. Struct. 1996. 25:137–62

THE SUGAR KINASE/HEAT SHOCK PROTEIN 70/ACTIN SUPERFAMILY: Implications of Conserved Structure for Mechanism

James H. Hurley

Laboratory of Molecular Biology, National Institute of Diabetes, Digestive and Kidney Diseases, National Institutes of Health, Bethesda, Maryland 20892-0580

KEY WORDS: protein structure, conformational change, X-ray crystallography, enzyme regulation, enzyme mechanism

ABSTRACT

Sugar kinases, stress-70 proteins, and actin belong to a superfamily defined by a fold consisting of two domains with the topology $\beta\beta\beta\alpha\beta\alpha\beta\alpha$. These enzymes catalyze ATP phosphoryl transfer or hydrolysis coupled to a large conformational change in which the two domains close around the nucleotide. The $\beta1$-$\beta2$ turns of each domain form hydrogen bonds with ATP phosphates, and conserved Asp, Glu or Gln residues coordinate Mg^{2+} or Ca^{2+} through bound waters. The activity of superfamily members is regulated by various effectors, some of which act by promoting or inhibiting the conformational change. Nucleotide hydrolysis eliminates interdomain bridging interactions between the second $\beta1$-$\beta2$ turn and the ATP γ-phosphate. This is proposed to destabilize the closed conformation and affect the orientation of the two domains, which might in turn regulate the activity of kinase oligomers, stress-70 protein-protein complexes, and actin filaments.

1056-8700/96/0610-137$08.00

CONTENTS

PERSPECTIVES AND OVERVIEW

One of the most common ways to regulate the activity of a protein is to couple its function to the hydrolysis of a nucleotide. Many of the proteins regulated in this way belong to a handful of structural classes. One such class, which comprises the 70-kDa heat shock–related proteins (stress-70), actin, and the sugar kinases, was first recognized on the basis of the similar three-dimensional folds of its members (22). The structural similarity of these proteins was quite unexpected, given their diversity of function and negligible overall sequence similarity. All these proteins have adenosine triphosphate (ATP) phosphotransferase or hydrolase activity, and all are thought to experience a large conformational change during their reaction cycle. The structural similarities of the sugar kinases, hsp70s, and actin include several short regions of conserved primary structure that are localized at the ATP binding site (6, 22, 37). It is now generally believed that these proteins diverged during ancient times from a common ancestor.

The past 4 years have brought a wave of studies targeting the mechanism of ATP cleavage and its coupling to the function and regulation of sugar kinases, actin, and the stress-70 proteins. Newly identified members have joined the superfamily, and the crystal structure of one of these, glycerol kinase (GK), has been determined. Several questions remain open regarding the relationships between these proteins. The central issue is whether the conserved structure implies not only a common ancestry but similar catalytic and regulatory mechanisms as well. The function and regulation of each of these proteins is of considerable interest individually. Insofar as these mechanisms are related to each other, the implications for individual superfamily members will be considered.

Superfamily Members

Hexokinase, GK, and other sugar kinases catalyze the committed step in the uptake of their carbohydrate substrates into metabolism, the ATP-dependent phosphorylation of the sugar. They are, therefore, key targets of metabolic regulation. Hexokinase and GK are oligomeric enzymes subject to both hetero- and homotropic allosteric regulation. Not all sugar kinases are related structurally to the superfamily. Known sugar kinase members of the superfamily include hexokinase, GK, glucokinase, fucokinase, fructokinase, rhamnulokinase, gluconokinase, xylulokinase, and ribulokinase (7).

The stress-70 proteins are central in cellular protein folding and in assembly and disassembly of supramolecular structures (12, 26, 27, 30). This family includes the eukaryotic heat-shock inducible proteins (hsp70s), the constituitively expressed heat-shock cognates (hsc70s), the endoplasmic reticulum immunoglobulin-binding protein, and the bacterial homologue DnaK. These proteins all bind ATP and couple ATP hydrolysis to their function. The stress-70 proteins contain separate peptide and ATP-binding domains. Intact stress-70 proteins hydrolyze ATP on stimulation by a target protein or peptide, but isolated ATP-binding domains hydrolyze ATP constituitively. Their biological roles differ, but the mechanism of action of these homologous proteins is very similar.

Actins are ubiquitous eukaryotic structural and contractility proteins that self-assemble into filaments (39, 58, 64). Together with myosin, actin forms the basic machinery for contractility in muscle cells. In other cell types, actin filaments play a central role in the cytoskeleton. Studies on actin isoforms from a range of organisms are referred to interchangeably in this review.

CONSERVED STRUCTURE AND CONFORMATIONAL CHANGE

Topology of the Conserved Core

The conserved core consists of two α/β subdomains related to each other by approximate dyad symmetry (Figure 1). The topology of each is of the form $\beta\beta\beta\alpha\beta\alpha\beta\alpha$ (Figure 2). The common core of structurally equivalent regions in hexokinase (66), hsc70 (21), actin (38), and GK (34) contains 127 residues (Figure 3). hsc70 and actin are more similar to each other than to the sugar kinases, and these two share a total of approximately 214 structurally equivalent residues. The structure of each superfamily member shows a distinct pattern of insertions at sev-

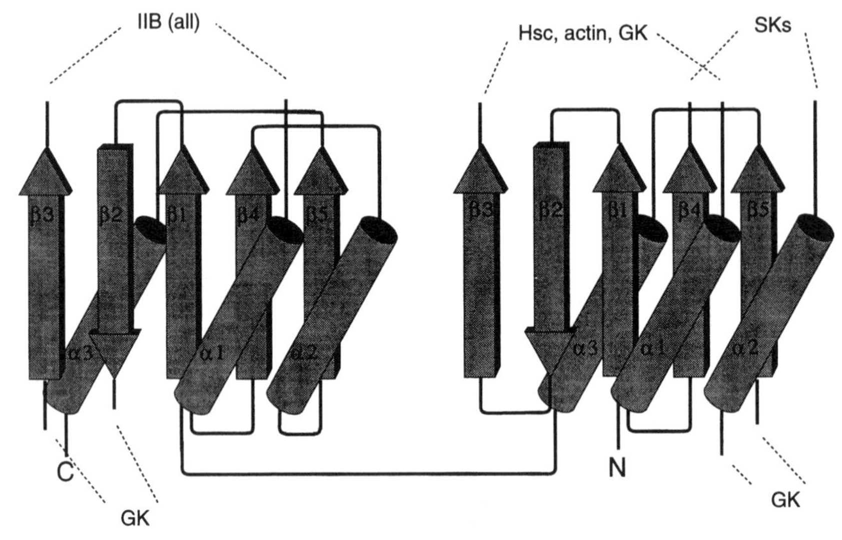

Figure 2 Schematic of the topology of the conserved core subdomains. Points of insertion for nonconserved regions are marked. SK, sugar kinases; all, all four family members.

```
                 IA  β1        β2           β3          α1                     β4
hexokinase       83  LAIDLGGTNLRVVLV 109 QSKY 123 NPDELWEFIADSLKAFIDEQ 152 PLGFTFS
GK                7  VALDQGTTSSRAVMD  31 SQRE  48 DPMEIWATQSSTLVEVLAKA  76 AIGITNQ
Hsc70             7  VGIDLGTTYSCVGVF  36 RTTP 116 PEEVSSMVLTKMKEIAEAYL 141 NAVTVPY
actin             8  LVCDNGSGLVKAGFA  29 AVFP  78 NWDDMEKIWHHTFYNELRVA 102 PTLLTEA

                      α2            β5        α3          IIA  β1         β2        β3
hexokinase      189  VPMLQKQIS 206 VALINDTTGTLVASY 227 KMGVIFGT-GVNGAY 267 NCEY
GK              165  VDTWLIWKM 240 SGIAGDQQAALFGQL 260 MAKNTYGT-GCFMLM 303 EGAV
Hsc70           157  ATKDAGTIA 170 LRIINEPTAAAIAYG 195 VLIFDLGGGTFDVSI 222 TAGD
actin           114  ANREKMTQI 132 MYVAIQAVLSLYASG 150 GIVLDSGDGVTHNVP 175 IMRL

                      α1              β4          α2            β5          α3
hexokinase      397  VCGIAAICQK 414 IAADGS 427 FKEKAA 452 KIVP 458 DGSGAGAAVIAAL
GK              388  TRDVLEAMQA 406 LRVDGG 420 QFQSDI 429 RVER 437 EVTALGAAYLAGL
Hsc70           313  TLDPVEKALR 334 IVLVGG 348 KLLQDF 358 ELNK 367 EAVAYGAAVQAAI
actin           274  IHETTYNSIM 297 NVMSGG 311 DRMQKE 328 KIIA 337 YSVWIGGSILASL
```

Figure 3 Sequences of structurally conserved regions of superfamily members. The hexokinase sequence is numbered according to the translated gene sequence of yeast hexokinase B (24). The alignment is based solely on three-dimensional structural equivalence. Only regions that are equivalent structurally in all four proteins are shown. Boxed residues are either identical in all sequences or have structurally equivalent roles in the nucleotide binding site.

eral different loci within the conserved fold. The two core subdomains and the insertions together make up the two domains of each structure. In the nomenclature first proposed for hsc70, the domains are denoted I and II; the conserved subdomains are IA and IIA; and the divergent subdomains are IB and IIB. The IA and IIA subdomains from different structures can be superimposed individually with approximately 2.5 Å rms deviation.

The unique biological functions are mediated by subdomains inserted at four different topologic positions (Figure 2). hsc70, actin, and GK have all or part of their IB subdomains inserted between $\beta3$ and $\alpha1$ of core subdomain IA. The $\beta3$-$\alpha1$ insertions comprise the entire α/β IB subdomains of hsc70 and actin. The GK $\beta3$-$\alpha1$ insertion consists of two antiparallel β strands that are part of a larger α/β structure. GK and hexokinase contain an insertion between $\beta4$ and $\alpha2$ of IA. In hexokinase, this insertion consists of two antiparallel β strands that comprise the small IB subdomain. In GK, the $\beta4$-$\alpha2$ insertion is larger and comprises much of its large α/β IB subdomain. GK has a unique insertion between $\alpha2$ and $\beta5$ that contributes to the IB subdomain and adds a β strand to the IA β sheet. All four proteins have insertions between $\beta3$ and $\alpha1$ of core subdomain IIA that comprise their entire IIB subdomains. hsc70, actin, and GK have topologically identical α/β IIB subdomains, but the IIB of GK is in a different position in three dimensions than are the other two. Hexokinase has a unique large α helical IIB subdomain. GK also has an additional β strand at each end of the IIA core β sheet. The first strand is inserted in the primary sequence between $\beta2$ and $\beta3$, and the second is inserted after the end of IIA. All four proteins have unique extensions at either their N- or C-termini, or both.

Nucleotide Binding Site

ATP binds in a deep cleft between the two domains (Figure 1). In the sugar kinases, the sugar substrate binds below the nucleotide at the bottom of the cleft. In hsc70 and actin, the cleft is closed below the nucleotide binding site. The position and conformation of the nucleotide in hsc70, actin, and GK (these coordinates are not available for hexokinase) are very similar. The base is in the *anti* conformation typical of enzyme-bound adenine nucleotides. The turn between $\beta1$ and $\beta2$ of subdomain IA plays a critical role in nucleotide binding, with main-chain nitrogens of the turn interacting with the phosphates. In GK, actin, and probably hexokinase, a basic side-chain from $\beta2$ binds the β phosphate of adenosine diphosphate (ADP). In subdomain IIA, the $\beta1$-$\beta2$ turn contributes main-chain amides to phosphate binding. In subdomain IA, the structure of the critical β1-β2 turn is nearly identical in all four proteins; in subdomain IIA, however, the turn is shortened by one residue in the sugar kinases compared with the ATPases. The adenosine base and ribose are bound by more diverse interactions by the domain IIs of each protein. The ribose interacts with residues from the IIB domains, which differ between family members. The adenine binds primarily through hydrophobic interactions with residues from IIA α1, β4, and the β4-α2 linker. The polar adenine N6 is solvent accessible in all three proteins.

Adenine nucleotides bind GK by relatively few interactions (Figure 4). The main-chain amide of residue 411 and the side-chain amide of Asn415 of IIA donate one hydrogen bond each to the α-phosphate and adenine moieties. The main-chain carbonyl of residue 310 hydrogen-bonds to the ribose. The main-chain amide of residue 267, on the first turn of the IIA β-sheet, donates a hydrogen bond to the β-phosphate.

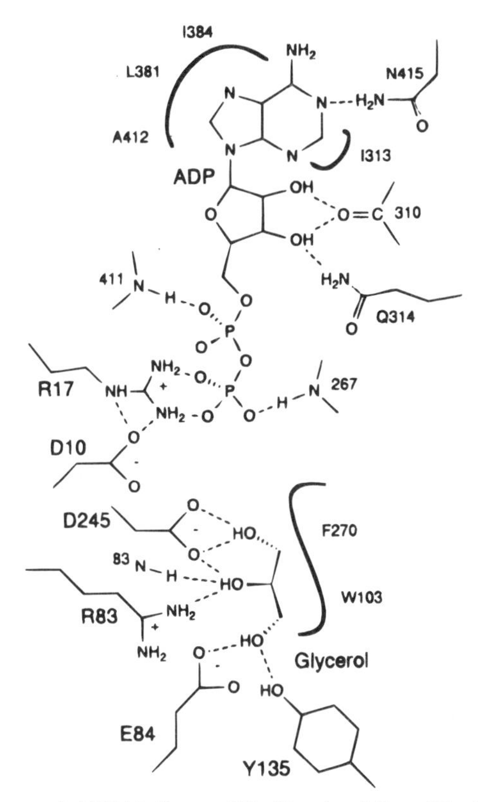

Figure 4 Sugar and ADP binding to GK. (Reprinted from 34 with permission.)

Arg17 of the second strand of the IA β-sheet forms a salt-bridge with the β-phosphate, the only direct interaction with domain I. The adenine base is surrounded partially by hydrophobic side chains of Ile313, Leu381, Ile384, and Ala412, and by the long first α-helix of IIA.

hsc70 and actin have more extensive contacts with bound nucleotide than do the sugar kinases. In the hsc70 structure, the ADP hydrogen bonds directly with four main-chain amides, in contrast to two in GK. There is an additional water-mediated hydrogen bond from the main-chain amide of the residue inserted in the IIA $\beta1$-$\beta2$ turns of hsc70 and actin relative to the sugar kinases. The other two additional amide hydrogen bonds are donated by the IA $\beta1$-$\beta2$ turn. The ribose participates in three hydrogen bonds (one water-mediated) with residues of the first two helices of the IIB domains, which are similar in the two ATPases but not in the sugar kinases. The adenine base accepts one hydrogen bond from a serine of IIB conserved only in hsc70 and actin.

The hsc70 complex lacks a basic residue analogous to the GK Arg17, but a K^+ ion forms an equivalent electrostatic interaction (75). A second K^+ ion bridges Asp199, Asp206, the carbonyl of Thr204, and the γ-phosphate of a model-built ATP. Whether other superfamily members have K^+ binding sites in the cleft is unknown but is unlikely. The hsc70 active site is more acidic than any of the others, and binding of the two monovalent cations leads to an active site charge distribution that is approximately equivalent to that of the rest of the superfamily. A unique interaction is found between either P_i or the ATP γ-phosphate and the side-chain of Thr13. The actin-ADP complex shows all of the amide-phosphate interactions seen in the hsc70 complex (Figure 5). There is also a salt bridge between the α and β phosphates and Lys 18. This interaction is analogous to that between Arg17 and the α and β phosphates of GK. The hsc70 and actin-ATP complexes show two more main-chain amide hydrogen bonds from the IIA $\beta1$-$\beta2$ turn to the γ-phosphate. In the actin-ATP complex, the interaction with Lys18 is lost to make these hydrogen bonds.

Divalent Metal Ion Binding Sites

ATP binding, hydrolysis, and phosphoryl transfer all depend on the presence of a divalent metal ion. Mg^{2+} and/or Mn^{2+} have been located in hsc70 (21, 23, 53) and GK (M Feese and SJ Remington, personal communication). Bound solvent plays a critical role in the active site, as it mediates metal coordination and serves as the phosphate acceptor for the hsc70 (Figure 6) and actin ATPases. In subdomain IA, an identically conserved aspartate at the end of $\beta1$ hydrogen bonds to two Mg^{2+}-coordinating water molecules. A second water-mediated metal ligand is contributed at the start of the crossover helix $\alpha3$ of domain IA. In

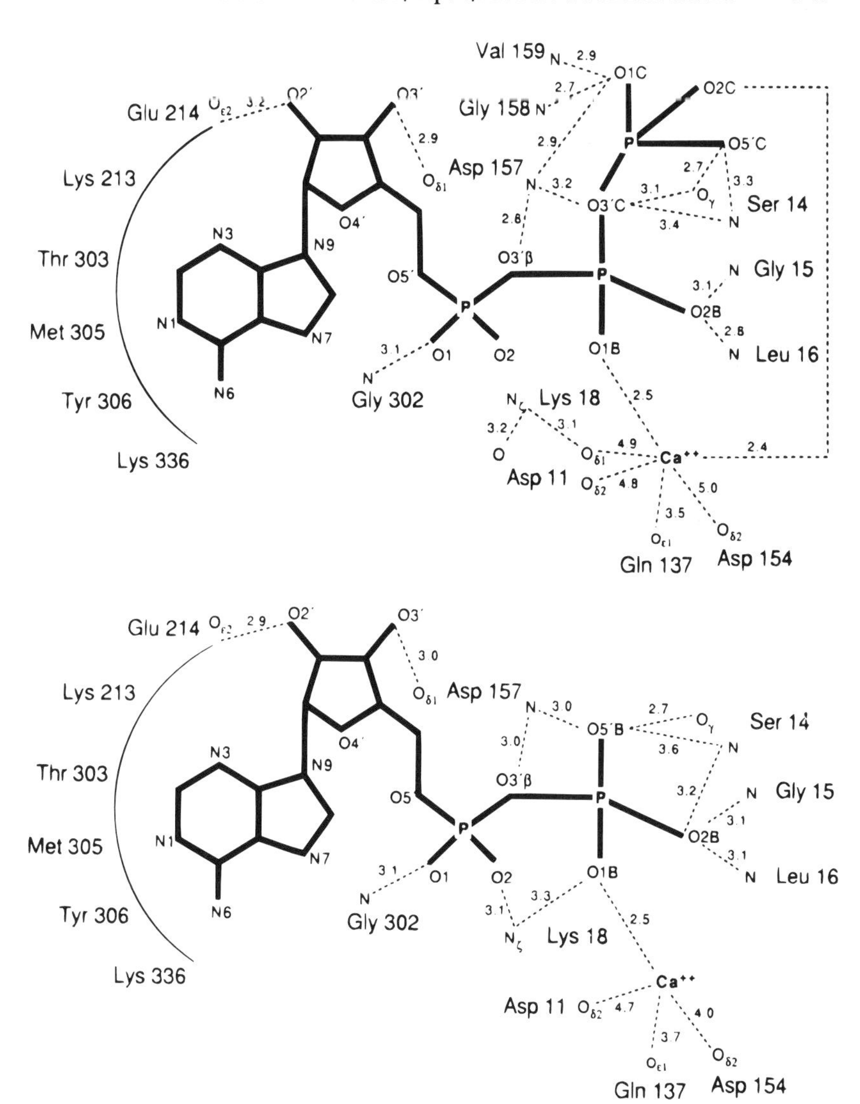

Figure 5 Schematic drawing of (*A*) ATP and (*B*) ADP bound to DNaseI-actin. (Reprinted from 38 with permission.)

hexokinase and GK, this position is an Asp that is almost certainly the catalytic base. In hsc70 and actin, the corresponding residue is a Glu or Gln, respectively. In hsc70 and actin, an Asp at the end of $\beta 1$, which is analogous to the conserved Asp at the end of $\beta 1$ of subdomain IA, contributes the third water-mediated metal ligand.

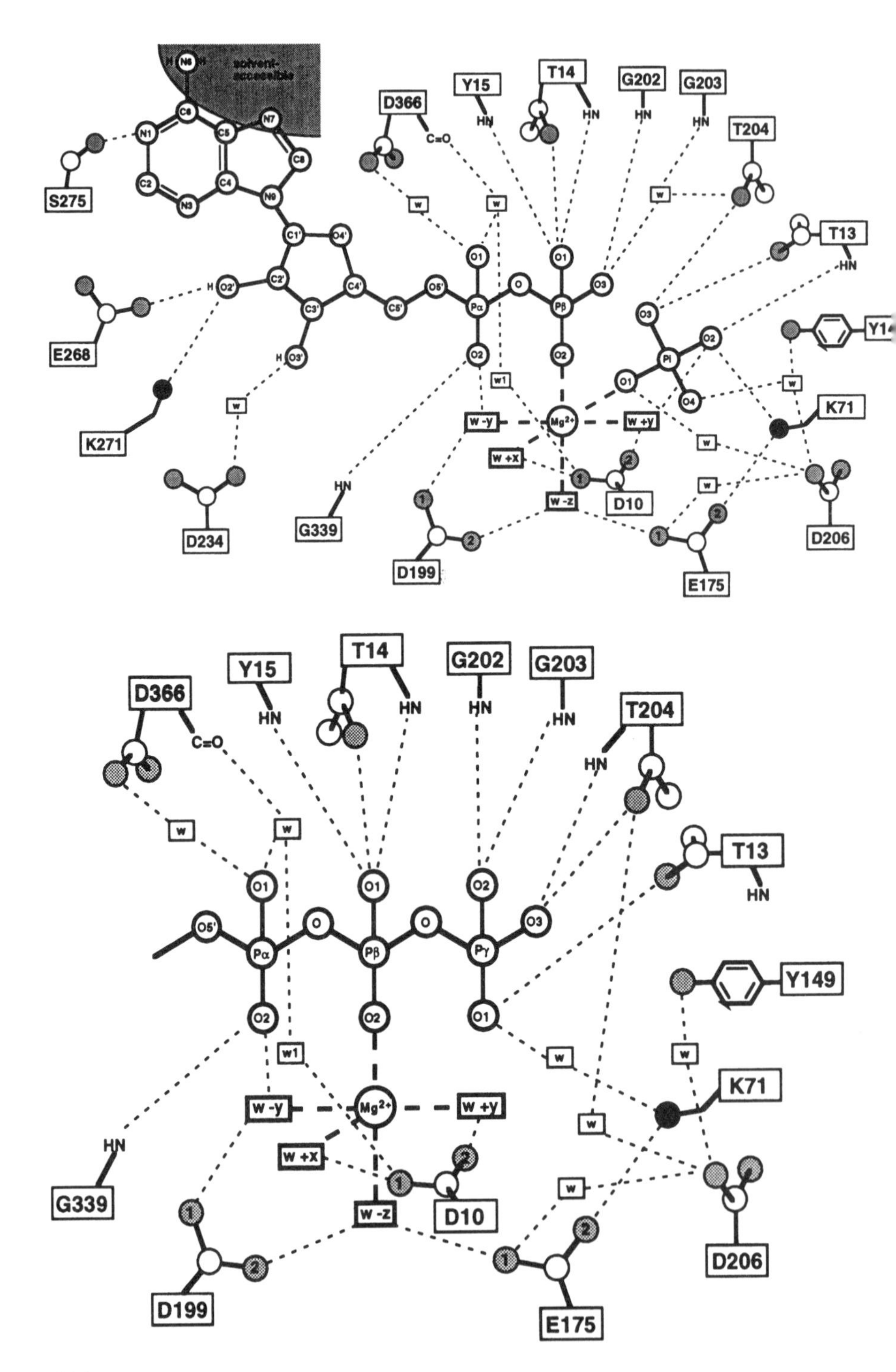
solvent-accessible
N6
N1
C6
C5
N7
C8
C2
C4
N3
N9
S275
C1'
O4'
C2'
C3'
C4'
C5'
O5'
O2'
O3'
E268
K271
D234
D366
C=O
Y15
HN
T14
G202
G203
T204
T13
Y149
K71
D206
E175
D199
G339
D10
O1
O2
O3
O4
O
Pα
Pβ
Pγ
Pi
Mg2+
w
w1
w -y
w +y
w +x
w -z

The structure of hsc70 bound to Mg^{2+} and the nonhydrolyzable ATP analogue AMPPNP (23; Figure 6) shows Mg^{2+} coordinated by one of the β phosphate oxygens and by four water molecules in an octahedral arrangement with one missing ligand. Two of the water molecules are bound to the first conserved ligating side-chain, Asp10. One water bridges the α phosphate and Asp199 (absent in sugar kinases, see Figure 3), and one bridges Asp199 to Glu175 (Asp or Gln in other members, see Figure 3). In the structure of the ADP and P_i product complex with hsc70, an oxygen of inorganic phosphate occupies the sixth octahedral Mg^{2+} ligand position. An analogous situation occurs in the ADP and 3-phosphoglycerol complex of GK (M Feese and SJ Remington, personal communication).

Bound Ca^{2+} has been located in actin and hsc70. Ca^{2+} binds to the actin-ATP complex in a manner closely analogous to the binding of Mg^{2+} to hsc70. Ca^{2+} is directly coordinated by the β and γ phosphates. Ca^{2+} binds within 3.5–5.0 Å but is not coordinated by Asp11, Gln137, or Asp154. These residues correspond to the acidic residues in the active site of hsc70. Ca^{2+} is within water-mediated interaction distance of these three residues and is positioned to form interactions analogous to those between Mg^{2+} and hsc70. Ca^{2+} binding to hsc70 reverses the situation found with the physiologically required Mg^{2+} and Mn^{2+} ions and with Ca^{2+} binding to actin. No direct interactions are formed with ATP, but the Ca^{2+} is coordinated directly by two of the acidic side chains in the active site. Ca^{2+} bound to hsc70 is coordinated in a distorted pentagonal bipyramid by Glu175, Asp199, and five waters (23). The waters bridge to the β and γ phosphates of ATP, Lys71, Asp199, and Asp206. The different modes of Ca^{2+} binding to the two enzymes are one of the few major differences in the structures of their active sites.

Domain Closure

All the well-characterized members of the superfamily undergo a conformational change in solution. Domain closure in hexokinase has been observed by small-angle X-ray scattering (48) and fluorescence spectroscopy (20, 31, 56). The susceptibility of GK to inactivation by sulfhydryl reagents is decreased dramatically in the presence of glycerol (69). Limited proteolysis has been used to demonstrate conformational

Figure 6 Schematic drawing of (*A*) ADP + P_i and (*B*) ATP in the active site of hsc70. Details of solvent structure and metal coordination are shown. (Reprinted from 23 with permission.)

changes in hsc70 (44). Conformational changes in actin have been detected with the use of fluorescence resonance energy transfer (51). Cysteine-reactivity studies (19) show that the conformational change of actin involves a change in the exposure of Cys10 at the nucleotide binding site.

The structure of hexokinase has been determined in both the open and closed forms (2, 3), thus providing the only direct view of large conformational changes in the superfamily (Figure 7). Apparently, no single residue or small set of residues serves as a "hinge" about which the domains rotate (4). The hexokinase conformational change appears to be best described as a combination of shear movements of the cross-over $\alpha3$ helices relative to each other and to the core IA and IIA β sheets (43). Individual secondary structure elements rotate 4–10 degrees and shift up to 1 Å after considering rigid body rotation of whole domains. The twist of the domain IA β sheet apparently is also altered. These conclusions are based on a 3.5-Å structure of glucose-bound hexokinase that was refined with the use of an "X-ray" sequence and should be considered tentative until the analysis is repeated with a glucose complex structure refined at higher resolution (T Steitz, personal communication). The open form of hexokinase A was obtained in the presence of a competitive inhibitor, o-toluoylglucosamine, whose toluoyl moiety blocks domain closure sterically. The open conformation thus crystallized probably represents the general features of the conformational change but is unlikely to be the only open conformation.

Structures of actin (38, 50, 63) and GK are now available in several crystal forms, all in closed conformations (34; HR Faber, JH Hurley, and SJ Remington, unpublished data; M Feese and SJ Remington, personal communication). Among these, the largest conformational difference is found between β-actin bound to profilin (63) relative to the other actin structures. Profilin-bound actin shows most of domain IB rotated by 5 degrees relative to its orientation in α-actin bound to DNaseI. The domain rotation can be largely accounted for by a torsional rotation at Ala 138. Gelsolin-actin (50) is in essentially the same conformation as DNaseI-actin. The nucleotide binding site, however, is similar in all the actin structures. Some specific hinges and a helical "spring" have been predicted in actin on the basis of modeling (70). The hinge model of Tiron & ben-Avraham (70) contrasts with the shear conformational change model of Lesk & Chothia (43). Rather than viewing the two models as contradictory, the hinge model may describe smaller-scale conformational changes accurately, whereas the shear model may describe large-scale domain opening and closing movements.

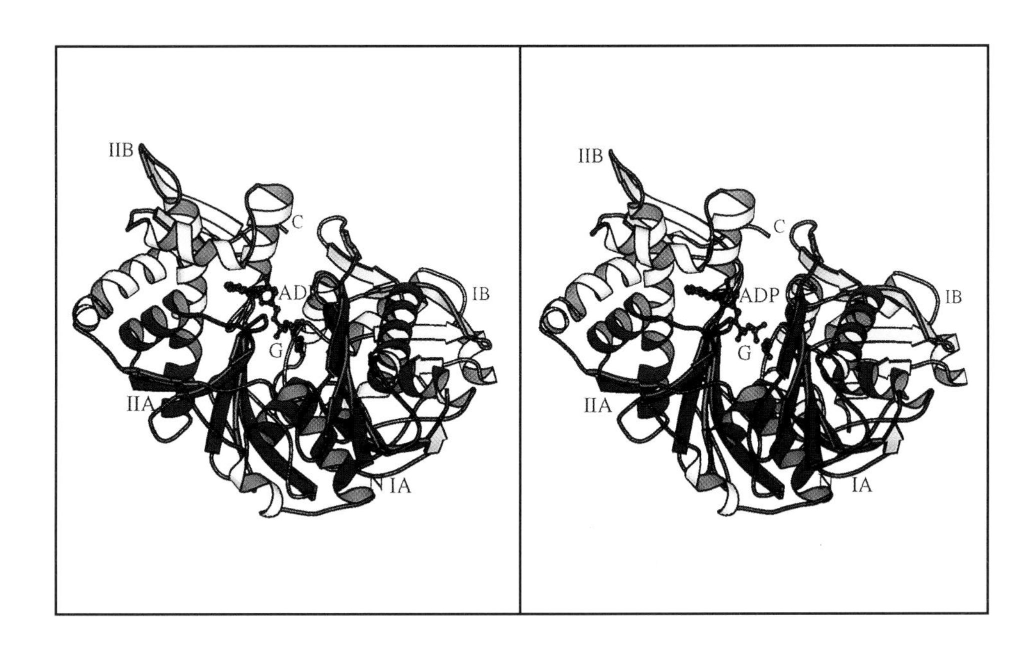

Figure 1 Glycerol kinase monomer with conserved elements of IA and IIA subdomains blue and other regions yellow. Subdomains, bound ADP, glycerol (G), and the N and C termini are labeled. (Generated with Molscript; P Kraulis.)

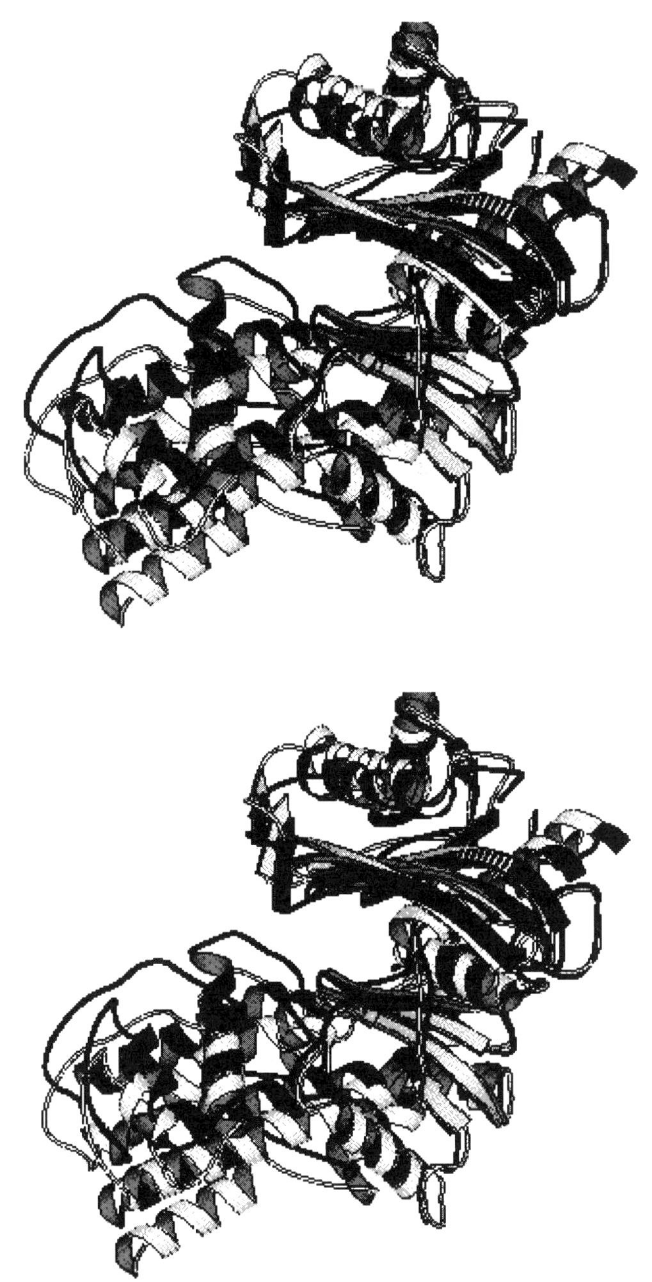

Figure 7 Open (*light*) and closed (*dark*) forms of hexokinase illustrating domain closure.

Sugar-Binding Site

Sugar binding and domain closure are coupled tightly in the sugar kinases. The open form of hexokinase cannot be crystallized in the presence of glucose (2, 3), and the closed form cannot be crystallized in its absence, at least in the known crystal forms. So far, GK has been crystallized only in the closed form in the presence of glycerol. Extended soaking of crystals in glycerol-free solutions fails to remove the bound glycerol (JH Hurley and SJ Remington, unpublished data). On the other hand, nucleotides can be diffused freely into crystals of the closed form of GK (34), thereby resulting in local structural changes only.

GK binds glycerol almost entirely with interactions that involve domain I (Figure 4). Hydrogen bonds are formed with Arg83 and Glu84 from a turn in the major IB β sheet, with Tyr135 in the IB helical bundle and Asp245 at the start of the $\alpha 3$ helix . Asp245 is hydrogen bonded to the phosphate-accepting 3-hydroxyl of glycerol. Chemical modification of GK with the substrate–site directed affinity label dibromoacetone labels Asp245 uniquely (DW Pettigrew, personal communication), which implicates this residue strongly in general base catalysis. Hydrophobic contacts are formed with Trp103 of the IB β-sheet and Phe270 of the IIA β-sheet.

It is remarkable that, even though glycerol binding drives domain closure, the only contact bridging to domain II directly is a single hydrophobic interaction. On its own, formation of a single hydrophobic contact would not appear to be an adequate driving force for domain closure. One possibility is that the structure has a built-in propensity to close, but closure in the absence of sugar would involve an unfavorable loss of hydration of the polar sugar-binding side-chains. In this model, sugar binding would replace the otherwise lost hydrogen bonds; it could be thought of as preventing the destabilization of the closed form rather than driving closure actively. Alternatively or additionally, a conformational change within subdomain IB might mediate sugar-induced domain closure.

Glucose binds at the bottom of the interdomain cleft of hexokinase, but all of the enzyme-substrate hydrogen bonds reported are with domain I (4). As for GK, the sugar does not form strong, direct bridging interactions between the domains. Two regions of hexokinase domain I, Asn164 and Thr175 (translated gene sequence numbering; 24), in the glucose-binding site show the largest deviations from the rigid-body model for domain closure. These structural changes are the only direct evidence in support of an allosteric mechanism that links the domain I sugar-binding site to domain closure.

MECHANISMS OF PHOSPHORYL TRANSFER AND ATP HYDROLYSIS

Sugar-Phosphoryl Transfer

Hexokinase catalyzes the transfer of the γ-phosphate of ATP to the 6-hydroxyl of a variety of furanose and pyranose sugars. The kinetic mechanism is random (14, 60), with a strong preference for binding of the sugar first (73). Phosphoryl transfer occurs with inversion of configuration, consistent with direct in-line transfer from ATP to the sugar without a phosphoenzyme intermediate (5). Transfer occurs by a dissociative mechanism in which the entering and leaving oxygens could maintain approximately fixed positions (36). Dissociation of the phosphorylated sugar product is much slower than chemical phosphotransfer or nucleotide dissociation under normal conditions, thereby making it the rate-limiting step (14, 59). Bound sugar is "sticky," as judged by isotope trapping (59). The pH profile of the reaction is consistent with a carboxylate with an elevated pKa serving as a general base in the forward reaction of the enzyme (72).

In addition to phosphoryl transfer, hexokinase catalyzes the slow hydrolysis (1×10^{-4} of k_{cat} for phosphate transfer) of ATP in the absence of substrate (15, 40). The rate of ATP hydrolysis is stimulated by the pyranoses xylose and lyxose, which lack the phosphorylatable 6-hydroxymethyl group (15). In the absence of sugar, the K_m for ATP with respect to hydrolysis is much higher than with respect to phosphoryl transfer. In the presence of xylose and lyxose, however, the K_m for ATP with respect to hydrolysis decreases to the same value as for the phosphotransferase reaction.

GK catalyzes the transfer of the γ-phosphate of ATP to the 3-hydroxyl of glycerol. Like hexokinase, GK has a random kinetic mechanism (42, 57). Typical kinetic parameters for GK are $k_{cat} = 14\ s^{-1}$ and $K_m = 4.9\ \mu M$ and $K_m = 8.4\ \mu M$ for glycerol and MgATP, respectively (57). GK has a slow substrate-independent MgATPase activity, with $K_m(ATP) = 35\ \mu M$ and V_{max} 6×10^{-6}-fold less than V_{max} for phosphate transfer (57).

Kinetic and structural data for the sugar kinases suggest the following picture: For either the forward or reverse reaction, binding of the sugar substrate induces domain closure, which greatly enhances affinity for the nucleotide and enables catalysis. Efficient phosphoryl transfer is probably the result of three factors. First, the conserved aspartate deprotonates the attacking sugar hydroxyl. Second, the Mg^{2+} ion stabilizes the buildup of negative charge in the pentavalent transition state. Finally, desolvation of the active site by domain closure enhances the

Mg^{2+}-dependent electrostatic stabilization of the transition state. Binding or release of the sugar is rate limiting. Sugar binding is both necessary and sufficient to stabilize the closed form of the enzyme; nucleotide binding is neither. Low-affinity nucleotide binding in the absence of sugar might occur by binding to only one of the two domains with the formation of a subset of the normal interactions. Domain closure is required for catalysis and is induced by sugar binding. The rate-limiting step probably corresponds in structural terms to domain opening coupled to the release of the sugar product.

Adenosine Triphosphate Hydrolysis

In the absence of stimulation, ATP cleavage by the hsc70 ATPase is much slower than by the sugar kinases. The overall rate for basal ATP hydrolysis by stress-70 proteins ranges from 0.0004 to 0.017 s^{-1} (29, 49). Release of P_i is the slow step for the N-terminal ATPase domain of hsc70, and chemical hydrolysis and P_i release are equally slow for intact hsc70 (29). Binding of substrate proteins and the accessory proteins DnaJ and GrpE, however, can stimulate hydrolysis by up to 200-fold (47). Hydrolysis is rate limiting for functional complexes of DnaK with accessory proteins and substrates (47). The kinetics of ATP hydrolysis by actin depend on the divalent cation bound. In the presence of Ca^{2+}, F-actin hydrolyzes ATP slowly, with an overall rate of 0.01 to 0.05 s^{-1} and releases P_i with a rate constant of 0.006 s^{-1} (11, 64). The rate of hydrolysis increases to 13 s^{-1} in the presence of Mg^{2+}. Nucleotides bind to these ATPases orders of magnitude more tightly than to the sugar kinases. hsc70 binds ATP and ADP with affinities of approximately 10^{-8} M (25). In the presence of divalent cations, actin binds ATP with an affinity of roughly 10^{-10} M (11, 41).

All the hsc70 side-chains in the immediate vicinity of the Mg^{2+} ion have been replaced by site-directed mutagenesis (33, 53, 74). Most structure-function studies of hsc70 have focused on the ATPase fragment, which is probably a good model for the basal activity of the intact protein. Replacements of Asp10 and Glu175 are the most deleterious mutations, as they affect both k_{cat} and K_m and reduce k_{cat} and K_m by approximately 500-fold and 1000-fold, respectively. Replacements of Asp199 and Asp206 reduce k_{cat} 50-fold and 10-fold, respectively but do not affect K_m. Replacement of Thr204, which is unique to hsc70, increases K_m 100-fold but increases k_{cat} twofold. The conserved Mg^{2+} ligands Asp10 and Glu175 clearly are very important, but none of the mutations abolish activity completely. The poorest mutant ATPase still hydrolyzes ATP at least three orders of magnitude faster than the rate of uncatalyzed hydrolysis in solution (74). Several explanations for this finding are possible, but one of the most likely is that no protein side-chain serves as a unique base. Glu175 has a position in the three-dimensional structure similar to that of the putative catalytic Asp of the sugar kinases. Glu175 does not form a hydrogen bond with the putative γ-phosphate–attacking water molecule (Figure 6), however. It is unlikely, therefore, that Glu175 deprotonates the attacking water directly. The conserved acidic residues serve primarily to bind, orient, and polarize water molecules, which in turn bind the catalytic metal ion.

The most likely attacking water molecule in hsc70 is bound to Lys71 (Figure 6). The lysine is in a salt bridge with Glu175 and is, therefore, presumably protonated, which makes it an improbable proton acceptor. The nucleophilic water, therefore, is more likely to transfer its proton to a second water molecule. This apparently suboptimal arrangement is not surprising in view of the low activity of the hsc70 ATPase fragment. The similar low ATPase activity of Ca^{2+}-actin and the apparent lack of a unique catalytic base in the Ca^{2+}-actin structure might also be related. The region of Lys71 in hsc70 is not conserved in actin, but the actin methylhistidine loop occupies a similar position in the three-dimensional structure. Gln137 of actin is the closest structural equivalent to the sugar kinase catalytic base, but Gln is a very poor general base. It is not yet clear whether the apparent lack of a unique, favorable catalytic base in the hsc70 ATPase fragment and Ca^{2+}-actin structures is inherent to these proteins, or whether the results are representative only of low activity states.

REGULATORY MECHANISMS

Allosteric Regulation of Sugar Kinases

The best-characterized hexokinase, yeast hexokinase B, is an allosterically regulated enzyme that can exist in solution as a monomer or dimer. Allosteric regulation is linked to the oligomeric state of the enzyme. ATP binding or the complete absence of ligands favors the dimer (65), whereas glucose and ADP binding favor the monomer (16). ATP stabilization of the dimer almost certainly occurs by binding to the dimer at an intersubunit site that is distinct from the catalytic site (1, 66). The other ligands bind only at the catalytic site, which is distant from the subunit interface.

The hexokinase dimer consists of two structurally nonequivalent monomers. The asymmetry of the dimer explains the observed half-of-the-sites reactivity of hexokinase (66). One subunit (''down'') contributes dimer contacts only from its domain II, whereas the other subunit (''up'') contributes contacts from both domains (66). Dimerization might be expected to restrict any conformational changes that would rearrange the two domains of the up, but not the down, subunit. Substrate-product exchange would be inhibited for one subunit but not the other. The less reactive site, therefore, most likely belongs to the up subunit.

Like hexokinase, GK is regulated by multiple allosteric effectors and is capable of subunit dissociation (57, 69). GK is inhibited allosterically by fructose 1,6-bisphosphate (68) and the phosphotransferase system protein factor III^{Glc} (52) and exhibits half-of-the-sites reactivity (57). Regulation of GK activity with respect to normal substrates by its allosteric effectors is purely velocity-modulated under physiologic conditions (57), although this does not hold with poorly reacting alternative substrates (DW Pettigrew, personal communication). GK is in equilibrium between dimeric and tetrameric forms (17, 18). At high enzyme concentrations and in the presence of fructose-1,6-bisphosphate (FBP), the tetrameric form of GK is favored. At low enzyme concentration, the tetramer dissociates into dimers. Only the tetramer is subject to allosteric regulation by FBP. Both the tetramer and dimer manifest negative cooperativity with respect to ATP, and both are inhibited by factor III^{Glc} of the phosphotransferase system.

The extensive GK dimer contact is formed entirely by the IIB subdomain. Tetramer contacts are much less extensive, formed by one α helix each from IA and IIB. Domain opening is constrained by the tetramer contacts but not by the dimer. The factor III^{Glc} binding site is within approximately 6 Å of the dimer interface but does not have direct contacts across the interface. The FBP site is formed at a tetramer interface (SJ Remington, personal communication). Destabilization of the tetramer by mutagenesis decreases both binding and inhibition of GK by FBP (45). Most, if not all, the inhibitory effect of FBP apparently can be accounted for by its stabilization of the tetramer, which is less active than the dimer. Tetramerization has no effect on binding and inhibition by III^{Glc}. Modeling of domain opening as a rigid-body motion with the dimer interface fixed does not suggest any new contacts with III^{Glc}. The exact or approximate symmetry of the GK tetramer structures contrasts with the asymmetric hexokinase B structure and offers no clue to the structural basis for half-of-the-sites reactivity in GK. Asymmetric forms of GK probably exist but have so far evaded crystallization. As the probable rate-limiting step, domain opening and closure is a likely target for regulation. Regulation targeting this step would be consistent with the unusual velocity-modulated regulation of GK. The abrogation of V-type regulation for poor substrates would be consistent with a change in the rate-limiting step, such as chemical hydrolysis instead of product release.

Heat Shock Protein Chaperone Activity

The relationship of ATP hydrolysis by hsp70s and their homologues to function has been investigated intensively for isolated stress-70 proteins and for their complexes with accessory proteins (47, 54, 55, 61, 62, 67). When protein-folding activity of DnaK is assayed in complex with the accessory proteins DnaJ and GrpE, a one-to-one correspondence is found between cycles of protein folding and ATP hydrolysis (47, 67). Given the occurrence of DnaJ and GrpE homologues in eukaryotes (13), this mechanism may well apply in all organisms. The reaction cycle of the DnaK-DnaJ-GrpE complex is as follows: 1. Unfolded substrate protein binds to ATP-bound DnaK. 2. Bound substrate and DnaJ promote ATP hydrolysis, and a stable DnaK-DnaJ-substrate protein complex forms. 3. GrpE causes ADP to dissociate from DnaK. 4. ATP binding to DnaK triggers release of substrate protein. ATP binding to hsp70s is coupled to a conformational change as judged by tryptic digestion (44).

To understand the mechanism of hsp70-mediated folding, the structural link between events at the nucleotide and substrate-protein binding sites must be determined at each step in the reaction cycle. At this writing, there is no three-dimensional structure of an intact hsp70, nor of any complex with accessory proteins. Some fascinating clues have emerged, however, through site-directed mutagenesis of the full-length DnaK (9, 10).

Mutation of the glutamate corresponding to Glu175 of hsc70 to Ala, Leu, or Lys decreases the affinity for nucleotide. This finding is consistent with that for the ATP-binding domain of hsc70 (74) alone. More surprisingly, these mutants actually uncouple ATP hydrolysis from peptide bonding and accelerate the velocity of basal ATP hydrolysis to the rate normally obtained after peptide stimulation (9). This glutamate is salt-bridged to the DnaK equivalent of the active site Lys71 of hsc70. In the hsc70-ATP complex, Lys71 binds the most likely hydrolytic water molecule. Given the observed salt-bridge to Glu175, the Lys presumably is protonated at neutral pH. A protonated Lys is a highly unfavorable ligand for a water required to attack the electrophilic γ-phosphorous. A disruption of the Lys-water interaction caused either by a mutagenesis of its salt-bridge partner or by a conformational change might be expected to enhance the rate of hydrolysis. Because both Lys71 and Glu175 are on domain I, the salt bridge cannot stabilize interdomain interactions directly. The lysine, however, occupies a position within approximately 2 Å of that corresponding to the phosphorylatable hydroxyl of the sugar-kinase substrates. By analogy with the role of sugars in triggering conformational changes in the sugar kinases, Lys71 could be poised to couple ATP hydrolysis to an allosteric change. In view of the suboptimal active site configuration described above, even a small-scale rearrangement of the water-Lys71-Glu175 network might result in substantial activation.

Buchberger et al (10) have also localized the binding site on DnaK for the ATP-exchange factor GrpE to Gly 32 (DnaK numbering), which is in a small insertion in $\beta 3$ of subdomain IA. The loop is found only in the stress-70 proteins. The readily testable proposal is made that Arg34 on this loop is salt-bridged to Glu369 of domain II, thereby stabilizing the closed conformation and decreasing the rate of nucleotide exchange. Binding of GrpE to the $\beta 3$ insertion might disrupt this salt bridge, thereby enhancing nucleotide exchange.

Actin Filament Assembly and Treadmilling

The self-assembly of actin monomers into a filament depends on the concentration of actin, on binding and hydrolysis of ATP (11, 42), and on regulatory proteins (71). The concentration above which monomers add to an end of a filament is referred to as the critical concentration. At steady state, the concentration of G-actin drops to the critical concentration. The filament is asymmetric, and the distinct ends of the filament are referred to as barbed and pointed. The critical concentration for monomer addition to the pointed end, which is enriched in ADP-actin, is higher than for addition to the barbed end, which is enriched in ATP- and ADP-P_i-actin. The mechanical stiffness of ADP-actin filaments is also less than that of ATP-actin filaments (35). The difference in the monomer-binding affinity at each end leads to simultaneous filament assembly at the barbed end and disassembly at the pointed end. This process is referred to as treadmilling.

The key mechanistic issues for control of filament assembly involve understanding the following: how differences in binding of ATP and ADP control the equilibrium between F-actin and G-actin; how ADP-P_i-actin stimulates hydrolysis on neighboring ATP-F-actin subunits; and how actin-binding regulatory proteins modulate these effects. These questions are not easy to address in structural terms because the critical biological effects are produced by small energetic differences. The critical concentration of ATP-actin for addition to the barbed and pointed ends differs by only a factor of 10–30 (11), which corresponds to a free energy difference no more than 2 kcal/mol. This difference corresponds to the interaction energy contributed by approximately two hydrogen bonds, or the burial of one or two large hydrophobic side-chains. The effects of ATP hydrolysis and P_i release on structure, therefore, need not be dramatic.

Understanding the relationship among mechanism, conformation, and ATP hydrolysis is complicated by the absence of an overall conformational change between the known structures of ADP-actin and ATP-actin. Localized changes occur at the nucleotide binding site (Figure 5), but these changes are not communicated to the filament or regulatory protein contact sites in the available crystal structures. The energetics of such a change may be so small, however, that the overall structure is held in place by crystal contacts. Apart from the actin-DNaseI contact itself, the nature of the crystal contacts is distinct from filament contacts, so effects of nucleotide binding on the energetics of the two sets of contacts need not be parallel. It is unappealing to invoke a conformational change that has not been observed directly, yet it is difficult to conceive of another model that would account for all of the effects of ATP hydrolysis on the actin filament.

An atomic model of the actin filament has been derived by fitting the crystal structure of DNaseI-actin to X-ray fiber diffraction data from the filament (8, 32, 46). With the exception of a rotation of subdomain IB, the structure of F-actin monomers is very close to that of DNaseI-actin. Indeed, several interactions in the filament, involving both domains, are mimicked in the actin-DNaseI complex, and DNaseI also inhibits nucleotide exchange by actin. The filament model shows extensive contacts involving both domains. The loop 264–273 of subdomain IIB forms a hydrophobic plug in a three-body interaction with the IB and IIA subdomains of two other monomers. The filament model shows that the relative orientation of the two domains is highly restricted in the filament. The tight restrictions placed on the relative orientations of all four subdomains explain the slow release of phosphate and the absence of ADP-ATP exchange on the filament.

In contrast to DNaseI, profilin increases the rate of nucleotide exchange when it binds actin (28). The profilin binding site spans subdomains IA and IIA, but profilin bridges the two below the cleft rather than across it (63). The exchange rate increase may be related to the 5-degree difference in interdomain angle in the profilin-actin structure compared with the other actin structures, or with some other as-yet unidentified coupling mechanism. Schutt et al (63) argue against the first possibility, because the nucleotide-binding sites in the different actin structures are essentially identical. These authors note that the 5-degree rotation either could have been an inherent feature of the β-actin structure or could have been induced by the ribbon-like packing in the profilin-actin crystal.

CONCLUDING REMARKS

In summary, the sugar kinases, stress-70 proteins and actins share a conserved two-domain fold and a nucleotide binding site that bridges both domains. There is clear primary sequence homology in a few localized regions, and these proteins almost certainly originate from a common ancestor. The catalytically essential metal ion binds to the ATP phosphates directly and to the enzyme by water-mediated interactions. After ATP hydrolysis, local conformational changes occur and interdomain bridging interactions are weakened.

The sugar kinases have a critical general base provided by a conserved Asp residue. In contrast, the ATPases appear to have no precise structural equivalent to this Asp, nor do they have any other uniquely required catalytic base. Product dissociation is rate limiting for the catalytically efficient sugar kinases. Either chemical hydrolysis or product dissociation may be rate limiting for the ATPases, depending on the regulatory and conformational state of the protein. The kinetics of basal ATP hydrolysis and phosphorylation of poor alternative substrates by sugar kinases more closely resembles the ATPases, which suggests an underlying mechanistic similarity.

All these proteins undergo a large conformational change involving domain closure. The conformational change involves shear movements within domains as well as movement of the two relative to each other. In the sugar kinases, sugar binding drives the conformational change, although not by bridging the two domains directly with strong interactions. The sugar kinases bind nucleotides weakly and do not require them for domain closure. In contrast, the actin and hsc70 ATPases bind nucleotides much more strongly. Nucleotide binding and hydrolysis are probably coupled to ATPase domain closure through changes in electrostatic and hydrogen-bonding interactions, which bridge the two domains.

The key regulatory responses of these proteins appear to be linked to their two-domain structure and domain closing conformational change by several different mechanisms. The hexokinase dimer, GK tetramer, actin filament, and DNaseI-actin complex all directly interfere, or potentially interfere, with domain opening, thereby limiting product dissociation. Peptide and DnaJ or DnaJ homologue binding to hsp70s and the ATP/ADP–actin boundary in actin filaments both appear to stimulate the velocity of hydrolysis allosterically. Regulation of DnaK and actin by GrpE and profilin, respectively, occurs by allosteric promotion of nucleotide exchange, probably by weakening domain-bridging interactions and promoting domain opening. Other regulatory mechanisms, including the mechanism of III^{Glc} inhibition of GK, remain to be explained. Given the complexity and variety of regulatory interactions in the superfamily, even the better understood cases should offer some interesting surprises in the coming years.

ACKNOWLEDGMENTS

I thank SJ Remington, DW Pettigrew, M Feese, and T Steitz for discussions and communicating unpublished results.

Literature Cited

1. Anderson CM, Steitz TA. 1975. Structure of yeast hexokinase IV. Low-resolution structure of enzyme-substrate complexes revealing negative co-operativity and allosteric interactions. *J. Mol. Biol.* 92:279–87
2. Anderson CM, Stenkamp RE, McDonald RC, Steitz TA. 1978. A refined model of the sugar binding site of yeast hexokinase B. *J. Mol. Biol.* 123: 207–19
3. Bennett WS, Steitz TA. 1980. Structure of a complex between yeast hexokinase A and glucose. I. Structure determination and refinement at 3.5 Å resolution. *J. Mol. Biol.* 140:183–209
4. Bennett WS, Steitz TA. 1980. Structure of a complex between yeast hexokinase A and glucose. II. Detailed comparisons of conformation and active site configuration with the native hexokinase B monomer and dimer. *J. Mol. Biol.* 140:211–30
5. Blattler WA, Knowles JR. 1979. Stereochemical course of phosphokinases. The use of adenosine [γ-(S)–$^{16}O,^{17}O,^{18}O$]triphosphate and the mechanistic consequences for the reactions catalyzed by glycerol kinase, hexokinase, pyruvate kinase, and acetate kinase. *Biochemistry* 18:3927–33
6. Bork P, Sander C, Valencia A. 1992. An ATPase domain common to prokaryotic cell cycle proteins, sugar kinases, actin, and hsp70 heat shock proteins. *Proc. Natl. Acad. Sci. USA* 89: 7290–94
7. Bork P, Sander C, Valencia A. 1993. Convergent evolution of similar enzymatic function on different protein folds: the hexokinase, ribokinase, and galactokinase families of sugar kinases. *Protein Sci.* 2:31–40
8. Bremer A, Aebi U. 1992. The structure of the F-actin filament and the actin molecule. *Curr. Opin. Cell Biol.* 4: 20–26
9. Buchberger A, Valencia A, McMacken R, Sander C, Bernd B. 1994. The chaperone function of DnaK requires the coupling of ATPase activity with substrate binding through residue E171. *EMBO J.* 13:1687–95
10. Buchberger A, Valencia A, McMacken R, Sander C, Bernd B. 1994. A conserved loop in the ATPase domain of the DnaK chaperone is essential for stable binding of GrpE. *Nat. Struct. Biol.* 1:95–101
11. Carlier M-F. Actin: protein structure and filament dynamics. 1991. *J. Biol. Chem.* 266:1–4
12. Craig EA, Weissman JS, Horwich AL. 1994. Heat shock proteins and molecular chaperones: mediators of protein conformation and turnover in the cell. *Cell* 78:365–72
13. Cyr DM, Lu X, Douglas MG. 1992. Regulation of Hsp70 function by a eukaryotic DnaJ homolog. *J. Biol. Chem.* 267:20927–31
14. Dannenburg KD, Cleland WW. 1975. Use of chromium-adenosine triphosphate and lyxose to elucidate the kinetic mechanism and coordination state of the nucleotide substrate for yeast hexokinase. *Biochemistry* 14: 28–39
15. DelaFuente G, Lagunas R, Sols A. 1970. Induced fit in yeast hexokinase. *Eur. J. Biochem.* 16:226–33

16. Derechin M, Rustum YM, Barnard EA. 1972. Dissociation of yeast hexokinase under the influence of substrates. *Biochemistry* 11:1793–97
17. deRiel JK, Paulus H. 1978. Subunit dissociation in the allosteric regulation of glycerol kinase from *Escherichia coli*. 1. Kinetic evidence. *Biochemistry* 17:5134–40
18. deRiel JK, Paulus H. 1978. Subunit dissociation in the allosteric regulation of glycerol kinase from *Escherichia coli*. 2. Physical evidence. *Biochemistry* 17:5141–45
19. Drewes G, Faulstich H. 1991. A reversible conformational transition in muscle actin is caused by nucleotide exchange and uncovers cysteine in position 10. *J. Biol. Chem.* 266:5508–13
20. Feldman I, Kramp DC. 1978. Fluorescence-quenching study of glucose binding by yeast hexokinase isoenzymes. *Biochemistry* 17:1541–47
21. Flaherty KM, DeLuca-Flaherty C, McKay DB. 1990. Three-dimensional structure of the ATPase fragment of a 70K heat-shock cognate protein. *Nature* 346:623–28
22. Flaherty KM, McKay DB, Kabsch W, Holmes KC. 1991. Similarity of the three-dimensional structures of actin and the ATPase fragment of a 70-kDa heat shock cognate protein. *Proc. Natl. Acad. Sci. USA* 88:5041–45
23. Flaherty KM, Wilbanks SM, DeLuca-Flaherty C, McKay DB. 1994. Structural basis of the 70-kilodalton heat shock cognate protein ATP hydrolytic activity. II. Structure of the active site with ADP or ATP bound to wild type and mutant ATPase fragment. *J. Biol. Chem.* 269:12899–907
24. Frohlick K-U, Entian KD, Mecke D. 1985. The primary structure of the yeast hexokinase PII gene (HXK2) which is responsible for glucose repression. *Gene* 36:105–11
25. Gao B, Greene L, Eisenberg E. 1994. Characterization of nucleotide-free uncoating ATPase and its binding to ATP, ADP, and ATP analogues. *Biochemistry* 33:2048–54
26. Georgopoulos C. 1992. The emergence of the chaperone machines. *Trends Biochem. Sci.* 17:295–99
27. Gething M-J, Sambrook J. 1992. Protein folding in the cell. *Nature* 355: 33–45
28. Goldschmidt-Clermont PJ, Machesky LM, Doberstein SK, Pollard TD. 1991. Mechanism of the interaction of human platelet profilin with actin. *J. Cell. Biol.* 113:1081–89
29. Ha J-H, McKay DB. 1994. ATPase kinetics of recombinant bovine 70 kDa heat shock cognate protein and its amino-terminal ATPase domain. *Biochemistry* 33:14625–35
30. Hendrick JP, Hartl FU. 1993. Molecular chaperone functions of heat-shock proteins. *Annu. Rev. Biochem.* 62: 349–84
31. Hoggett JG, Kellett GL. 1976. Yeast hexokinase: substrate-induced association-dissociation reactions in the binding of glucose to hexokinase P-II. *Eur. J. Biochem.* 66:65–77
32. Holmes KC, Popp D, Gebhard W, Kabsch W. 1990. Atomic model of the actin filament. *Nature* 347:44–49
33. Huang S-P, Tsai M-Y, Tzou Y-M, Wu W-G, Wang C. 1993. Aspartyl residue 10 is essential for ATPase activity of rat hsc70. *J. Biol. Chem.* 268:2063–68
34. Hurley JH, Faber HR, Worthylake D, Meadow ND, Roseman S, et al. 1993. Structure of the regulatory complex of *Escherichia coli* III^{Glc} with glycerol kinase. *Science* 259:673–77
35. Janmey PA, Hvidt S, Oster GF, Lamb J, Stossel TP, et al. 1990. Effect of ATP on actin filament stiffness. *Nature* 347:95–99
36. Jones JP, Weiss PM, Cleland WW. 1991. Secondary ^{18}O isotope effects for hexokinase-catalyzed phosphoryl transfer from ATP. *Biochemistry* 30: 3634–39
37. Kabsch W, Holmes KC. 1995. The actin fold. *FASEB J.* 9:167–74
38. Kabsch W, Mannherz HG, Suck D, Pai EF, Holmes KC. 1990. Atomic structure of the actin: DNase I complex. *Nature* 347:37–44
39. Kabsch W, Vandekerckhove J. 1992. Structure and function of actin. *Annu. Rev. Biophys. Biomol. Struct.* 21: 49–76
40. Kaji A, Colowick SP. 1965. Adenosine triphosphatase activity of yeast hexokinase and its relation to the mechanism of the hexokinase reaction. *J. Biol. Chem.* 240:4454–62
41. Korn ED, Carlier M-F, Pantaloni D. 1987. Actin polymerization and ATP hydrolysis. *Science* 238:638–44
42. Knight WB, Cleland WW. 1989. Thiol and amino analogues as alternate substrates for glycerokinase from *Candida mycoderma*. *Biochemistry* 28: 5728–34
43. Lesk AM, Chothia C. 1984. Mechanisms of domain closure in proteins. *J. Mol. Biol.* 174:175–91
44. Liberbek K, Skowyra D, Zylicz M, Johnson C, Georgopolous C. 1991.

The *Escherichia coli* DnaK chaperone, the 70-kDa heat shock protein eukaryotic equivalent, changes conformation upon ATP hydrolysis, thus triggering its dissociation from a bound target protein. *J. Biol. Chem.* 266:14491–96
45. Liu WZ, Faber HR, Feese M, Remington SJ, Pettigrew DW. 1994. *Escherichia coli* glycerol kinase: role of a tetramer interface in regulation by fructose 1,6-bisphosphate and phosphotransferase system regulatory protein III^{Glc}. *Biochemistry* 33:10120–26
46. Lorenz M, Popp D, Holmes KC. 1993. Refinement of the F-actin model against x-ray fiber diffraction data by the use of a directed mutation algorithm. *J. Mol. Biol.* 234:826–36
47. McCarty JS, Buchberger A, Reinstein J, Bukau B. 1995. The role of ATP in the functional cycle of the DnaK chaperone system. *J. Mol. Biol.* 249: 126–37
48. McDonald RC, Steitz TA, Engelman DM. 1979. Yeast hexokinase in solution exhibits a large conformational change upon binding glucose or glucose 6-phosphate. *Biochemistry* 18: 338–42
49. McKay DB. 1993. Structure and mechanism of 70-kDa heat-shock-related proteins. *Adv. Protein Chem.* 44: 67–98
50. McLaughlin PJ, Gooch JT, Mannherz HG, Weeds HG. 1993. Structure of gelsolin segment 1-actin complex and the mechanism of filament severing. *Nature* 364:685–92
51. Miki M, Kouyama T. 1994. Domain motion in actin observed by fluorescence resonance energy transfer. *Biochemistry* 33:10171–77
52. Novotny MJ, Frederickson WL, Waygood EB, Saier MH. 1985. Allosteric regulation of glycerol kinase by enzyme III^{Glc} of the phosphotransferase system in *Escherichia coli* and *Salmonella typhimurium*. *J. Bacteriol.* 162: 810–16
53. O'Brien MC, McKay DB. 1993. Threonine 204 of the chaperone protein Hsc70 influences the structure of the active site, but is not essential for ATP hydrolysis. *J. Biol. Chem.* 268: 24323–29
54. Palleros DR, Reid KL, Shi L, Welch WJ, Fink AL. 1993. ATP-induced protein-Hsp70 complex dissociation requires K^+ but not ATP hydrolysis. *Nature* 365:664–66
55. Palleros DR, Welch WJ, Fink AL. 1991. Interaction of hsp70 with unfolded proteins: effects of temperature and nucleotides on the kinetics of binding. *Proc. Natl. Acad. Sci. USA* 88:5719–23
56. Peters BA, Neet KE. 1978. Yeast hexokinase PII. Conformational changes induced by substrates and substrate analogues. *J. Biol. Chem.* 253:6826–31
57. Pettigrew DW, Yu GY, Liu Y. 1990. Nucleotide regulation of *Escherichia coli* glycerol kinase: initial-velocity and substrate binding studies. *Biochemistry* 29:8620–27
58. Pollard TD. 1990. Actin. *Curr. Opin. Cell Biol.* 2:33–40
59. Rose IA, O'Connell EL, Litwin S, Bar Tana J. 1974. Determination of the rate of hexokinase-glucose dissociation by the isotope-trapping method. *J. Biol. Chem.* 249:5163–68
60. Rudolph FB, Fromm HJ. 1971. Computer simulation studies with yeast hexokinase and additional evidence for the random bi bi mechanism. *J. Biol. Chem.* 246:6611–19
61. Sadis S, Hightower LE. 1992. Unfolded proteins stimulate molecular chaperone Hsc70 ATPase by accelerating ADP/ATP exchange. *Biochemistry* 31:9407–12
62. Schmid D, Baici A, Gehring H, Christen P. 1994. Kinetics of molecular chaperone action. *Science* 263:971–73
63. Schutt CE, Myslik JC, Rozycki MD, Goonesekere NCW, Lindberg U. 1993. The structure of crystalline profilin-β-actin. *Nature* 365:810–16
64. Sheterline P, Sparrow JC. 1994. Actin. *Protein Profile* 1:1–100
65. Shill JP, Neet KE. 1975. Allosteric properties and the slow transition of yeast hexokinase. *J. Biol. Chem.* 250: 2259–68
66. Steitz TA, Fletterick RJ, Anderson WF, Anderson CM. 1976. High resolution x-ray structure of yeast hexokinase, an allosteric protein exhibiting a non-symmetric arrangement of subunits. *J. Mol. Biol.* 104:197–222
67. Szabo A, Langer T, Schroder H, Flanagan J, Bukau B, et al. 1994. The ATP hydrolysis-dependent reaction cycle of the *Escherichia coli* Hsp70 system-DnaK, DnaJ, and GrpE. *Proc. Natl. Acad. Sci. USA* 91:10345–49
68. Thorner JW, Paulus H. 1971. Composition and subunit structure of glycerol kinase from *Escherichia coli*. *J. Biol. Chem.* 246:3885–94
69. Thorner JW, Paulus H. 1973. Catalytic and allosteric properties of glycerol kinase from *Escherichia coli*. *J. Biol. Chem.* 248:3922–32

70. Tirion MM, ben-Avraham D. 1993. Normal mode analysis of G-actin. *J. Mol. Biol.* 230:186–95
71. Vanderkerckove J. 1990. Actin-binding proteins. *Curr. Opin. Cell Biol.* 2: 41–50
72. Viola RE, Cleland WW. 1978. Use of pH studies to elucidate the chemical mechanism of yeast hexokinase. *Biochemistry* 17:4111–17
73. Viola RE, Raushel FM, Rendina AR, Cleland WW. 1982. Substrate synergism and the kinetic mechanism of yeast hexokinase. *Biochemistry* 21: 1295–1302
74. Wilbanks SM, DeLuca-Flaherty C, McKay DB. 1994. Structural basis of the 70-kilodalton heat shock cognate protein ATP hydrolytic activity. I. Kinetic analyses of active site mutants. *J. Biol. Chem.* 269:12893–98
75. Wilbanks SM, McKay DB. 1995. How potassium affects the activity of the molecular chaperone Hsc70. II. Potassium binds specifically in the ATPase active site. *J. Biol. Chem.* 270: 2251–57

Annu. Rev. Biophys. Biomol. Struct. 1996. 25:163–95

USE OF ^{19}F NMR TO PROBE PROTEIN STRUCTURE AND CONFORMATIONAL CHANGES

Mark A. Danielson and Joseph J. Falke

Department of Chemistry and Biochemistry, University of Colorado, Boulder, Colorado 80309–0215

KEY WORDS: fluorine, paramagnetic broadening, spin-label, dynamics, X-ray crystallography

ABSTRACT

^{19}F NMR has proven to be a powerful technique in the study of protein structure and dynamics because the ^{19}F nucleus is easily incorporated at specific labeling sites, where it provides a relatively nonperturbing yet sensitive probe with no background signals. Recent applications of ^{19}F NMR in mapping out structural and functional features of proteins, including the galactose-binding protein, the transmembrane aspartate receptor, the CheY protein, dihydrofolate reductase, elongation factor-Tu, and D-lactose dehydrogenase, illustrate the utility of ^{19}F NMR in the analysis of protein conformational states even in molecules too large or unstable for full NMR structure determination. These studies rely on the fact that the chemical shift of ^{19}F is extremely sensitive to changes in the local conformational environment, including van der Waals packing interactions and local electrostatic fields. Additional information is provided by solvent-induced isotope shifts or line broadening of the ^{19}F resonance by aqueous and membrane-bound paramagnetic probes, which may reveal the proximity of a ^{19}F label to bulk solvent or a biological membrane. Finally, the effect of exchanging conformations on the ^{19}F resonance can directly determine the kinetic parameters of the conformational transition.

1056–8700/96/0610–0163$08.00

CONTENTS

PERSPECTIVES AND OVERVIEW

Since the initial application of solution ^{19}F NMR to protein systems in the late 1960s (80), the technique has evolved to become a versatile tool in the study of protein structure and dynamics and, thus, has provided an approach that complements established structure determination techniques. When high-resolution structural methods cannot be used, ^{19}F NMR generates valuable low-resolution structural information. Perhaps the most powerful application of ^{19}F NMR, however, is to probe proteins of known structure in their native solution environment, where the extreme sensitivity of the ^{19}F resonance to its molecular surroundings can reveal important structural and kinetic features of protein conformational changes.

The rapidly expanding number of high-resolution protein crystal structures has emphasized the need for alternative approaches that can be used to monitor specific locations within a known structural framework in solution. For smaller proteins ($M_r < 30$ kDa), whose solution structures can be probed effectively by multidimensional NMR, such techniques clearly provide the maximum obtainable information and represent the method of choice to complement crystallographic data. Many proteins of known crystallographic structure, however, are too large for multidimensional NMR but still fall within the range of molecular weights accessible to solution ^{19}F NMR ($M_r < 100$ kDa). For such a protein, one or more fluorine labels can be introduced, either biosynthetically or chemically, into specific sites, thereby enabling the use of ^{19}F NMR to map out the regions of the molecule involved in a conformational change, characterize the kinetics of the event, and, in some cases, measure the intramolecular distance changes that result.

In this review, we use recent ^{19}F NMR studies to illustrate the analysis of both proteins of known crystal or NMR structure and proteins for which no high-resolution structural information exists. We focus primarily on studies published during 1993–1995, in which the fluorine probes are introduced covalently into the protein itself. For a comprehensive summary of earlier applications, including theoretical aspects and studies that introduce fluorine into a small molecule or peptide ligand, see the excellent reviews by Gerig (30, 31), de Dios et al (21), Ho et al (37), and Sykes & Hull (85).

BACKGROUND AND METHODS

Useful Properties of the ^{19}F Nucleus

Several factors contribute to the power of ^{19}F NMR (20, 30, 31, 51):

1. The spin ½ ^{19}F nucleus occurs at 100% natural abundance and has 83% the sensitivity of ^{1}H.
2. ^{19}F does not occur naturally in proteins; thus, there are no background signals with which to contend.
3. Fluorine incorporation is generally nonperturbing, particularly when substituted for hydrogen in an amino acid sidechain (see below).
4. Although the large anisotropy of the chemical shift tensor leads to broader linewidths at high field strengths, the ^{19}F chemical shift range is 100-fold larger than that of ^{1}H. This resolution, coupled with the high detection sensitivity and absence of background signals, generally yields well-resolved ^{19}F resonances in one-dimensional spectra.
5. One-dimensional ^{19}F NMR studies generally require lower protein concentrations and shorter spectral acquisition times than do multidimensional NMR techniques.
6. The ^{19}F chemical shift is controlled primarily by the fluorine lone-pair electrons, which provide a large paramagnetic term in the shielding formula. The chemical shift, therefore, is exquisitely sensitive to changes in the local van der Waals environment, as well as to local electrostatic fields.
7. The exposure of specific fluorine labels to paramagnetic centers, such as a bound or aqueous metal ion, a spin-labeled analogue of a ligand, or a spin-labeled lipid probe, can be easily detected.

Incorporation of ^{19}F Labels

A wide variety of fluorine labels, including fluorinated amino acids, fluorinated reagents that react covalently with specific sidechains in proteins, and fluorinated ligands, are available. The most commonly used fluorinated amino acids are analogues of the aromatic amino acids, all of which are available commercially as mixtures of the D and L enantiomers (30). These amino acids include the ortho, meta, and para derivatives of phenylalanine (2-, 3-, and 4-F-Phe), the meta derivative of tyrosine (3-F-Tyr), and tryptophan fluorinated at specific indole ring positions (4-, 5-, and 6-F-Trp).

Several methods exist for incorporating fluorinated amino acids into a selected protein. Although chemical synthesis is impractical for any protein of appreciable size, several proteins have been labeled using a "semi-synthetic" approach (11, 46, 88). In this method, a peptide that contains a fluorinated amino acid is synthesized and combined with the remainder of the protein that has been produced biosynthetically, thereby producing an active, labeled protein. Recent advances in protein splicing techniques may soon revive this approach, which has the advantage of labeling a unique region of the protein (98).

More commonly, fluorinated amino acids are incorporated biosynthetically by microbial protein expression in the presence of the desired amino acid analogue. A fluorinated aromatic amino acid can be incorporated into a bacterial expression system easily, for example, by including it in the growth medium and eliminating the ability of the cell to synthesize that amino acid endogenously. The latter is accomplished either by using a bacterial strain auxotrophic for the amino acid of interest (85) or by adding glyphosate, which inhibits the synthesis pathways of all three aromatic amino acids (44). Depending on the application, the level of incorporation can be adjusted from less than 5% (22) to greater than 90% (50). Similarly, biosynthetic incorporation can be carried out by using a yeast expression system (96) or by including the amino acid analogue in the diet of such vertebrates as rabbit (95), avians (51), or primates (28).

When biosynthetic incorporation of fluorinated amino acids is impractical, or when alternative probe locations are desired, a fluorinated modification reagent can be used to label specific protein sidechains covalently. The most popular of these reagents are the cysteine-reactive probes, which target the unique chemistry of the cysteine thiol. Conveniently, the low frequency of cysteine usage in natural protein sequences generally guarantees a manageably small number of labeling sites or enables the engineering of labeling sites into important, selected locations in the protein structure. Other amino acids that may be labeled with fluorinated reagents include lysine (as well as the N-terminus), serine, tyrosine, and histidine. These reagents have been reviewed recently (30).

Finally, a wide variety fluorinated analogues can be used to substitute for natural ligands, substrates, and intermediates. For example, several fluorinated compounds have been found to be effective enzyme inhibitors (reviewed in 1). Other sources provide further information regarding these small molecules, as well as nucleic acid applications (16–18, 23, 36, 43, 58, 73–75, 81, 86).

Effect of ^{19}F Labels on Protein Structure and Activity

Fluorine incorporation, particularly at a ring position of an aromatic sidechain, generally has little effect on the structure and function of a protein. *Escherichia coli*, for example, has been labeled with 5-F-Trp to a level of 80% at every tryptophan position within the cell without affecting growth seriously (44, 54), whereas higher order animals can tolerate at least 25% incorporation of 4-F-Phe (95). Similarly, at least 15 proteins have been shown to be structurally or functionally unperturbed by incorporation of fluorine (2, 11, 19, 22, 29, 33, 42, 49, 53, 54, 64, 71, 72, 77, 87, 91, 97). In contrast, detectable perturbations have been observed for a pentafluoro analogue of phenylalanine (8), as well as for monofluoro derivatives of tyrosine (47) and histidine (89), in which the ring fluorine alters the sidechain pK_A substantially. Such perturbations, however, appear to be the exception. More caution must be used when fluorine is introduced by chemical modification, because the relatively large size of the modification reagent itself can generate a significant structural perturbation.

Several factors likely contribute to the surprisingly nonperturbing nature of fluorine when it is substituted for hydrogen within amino acid sidechains. The fluorine atom is similar in size to hydrogen [covalent radii of 1.35 Å and 1.2 Å, respectively (68)], such that fluorine substitution has little steric effect. Incorporation of a single fluorine atom into phenylalanine, for example, increases the volume of the sidechain only 0.7%. Moreover, the aliphatic C-F bond is only moderately polar, such that the fluorine atom is a weak hydrogen bond acceptor, at best (66). Aromatic C-F bonds, in which the fluorine electron density is significantly reduced by backbonding to the ring (68), are even less polar and, therefore, are correspondingly less able to accept a hydrogen bond. In fact, no case of hydrogen bond formation that involves aromatic fluorine has been documented. Finally, extensive mutagenic studies have shown that protein structures are relatively plastic and can accommodate a variety of amino acid substitutions, except for those targeted to the active site or locations critical for folding. Substitution of fluorine into a sidechain for hydrogen represents perhaps the most subtle type of "single-atom mutagenesis" possible.

To draw conclusions regarding the native protein from ^{19}F NMR studies, however, it remains essential to establish that the fluorine label is nonperturbing. Two classes of approach have been used to test for perturbations in different types of applications. For those cases in which the level of fluorine incorporation is high enough, the effect of fluorine on protein structure and activity has been assayed directly (2, 11, 29, 33, 42, 49, 53, 54, 64, 71, 72, 77, 87, 91, 97). When the level of incorporation is low, however, an alternative mutagenic approach has been used (19, 22). The latter approach substitutes a natural amino acid, rather than a fluorinated analogue, at a position to be probed by fluorine incorporation. For example, tyrosine has been substituted at phenylalanine positions targeted for labeling with 4-F-Phe, thereby enabling the effect of substituents at the para position of specific phenylalanine residues to be measured directly (19, 22). In this example, the substitution of tyrosine is expected to overestimate any perturbation caused by the incorporation of 4-F-Phe, because the hydroxyl group is considerably larger, more polar, and better able to form hydrogen bonds than is fluorine.

Assignment of Resonances

Several methods have been used to assign ^{19}F NMR resonances. The most rigorous assignments have used site-directed mutagenesis either to replace or to "nudge" each specific fluorine label. In the first method, the residue targeted for assignment is substituted with the most similar sidechain available, thereby causing the corresponding resonance in the ^{19}F NMR spectrum to disappear while the remaining resonances are unperturbed (19, 22, 39, 50, 54, 76, 77). The direct replacement method fails occasionally, because each attempted substitution has perturbed the nontarget resonances. In proteins of known structure, the nudge method can then be used (19, 22). This approach identifies a sidechain in van der Waals contact with the target residue for conservative replacement, thereby generating a local nudge. In favorable cases, this nudge is not transmitted to the remaining resonances. Figure 1 illustrates both types of assignments in the 4-F-Phe-labeled CheY protein (22). A nonmutagenic, but similar, method of assignment has been illustrated in the egg white lysozyme system, for which ^{19}F NMR spectra were obtained with the use of nonidentical lysozymes from several avian species (51).

Additional information, including solvent exposure (see below), perturbation of a resonance by a small molecule ligand or a substrate that binds to a specific site, or chemical modification of an amino acid near a fluorinated residue, can be used to confirm or tentatively assign

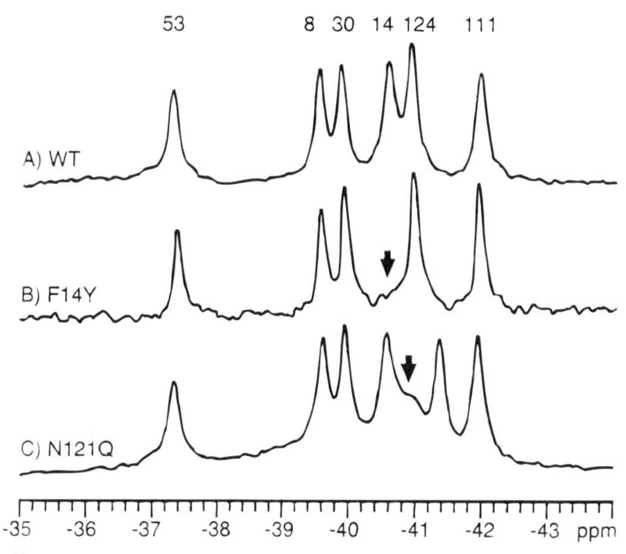

Figure 1 ^{19}F NMR spectra illustrating the two protein engineering approaches used to assign resonances of the phospho-signaling protein CheY (22). (*A*) The protein is labeled at its six phenylalanine positions with 4-F-Phe, thereby giving rise to a well-resolved ^{19}F NMR spectrum at 470 MHz. (*B*) The assignment of the Phe14 resonance by direct replacement with tyrosine. (*C*) The nudge assignment of Phe124 by replacement of the adjacent Asn121. The same two methods have also assigned the remaining resonances, as indicated.

resonances. These approaches can be complicated, however, by the ability of ^{19}F NMR to detect the local structural and electrostatic effects triggered by a long-range conformational change that originates in a different region of the protein.

Detection of Solvent Exposure

Solvent exposure can be detected either by a solvent-induced isotopic shift (SIIS) or by paramagnetic broadening that stems from a phase-specific probe. The SIIS effect generates up to a 0.25-ppm chemical shift of a ^{19}F resonance exposed to aqueous solution when the solvent is changed from H_2O to D_2O (30, 42). In contrast, a ^{19}F resonance that emanates from the protein interior does not experience solvation and the accompanying chemical shift change, unless the residue interacts with buried water molecules. The alternative method uses an aqueous or hydrophobic paramagnetic probe to broaden the resonances of water-exposed and lipid-proximal fluorine labels, respectively. The aqueous reagent Gd^{3+} $EDTA^{4-}$, in which the metal ion possesses seven unpaired electrons, and the membrane probe 8-doxylpalmitic acid have proved useful in many studies (19, 20, 70, 77, 84, 90). Finally, a ^{19}F NMR method has been developed recently to detect bound water molecules within a macromolecule (16).

Interpretation of Chemical Shifts

The large covalent chemical shift range of ^{19}F, which approaches 1000 ppm, stems from the paramagnetic shielding generated by the fluorine lone-pair electrons. This paramagnetic term of the shielding equation also causes the chemical shift to be highly sensitive to the noncovalent tertiary interactions in a folded protein structure. In an unfolded protein that is labeled homogeneously with a given fluorine probe, the resonances from multiple probe positions collapse to give a single peak, because of the equivalent solvent environment experienced by each probe position. In a folded protein, however, the local packing and electrostatic fields perturb the lone-pair electrons of each probe differently, thereby providing a noncovalent chemical shift range as large as 17 ppm (69).

Several theoretical descriptions have been offered to explain observed range of environmental ^{19}F chemical shifts in proteins. Millet & Raftery (61) divided fluorine shielding into contributions from (*a*) local magnetic fields that arise from electronically anisotropic groups, (*b*) hydrogen bond formation, (*c*) electrostatic fields from dipoles or formal charges, and (*d*) van der Waals interactions. Local magnetic fields from aromatic rings, carbonyls, and other electronically anisotropic groups generate chemical shift changes less than 2 ppm in magnitude and, therefore, cannot explain the observed protein chemical shift range (32). The contribution of hydrogen bonding is uncertain, because fluorine hydrogen bonds in proteins are weak or nonexistent. The two forces that dominate environmental ^{19}F chemical shifts in proteins, therefore, are likely to be van der Waals interactions and electrostatic fields.

Considerable evidence suggests that van der Waals interactions play an important role. Hull & Sykes (42) observed a rough correlation between chemical shift and T_1 relaxation for fluorines in alkaline phosphatase labeled with 3-F-Tyr. Because T_1 relaxation and the van der Waals force have the same r^{-6} dependence on the distance to nearest neighbor protons, T_1 relaxation was regarded as an indicator of the extent of van der Waals contacts or the extent of burial within the protein interior. The latter relationship was verified by solvent exposure measurements with the use of SIIS. These authors therefore concluded that van der Waals contacts were the primary factor influencing the chemical shift. Gregory & Gerig (32) reached a similar conclusion in a study comparing the results of molecular dynamics calculations with experimental data obtained for fluorinated ribonuclease-S (11). van der Waals interactions were proposed to be the dominant factor controlling ^{19}F chemical shift, together with smaller contributions from local magnetic and electric fields. This study demonstrated the importance of molecular dynamics in chemical shift calculations, because calculations based on static structures yielded incorrect chemical shifts. More recently, van der Waals interactions within a molecular dynamics simulation have been used to explain the ^{19}F chemical shifts of the *E. coli* galactose-binding protein in terms of its known high-resolution structure (12).

If van der Waals interactions were the sole effector of fluorine chemical shift changes, however, one would expect all ^{19}F NMR resonances from buried probe positions to be shifted downfield from the denatured protein resonance. The fact that many ^{19}F NMR resonances are observed upfield of the denatured protein resonance (30) indicates that another factor must also contribute to the observed chemical shifts. On the basis of electric field effects alone, Augspurger et al (3) could predict accurately chemical shift ranges in several nuclei, including ^{19}F. Subsequently, a self-consistent field approach, denoted "charge field perturbation-gauge including atomic orbital," was used to predict the chemical shifts of the *E. coli* galactose-binding protein solely on the basis of electric field effects (21, 69). A strong correlation was seen between the predicted and experimental chemical shifts, which led to the conclusion that weak electrical interactions dominate the fluorine chemical shielding term. Surface charge does not significantly contribute to the shielding (51, 69).

Overall, it appears likely that van der Waals and electrostatic interactions together control the environmental ^{19}F chemical shifts observed in proteins. Significantly, such chemical shifts are regulated by local, rather than long-range, factors. Thus, although it remains difficult to convert either the magnitude or the sign of a chemical shift into a structural parameter, the observation of a chemical shift change indicates that the protein structure or electrostatics has been altered in the vicinity of that probe. In fact, the extreme sensitivity of the ^{19}F chemical shift to its environment provides what is perhaps the most sensitive physical method available for the detection of local conformational changes.

PROBING CONFORMATIONAL CHANGES IN PROTEINS OF KNOWN STRUCTURE

X-ray crystallography and ^{19}F NMR have proven to be splendidly complementary techniques. Starting with a high-resolution structure solved under relatively static conditions, ^{19}F NMR allows us to observe the dynamics of the protein, as well as to map out areas where changes occur as the protein enters a conformation inaccessible to crystallography. Studies to date have primarily used proteins labeled with fluorinated aromatic amino acids.

Because of the importance of aromatic sidechains in many aspects of protein function, aromatic fluorine labels are often located in critical regions of the target protein, including most ligand and effector protein binding sites. For the most part, two ^{19}F NMR approaches have been used to probe conformational changes. The first approach maps out the regions involved in the conformational change by observing which ^{19}F NMR resonances undergo chemical shift changes. The second approach uses paramagnetic broadening to measure distances and changes in solvent exposure.

In addition to providing spatial maps of conformational changes in proteins, ^{19}F NMR has been used to characterize the kinetic parameters of these conformational changes. The standard relationships of NMR exchange averaging places limits on the frequency of a conformational change, or, in favorable cases, can determine this rate directly (94):

$$\text{Slow exchange: distinct resonances} \quad (\nu_{ic} \ll |\nu_A - \nu_B|) \qquad 1.$$

$$\text{Intermediate exchange: broad resonances} \quad (\nu_{ic} \sim |\nu_A - \nu_B|) \qquad 2.$$

$$\text{Fast exchange: single averaged resonance} \quad (\nu_{ic} \gg |\nu_A - \nu_B|) \qquad 3.$$

When two conformations, A and B, which yield NMR frequencies ν_A and ν_B, interconvert over time at a frequency of ν_{ic}, two distinct resonances are observed if the interconversion is slow relative to the difference in their NMR frequencies (Equation 1). If, however, the interconversion rate is rapid relative to the NMR frequency difference, then a single exchange-averaged resonance is observed (Equation 3). In the intermediate exchange limit, where the interconversion rate approximately equals the NMR frequency difference, the two resonances become very broad and disappear into the baseline. Observation of the latter intermediate exchange limit enables direct determination of the interconversion rate, whereas the slow and rapid exchange cases provide upper and lower limits, respectively, on the interconversion rate. Spectral simulation can provide even more precise information regarding both interconversion kinetics and the relative populations of different conformational states.

The Proteins of Bacterial Chemotaxis

The chemotaxis pathway of *E. coli* and *Salmonella typhimurium* provides a well-suited system for the study of conformational changes involved in signal transduction (reviewed in 6, 35, 67, 83). The protein components of the pathway are defined fully, and several have been characterized structurally by crystallography or NMR. Such structural studies have helped define key mechanistic questions, the answers to which will provide a molecular understanding of signaling protein activation. To address these questions, ^{19}F NMR has been utilized to augment high-resolution structures. Studies of the galactose-binding protein (54–56) represent the earliest application of ^{19}F NMR to probe intrinsic sidechain labels in a protein of known crystal structure. The same approach has been extended to studies of the aspartate receptor (19, 26) and CheY (7, 22). Together, these three proteins illustrate how solution ^{19}F NMR can complement the information provided by high-resolution structure determination. In some cases, the power and limitations of the ^{19}F NMR method have been defined by direct comparison with results from independent approaches.

GALACTOSE-BINDING PROTEIN Chemosensing in bacteria often begins with a soluble periplasmic receptor, also termed a binding protein, which engulfs a specific small molecule ligand, then docks to and activates one of several transmembrane receptors that span the cytoplasmic membrane. In general, the soluble receptor also transports its ligand by associating with a membrane-bound transport system distinct from the sensory pathway. As a representative member of the soluble receptor family, the galactose-binding protein plays both a sensory and transport role for its two ligands, D-galactose and D-glucose. The crystal structure of the monomeric 32-kDa protein has been solved to 1.9 Å resolution for the conformation that contains bound D-glucose, as illustrated in Figure 2 (93), but the apo-structure is undefined. Thus, little is known about the conformational change that the protein undergoes when it binds ligand. Overall, the structure consists of two homologous domains, each consisting of a β-sheet sandwiched between layers of α-helices. The two domains are connected by three polypeptide strands, and the sugar-binding site lies within the cleft between the domains. The protein also possesses an EF-hand–like calcium-binding site, which stabilizes the protein structure. Luck & Falke (54–56) have fluorine-labeled the protein with 5-F-Trp and 3-F-Phe to investigate the conformational changes induced by sugar and calcium binding.

As seen in Figure 2, the targeted 12 probe locations (five tryptophans and seven phenylalanines) are well distributed throughout the molecule, including the sugar- and calcium-binding site regions. Biosynthetic incorporation of fluorinated amino acids into the appropriate *E. coli* auxotrophs yields 65–80% incorporation of 5-F-Trp or 20% incorporation

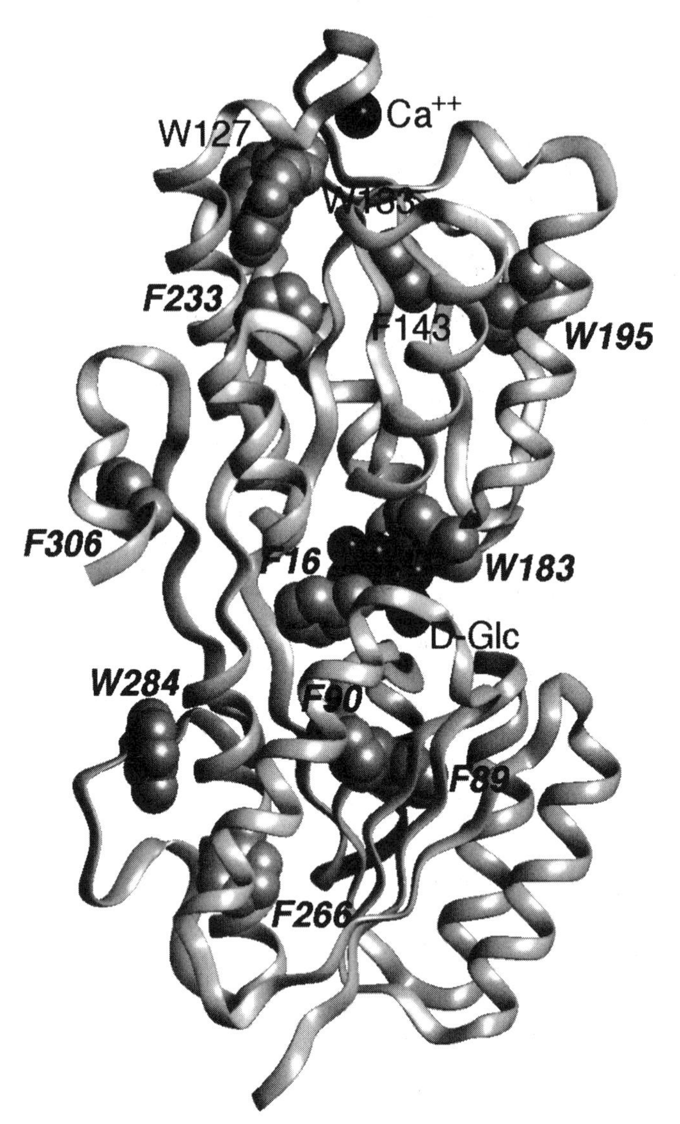

Figure 2 Backbone ribbon structure of the galactose-binding protein (93), which illustrates bound ligands (Ca^{2+} and D-glucose, both in black) and the 12 fluorine probe positions. The protein has been labeled at its five tryptophan positions with 5-F-Trp, or its seven phenylalanine positions with 3-F-Phe (gray VDW surfaces), thereby enabling the use of ^{19}F NMR to probe ligand-induced conformational changes (54–56). The binding of D-galactose or D-glucose generates a long-range conformational change, thus yielding significant chemical shift changes for the nine probe positions indicated by italics (see Figure 5). In contrast, the three remaining probes, all in the vicinity of the Ca^{2+} binding site, are unperturbed. When Ca^{2+} binds, the converse pattern of frequency changes is observed, such that only the resonances of the latter three probes are observed to shift.

of 3-F-Phe (54). The resonance that corresponds to each 5-F-Trp probe has been assigned by the direct-replacement mutagenesis method (see above).

The effect of fluorine labeling on the structure has been assayed directly, thereby revealing relatively minor effects caused by fluorine substitution (54). The estimated upper limits of these effects, calculated for the fully labeled protein, are as follows: (*a*) a 2.2- to 3-fold decrease in the D-galactose affinity caused by 5-F-Trp substitution, or a 1.5- to 4-fold affinity decrease caused by 3-F-Phe substitution, respectively; (*b*) a 1.5- to 2.5-fold increase in the calcium dissociation rate caused by 5-F-Trp substitution, whereas 3-F-Phe substitution has no detectable effect; and (*c*) a measurable decrease in protein stability, as revealed in urea denaturation curves yielding $\Delta\Delta G_U$ (H_2O) = -1.2 kcal mol^{-1} for 5-F-Trp, or -0.8 kcal mol^{-1} for 3-F-Phe. For this protein, then, the effects of fluorine incorporation on protein structure and function are quite small, especially considering that bound ligand is stabilized by direct van der Waals interactions with tryptophan and phenylalanine sidechains (one of each).

At the time the ^{19}F NMR studies were carried out, it was proposed that a key feature of binding protein activation was closure of the ligand-binding cleft through bending of the interdomain linkage (57a). Such a hinged-cleft mechanism, however, had not been demonstrated experimentally for any member of the binding protein family. To test this mechanism, and to measure the minimum angle of cleft opening on removal of sugar, ^{19}F NMR paramagnetic broadening measurements were carried out with the use of the aqueous paramagnet Gd^{3+} $EDTA^{4-}$(56). With this approach, the inverse sixth power dependence of paramagnetic broadening is used to estimate the separation between the paramagnet and the ^{19}F nucleus of a 5-F-Trp probe located at position 183 within the cleft. Addition of the paramagnet to the sugar-empty protein causes selective broadening of the probe resonance, as shown in Figure 3, whereas the paramagnet has no effect on the probe resonance when the cleft is closed with D-glucose trapped inside. The broadening of the probe resonance can be used to place an upper limit on the probe–paramagnet separation for the empty cleft. These results demonstrate that the cleft can open by at least 18 degrees in the absence of sugar, thereby enabling the paramagnetic probe to enter, as illustrated schematically in Figure 4 (56). Subsequent crystallographic studies have confirmed large, hinged-cleft motions in related binding proteins (27, 79).

More subtle conformational changes are also revealed by ^{19}F NMR chemical shift changes within the individual protein domains (Figure

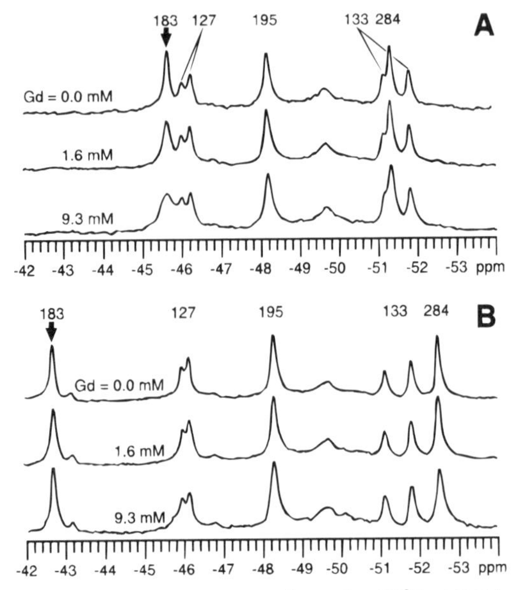

Figure 3 Effect of the aqueous paramagnetic probe Gd^{3+} EDTA on the ^{19}F NMR spectrum of the galactose-binding protein (470 MHz) (56). The resonance of the 5-F-Trp probe at the Trp183 position within the sugar-binding cleft is broadened selectively by increasing concentrations of the paramagnet when the cleft is empty (*A*), whereas the same probe resonance in the closed cleft containing bound D-glucose is unaffected (*B*). The observed broadening of the Trp183 resonance in the empty cleft has been used to place an upper bound on the closest approach of the paramagnet (see Figure 4). The duplicity observed for the Trp127 and 133 resonances stems from two kinetically stable conformations of their environment.

5). Inside the sugar cleft itself, the binding of D-galactose causes a dramatic chemical shift change (+3.8 ppm) in the resonance that corresponds to Trp 183, which contacts the bound sugar. The binding of D-glucose, which differs from D-galactose by the stereochemistry about a single carbon atom, induces a slightly smaller chemical shift change (+2.8 ppm) in the same resonance. In contrast, at more distal 5-F-Trp or 3-F-Phe probe positions, the two sugars induce identical chemical shift changes, with the exception of resonances in the vicinity of the calcium-binding site. These resonances are unaffected by sugar binding (Figure 2). Conversely, calcium binding alters the chemical shifts of only those residues located in the vicinity of its binding site and does not affect the resonances coupled to the sugar site (55). Calcium binding, therefore, induces only a local structural change, whereas sugar binding generates a long-range conformational change; the two sites are allosterically independent of one another. The long-range nature of the sugar-

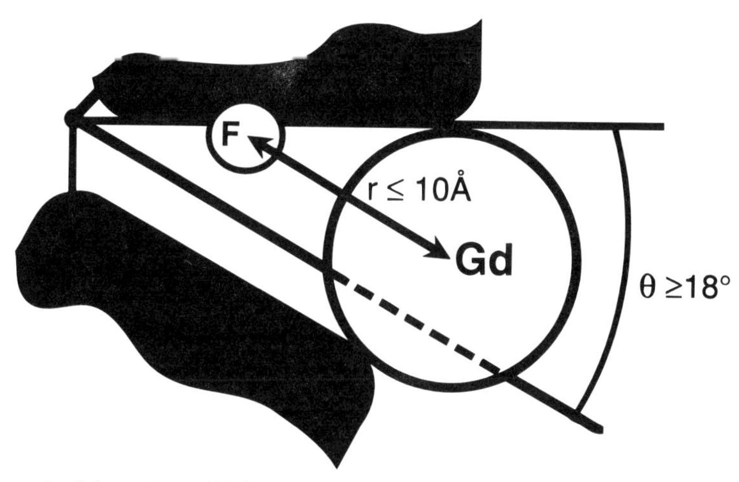

Figure 4 Schematic model for the empty sugar cleft of the galactose-binding protein (56). Addition of the aqueous paramagnet Gd^{3+} EDTA (Gd) to the empty cleft is observed to broaden the ^{19}F NMR resonance of the 5-F-Trp probe (F) located at the Trp183 position within the cleft (see Figure 3). The resulting broadening indicates that the paramagnet can approach the fluorine probe at a distance of 10 Å or less, which places the paramagnet within the cleft. To accommodate the paramagnet, the known structure of the closed cleft must open by an angle of at least 18 degrees.

induced conformational change is consistent with its function, which is to regulate protein-protein interactions that involve large regions of the binding protein surface.

Further analysis of ^{19}F NMR spectra places a lower limit on the kinetics of sugar release. At substoichiometric concentrations of D-glucose and D-galactose, the spectrum of the 5-F-Trp–labeled protein shows distinct resonances that correspond to each apo- and sugar-bound state (54), thereby placing sugar release in the slow exchange limit. Equation 1 indicates that the lifetime of the D-galactose-bound state is $\tau_{ic} \gg 0.6$ ms, whereas the lifetime of the D-glucose–bound state is $\tau_{ic} \gg 0.8$ ms, which is consistent with direct kinetic measurements that yield bound sugar lifetimes of 220 ms for D-galactose and 710 ms for D-glucose (60).

Altogether, the ^{19}F NMR data indicate that the sugar-binding cleft does, indeed, close on sugar binding, thereby trapping the sugar inside until the complex can diffuse to the appropriate membrane receptor. Moreover, subtle, long-range conformational changes are induced in both domains by sugar, but not calcium, binding. These intradomain conformational changes, which are not easily detected by crystallography, likely stem from the conserved architecture of the binding protein

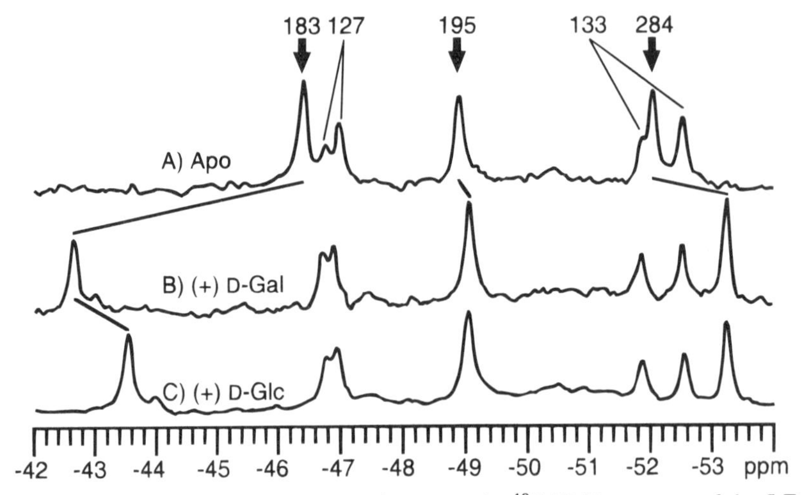

Figure 5 Effect of D-galactose and D-glucose on the ^{19}F NMR spectrum of the 5-F-Trp labeled galactose-binding protein (470 MHz) (54). (*A*) The spectrum of the sugar-empty protein. (*B*) The protein saturated with D-galactose. (*C*) The protein saturated with D-glucose. (*bold arrows*) Resonances for which significant ligand-induced chemical shift changes are observed. The greatest changes are observed for the resonance from the Trp183 position, which lies in van der Waals contact with the bound sugar molecule.

family, in which most secondary structure elements originate or terminate in the substrate cleft. Such an architecture may transmit conformational information from the buried ligand-binding site to the protein surface, where it can be used to regulate the critical membrane docking sites (45).

ASPARTATE RECEPTOR The transmembrane aspartate receptor is activated either directly, by the binding of the chemoattractant aspartate to its periplasmic ligand-binding domain, or indirectly, by association with the maltose binding protein. The resulting transmembrane signal regulates an associated cytoplasmic histidine kinase, which ultimately controls the swimming behavior of the cell. The receptor is a homodimer of identical 60-kDa subunits in both the presence and the absence of ligand (25, 62). The structure of the isolated periplasmic domain has been determined to 2.0 Å resolution by X-ray crystallography, as displayed in Figure 6 (59). This domain consists of a homodimer of four-helix bundles, with two symmetric, nonoverlapping attractant binding sites at the dimer interface. In the intact receptor, the structure of the transmembrane domain has been defined by disulfide mapping, which indicates that the N- and C-terminal α-helices of the ligand-binding domain continue uninterrupted through the membrane.

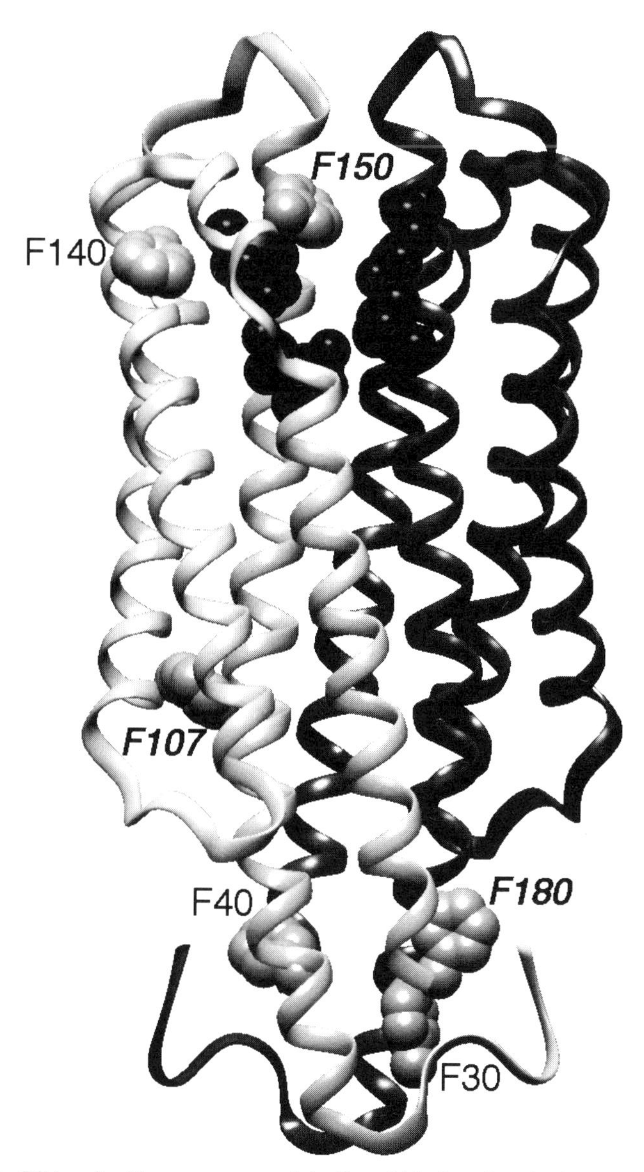

Figure 6 Ribbon backbone structure of the ligand-binding domain from the aspartate receptor (59), which shows the residues (*black*) in one of the two symmetric ligand-binding sites, as well as the six fluorine probe positions in one of the two symmetric subunits. The isolated domain has been labeled with 4-F-Phe at all six phenylalanines (gray VDW surfaces), thereby enabling the use of ^{19}F NMR to detect ligand-induced chemical shift changes (see Figure 7) (19). Aspartate is observed to induce a long-range conformational change, which yields significant changes in chemical shift for the three probe positions indicated by italics. In the intact receptor, two transmembrane helices from each subunit would extend downward into the bilayer located below the figure, thus carrying the ligand-induced signal across the membrane.

To map out the spatial and kinetic features of the aspartate-induced conformational change, both the intact receptor and the isolated ligand-binding domain have been probed by solution ^{19}F NMR. The intact receptor has been labeled with 5-F-Trp in its native membrane (26). Because of the very slow tumbling of the protein–membrane complex, all but one of the receptor 5-F-Trp resonances are too broad to detect. The lone resonance observed lies near the mobile C-terminus of the receptor and is exposed to the cytoplasmic compartment, where it possesses considerable local motion. Aspartate binding to the periplasmic ligand-binding site induces a long-range, transmembrane conformational change that is transmitted to the C-terminal 5-F-Trp probe located more than 90 Å away, thereby perturbing the chemical shift of this resonance.

A higher resolution picture of the conformational change has been obtained with the use of the 36-kDa isolated ligand-binding domain, which has been cloned and shown to retain native aspartate binding (19, 63). The homodimeric domain has been biosynthetically labeled with 4-F-Phe, with the use of glyphosate to inhibit endogenous synthesis of phenylalanine (19). Each identical subunit possesses six phenylalanine labeling positions, which are distributed throughout the known structure (Figure 6). Significantly, two probe positions lie in or near the ligand-binding pocket, whereas the other four probes lie at positions distal to the bound aspartate, including distal sites on the N- and C-terminal α-helices. Assignment of the resulting resonances has been carried out with the use of a combination of the direct replacement and the nudge mutagenic techniques.

The chosen low level of fluorine incorporation (7%) ensures that the labeled domain typically possesses no more than one fluorine atom, thereby preventing pairwise or higher order interactions. To estimate the effects of such fluorine incorporation on domain structure and function, each of the six phenylalanine residues has been individually mutated to tyrosine (19). The resulting para-hydroxyl moiety is expected to be even more perturbing than the corresponding para-fluorine (see above), yet the tyrosine substitutions yield little effect on the aspartate binding site, which retains glutamate affinities ranging from 4-fold to 0.6-fold of native. These effects are quite small on a free energy scale ($\leq$0.8 kcal mol^{-1}), which suggests that hydroxyl substitutions at the para position of each phenylalanine, as well as the corresponding fluorine substitutions, cause little or no perturbation of ligand binding and the ensuing conformational change.

The ^{19}F NMR spectrum of the ligand-binding domain has been obtained for the apo-, aspartate-bound, and glutamate-bound states (19).

Only six resonances are observed for the 12 4-F-Phe positions in the dimer, which indicates that the monomers are equivalent, on average, owing to rapid exchange of ligand between the two symmetric binding sites. As observed for the galactose-binding protein, different ligands elicit contrasting chemical shift changes for a resonance within their shared binding pocket (4-F-Phe150) but yield the same ligand-induced pattern of chemical shift changes at more distant probe positions. These ligands therefore generate similar or identical long-range conformational changes outside the binding site itself.

Figure 6 summarizes the probe positions for which significant ligand-induced chemical shift changes have been detected (19). Importantly, chemical shift changes are observed for probe positions lying on the C-terminal helix but not for those on the N-terminal helix, which implicates the former helix as the transmitter of the transmembrane signal. Subsequent analyses of data from X-ray crystallography and disulfide engineering have confirmed this identification of the C-terminal helix as the key transmembrane signaling element, whereas the N-terminal helix is proposed to play a structural role in stabilizing the dimer (13–15, 48).

Additional stoichiometric and kinetic information has been extracted from a titration of the ^{19}F NMR spectrum with ligand (19), as illustrated in Figure 7. Note that the maximum effect of ligand on the resonances is attained at a ligand:dimer mole ratio of 1:1, although there are two symmetric binding sites in the apo-dimer. This result confirms the earlier conclusion of the crystallographic and direct binding measurements (5, 59), which indicated strong negative cooperativity between the two aspartate binding sites. The titration also reveals that one of the resonances, which arises from 4-F-Phe150 in the ligand-binding site, disappears because of exchange broadening at a nonsaturating aspartate concentration, thereby enabling the kinetics of aspartate binding and dissociation to be measured. The resulting calculation indicates that aspartate binding and release is rapid, approaching the diffusion controlled limit observed for the fastest enzymes (19). Ligand binding and release, therefore, are not rate limiting in the chemosensory pathway, because subsequent phosphorylation and/or protein diffusion events in the cytoplasm are significantly slower (78).

CHEY CheY is the response regulator protein of the chemotaxis pathway. It receives a phosphate from the histidine kinase associated with the transmembrane receptors and subsequently docks to and regulates the behavior of the flagellar motor. The structure of the unphosphorylated 13-kDa protein has been determined crystallographically by

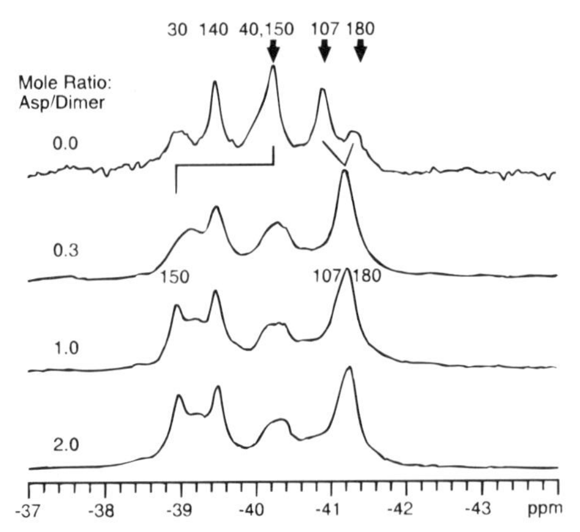

Figure 7 The ligand-binding domain of the aspartate receptor: titration of the ^{19}F NMR spectrum with aspartate (470 MHz) (19). (*upper*) The spectrum of the 4-F-Phe-labeled apo-domain to which is added increasing concentrations of ligand aspartate. (*bold arrows*) The resonances that undergo chemical shift changes on ligand binding. The final chemical shifts are observed at a mole ratio of one aspartate molecule per dimer, which indicates half-of-sites occupancy. In addition, the Phe150 resonance is observed to disappear at an intermediate loading (mole ratio = 0.3), then reappears at a new frequency on saturation (mole ratio = 1.0), thus demonstrating that this resonance passes through the intermediate exchange limit.

several groups (4, 82, 92), and independent solution NMR structures have also been completed (9, 52, 65). CheY folds as a compact $(\alpha/\beta)_5$ motif illustrated in Figure 8, with a central antiparallel β-sheet sandwiched between layers of α-helices. The activation site, where phosphorylation occurs and activating mutantations are found, lies at one end of the β-sheet and includes four residues that exhibit 70–100% conservation across the response regulator family: Asp12, Asp13, Asp57, and Lys109. The Asp57 sidechain is the site of phosphorylation and provides one of the coordinating oxygens for a bound magnesium ion, which serves as a cofactor in phosphorylation and dephosphorylation of Asp57. A salt bridge has been proposed to exist between the highly conserved Asp57 and Lys109 residues (92). This salt bridge could provide a structural on-off switch in response regulator activation; alternatively, the Lys109 sidechain could play a role in the active site chemistry.

The short lifetime of the phospho-Asp57 linkage ($\tau_{1/2} \sim 10$ s) has prevented the crystallization of the protein in the phosphorylated state. Instead, ^{19}F NMR has been used to probe CheY labeled with 4-F-Phe, thereby enabling the conformational changes triggered by both

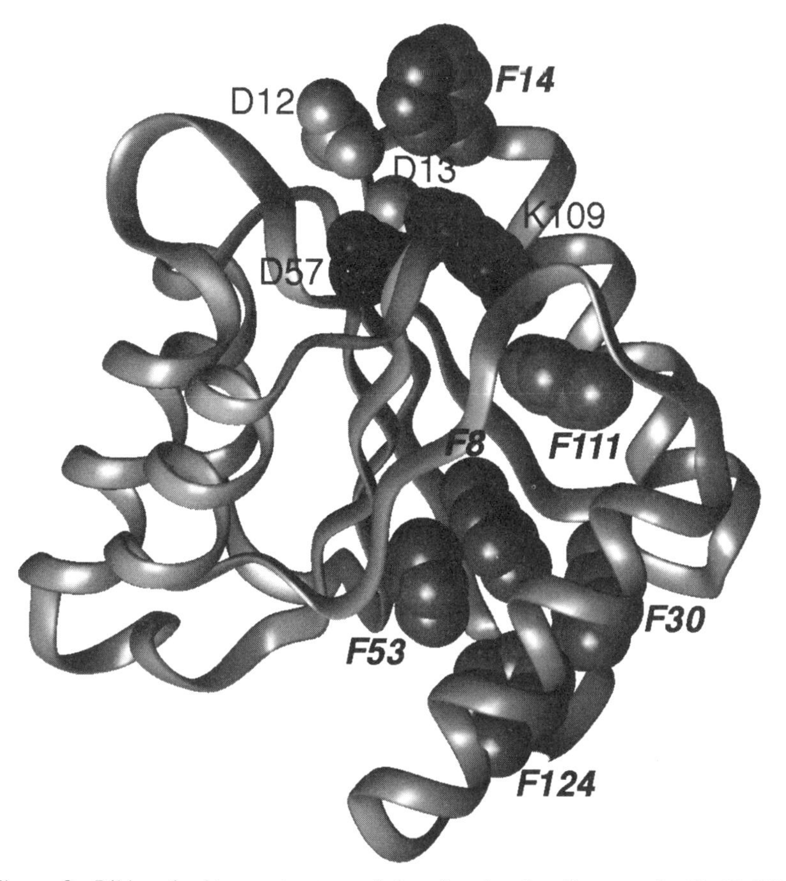

Figure 8 Ribbon backbone structure of the phospho-signaling protein CheY (92). Shown are Asp57, the site of phosphorylation, the adjacent highly conserved Lys109 (*black*), two other conserved active site residues (Asp12,13) and the fluorine probe positions. The protein has been labeled with 4-F-Phe at all six phenylalanines (gray VDW surfaces), thereby enabling the use of ^{19}F NMR to detect activation-induced chemical shift changes (7, 22). Phosphorylation of Asp57 is observed to generate significant chemical shift changes at all six probe positions, indicated by italics. In contrast, constitutive activation of the protein by the D13K mutation gives rise to large chemical shift changes for only the Phe14 and Phe111 resonances, both of which arise from the vicinity of the active site.

phosphorylation and activating mutations to be mapped out (22). Using an *E. coli* strain auxotrophic for phenylalanine, 4-F-Phe has been incorporated to a level of 5%. The resulting six fluorine probe positions (Figure 8) are found at interesting locations, including the active site (Phe14), the loop containing the putative "switch" residue Lys109

(Phe111), and a phenylalanine cluster on the opposite side of the molecule from the active site, which can serve as an antenna for long-range conformational changes. The six resonances have been assigned with the use of direct replacement and the nudge method (Figure 1), and the maximal effect of fluorine incorporation at each phenylalanine position has been estimated by the corresponding tyrosine substitutions (see above). Each of these tyrosine substitutions has a negligible effect on the chemotaxis pathway in vivo, which indicates that perturbations of CheY caused by para substituents on its phenylalanine rings are minimal or nonexistent (22).

To probe the conformational change induced by phosphorylation, the short lifetime of the phosphorylated state has been overcome by chemically phosphorylating Asp57 with the small molecule phosphodonor, acetyl phosphate (57). Excess acetyl phosphate can maintain a high steady-state level of phosphorylation for 10 min—sufficient time to obtain a one-dimensional ^{19}F NMR spectrum. Phosphorylation induces a global conformational change in the CheY molecule, which extends from the active site to the distant phenylalanine cluster (Figure 8). Subsequent multidimensional solution NMR studies have confirmed this long-range conformational change and have provided a higher resolution map of its extent (52). The long-range nature of the phosphorylation-induced structural change may play an important role in the large family of two-domain response regulator proteins (67), in which phosphorylation of the regulatory domain could transmit an allosteric signal to the effector domain.

In contrast to the long-range structural change triggered by phosphorylation, activation of CheY by specific "lock-on" mutations generates a much more localized conformational perturbation (7). Six single or double mutations that activate CheY constitutively all generate an upfield chemical shift change for the resonance corresponding to Phe111, which lies in the same loop as Lys109 (Figure 8). The structural change detected for this loop could represent a key component of the on-off "switch." On the basis of circumstantial evidence, it has been proposed (7) that the Asp57-Lys109 salt bridge is formed in the off state and broken in the on state; this conclusion remains controversial, however, and must be tested by other approaches. The localized nature of the conformational change triggered by one of the lock-on mutations has recently been confirmed by multidimensional NMR (FW Dahlquist, personal communication).

Nuclear magnetic resonance methods, including ^{19}F NMR, have begun to reveal key aspects of the protein–protein interaction between CheY and its kinase protein, CheA. Interestingly, the high-affinity kinase binding site on CheY appears to be located at a surface distal from the phosphorylated residue Asp57. Thus, when the CheA fragment responsible for CheY docking is added to the labeled protein, large chemical shift changes at positions distant from the active site have been observed in both ^{19}F (20; TB Morrison, M Welch, Y Blat, SL Butler, JJ Falke, M Eisenbach, and JS Parkinson, submitted for publication) and multidimensional (84a) NMR studies. The phosphorylation domain of the kinase, therefore, must be large enough to bridge this distance.

Other Examples

DIHYDROFOLATE REDUCTASE The dihydrofolate reductase (DHFR) enzyme of *E. coli* is an 18-kDa monomeric protein that catalyzes the NADPH-dependent reduction of 7,8-dihydrofolate to 5,6,7,8-tetrahydrofolate, an important cofactor for several biosynthetic pathways. Because of its small size, well-characterized enzymatic mechanism, well-refined structure, and the reversibility of its folding reaction in the presence of chemical denaturants, Hoeltzi & Frieden (39) have deemed this a good model for studies of protein folding. The use of ^{19}F NMR provides an advantage over fluorescence, absorbance, or circular dichroism in that it can provide information regarding specific sites within the protein. The five tryptophan residues, which have been labeled with 6-F-Trp, are distributed throughout the molecule illustrated in Figure 9 and give rise to well-resolved resonances that have been assigned by direct replacement mutagenesis (39).

Dihydrofolate reductase possesses two ligand-binding sites, one for NADPH and the other for 7,8-dihydrofolate. The latter site also binds methotrexate, a clinically important anticancer drug. A comparison of the spectra observed for the apo, NADPH-bound, methotrexate-bound, and doubly liganded (NADPH and methotrexate) forms of the protein reveal that NADPH significantly shifts two of the five probe resonances in or near its binding site (Trp74 and Trp22), whereas methotrexate triggers a longer range conformational change revealed by large chemical shift changes of three probe resonances (Trp22, Trp133, and Trp47; see Figure 9). Moreover, methotrexate sharpens the resonances of two probe resonances (Trp22, Trp30) considerably, whereas NADPH broadens two resonances (Trp22, Trp74) relative to the apo state. The doubly liganded protein exhibits the maximum chemical shift dispersion and sharpest resonances, which suggests that this conformation is the most stable or least flexible and possesses less structural heterogeneity than the lower ligation states. Of the two ligands, the methotrexate appears to generate the greater increase in structural homogeneity (39).

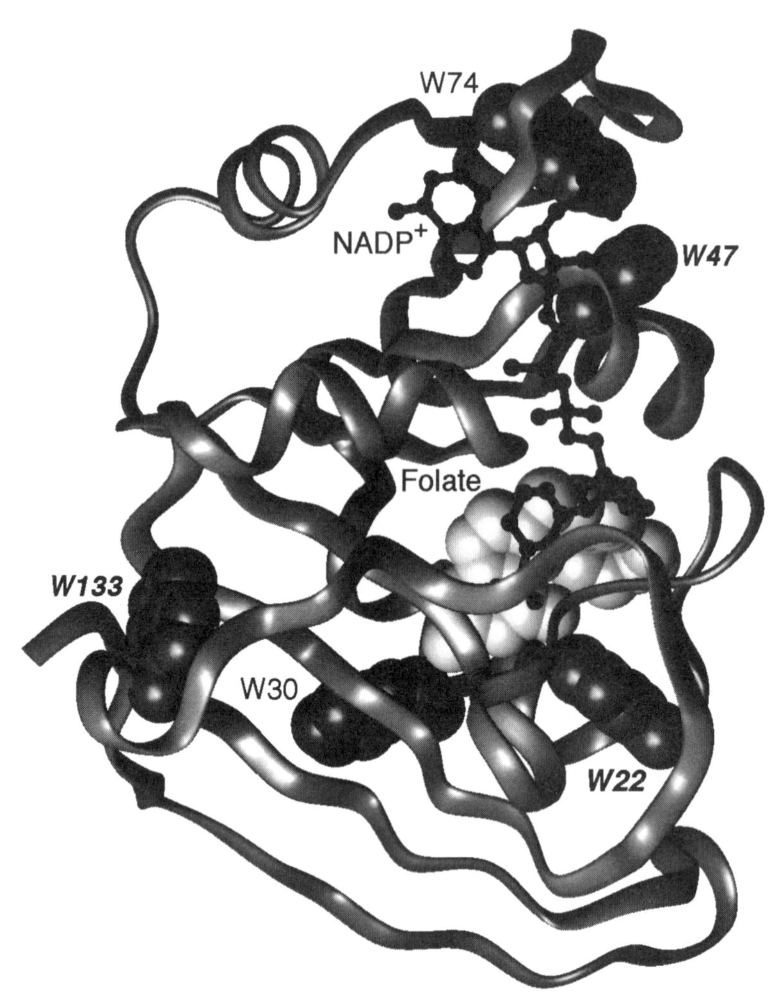

Figure 9 Ribbon backbone structure of the biosynthetic enzyme dihydrofolate reductase (10). Shown are the bound ligands $NADP^+$ (*black ball* and *stick*) and folate (*light VDW surface*), as well as the fluorine probe positions. The protein has been labeled with 6-F-Trp at its five tryptophan positions (*gray VDW surfaces*), thereby enabling ^{19}F NMR studies of ligand-induced chemical shift changes (40). Binding of methotrexate to the site that contains folate triggers a long-range conformational change, thereby yielding chemical shift changes at the probe positions indicated by italics and increasing the structural homogeneity of the protein (see text). In contrast, the binding of NADPH generates a more localized conformational change, detected by chemical shifts changes of the Trp22 and Trp74 resonances that arise from the NADPH binding pocket itself.

Crystallographic evidence also supports the ordering effect of this ligand, which is observed to order the loop containing one of the fluorine probe positions, Trp22 (10).

To study the unfolding process of this protein further, Hoeltzi et al (40, 41) have constructed a stopped-flow ^{19}F NMR device to monitor unfolding in real time with a resolution of 1.5 s. They observe that immediately after the addition of urea, the resonances from the native protein disappear, but the denatured resonances exhibit only 20% of their final intensity. Together with fluorescence and circular dichroism data, these stopped-flow ^{19}F NMR results indicate that the protein unfolds by two pathways. Approximately 20% of the population denatures rapidly, whereas 80% of the protein forms a molten globule intermediate that retains native-like secondary structure but loses tertiary contacts, thereby causing the tryptophan sidechains to exhibit considerable structural heterogeneity (40). Interestingly, when the protein is titrated with increasing concentrations of urea in a separate equilibrium experiment, the resonance corresponding to Trp22 narrows and moves toward its denatured chemical shift at urea concentrations well below the denaturation midpoint (39). Thus, as suggested by the ^{19}F NMR ligand-binding and crystallographic results, the region of the methotrexate binding site, including the Trp22 residue, appears to be less folded than the rest of the molecule in the absence of its ligand.

ELONGATION FACTOR TU In the GTP-bound form, elongation factor Tu (EF-Tu) delivers aminoacyl-transfer RNA to the ribosome. After hydrolysis of the GTP to GDP, EF-Tu dissociates, and the GTP-bound form is regenerated by a different elongation factor. The structure of each of the three domains of trypsin-treated EF-Tu have been determined crystallographically. Subsequently, ^{19}F NMR has been used to monitor the effects of various ligands on the conformation of full-length EF-Tu and to examine the individual domains for perturbations that arise from their isolation through trypsin cleavage (24). Elongation factor Tu has been labeled with 3-F-Tyr at its ten tyrosine positions, and tentative assignments are postulated for two of the resulting resonances that exhibit the largest upfield shifts for the GDP-bound state relative to other conformations. One of the assigned resonances stems from a buried tyrosine in domain III of the known structure, which suggests that this domain interacts with domain I containing the GDP/GTP binding site. No major chemical shift changes are seen between the native and trypsin-treated forms of the protein, which indicates that, in all likelihood, the crystal structures represent native domain conformations.

INVESTIGATING A PROTEIN OF UNKNOWN STRUCTURE

We have seen how ^{19}F NMR can complement a high-resolution protein structure in probing conformational changes. Even for proteins that have not been characterized structurally, however, ^{19}F NMR can be used to define the solvent exposure or separations of specific probe positions. The best example of such an application is *E. coli* D-lactate dehydrogenase (D-LDH), a peripheral membrane protein for which Ho and coworkers have developed a low-resolution structural model on the basis of extensive ^{19}F NMR data (38, 70, 77, 84, 90).

Rule et al (77) have incorporated 4-, 5-, and 6-fluorotryptophan into the protein to levels reaching 90% and assigned the ^{19}F NMR resonances by direct replacement mutagenesis. None of the substitutions alter the kinetic parameters of the protein substantially, and stability studies at 55°C have demonstrated that both the 4- and 6-fluoro-Trp labels retain the native activity decay constant of 10 s. Moreover, 5-F-Trp labeling actually increases this decay constant to 40 s (77). The secondary structure of the 4- and 5-F-Trp-labeled D-LDH appears, by circular dichroism measurements, to be unaltered from native, although the 6-F-Trp-labeled D-LDH shows a minor perturbation (77). Because the 5-F-Trp resonances exhibit wider dispersion than the 4- or 6-F-Trp resonances (7 ppm vs 6 or 3 ppm, respectively), studies have focused on this analogue. To increase the molecular tumbling rate and thereby minimize the linewidths of the probe resonances, the protein was solubilized from its native membrane to yield a protein–detergent micellar complex, in which the resonances were resolved easily. Although this solubilization narrowed the linewidths significantly, the chemical shifts of the resonances were unperturbed, which indicated that detergent micelles did not induce a conformational change (77). This finding was verified later by repeating the studies in a small unilamellar vesicle system, which led to the same conclusions drawn in the detergent-solubilized system (90).

Of the five labeled tryptophan positions, two (Trp384 and 567) are revealed by their solvent-induced isotopic shifts to be near the aqueous surface of the molecule. One position (Trp369) displays an altered chemical shift and line broadening on substrate binding and the accompanying reduction of the bound flavin cofactor (FAD), which suggests that this position may lie in or near the lactate-binding site (77). The observed line broadening is attributed to exchange of the probe between multiple environments, presumably owing to fluctuations of the local structure (38). In further studies, the proximity of the fluorine positions to the detergent micelle or membrane has been examined with the use of the nitroxide-containing fatty acid 8-doxylpalmitic acid, a paramagnetic probe that will broaden any resonance within approximately 15 Å. Because the nitroxide lies in the middle of the fatty acid, 7 Å from either end, any resonance broadened would lie in or near the hydrocarbon phase. None of the intrinsic tryptophan positions are broadened by this probe, which suggests that they are not located near the membrane-binding surface (77).

To elaborate on these results, the repertoire of labeling positions has been expanded by substituting tryptophan for other aromatic sidechains within D-LDH through site-directed mutagenesis (70, 84). Only those mutants that retain significant enzymatic activities were used in the subsequent structural analysis. Figure 10 summarizes the resulting engineered 5-F-Trp positions, as well as the intrinsic 5-F-Trp probes, located within in a region of D-LDH proposed to contain the membrane-binding site. Also summarized are three types of structural information obtained by ^{19}F NMR: (*a*) chemical shift changes induced by substrate binding and accompanying FAD reduction, (*b*) line broadenings caused by the paramagnetic fatty acid probe placed in the associated micelle or bilayer, and (*c*) solvent-induced isotopic shifts. The identified membrane-binding region extends from Tyr226 to Trp384. Within this region,

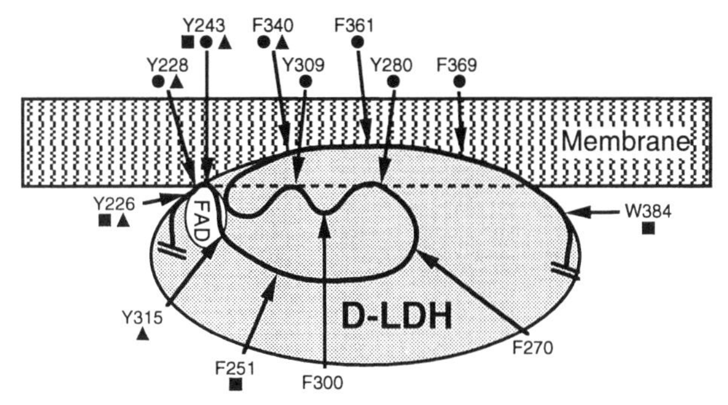

Figure 10 Schematic low-resolution structural model for the membrane-binding domain of D-LDH, developed by ^{19}F NMR studies (84). Shown are the predicted locations of the FAD binding site and the locations of fluorine probes within the proposed membrane-binding domain, spanning residues 226–384. The protein has been labeled with 5-F-Trp at its native tryptophan sidechains. In addition, engineered tryptophans have been substituted for other aromatic sidechains and labeled with 5-F-Trp as indicated. (*filled squares*) Probe positions judged to be exposed to aqueous solvent by their solvent isotope-induced shifts; (*filled triangles*) positions yielding FAD-induced chemical shift changes; and (*filled circles*) positions for which paramagnetic line broadening is observed on addition of a spin-labeled fatty acid, which implies close proximity to the membrane-binding surface.

the ^{19}F NMR data have allowed the development of a low-resolution structural model (84), as illustrated in Figure 10.

In regions of D-LDH lying outside the proposed membrane-binding domain, limited structural information has been obtained from the chemical shift and linewidth changes induced by sidechain substitutions. In general, the effects of such point mutations stem from local structural perturbations of the region surrounding the substitution, thereby identifying fluorine probes in the vicinity of the altered sidechain (70). Such identifications must be regarded as tentative, however, because of the ability of ^{19}F NMR to detect long-range conformational changes triggered at a distant site.

CONCLUSIONS AND FUTURE DIRECTIONS

In summary, the systematic use of solution ^{19}F NMR has shed light on the kinetics and spatial ranges of conformational changes in several proteins of known structure and has been used to develop a low-resolution conformational map for one protein of unknown structure. In cases where the conclusions of ^{19}F NMR studies can be checked against newer information from crystallographic or multidimensional NMR structures, it generally has been observed that ^{19}F NMR provides an accurate and powerful method for probing protein conformations. As with any method, the main caution is to avoid the overinterpretation of a limited data set.

Challenges for the future include the use of ^{19}F NMR to probe specifically labeled proteins in vivo. One example of such an application has already been described in living yeast cells (96). Further, it should be possible to use paramagnetic probes that are attached covalently to a protein functionality, or paramagnetic metal ions bound in an intrinsic site, to map out multiple fluorine-paramagnet distances, thereby facilitating the development of a low-resolution structure for macromolecules inaccessible to other structural techniques. Such an approach has been used successfully in ^{1}H NMR studies (34) but still needs to be tested for ^{19}F NMR in a protein of known structure. Finally, the assignment of ^{19}F NMR resonances in proteins of known structure could be facilitated by ^{19}F-^{1}H nuclear Overhauser effect experiments. Such approaches could, in favorable cases, allow full assignment of the ^{19}F NMR spectrum without the aid of protein engineering. Overall, as additional high-resolution protein structures become available, and as the techniques of molecular cloning make new proteins of unknown structure accessible, ^{19}F NMR will continue to be a useful technique in the toolbox of the protein chemist.

ACKNOWLEDGMENTS

The authors thank a large number of colleagues in the field for sending reprints and preprints. We apologize to those whose work was, unfortunately, outside the scope of the present review. Special thanks are due to Steven Drake, Dr. Linda Luck, and Dr. Olve Peersen for helpful discussions and important experimental contributions, and to the National Institutes of Health for funding (GM40731 and GM48203 to JJF).

Literature Cited

1. Abeles RH, Alston TA. 1990. Enzyme inhibition by fluoro compounds. *J. Biol. Chem.* 265:16705–8
2. Arseniev AS, Kuryatov AB, Tsetlin VI, Bystrov VF, Ivanov VT, Orchinnikov YA. 1987. ^{19}F NMR study of 5-fluorotryptophan-labeled bacteriorhodopsin. *FEBS Lett.* 213:283–88
3. Augspurger J, Pearson JG, Oldfield E, Dykstra CE, Park KD, Schwartz D. 1992. Chemical-shift ranges in proteins. *J. Magn. Reson.* 100:342–57
4. Bellsolell L, Prieto J, Serrano L, Coll M. 1994. Magnesium binding to the bacterial chemotaxis protein CheY results in large conformational changes involving its functional surface. *J. Mol. Biol.* 238:489–95
5. Biemann H-P, Koshland DE Jr. 1994. Aspartate receptors of *Escherichia coli* and *Salmonella typhimurium* bind ligand with negative and half-of-sites cooperativity. *Biochemistry* 33: 629–34
6. Bourret RB, Brokovich KA, Simon MI. 1991. Signal transduction pathways involving protein phosphorylation in prokaryotes. *Annu. Rev. Biochem.* 60:401–41
7. Bourret RB, Drake SK, Chervitz SA, Simon MI, Falke JJ. 1993. Activation of the phosphosignaling protein CheY II. Analysis of activated mutants by ^{19}F NMR and protein engineering *J. Biol. Chem.* 268:13089–96
8. Bovy PR, Getman DP, Matsoukas JM, Moore GJ. 1991. Influence of polyfluorination of the phenylalanine ring of angiotensin II on conformation and biological activity. *Biochim. Biophys. Acta* 1079:23–28
9. Bruix M, Pascual J, Santoro J, Prieto J, Serrano L, Rico M. 1993. ^{1}H and ^{15}N-NMR assignment and solution structure of the chemotactic *Escherichia coli* Che Y protein. *Eur. J. Biochem.* 215:573–85
10. Bystroff C, Oatley SJ, Kraut J. 1990. Crystal strucutres of *Escherichia coli* dihydrofolate recuctase: the NADP^{+} holoenzyme and the folate NADP^{+} ternary complex. Substrate binding and a model for the transition state. *Biochemistry* 29:3263–77
11. Chaiken IM, Freedman MH, Lyerla JR Jr, Cohen JS. 1973. Preparation and studies of ^{19}F-labeled and enriched ^{13}C-labeled semisynthetic ribonuclease-S′ analogues. *J. Biol. Chem.* 248: 884–91
12. Chambers SE, Lau EY, Gerig JT. 1994. Origins of fluorine chemical shifts in proteins. *J. Am. Chem. Soc.* 116:3603–4
13. Chervitz SA, Falke, JJ. 1996. Molecular mechanism of transmembrane signaling by the aspartate receptor. *Proc. Natl. Acad. Sci. USA.* In press
14. Chervitz SA, Falke JJ. 1995. Lock on/off disulfides identify the transemembrane signaling helix of the aspartate receptor. *J. Biol. Chem.* 41:24043–53
15. Chervitz SA, Lin CM, Falke JJ. 1995. Transmembrane signaling by the aspartate receptor: engineered disulfides reveal static regions of the subunit interface. *Biochemistry* 34:9722–33
16. Cistola DP, Hall KB. 1995. Probing internal water molecules in proteins using two-dimensional ^{19}F–^{1}H NMR.

J. Biomol. Nucl. Magn. Reson. 5: 415–19
17. Colmenares LU, Liu RSH. 1996. Fluorinated phenylrhodopsin analogs-binding selectivity, restricted rotation, and ^{19}F NMR studies. *Tetrahedron*. In press
18. Connick TJ, Reilly RT, Dunlap RB, Ellis PD. 1993. Fluorine-19 nuclear magnetic resonance studies of binary and ternary nuclear magnetic resonance studies of binary and ternary complexes of thymidilate synthase utilizing a fluorine-labeled folate analogue. *Biochemistry* 32:9888–95
19. Danielson MA, Biemann H-P, Koshland DE Jr, Falke JJ. 1994. Attractant- and disulfide-induced conformational changes in the ligand binding domain of the chemotaxis aspartate receptor: a ^{19}F NMR study. *Biochemistry* 33: 6100–9
20. Danielson MA, Falke JJ. 1996. Fluorine NMR of proteins involved in chemotaxis. In *Encyclopedia of Nuclear Magnetic Resonance*, ed. DM Grant, RK Harris, pp. 855–61 Chichester: Wiley.
21. de Dios AC, Pearson JG, Oldfield E. 1993. Secondary and tertiary structural effects on protein NMR chemical shifts: an ab initio approach. *Science* 260:1491–96
22. Drake SK, Bourret RB, Luck LA, Simon MI, Falke JJ. 1993. Activation of the phosphosignaling protein CheY I. Analysis of the phosphorylated conformation by ^{19}F NMR and protein engineering. *J. Biol. Chem.* 268: 13081–88
23. Dubois BW, Cherian SF, Evers AS. 1993. Volatile anesthetics compete for common binding sites on bovine serum albumin: a ^{19}F NMR study. *Proc. Natl. Acad. Sci. USA* 90: 6478–82
24. Eccleston JF, Molloy DP, Hinds MG, King RW, Feeney J. 1993. Conformational differences between complexes of elongation factor Tu studied by ^{19}F-NMR spectroscopy. *Eur. J. Biochem.* 218:1041–47
25. Falke JJ, Koshland DE Jr. 1987. Global flexibility in a sensory receptor: a site-directed cross-linking approach. *Science* 237:1596–600
26. Falke JJ, Luck LA, Scherrer J. 1992. ^{19}F nuclear magnetic resonance studies of aqueous and transmembrane receptors. Examples from the *Escherichia coli* chemosensory pathway. *Biophys. J.* 62:82–86
27. Flocco MM, Mobray SL. 1994. The 1.9 Å X-ray structure of a closed unliganded form of the glucose/galactose receptor from *Salmonella typhimurium*. *J. Biol. Chem.* 269:8931–36
28. Gamcsik MP, Gerig JT, Swenson RB. 1986. Fluorine-NMR studies of chimpanzee hemoglobin. *Biochim. Biophys. Acta* 874:372–74
29. Gammon KL, Smallcombe SH, Richards JH. 1972. Magnetic resonance studies of protein-small molecule interactions. Binding of N-trifluoroacetyl-D-(and L-)-p-fluorophenylalanine to chymotrypsin. *J. Am. Chem. Soc.* 94:4573–80
30. Gerig JT. 1994. Fluorine NMR of proteins. *Prog. Nucl. Magn. Reson. Spectrosc.* 26:293–370
31. Gerig JT. 1989. Fluorine nuclear magnetic resonance of fluorinated ligands. *Methods Enzymol.* 177:3–23
32. Gregory DH, Gerig JT. 1991. Prediction of fluorine chemical shifts in proteins. *Biopolymers* 31:845–58
33. Gregory DH, Gerig JT. 1991. Structural effects of fluorine substitution in proteins. *J. Comp. Chem.* 12:180–85
34. Guiles RD, Basus VJ, Sarma S, Malpure S, Fox KM, et al. 1993. Novel heteronuclear methods of assignment transfer from a diamagnetic to a paramagnetic protein: application to rat cytochrome b5. *Biochemistry* 32: 8329–40
35. Hazelbauer GL, Berg HC, Matsumura P. 1993. Bacterial motility and signal transduciton. *Cell* 73:15–22
36. Henry GD, Maruta S, Ikebe M, Sykes BD. 1993. Observation of multiple myosin subfragment 1-ADP-fluoroberyllate complexes by ^{19}F NMR spectroscopy. *Biochemistry* 32:10451–56
37. Ho C, Dowd SR, Post JFM. 1985. ^{19}F NMR investigations of membranes. *Curr. Top. Bioenerg.* 14:53–95
38. Ho C, Pratt EA, Wu C-S, Yang JT. 1989. Membrane-bound D-lactate dehydrogenase of *Escherichia coli*: a model for protein interactions in membranes. *Biochim. Biophys. Acta* 988: 173–84
39. Hoeltzi SD, Frieden C. 1994. ^{19}F NMR spectroscopy of [6–^{19}F]tryptophan-labeled *Escherichia coli* dihydrofolate reductase: equilibrium folding and ligand binding studies. *Biochemistry* 33:5502–9
40. Hoeltzi SD, Frieden C. 1995. Stopped-flow NMR spectroscopy: real-time unfolding studies of 6-^{19}F-tryptophan labeled *E. coli* dihydrofolate reductase. *Proc. Natl. Acad. Sci. USA* 92: 9318–22

41. Hoeltzi SD, Ropson IJ, Freiden C. 1994. Application of equilibrium and stopped-flow ^{19}F NMR spectroscopy to protein folding: studies of *E. coli* dihydrofolate reductase. *Tech. Protein Chem.* 5:455–65
42. Hull WE, Sykes, BD. 1976. Fluorine-19 nuclear magnetic resonance study of fluorotyrosine alkaline phosphatase: the influence of zinc on protein structure and a conformational change induced by phosphate binding. *Biochemistry* 15:1535–46
43. Johnson CD, Gerig JT. 1995. Dynamics at the active site of N-(4-fluorophenyl),N-(2,6-difluorophenyl)carbamoyl-α-chymotrypsin. *Magn. Res. Chem.* In press
44. Kim H-W, Perez JA, Ferguson SJ, Campbell ID. 1990. The specific incorporation of labelled aromatic amino acids into proteins through growth of bacteria in the presence of glyphosate. *FEBS Lett.* 272:34–36
45. Kossman M, Wolff C, Manson MD. 1988. Maltose chemoreceptor of *Escherichia coli* interaction of maltose-binding protein and the tar signal transducer. *J. Bacteriol.* 170:4516–21
46. Koul AK, Wasserman GF, Warme, PK. 1979. Semi-synthetic analogs of cytochrome C at positions 67 and 74. *Biochem. Biophys. Res. Commun.* 89: 1253–59
47. Labroo VM, Hebel D, Krik KL, Cohen LA, Lemieux C, Schiller PW. 1991. Direct electrophilic fluorination of tyrosine in dermorphin analogs and its effect on biological activity, receptor affinity, and selectivity. *Int. J. Pept. Protein Res.* 37:440–49
48. Lee GF, Dutton DP, Hazelbauer GL. 1995. Identification of functionally important helical faces in transmembrane segments by scanning mutagenesis. *Proc. Natl. Acad. Sci. USA* 92: 5416–20
49. Li E, Qian S-J, Nader L, Young N-CC, d'Avignon A, et al. 1989. Nuclear magnetic resonance studies of 6-fluorotryptophan-substituted rat cellular retinol-binding protein II produced in *Escherichia coli. J. Biol. Chem.* 264: 17041–48
50. Li E, Quian S-J, Yang N-CC, d'Avignon A, Gordon JI. 1990. Nuclear magnetic resonance studies of 6-fluorotryptophan-substituted rat cellular retinol binding protein II produced in *Escherichia coli. J. Biol. Chem.* 265: 11549–54
51. Lian C, Le H, Montez B, Patterson J, Harrell S, et al. 1994. Fluoine-19 nuclear magnetic resonance spectroscopic study of fluorophenylalanine-and flurortryptophan-labeled avian egg white lysozymes. *Biochemistry* 33:5238–45
52. Lowry DF, Roth AF, Rupert AF, Rupert PB, Dahlquist FW, et al. 1994. Signal transduction in chemotaxis. A propogating conformation change upon phosphorylation of cheY. *J. Biol. Chem.* 269:26358–62
53. Lu P, Jarema M, Mosser K, Daniel WE. 1976. Lac repressor: 3-fluorotyrosine substitution for nuclear magnetic resonance studies. *Proc. Natl. Acad. Sci. USA* 73:3471–75
54. Luck LA, Falke JJ. 1991. ^{19}F NMR studies of the D-galactose chemosensory receptor. 1. Sugar binding yields a global structural change. *Biochemistry* 30:4248–56
55. Luck LA, Falke JJ. 1991. ^{19}F NMR studies of the D-galactose chemosensory receptor. 2. Ca(II) Binding yields a local structural change. *Biochemistry* 30:4257–61
56. Luck LA, Falke JJ. 1991. Open conformation of a substrate-binding cleft: ^{19}F NMR studies of cleft angle in the D-galactose chemosensory receptor. *Biochemistry* 30:6484–90
57. Lukat GS, McCleary WR, Stock AM, Stock JB. 1992. Phosphorylation of bacterial response regulator proteins by low molecular weight phospho-donors. *Proc. Natl. Acat. Sci. USA* 89: 718–22
57a. Mao B, Pear MR, McCammon JA, Quiocho FA. 1982. Hinge-bending in L-arabinose binding protein: the "venus-flytrap" model. *J. Biol. Chem.* 257:1131–33
58. Maruta S, Henry GD, Sykes BD, Ikebe M. 1993. Conformation of the stable myosin-ADP-aluminum fluoride and myosin-ADP-beryllium fluoride complexes and their analysis using ^{19}F NMR. *J. Biol. Chem.* 268:7093–100
59. Milburn MV, Prive GG, Milligan DL, Scott WG, Yeh J, et al. 1991. Three-dimensional structures of the ligand-binding domain of the bacterial aspartate receptor with and without a ligand. *Science* 254:1342–47
60. Miller DM, Olson JS, Quiocho FA. 1980. The mechanism of sugar binding to the periplasmic receptor for galactose chemotaxis and transport in *Escherichia coli. J. Biol. Chem.* 255: 2465–71
61. Millet F, Raftery MA. 1972. An NMR method for characterizing conforma-

tion changes in proteins. *Biochem. Biophys. Res. Commun.* 47:625–32
62. Milligan DL, Koshland DE Jr. 1988. Site-directed crosslinking. Establishing the dimeric structure of the aspartate receptor of bacterial chemotaxis. *J. Biol. Chem.* 263:6268–75
63. Milligan DL, Koshland DE Jr. 1993. Purification and characterization of hte periplasmic domain of the aspartate chemoreceptor. *J. Biol. Chem.* 268:19991–97
64. Mirmira RG, Tager HS. 1991. Disposition of the phenylalanine-B25 side-chain during insulin-receptor and insulin-insulin interactions. *Biochemistry* 30:8222–29
65. Moy FJ, Lowry DF, Matsumura P, Dahlquist FW, Krywko JE, Domaille PJ. 1994. Assignments, secondary structure, global fold, and dynamics of chemotaxis Y protein using three- and four-dimensional heteronuclear (13C, 15N) NMR spectroscopy. *Biochemistry* 33:10731–42
66. Murray-Rust P, Stallings WC, Monti CT, Preston RK, Glusker JP. 1983. Intermolecular interactions of the C-F bond: the crystallographic environment of fluorinated carboxylic acids and related structures. *J. Am. Chem. Soc.* 105:3206–14
67. Parkinson JS, Kofoid EC. 1992. Communication modules in bacterial signaling proteins. *Annu. Rev. Genet.* 26: 71–112
68. Pauling L. 1960. *The Nature of the Chemical Bond.* Ithaca: Cornell Univ. Press. 644 pp. 3rd ed.
69. Pearson JG, Oldfield E, Lee FS, Warshell A. 1993. Chemical shifts in proteins: a shielding trajectory analysis of the fluorine nuclear magnetic resonance spectrum of the *Escherichia coli* galactose binding protein using a multipole shielding polarizability-local reaction field-molecular dynamics approach. *J. Am. Chem. Soc.* 115: 6851–62
70. Peersen OB, Pratt EA, Truong H-TN, Ho C, Rule GS. 1990. Site-specific incorporation of 5-fluorotryptophan as a probe of the structure and function of the membrane-bound D-lactate dehydrogenase of *Escherichia coli*: a ^{19}F nuclear magnetic resonance study. *Biochemistry* 29:3256–62
71. Post JF, Cottam PF, Simplaceanu V, Ho C. 1984. Fluorine-19 nuclear magnetic resonance study of 5-fluorotryptophan-labeled histidine-binding protein J of *Salmonella typhimurium*. *J. Mol. Biol.* 179:729–43
72. Pratt EA, Ho C. 1975. Incorporation of fluorotryptophans into proteins of *Escherichia coli*. *Biochemistry* 14: 3035–40
73. Rastinejad F, Artz P, Lu P. 1993. Origin of the asymmetrical contact between *lac* repressor and *lac* operator DNA. *J. Mol. Biol.* 233:389–99
74. Rastinejad F, Evilia C, Lu P. 1995. Studies of nucleic acids and their protein interactions by ^{19}F NMR. *Methods Enzymol.* 261:560–75
75. Rastinejad F, Lu P. 1993. Bacteriophage T7 RNA polymerase: ^{19}F nuclear magnetic resonance observations at 5-fluorouracil-substituted promotor DNA and RNA transcript. *J. Mol. Biol.* 232:105–22
76. Ropson IJ, Frieden C. 1991. Dynamic NMR spectral analysis and protein folding: identification of a highly populated folding intermediate of rat intestinal fatty acid-binding protein by ^{19}F NMR. *Proc. Natl. Acad. Sci. USA* 89:7222–26
77. Rule GS, Pratt EA, Simplaceanu V, Ho C. 1987. Nuclear magnetic resonance and molecular genetic studies of the membrane-bound D-lactate dehydrogenase of *Escherichia coli*. *Biochemistry* 26:549–56
78. Segall JE, Manson MD, Berg HC. 1982. Signal processing times in bacterial chemotaxis. *Nature* 296:855–57
79. Sharff AJ, Rodseth LE, Spurlino JC, Quiocho FA. 1992 Crystallographic evidence of a large ligand-induced hinge-twist motion between the two domains of the maltodextrin binding protein involved in active transport and chemotaxis. *Biochemistry* 31: 10657–63
80. Spotswood TM, Evans JM, Richards JH. 1967. Enzyme-substrate interaction by nuclear magnetic resonance. *J. Am. Chem. Soc.* 89:5052–54
81. Stirtan W, Withers SG. 1993. Direct ^{19}F NMR titration of phosphorylase molecules binding to fluorine-labeled glycogen particles. *Carbohydr. Res.* 249:253–58
82. Stock AM, Martinez-Hackert E, Rasmussen BF, West AH, Stock JB, et al. 1993. Structure of the Mg^{2+}-bound form of CheY and mechanism of phosphoryl transfer in bacterial chemotaxis. *Biochemistry* 32:13375–80
83. Stock JB, Lukat GS. 1991. Bacterial chemotaxis and the molecular logic of intracellular signal transduction networks. *Annu. Rev. Biophys. Biophys. Chem.* 20:109–36
84. Sun Z-Y, Truong H-TN, Pratt EA,

Sutherland DC, Kulig CE, et al. 1993. A ^{19}F-NMR study of the membrane-binding region of D-lactate dehydrogenase of *Escherichia coli*. *Protein Sci.* 2:1938–47

84a. Swanson RV, Lowry DF, Matsumura P, McEvoy MM, Simon MI, Dahlquist FW. 1995. Localized perturbations in CheY structure monitored by NMR identify a CheA binding interface. *Nat. Struct. Biol.* 2:906–10

85. Sykes BD, Hull, WE. 1978. Fluorine nuclear magnetic resonance studies of proteins. *Methods Enzymol.* 49: 270–95

86. Sylvia LA, Gerig JT. 1993. NMR studies of the a-chymotrypsin-⟩–1-acetamido – 2-(4-fluorophenyl)ethane – 1-boronic acid complex. *Biochim. Biophys Acta* 1163:321–34

87. Tanaka H, Osaka F, Ohashi M, Shiraki M, Munekata E. 1986. Substance-P analogues containing para-fluoro-L-phenylalanine. *Chem. Lett.* 391–94

88. Taylor HC, Komoriya A, Chaiken IM. 1985. Crystallographic structure of an active, sequence-engineered ribonuclease. *Proc. Natl. Acad. Sci. USA* 82:6423–26

89. Taylor HC, Richardson DC, Richardson JS, Wlodawer A, Komoriya A, Chaiken IM. 1981. Active conformation of an inactive semi-synthetic ribonuclease-S. *J. Mol. Biol.* 149:313–17

90. Truong T-TN, Pratt EA, Ho C. 1991. Interaction of the membrane-bound D-lactate dehydrogenase of *Escherichia coli* with phospholipid vesicles and reconstitution of activity using a spin-labeled fatty acid as an electron acceptor: a magnetic resonance and biochemical study. *Biochemistry* 30: 3893–98

91. Vine WH, Brueckner DA, Needleman P, Marshall GR. 1973. Synthesis, biological activity, and ^{19}F nuclear magnetic resonance spectra of angiotensin II analogs containing fluorine. *Biochemistry* 12:1630–37

92. Volz K, Matsumura P. 1991. Crystal structure of *Escherichia coli* CheY refined at 1.7 Å resolution. *J. Biol. Chem.* 266:15511–19

93. Vyas NK, Vyas MN, Quiocho FA. 1987. A novel calcium binding site in the galactose-binding protein of bacterial transport and chemotaxis. *Nature* 327:635–38

94. Wagner G, Wütrich K. 1986. Observation of internal motility of proteins by nuclear magnetic resonance in solution. *Methods Enzymol.* 131:307–26

95. Westhead EW, Boyer PD. 1961. The incorporation of *p*-fluorophenylalanine into some rabbit enzymes and other proteins. *Biochim. Biophys. Acta* 54:145–50

96. Williams S-P, Fulton AM, Brindle KM. 1993. Estimation of the intracellular free ADP concentration by ^{19}F NMR studies of fluorine-labeled yeast phosphoglycerate kinase in vivo. *Biochemistry* 32:4895–902

97. Wilson ML, Dahlquist FW. 1985. Membrane protein conformational change dependent on the hydrophobic environment. *Biochemistry* 24: 1920–28

98. Wuttke DS, Gray HB, Fisher SL, Imperaili B. 1993. Semisynthesis of bipyridyl-alanine cytochrome c mutants: novel proteins with enhanced electron-transfer properties. *J. Am. Chem. Soc.* 115:8455–56

Annu. Rev. Biophys. Biomol. Struct. 1996. 25:197–229

ANTIBODIES AS TOOLS TO STUDY THE STRUCTURE OF MEMBRANE PROTEINS: The Case of the Nicotinic Acetylcholine Receptor

Bianca M. Conti-Fine and Sijin Lei*

Department of Biochemistry, University of Minnesota, St. Paul, Minnesota 55108, and Department of Pharmacology, University of Minnesota, Minneapolis, Minnesota 55455

Kathryn E. McLane

Department of Chemistry and Department of Biochemistry and Molecular Biology, University of Minnesota, Duluth, Minnesota 55812

KEY WORDS: transmembrane topology, ton channel superfamily, cholinergic receptor

ABSTRACT

The nicotinic acetylcholine receptor is the prototype of the ionotropic receptor superfamily of proteins, which includes the closely related γ-aminobutyric acid type A and glycine receptors, and more distantly related serotonin type-3 and glutamate receptors. Several models of the transmembrane topology of the nicotinic acetylcholine receptor subunits were originally proposed based on hydropathy analysis of their deduced amino acid sequences. Antibodies specific to different epitopes of the nicotinic acetylcholine receptor have proven to be valuable probes for examining the validity of those models. Despite important caveats, a viable model for the transmembrane structure and functional topology

*previous literature cited under Bianca M. Conti-Tronconi

1056–8700/96/0610–0197$08.00

of the nicotinic acetylcholine receptor subunits has been obtained from the antibody mapping studies. This model, and the associated methodological shortcomings and obstacles that were overcome in the process of its formulation, can legitimately be extended to other members of the ionotropic receptor superfamily and to other membrane proteins as well.

CONTENTS

INTRODUCTION

Although X-ray crystallographic studies of scarce membrane proteins are still daunting, the sequences of many membrane proteins are available, and sequence-specific antibodies (Abs) can be developed easily with the use of synthetic or biosynthetic peptides (7, 104). In spite of their wide use, however, there is a great deal of controversy as to whether Abs raised against denatured forms of a protein can be used legitimately for structural studies of the cognate native protein.

The nicotinic acetylcholine receptor (AChR) of *Torpedo* electric organ (TAChR), the prototype of the ligand-gated ion channel superfamily (26), can be purified in large amounts. This feature makes TAChR amenable to structural studies that are impossible for any other neurotransmitter receptor and, in general, for most membrane proteins. Those studies included Ab-based investigations, which could be verified in most cases by different experimental approaches. The TAChR, therefore, is the benchmark to define the legitimacy, limits, and caveats of the use of Abs for structural studies of membrane proteins.

The AChR is a complex transmembrane protein formed by five homologous or identical subunits, which are symmetrically arranged around a central cation channel (3, 25, 26, 59, 82, 90). In peripheral tissues, such as vertebrate muscle and fish electric organ, the AChR is composed of four types of subunits—α, β, γ (or ϵ), and δ—in a stoichiometry $\alpha_2\beta\gamma\delta$. Neurons express AChRs that may be formed by only three (one α_X, one α_Y, and three β subunits) or two (two α and three β) types of homologous subunits. Neuronal AChRs may also exist as homo-oligomers of five identical α subunits. Peripheral tissues express only one or two isoforms of the AChR subunits. Neuronal tissues express multiple subtypes of α (from $\alpha 2$ to $\alpha 9$) and β (from $\beta 2$ to $\beta 5$) subunits, with characteristic localization within the nervous system. A great variety of neuronal AChR subtypes results from the combination of different α and β subunits.

All AChR subunits have a high degree of sequence identity. Several ligand-gated ion channels share sequence homology with the AChR subunit and are considered members of the same protein superfamily (26). The subunits of the glycine and γ-amino butyric acid type A ($GABA_A$) receptors have 22–34% amino acid sequence identity with the AChR subunits, whereas more distant relatives, such as the serotonin type-3 (5HT-3) receptor and, possibly, the glutamate receptors, share less than 20% sequence identity. Multiple duplications of a common ancestral gene, and the divergence of the resulting new genes, likely created the different subunits of a single receptor complex, various subtypes of the same receptor, and ultimately receptors that bind different ligands and have different ion selectivity.

Given their sequence similarity, the different members of the ionotropic receptor superfamily may share a conserved structural framework. Conclusions obtained for the easily studied TAChR, therefore, might be extrapolated to the structure and transmembrane folding of other members of the superfamily. This possibility is supported by the demonstration that functional, ligand-gated homomeric ion channels are obtained by expressing a chimeric subunit that combines the extracellular domain of the $\alpha 7$ AChR subunit, which contains the ligand-binding site, with the carboxyl-terminal region of the 5HT-3 receptor, which forms the transmembrane and intracellular domains. The resulting chimeric receptors have the pharmacology of the $\alpha 7$ AChR and the ion channel properties of the 5HT-3 receptor (35). The glutamate receptors, however, appear to have distinctly different topologic features, compared with the other, more closely related members of this protein superfamily.

Sequence characteristics shared by all the AChR, $GABA_A$, glycine, and 5HT-3 receptor subunits include the following: (*a*) a putative extracellular aminoterminal domain that contains two cysteine residues separated by approximately 15 amino acids ("Cys-Cys loop"); (*b*) four putative transmembrane regions, designated M1–M4; (*c*) conservation of a proline in the M1 segment; (*d*) an abundance of serine, threonine, and small aliphatic amino acids in the M2 segment; and (*e*) a putative cytoplasmic, nonconserved region between M3 and M4.

In this review, we first summarize the characteristics and caveats of the Abs used for structural studies of the AChR. We then review studies, carried out with the use of Abs or other approaches, on the transmembrane topology of the AChR subunits and the sequence regions of the AChR that form surface domains involved in AChR function and dysfunction. Finally, we discuss evidence that the $GABA_A$, glycine, and glutamate receptors have both conserved and diverged structural features.

USE OF ANTIBODIES FOR STRUCTURAL STUDIES OF THE ACETYLCHOLINE RECEPTOR: CHARACTERIZATION AND CAVEATS

Most of these Abs fall into one of two groups, which have different advantages and shortcomings. One group includes Abs raised against native AChR: The sequence regions and residues that form their epitopes were identified with the use of synthetic AChR sequences. The other group includes Abs raised against short synthetic AChR sequences, which were used to study the native AChR molecule.

A different type of Ab used for structural studies of the AChR (4) is against a "reporter epitope" unrelated to the AChR, which is inserted by genetic engineering at a defined position in the sequence of a TAChR subunit. The altered subunit is expressed as part of a native TAChR molecule, which is used for immunolocalization studies of the transmembrane disposition of the reporter epitope and, therefore, of the TAChR residues surrounding it.

Antibodies Raised Against Native Acetylcholine Receptor: Mapping of Their Epitopes with the Use of Synthetic Sequences

Synthetic peptides have been used extensively as representative structural elements of the cognate protein to identify sequence(s) and residues recognized by Abs raised against the native cognate protein, including the AChR. Experimental strategies that use a sequence region excised from the structural context of the cognate native protein are laden with important caveats (63). The legitimacy of these strategies needs to be investigated carefully, because most Ab epitopes are formed by discontinuous sequence regions (8, 29). It is unlikely, therefore, that a short peptide, excised from the structural context of the native protein, accurately represents the surface domain to which that sequence contributes.

To map Ab epitopes successfully with the use of synthetic peptides, a substantial portion of the surface domain recognized by the Ab must be formed by residues within a continuous sequence region. This region may then be represented by a synthetic peptide. Synthetic peptides may be used more confidently to study domains that interact with protein ligands, such as Abs, because high-affinity protein–protein binding involves interaction of large surfaces, which may include several residues contained in continuous sequence segments.

Sequence segments that contain several residues crucial to the formation of epitopes may fold in a manner incompatible for interaction with the specific Ab when they are released from the structural constraints of the native protein. This occurrence could lead to false negative conclusions.

A less frequent, but possible, shortcoming of short synthetic sequences is that Abs, or other large ligands, might bind to a peptide whose corresponding sequence in the native protein is inaccessible, because of obstruction by surrounding residues. This possibility has been verified in studies of the AChR. For example, some monoclonal Abs (mAbs) against the main immunogenic region (MIR), which is the epitope(s) that dominates the anti-AChR Ab response in the autoimmune disease myasthenia gravis, specifically recognize the synthetic sequence 67–76 of the human muscle AChR α-subunit—a sequence region which is known to contain important constituent elements of the MIR but does not cross-react with native human AChR (95, 97).

Despite their potential pitfalls, Ab-based studies of the AChR that used synthetic peptides have had reliable predictive value, as discussed below. Synthetic peptides allowed the rapid screening of the sequence of AChR subunits for identification of sequence regions and residues that are likely to contribute to the structure of individual domains.

Antibodies Against Short Synthetic Acetylcholine Receptor Sequences: Determination of Their Ability to Bind Nondenatured Acetylcholine Receptor

Although Abs against synthetic sequences of scarce membrane receptors have been used extensively to study their cellular localization, subunit composition, transmembrane topology, and structure of their functional domains (7), it is unclear whether Abs raised against synthetic peptides recognize the native cognate protein. It has been argued that because Ab epitopes on native proteins are formed by residues from different sequence regions (8, 29), any cross-reactivity between an antipeptide Ab and the cognate native protein must be caused by the presence of denatured molecules in preparations of "native" protein. For example, antipeptide Abs believed to recognize the cognate native protein—to which they bound in enzyme-linked immunosorbent assays (ELISA), where the protein may be partially denatured because of its interaction with the plastic surface—did not recognize the same native protein in solution (88).

The ability of different antipeptide mAbs to recognize the native TAChR molecules could be determined accurately with the use of quantitative immunoprecipitation of solubilized TAChR and immunoelectron microscopic analysis of postsynaptic membrane fragments. These methods also indicated the transmembrane topology of the corresponding sequence regions (55). Only mAbs against the synthetic sequence α304–322, α332–350, and α360–378 cross-reacted fully with solubilized TAChR molecules and bound to the cytoplasmic surface of TAChR in membrane fragments. These sequence segments are exposed largely on the TAChR surface, and the corresponding synthetic peptides must be able to fold in a conformation reminiscent of that assumed by the same sequence on the surface of the native TAChR.

The epitopes recognized by the mAbs that are cross-reactive with native TAChR need not be formed by linear sequences of residues. Antipeptide mAbs may recognize discrete side chains or clusters of side chains along the peptide sequence, which are brought together by the folding of the peptide bound to the mAb, resulting in a similar conformation for the same sequence in both the peptide and the native cognate protein (6, 24).

A second mechanism that contributes to the ability of antipeptide mAbs to cross-react with the cognate protein is the flexibility of the Ab binding site. This flexibility, which allows for an induced fit upon antigen binding, was demonstrated for complexes of antipeptide mAbs cross-reactive with the cognate protein [influenza virus hemagglutinin (IVH)], with the peptide immunogen. The dual recognition by the mAbs of the peptide and the IVH occurred both because the folding of the peptide was similar to that of the same sequence in the native IVH, and a conformational change occurred in the mAb binding site upon antigen binding, which resulted in a better fit with the peptide (77).

Antipeptide mAbs that recognize the synthetic aminoterminal segment of the TAChR α subunit, α1–20, did not cross-react with native TAChR (55). Those mAbs bound to TAChR in ELISA, which confirmed that interaction with the plastic surface denatures the TAChR and that ELISA is not a reliable method to assess cross-reactivity of Abs with native proteins (88).

Some TAChR synthetic sequences that had substantial stretches of hydrophobic residues resulted in induction of mAbs that cross-reacted widely with unrelated synthetic peptides (55). Although no data are available regarding the epitope structure recognized by those mAbs, those data strongly advise avoiding sequences that contain hydrophobic regions for antipeptide Abs induction.

STUDIES ON THE STRUCTURE AND TOPOLOGY OF THE ACETYLCHOLINE RECEPTOR SUBUNITS

All AChR subunits form extracellular and cytoplasmic domains (25, 26, 90). Hydropathy analysis of a "typical" AChR subunit (25, 90) identifies a long aminoterminal region of approximately 200 amino acids rich in hydrophilic residues which could form an extracellular region. The putatuve extracellular domain is followed by four hydrophobic segments approximately 20 amino acids long, which potentially could be transmembrane α-helices (Figure 1), referred to as M1–M4. Between M3 and M4 is the most variable region. This region contains a segment, called MA, that has the periodicity of an amphipatic α helix (37). M4 is followed by a short carboxyl terminal region.

Models of the transmembrane folding of the AChR subunits have been proposed, with four (M1–M4), five (M1–M4 and MA), or six (M1–M6) transmembrane segments. The carboxyl terminus is extracellular in the four-transmembrane segment model and is cytoplasmic in the five-transmembrane segment model; the whole region preceding M1 is extracellular in both models. The six-transmembrane segment model includes two additional transmembrane domains formed by sequence segments preceding M1; in this model, both the amino and the carboxyl termini are extracellular.

We review the experimental evidence that supports each model. We emphasize the evidence obtained by using Abs and compare the results of nonimmunologic studies.

The Extracellular Aminoterminal Domain

STUDIES WITH ANTIBODIES Antibodies specific for the aminoterminus of AChR subunits do not bind native AChR. This finding suggests that

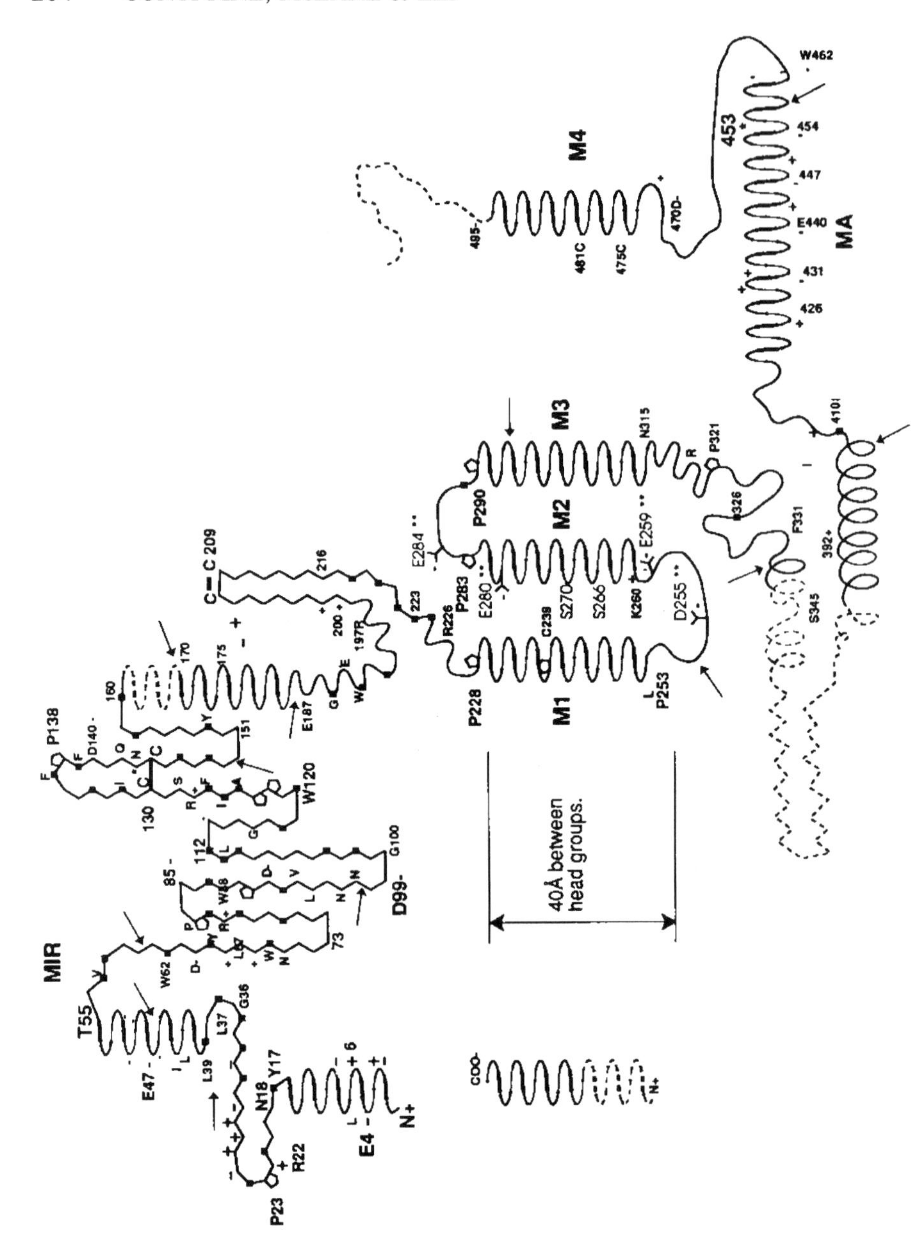
MIR
M1
M2
M3
M4
MA
40Å between
head groups.
C — C 209
P228
P253
P283
P290
N315
P321
E440
W462
W120
P138
T55
E47
P23
R22
E4
N+
D99-
G100

the aminoterminus of native AChR, although extracellular as demonstrated by nonimmunologic studies, is inacccssible to Abs (55).

Results obtained with two Abs against the synthetic sequence α152–167 of TAChR suggested the existence in this region of two unsuspected transmembrane segments (M6 and M7) (28). Although M7 (residues 160–185) could be modeled as an amphipathic α helix long enough to span the membrane, M6 includes only ten residues (residues 142–151) and would cross the membrane in some extended conformation. Another study that used anti-AChR Abs suggested that the M7 sequence is not transmembrane and may be extracellular (67).

At least part of the sequence region between the aminoterminus and M1 must be extracellular because mAbs and other probes that recognize the MIR and the cholinergic binding site (see below), which are on the extracellular surface, bind to sequence segments within this region.

Anti–MIR mAbs recognize the sequence regions 67–76 or 61–76 of the electroplax and muscle α subunits (26). This sequence region is predicted to form a hydrophilic hairpin loop, whose negatively charged apex is a type I β-turn formed by residues α68–71 and whose arms are β-strand (26). Residues that comprise the apex of the loop likely interact with anti-MIR mAbs, because substitutions of individual residues at positions α68–72—most notably Asn68, Pro69, Asp71, and Tyr72—affected mAb binding strongly (26). Asn68 and Asp71 are

Figure 1 Consensus model of a possible topology of a peripheral AChR subunit and secondary structure as predicted by amphipatic analysis. The four nonpolar, hydrophobic, putative transmembrane regions M1–M4 are indicated as α helices, although M2 and conceivably other transmembrane segments may have, partially or entirely, a β sheet configuration. Letters identify the amino acid residue of highly conserved positions. *Black squares* (single letter codes) identify conservation of hydrophobic residues in the aminoterminal putative extracellular domain, which would tend to fold toward the inside. Residues in M2 whose mutation causes charge in conductance are indicated in narrow letters; **, those that are charged. Positive and negative signs identify regions that generally carry charged side chains. The conserved cysteines, known to form a disulfide-linked loop in all four neuromuscular AChR subunits (indicated as C130–C144, according to the numbering of the consensus aligment and corresponding to Cys128 and Cys142 of the *Torpedo* α subunit), enclose the conserved site of N-linked glycosylation site, found in all neuromuscular AChR subunits. The adjacent, disulfide-linked cysteines at the agonist binding site, found only in AChR α subunits, are labeled C208 and 209, following the numbering of the consensus alignment (corresponding to Cys192 and Cys193 of the TAChR α subunit). The MIR is located on the aminoterminal extracellular domain. *Dashed lines* indicate sequence regions that are not conserved between species, and *arrows* indicate common intron boundaries. (From Reference 90, with permission.)

highly conserved in all AChRs that bind anti-MIR mAbs and are substituted nonconservatively in a frog AChR isomer, which is the only known muscle AChR that cannot bind anti-MIR mAbs (42). Some anti-MIR mAbs bind to the synthetic sequence α55–74, which overlaps α67–76 and might contribute to the cholinergic site (see below). Within this sequence, Trp60 and Asp62 were identified as involved in MIR formation (26).

A recent elegant study of the tridimensional localization of the MIR, in which electron microscopic investigation of the complexes between TAChR and an anti-MIR mAb was used, located the two MIR on each AChR complex at the extreme synaptic end of each α subunit (10). The end of each α subunit forms a distinct protrusion that seems well separated from those of the neighboring subunits. Subunits other than the α, therefore, do not seem involved in MIR formation (10).

STUDIES WITH OTHER APPROACHES The TAChR δ subunit contains a processed signal peptide (5), which indicates that the aminoterminus of the mature subunit is extracellular. The aminotermini of AChR subunits expressed in vitro are translocated into the lumen of microsomal vesicles, which are equivalent topologically to the extracellular space (22 and references therein). Studies that investigated the transmembrane orientation of the TAChR δ subunit by selective proteolysis also concluded that the aminoterminus is extracellular (108).

The sequence region surrounding positions 30 and 141 must be extracellular, because Asn residues at these positions are N-glycosylated (25). The transmembrane disposition of the aminoterminus of the α subunit has been determined also by introducing novel glycosylation sites (22). An α subunit fragment that terminated at position α_{207} was found to be a nonintegral membrane protein. Glycosylation sites introduced at position α_{154} and α_{200} were on the lumenal side of microsomal vesicles, which suggests that the entire domain preceding M1 is extracellular (22).

Studies on the tridimensional structure of TAChR that use low-dose electron microscopy and X-ray diffraction concluded that the volume of protein protruding toward the extracellular space is 215,000 Å^3, which corresponds to 157 kDa (26). This finding corresponds precisely to the predicted molecular mass formed by the complete aminoterminal region, up to M1, of all five glycosylated subunits plus the molecular mass of the oligosaccharide moieties present in the AChR (26).

SEQUENCE REGIONS AND INDIVIDUAL RESIDUES FORMING THE CHOLINERGIC BINDING SITE The AChR has at least two cholinergic ligand-binding sites. In vertebrate muscle and fish electroplax AChRs, one site is at the interface between one α subunit and the γ subunit, and the other is at the interface between the other α subunit and the δ subunit. In neuronal AChRs, the sites are at the interfaces between α and β subunits (26). The α subunits have a crucial role in the formation of binding sites, as indicated by the labeling with cholinergic affinity probes and the snake toxin, α-bungarotoxin (αBTX), and the influence of the α subunit subtype on the pharmacologic properties of the corresponding AChR complex (26).

Monoclonal antibodies that can compete with cholinergic ligands and can recognize the cholinergic binding site have been used to demonstrate that this region is extracellular. The structural elements that form the binding site for cholinergic ligands also have been studied with the use of α and κ neurotoxins. These neurotoxins are high-affinity competitive antagonists, specific for peripheral AChR and selected neuronal AChRs (the α toxins) and for selected neuronal AChRs (the κ toxins). Because of their polypeptide nature and the large area of contact with the AChR molecule, their binding properties are reminiscent of those of Abs: very high affinity, slow reversibility or virtual irreversibility of their binding to the AChR, and ability to recognize denatured forms of the α subunits. The sequence segments and residues that form the cholinergic site were also identified by affinity labeling and mutation approaches. Those studies have been reviewed extensively elsewhere (26, 56, 63). Here we summarize only the results relevant to identification of sequence regions and residues exposed on the extracellular surface of the TAChR.

The combined results of studies that used cholinergic competitive mAbs, α and κ neurotoxin, and synthetic or biosynthetic sequences of different α subunit subtypes as representative structural elements of the corresponding AChRs indicate that two or more sequence regions, within the aminoterminal domain before M1, contribute to the cholinergic binding site.

In both TAChR and muscle AChRs, the sequence region that surrounds the vicinal Cys/Cys pair at positions approximately 192 and 193, found in all α subunits, binds αBTX (26, 56). In the TAChR α subunit, the sequence segment α55–74 is also recognized by αBTX and three different cholinergic competitive mAbs, W2, WF5, and WF6 (26).

Residues within the sequence α181–200, critical for binding of αBTX and mAb WF6, were determined with the use of single-residue substituted peptide analogues (26). Although WF6 and αBTX compete for TAChR binding mutually, the amino acid residues that form their binding site are not the same. The contact points identified for αBTX (Val188, Tyr189, Tyr190, Cys192, Cys193, and Phe194) and WF6

(Trp187, Thr191, Pro194, Asp195, and Tyr198) must be exposed on the TAChR surface. Other studies that used synthetic or biosynthetic peptides indicated that Asp195, Trp187, Tyr189, Tyr190, Thr191, and Pro194 were involved in interaction with αBTX (21, 93, 98).

Studies on the binding of α-neurotoxins to single-residue substituted synthetic analogues of the sequence region α55–74 of TAChR concluded that Arg55, Arg57, Trp60, Arg64, Leu65, Arg66, Trp67, and Asn68 also are involved in the interaction with α-neurotoxins (26).

Physostigmine is a representative member of a new class of cholinergic agonists. These agonists bind to a site that includes residue Lys125, which is not recognized by αBTX or mAb WF6 but is recognized by a specific mAb, FK1 (85 and references therein). Monoclonal antibodies FK1 and WF6 partially compete for binding to TAChR, which suggests that they bind to distinct epitopes that share a limited overlap (85). Studies in which biosynthetic and synthetic sequences of the TAChR α subunit were used demonstrated that structural components of the epitope for FK1 are contributed by α118–145. This epitope includes Lys125 and the "Cys-Cys loop" (formed by Cys128 and Cys142) and, to a much lesser extent, α181–216 (85 and references therein). Binding of FK1 to single-residue substituted analogues of α181–200 indicated that, as expected, FK1 has different attachment points than αBTX and WF6. Only Cys192 is important for binding of both KF1 and αBTX, and Trp187 and Asp195 are important for binding of both KF1 and WF6 (85). Tyr181, Trp184, and Asp200 are involved in KF1 binding only (85).

Affinity labeling of TAChR with cholinergic ligands confirmed that both the sequence region surrounding the vicinal Cys/Cys pair and that containing the "Cys-Cys loop" of the α subunit contribute to cholinergic binding sites (26). The cholinergic affinity labels 4-(N-malemido)benzyltri[^{3}H] methylammonium ([^{3}H]MBTA) and [^{3}H]*p*-(dimethylamino)-benzenediazonium fluoroborate (DDF) specifically labeled αCys192 and αCys193; DDF also labels Tyr190 and, to a lesser extent, Trp149 and Tyr93 (30). [^{3}H]ACh and [^{3}H]nicotine specifically label αTyr93 and αTyr198, respectively; lophotoxin, a cholinergic antagonist, labels αTyr190.

The region that surrounds αCys192/Cys193 can be cross-linked with δ164–224, which indicates proximity of these different subunit regions (26, 48). A *d*-tubocurare analogue labels a region that contains the Cys-Cys loop of the γ and δ subunits, and γTrp55 and δTrp57 (26). Mutation of γTrp55 to a Leu results in an eightfold decrease in ACh affinity and *d*-tubocurare inhibitory potency (70). These findings are consistent with the notion that the α and δ subunits interact with the α subunit in the

AChR complex to form homologous binding sites for cholinergic ligands (26).

Because mAbs against different neuronal AChR isotypes that compete with cholinergic ligands were not available, peptide-based studies that map the cholinergic sites on neuronal α subunits primarily have used snake toxins (26). In all cases, the sequence region that surrounds the vicinal Cys/Cys pair bound αBTX or κBTX. Snake κ toxins also bind the sequence regions 51–70 of the α3 subunit, which is homologous to α55–70 of TAChR recognized by αBTX and by cholinergic competitive mAbs.

The αBTX binding sequence 180–200 of the rat α5 subunit binds αBTX, although it is relatively divergent compared with TAChR and muscle AChRs (26). Residues critical for αBTX binding were identified by testing the effect of single residue substitutions: Lys184, Arg187, Cys191, and Pro195 were required for αBTX binding, and substitution of Gly185, Asn186, Asp189, Trp193, Tyr194, and Tyr196 lowered its affinity. Several aromatic amino acids were found to be critical for αBTX binding to the α5 peptide, similar to the TAChR sequence α180–200. Thus, despite the apparent divergence of the α5 sequence from other αBTX binding α subunits, some structural features have been conserved.

Single-residue substituted analogues of the sequence segment 50–71 of the α3 subunit identified residues that may form the binding site of the κ toxins, κ bungarotoxin (κBTX) and κ-flavitoxin (κFTX) (26). Several aliphatic and aromatic residues were important for κ-toxin binding (Leu54, Leu56, and Tyr63 for both κ toxins, and Trp at positions 55, 60, and 67 for κFTX). Two negatively charged residues (Glu51 and Asp62) were important for κBTX binding, whereas positively charged residues appeared to mediate electrostatic interactions with κFTX (Lys at positions 57, 64, 66, and 68). These differences in amino acid specificity correlated with sequence differences of κBTX and κFTX, thereby providing clues regarding the residue interactions at the κ-toxin/AChR interface.

Mutation of Trp54 in the α7 subunit resulted in decreased binding affinity and responses for nicotinic cholinergic agonists, without affecting αBTX binding (27). This residue is homologous to γTrp55 and δTrp57, which are labeled by *d*-tubocurare, and is in a region homologous to α50–75, which is involved in ligand binding in both TAChR and neuronal α3 AChR.

The sequence regions α181–200 and α50–75 are unusually rich in aromatic residues, and several aromatic residues contribute to cholinergic sites in both peripheral and neuronal AChRs. These findings suggest that the anionic cholinergic binding site of the AChR is formed not by a single negatively charged residue but rather by interaction of the π electrons of aromatic rings (32), as demonstrated for the cholinergic site of acetylcholinesterase (91).

Potential Transmembrane Segments (M1–M4 and MA)

Most of the information on the topology of putative transmembrane domains was obtained with the use of nonimmunologic approaches. We therefore review together the results obtained in both immunologic and nonimmunologic studies.

SEGMENTS M1–M4 Studies in which sequence-specific mAbs were used indicate that the sequence α235–242, which separates M1 and M2, is cytoplasmic (28). Because Cys192–193 and the immediately flanking region are extracellular, as part of a cholinergic binding site, M1 must cross the membrane. M1 and M2 are likely to be transmembrane, because they contribute to the ion channel (see below). A transmembrane location of M3 and M4 is supported by a study in which the disposition of these sequence segments was deduced with the use of proteolysis protection assays of fusion proteins that contain a reporter group (23). By labeling Lys residues of TAChRs in sealed microsacs, a transmembrane disposition of M4 was suggested: Lysα380, which is in a sequence region immediately aminoterminal to M4, is cytoplasmic, and Lysγ486, which is carboxyl-terminal to M4, is extracellular (33).

3-Trifluoromethyl-3-(m-[^{125}I]iodophenyl)diazirine ([^{125}I]TID, which is a hydrophobic, photoactivatable probe that partitions in the lipid bilayer and labels residues exposed to the lipid interface, reacted with both M3 and M4 (15). The distribution of the labeled residues in the M3 region suggested an α-helical conformation and defined a strip in contact with the lipids that includes three (β and γ subunits) or four (δ subunits) helical turns. [^{125}I]TID labeled five residues in the M4 segment of the α subunit, at the positions expected for an α helix, formed by five turns.

The finding that M3 and M4 are exposed to the lipid interface and may have an α helical structure has been verified by the use of [^{3}H]diazofluorene, another photoactivatable probe that generates carbenes and labels the surface of membrane proteins exposed to the hydrophobic core of the membrane (16).

The secondary structure of the transmembrane domains of TAChR has been investigated with the use of Fourier-transform infrared spectroscopy of TAChR-rich membrane preparation treated with proteinase K to remove the extramembrane part of the AChR (41). That study concluded that the transmembrane part of the TAChR contains considerable amounts of β structure, which did not seem to be uniaxially oriented, in addition to an uniaxially oriented α-helical component. That study suggested that M1 and M3 could form primarily β strands, whereas M2 and M4 may be primarily α-helical.

SEGMENTS M1 AND M2 LINE THE ION CHANNEL Hydropathy analysis of the TAChR subunit sequences first suggested that M1 and M2 may be involved in ion channel formation (25, 26). The main experimental approaches used to investigate the structure of the AChR ion channel have not used Abs, which are too large probes. These approaches are studies of mutated or chimeric AChRs (25, 26), photoaffinity labeling in which channel-specific blockers are used (25, 26, 39, 90), and high-resolution studies of the tridimensional structure of the lining of the channel in the closed and in the open state (102, 103).

Chimeric *Torpedo*-calf AChRs expressed in *Xenopous* oocytes were used to investigate the effects on conductance of mutations in the δ subunits. Mutations in the M2 region and the adjacent segment that links M2 and M3 affected channel conductance profoundly and conferred *Torpedo*-like or bovine-like channel properties to the resulting chimeric or mutated AChRs (25).

A study investigated the effect of mutation of different Ser residues in the M2 segments on the binding of QX-222, an open channel blocker (57). The apparent affinity of QX-222 decreased with the number of mutated Ser residues, which implicated M2 in channel formation. That study suggested a model in which QX-222 binds within an ion channel formed by M2 helices contributed by each of the subunits (20, 57). Another mutation study altered the size and polarity of uncharged polar amino acids between the putative cytoplasmic and extracellular negatively charged rings at either end of M2 (47). That study concluded that Thr and Ser residues in the M2 segments of the α, β, γ, and δ subunits line a short, narrow channel close to the cytoplasmic side of the membrane (47). Spontaneous mutations of the M2 segment in human muscle AChR cause a change in the ion-gating properties of the AChR, with an increase in channel open time (69).

The M2 domain has been implicated in channel formation in homo-oligomeric neuronal AChRs formed by $\alpha 7$ subunits. Mutation of Leu247 to Thr altered activation and desensitization properties and the pharmacologic profile of the resulting AChR (11, 76). This finding suggests that Leu247 blocks the ion channel in the desensitized AChR, but that mutation to a Thr renders this state conductive. Mutation of residues within the M2 segment or flanking its cytoplasmic end altered the calcium permeability of the $\alpha 7$ AChR (12).

Mutation to Cys of amino acid residues in the M2 region of the mouse muscle α subunit, followed by chemical modification of the new thiol groups, affected the ion-gating properties of the resulting AChR, which suggests that M2 is part of the channel. This approach identified residues exposed on the channel surface from their accessibility to polar sulfhydryl-specific reagents (1, 2). In one study (2), residues at positions 241–250 were mutated individually to Cys. The results obtained suggested that the alternating residues Ser248, Leu250, Ser252, and Thr254 are exposed in the closed channel, thus leading to the conclusion that this region of M2 has a β-strand structure rather than an α-helix, as was generally assumed. A channel (partly or entirely) comprised of M2 segments from five subunits, therefore, would consist of or include some form of β-barrel structure. In another study (1), 22 consecutive residues, Glu241–262, were mutated individually to Cys. All but the Lys242 mutant had vigorous ACh-evoked currents. The results suggested that Glu241, Thr244, Leu245, Ser248, Leu250, Leu251, Ser252, Val255, L258, and Glu262 contribute to channel lining. Residues at several of these positions in the α or in other AChR subunits were suggested previously to be involved in formation or function of the ion channel (26, 48). The proposed structure of M2 remains controversial, because the pattern of accessibility of residues 241–248 and 243–262 suggests an α-helical conformation, whereas the accessibility of residues Leu250, Leu251, and Ser252 supports a β strand or some other extended conformation. The segment M2 may form a broken helix interruped by a short link.

The possibility that the lining of the channel and presumably M2 is a broken α helix is supported by high-resolution (9 Å) images of TAChR in closed and open channel forms (102, 103). The membrane-spanning region of the AChR includes a pore lined by five rods (one for each subunit) of dimensions consistent with an α helix, surrounded by a continuous rim of density, which could be formed by a β sheet. The rods are bent near their midpoint, where they are closest to the axis of the pore, and tilt gradually outward on either side. On the basis of correlation between the appearance of the rods and the known M2 sequence, a Leu residue conserved in an all AChR subunit (at position 251 of the α subunit) was predicted to align with the bend, whereas the charged groups at either ends of M2 were placed symmetrically on either side of the bilayer. The side chain of the conserved Leu would project into the lumen, thereby forming a hydrophic "plug" impermeable to hydrated ions when the channel is closed (103).

The structure of the open state of the AChR channel also was identified (103). The rods surrounding the pore rotate, thereby yielding a right-handed barrel of α helixes and resulting in opening of the restriction which in the closed state results from the proximity of the rod bends. Experiments that investigated the change in accessibility of M2 residues in the closed and in the open state of the AChR suggest that opening of the channel is associated with structural changes along its entire length. Those experiments found that the changes were distributed over the entire length of M2 (1).

On the extracellular end, the ion channel is wider than in its more cytoplasmic region and seems to be formed by five α-helices (presumably M2 segments) that alternate with five other segments. Photolabeling studies (31) support the possibility that M1 contributes to the channel lining at its extracellular end.

Studies that used as photoaffinity labels noncompetitive inhibitors that bind within the ion channel also implicated M2 as a channel-forming sequence (26). The local anesthetic [^{3}H]triphenylmethylphosphonium ([^{3}H]TPMP^{+}) labels Ser262 of the δ subunit, Ser254 of the β subunit, and Ser248 of the α subunit, all within M2; [^{3}H]chlorpromazine labels Ser residues within M2 in all subunits (at positions δ262, β254, α248, and γ257), in addition to Leu residues at positions β257 and γ260 and Thr at γ253. [^{125}I]TID, which is also a potent noncompetitive inhibitor, specifically photolabels Leu257 and Val261 in the M2 segments of the β subunit and the homologous residues of the other subunits when the TAChR is in the resting state. These residues should contribute to the binding site(s) of closed channel blockers, which suggests that the permeability barrier in the closed AChR might be an hydrophobic ''plug'' made up by the aliphatic site chains of those residues. The conserved Leu residue identified in these studies is the same as that proposed to be at the apex of the observed kink in the ion-channel lining, projecting into the lumen of the channel (103).

Photoaffinity labeling by [^{3}H]quinacrine azide has been used to label TAChR in the open channel conformation (31). The segment α208–243 was labeled, which encompasses the M1 domain, but not M2.

In the presence of agonists, [^{125}I]TID specifically labels M1 residues of the γ and δ subunits (109). In the δ subunit, residues both preceding (Phe232, Ile233) and following (Cys236) a conserved Pro in the middle of M1 were labeled. The residues labeled in the γ subunit were not identified. The nonequivalence of the labeling pattern of the M1 segments of different subunits suggest that agonist-induced changes occur in the structure of the M1 segment in the δ and possibly in the γ subunit. These changes likely are coupled to the agonist-induced structural changes of the M2 domain, which are also revealed by the pattern of specific labeling by [^{125}I]TID (109).

SEGMENT MA The segmet MA was suspected initally to be involved in lining the ion channel (90). A synthetic peptide corresponding to the MA of the TAChR β subunit forms ion channels in artificial phospholipid bilayers (40). MA can be deleted from the TAChR subunits, however, without affecting formation of the ion channel expressed in *Xenopus* oocytes (90). Antibodies that recognize sequences within MA (α378–391, α379–385, α395–401, and α389–408) bind to the cytoplasmic surface of AChR (26). In addition, trypsinization of AChR causes disappearance of Ab epitopes within MA (80), and proteolysis protection assays of fusion proteins that contain a reporter group and the MA segment suggested that MA is not transmembrane (23). Pyridoxamine-phosphate labeling of membrane bound TAChR in sealed microsacs demonstrated that αLys380, within MA, is cytoplasmic (33). In light of those results, MA might form an intracellular domain with regulatory function, rather than the ion channel.

Sequence Regions Involved in Formation of the Ion Filter

M1 and M2, which likely line the AChR ion channel, are uncharged and highly homologous to those of other members of the ionotropic receptor superfamily. These members include proteins, such as the $GABA_A$ and the glycine receptors, which transport anions (13). Several Ab-based or mutation-based studies have searched for sequence regions and individual residues, outside M1 and M2, which may confer charge selectivity to the AChR channel.

Some mAbs block TAChR without interfering with binding of cholinergic ligands (100 and references therein). They all recognize epitopes on the cytoplasmic surface of the TAChR, on the α, β, or γ subunit, within residues 368–399 (as numbered in the consensus alignment of the TAChR subunits), in the putative cytoplasmic domain, and shortly after M3 (98, 100). All the epitopes include, or are very close to, a phosphorylation site (100). One mAb recognized an epitope formed by the sequence region α332–350 (380–399 in the consensus alignement) (98), which is unusually rich in negatively charged residue, as expected for a cation filter.

Expression of TAChRs that carry mutations of charged and polar amino acids in three anionic rings on either side of the M2 segment (extracellular, intermediate, and cytoplasmic) suggested that amino acids in the intermediate ring may be part of a cation selectivity filter (50). Mutations in the segment that links M2 and M3 profoundly affected channel conductance and conferred *Torpedo*-like or bovine-like channel properties to the resulting chimeric or mutated AChRs (46). Removal of negative charges at either end of M2 affected the conductance properties of the mutated TAChRs (46). Rings of negative charge at either end of M2, therefore, are important determinants of channel conductance.

By replacing selected amino acids of M2 and its flanking regions with residues found in anion-selective channels, it was found that the insertion or deletion of a neutral amino acid residue in the segment between M1 and M2 was critical for ion selectivity. This finding suggests that ion selectivity is influenced strongly by the geometry of the channel (38). In addition, mutation of two rings of conserved Leu residues at the synaptic end of M2 abolished the calcium permeability of the neuronal α7 AChR, and mutations within the "intermediate" ring of negatively charged residues, on the cytoplasmic side of M2, reduced calcium permeability (12). These mutations did not affect the relative permeability for sodium and potassium (12).

The Putative Cytoplasmic Domain Between the Transmembrane Segments M3 and M4

This sequence region is highly divergent among different AChR subunits and may be involved in the differential functional characteristics that each subunit confers to the resulting AChR complexes (25).

STUDIES WITH ANTIBODIES The aminoterminal part of this sequence region is cytoplasmic. Several studies that used sequence-specific Abs that recognize parts of this region in different AChR subunits (sequences α304–322, α330–346, α332–350, α339–378, α349–364, α360–378, β350–358, β368–406, and γ360–377) in *Torpedo* and muscle AChRs consistently found a cytoplasmic location of Ab binding (26).

Antibody-based studies indicated that parts of the sequence region between M3 and M4 are exposed largely on the TAChR cytoplasmic surface. One study found that mAbs obtained by immunization with the synthetic sequences α304–322, α332–350, and α360–378 cross-reacted fully with native TAChR (55). Those sequences, therefore, must be exposed largely on the TAChR surface (55). A similar conclusion was reached by another study, which identified the sequence region between M3 and M4 as highly immunogenic when denatured TAChR was used for immunization and able to induce formation of Abs that were cross-reactive with native TAChR (74). In addition, a study that used the "reporter epitope" technique demonstrated that a reporter epitope inserted at position α347 was cytoplasmic (4).

Antibody-based and mutation studies have shown that this sequence region, in different TAChR subunits, contains residues that are phosphorylated and, therefore, cytoplasmic. Their phosphorylation is important in modulating AChR function (25; see also 96, 100).

STUDIES WITH OTHER APPROACHES Proteolysis protection assays of fusion proteins that contain a reporter group fused after the nucleic acid–sequence encoding each putative transmembrane domain demonstrated a cytoplasmic disposition of most or all the sequence region between M3 and M4 (23). Another study that used incorporation of pyridoxamine phosphate into membrane-bound TAChR in sealed microsacs, in the presence and in the absence of saponin, concluded that residue αLys380, which is just aminoterminal to M4, has cytoplasmic location (33).

In conflict with the conclusion that the whole region between M3 and M4 is cytoplasmic, a study that determined the sequence of AChR fragments released on brief proteolytic treatment of sealed AChR-rich membrane vesicles found that the sequences α341–380, β351–385, γ353–414, and δ328–341, which are part of the sequence region between M3 and M4, were quickly released by trypsin treatment (66). This finding suggests that these sequences are exposed on the extracellular surface (66).

The Carboxyl Terminal Domain

The segment M4 is followed by a carboxyl-terminal segment, which is very short in the α subunit and increasingly longer in the β, γ, and δ subunits (25, 26, 90). We review here the results of several studies on the transmembrane topology of this domain. These studies have used immunologic, biochemical, and genetic approaches, which yielded conflicting results. We review separately a recent immunologic study that sought structural reasons of the disparate results of those experimental efforts: the results of that study suggested the unexpected conclusion that the carboxyl terminus of at least the δ subunit might be exposed on both sides of the membrane.

STUDIES THAT USE ANTIBODIES GENERALLY SUGGEST A CYTOPLASMIC LOCATION OF THIS SEQUENCE REGION, WHEREAS STUDIES THAT USE NONIMMUNOLOGIC APPROACHES SUPPORT AN EXTRACELLULAR LOCATION Several studies that used Abs against the carboxyl terminus of different TAChR subunits supported a cytoplasmic location of this sequence region, because the Abs bound to the cytoplasmic side of the postsynaptic membrane (25, 26, 39). On the other hand, a reporter epitope inserted at position α429 (i.e. after M4) was located on the extracellular surface of the TAChR (4).

Monoclonal antibodies against the AChR δ subunit, whose epitope was mapped to a sequence segment between M4 and the carboxyl terminus, bound solubilized TAChR and inhibited the binding of other mAbs known to bind the cytoplasmic surface the γ and δ subunits (99).

These mAbs, however, did not bind to TAChR in sealed membrane fragments (99).

Several nonimmunologic studies found the carboxyl terminus to be extracellular. Some studies took advantage of the fact that TAChR exist as dimers, formed by a disulphide bridge between the penultimate Cys residue of the δ subunit of two TAChR monomers (25, 39). Studies that used membrane-impermeable reducing agents and sealed TAChR-rich vesicles consistently found that TAChR dimers can be reduced from the extracellular surface (25, 26, 39). In addition, Lys486 of the γ subunit, which is on the carboxyl terminal side of M4, could be labeled with membrane-impermeable reagents in sealed TAChR-rich vesicles, which suggests that the carboxyl terminus of the γ subunit is extracellular (33).

The location of the carboxyl terminus of the α and δ subunits of mammalian muscle AChR has been studied with the use of fusion proteins, which contain a reporter sequence attached downstream M4 (23). By using proteolysis protection assays to determine the orientation relative to the microsomal membrane, the carboxyl termini were found to be extracellular.

THE CARBOXYL TERMINAL DOMAIN MIGHT HAVE AN UNUSUAL TRANSMEMBRANE DISPOSITION AND BE EXPOSED ON BOTH SIDES OF THE MEMBRANE

The failure of several different investigations to reach a consensus regarding the topology of the carboxyl termini of AChR subunits might be the result of identifiable structural characteristics of the AChR. We sought to answer this dilemma by using a large panel of mAbs specific for overlapping epitopes within the carboxyl terminal region of the TAChR δ subunit. One mAb was obtained from a mouse immunized against a synthetic TAChR sequence; the others were obtained by immunization with native TAChR. All mAbs cross-reacted fully with nondenatured TAChR and recognized overlapping epitopes within the sequence segment δ485–493 in the carboxyl terminal region of the δ subunit.

The transmembrane localization of their epitopes was identified by immunoelectronmicroscopic analysis of TAChR-rich membrane fragments. Some mAbs, similar to those used in previous studies (55, 96), recognized only the cytoplasmic side of TAChR-rich membrane fragments. Others bound to both sides to a similar extent or bound primarily to the extracellular side. Binding of all mAbs was blocked specifically by synthetic peptides that contained the carboxyl terminal region of the δ subunit. The δ subunit of the AChR, therefore, might have alternative conformations, thus leading to exposure of the same sequence region on the extracellular or the cytoplasmic surface.

Conflicting findings with previous studies that used nonimmunologic approaches could be explained as follows. Chavez & Hall (23) and Dwyer (33) used a qualitative approach in their elegant studies. These authors demonstrated that a substantial fraction of the carboxyl termini of different subunits is extracellular. However, the presence of undetected AChR isomers that have the carboxyl termini of one or more subunits exposed on the cytoplasmic surface could not be excluded.

In addition, studies that concluded that the disulfide bridge holding together TAChR dimers is extracellular (25, 26, 39) can be reconciled, considering that the carboxyl terminal region of the AChR δ subunit in TAChR dimers must be long and flexible and able to assume more than one conformation. This conclusion was reached by studies on the motility relative to each other of the two monomers within a TAChR dimer, which demonstrated that monomers could rotate relative to the other up to almost 180 degrees (36). The ability of this region in mature TAChR molecules to interconvert spontaneously to different isomeric forms would well explain how exposure to extracellular reducing agents for relatively long periods of time may result ultimately in complete reduction of the disulfide bond that holds two AChR monomers within a dimer.

The carboxyl terminal region of the δ subunit is rich in Pro residues, most of which are highly conserved in AChRs from different species and tissues. A *cys* or *trans* conformation of one or more Pro residue could be the structural basis of the different isomeric forms of the carboxyl terminal region of the δ subunit. A striking abundance of highly conserved Pro residues is common to subunits of both AChRs and to other members of the ligand-gated ion channel superfamily.

The presence of AChR subunit isomers that are different in isomerization of one or more Pro residues, and the necessity of different isomeric forms for assembly of functional AChRs, has been demonstrated for homo-oligomeric neuronal AChRs formed by $\alpha 7$ subunits (43). That study showed that the peptidyl-prolyl isomerase cyclophilin is needed for assembly of both a functional $\alpha 7$ AChR and the 5HT-3 receptor. Block of cyclophilin reduced the expression of functional $\alpha 7$ AChR strongly but did not affect that of heteroligomeric muscle AChRs. In the presence of cyclophilin, expression of functional AChRs containing the $\alpha 7$ subunit could be restored by coexpression of muscle type non-α AChR subunits. That a muscle non-α subunit can substitute for the missing $Pro_{cys/trans}$ $\alpha 7$ isomer suggests that muscle-type non-α subunits may fold in different $Pro_{cis/trans}$ isomeric forms without intervention of cyclophilin, and possibly spontaneously.

The hydrophobic segment M4 is very nonpolar, which could allow it to assume different positions relative to the plane of the membrane. Topographic excursions of the carboxyl terminal "tail" of the δ subunit, therefore, could be accommodated.

The existence of different foldings of mature proteins that have identical primary sequences has been demonstrated in crystallographic studies of the structure of the constituent subunits of the pentameric protein VP1, which forms the outer shell of simian virus 40 (SV40) (58). These pentamers in SV40, like the pentameric AChR molecules in the tightly packed postsynaptic membrane, assemble in hexagonal lattices. This arrangement raises the puzzle of fitting pentamers into hexavalent holes, which requires unexpected flexibility and alternative bonding to account for the mismatch of symmetries. This structural dilemma is solved by assembling pentamers whose subunits have identical conformations, except for their carboxyl terminal segment. These segments may assume alternative stable conformations and take different directions as they emerge from the core of the corresponding subunit and head toward a neighbor pentamer, thus accommodating the required variability in the geometry of contacts. The folding of the constituent subunits and the packing patterns of mature pentamers might be similar in the AChR and in the SV40 VP1. This possibility is supported by the fact that, in both cases, the pentamers form disulfide stabilized dimers, which can further pack to yield tubular aggregates (17).

The SV40 shell, therefore, is formed by dimers of VP1 pentamers, linked by flexible carboxyl terminal arms of two neighbors, stabilized by disulfide bonds, and loosely tied together instead of locked in a tight fit between pentameric units. Liddington et al (58) have suggested that this is "one general solution to the problem of how to generate a structure with several kinds of contacts among standard building blocks." Further, they suggested "that subcellular assemblies frequently will be found to exhibit these sorts of linkages. Many such assemblies seem to be flexible yet highly specific. Interaction through arms can ensure specificity without requiring a rigid geometry and without imposing strong restrictions on symmetry" (58). The TAChR may be the first demonstration of the truth of their prediction.

STUDIES ON THE TRANSMEMBRANE FOLDING OF OTHER MEMBERS OF THE ACETYLCHOLINE RECEPTOR SUPERFAMILY: SIMILARITIES AND DIFFERENCES

The validity of extending models of the stoichiometry and topology of subunits based on the AChR to other members of the superfamily is indicated by the common phosphorylation, N-glycosylation, and ligand-binding regions of the AChR, $GABA_A$, and glycine receptors. In contrast, the glutamate receptor subunits appear to differ in transmembrane folding from the models developed for the TAChR.

Conserved Structural Features and Topology of the $GABA_A$ and the Glycine Receptor Subunits

SUBUNIT COMPOSITION AND PENTAMERIC STRUCTURE OF THE $GABA_A$ RECEPTORS Sixteen $GABA_A$ receptor subunit cDNAs have been isolated so far: $\alpha1-\alpha6$, $\beta1-\beta4$, $\gamma1-\gamma3$, $\delta1$, and $\rho1-\rho2$. Additional heterogeneity is generated by alternative messenger RNA splicing of the $\gamma2$ subunit (61). Subunit-specific Abs showed that $\alpha1$ colocalizes preferentially with the $\gamma2$ and either $\beta2$ or $\beta3$ subunits, whereas $\alpha6$ colocalizes preferentially with either $\beta2$ or $\beta3$ subunits and either the δ or $\gamma2$ subunits (19, 71). Antibodies specific for $\gamma2$ subunit splicing isoforms provided evidence that two $\gamma2$ subunit isoforms can be found in the same $GABA_A$ receptor complex (49). The molecular pharmacology of $\alpha/\beta/\gamma$ heteromers has been studied systematically in expression studies (34). As in AChRs, $GABA_A$ receptor α subunit subtype plays an important role in determining the pharmacology of the receptor complex (111). The γ subunit is required for benzodiazepine modulation, which indicates that ligand-binding sites reside at the α/γ subunit interface, analogous to the agonist sites of AChRs (111). Other ligand-binding sites appear to be influenced by the β subunit subtype (110). Homo-oligomers of $\rho1$ or $\rho2$ subunits form bicuculline-insensitive $GABA_A$ receptors (106). Rotational symmetry analysis of electron micrographs of several different native $GABA_A$ receptors indicates that they are pentameric structures (68).

EXTRACELLULAR AMINOTERMINAL DOMAIN OF THE $GABA_A$ RECEPTOR Site-directed mutagenesis of potential N-glycosylation sites (Asn10 and Asn110) of the $GABA_A$ $\alpha1$ subunit suggest that the aminoterminus is extracellular (18).

Muscimol labels $GABA_A$ receptors at αPhe65 (87). Mutation of either αHis101 or αPhe64, or deletion of $\alpha57-66$ (a region homologous to a ligand-binding segment of the TAChR cholinergic site), decreases the affinity for agonists (61). Substitution of Thr142 on the $\gamma2$ subunit (64) or Arg119 on the $\alpha6$ subunit (107) affects the response to benzodiazepines, which indicates that these residues are on the extracellular surface.

THE PUTATIVE CYTOPLASMIC DOMAIN BETWEEN M3 AND M4 OF THE $GABA_A$ RECEPTOR A cytoplasmic disposition of this segment is supported by the findings that antipeptide Abs against this sequence region block phosphorylation of the β subunit. In addition, substitutions of Ser residues in this region reduce the response of $GABA_A$ receptors to phosphorylation markedly (60).

SUBUNIT COMPOSITION AND PENTAMERIC STRUCTURE OF THE GLYCINE RECEPTOR Several glycine receptor subunits, which are designated $\alpha 1$, $\alpha 2$, $\alpha 2^*$, $\alpha 3$, $\alpha 4$, and β, have been isolated (14). Alternate RNA splicing of α subunits adds to glycine receptor diversity. Functional glycine receptors are expressed as α subunit homomers or α/β heteromers (14). A subunit stoichiometry, $\alpha_3\beta_2$, and a pentameric structure of the glycine receptor are suggested by cross-linking (14) and expression studies (51).

THE EXTRACELLULAR AMINOTERMINAL DOMAIN OF THE GLYCINE RECEPTOR An extracellular location of this sequence region is supported by the identification of sequence and individual residues within this domain, which contribute to ligand-binding sites. The $\alpha 2^*$ subunit differs from $\alpha 2$ by substitution of Gly167 with Glu, which confers neonatal resistance to strychnine and a lower affinity for agonists (14). Mutation studies identified other residues of the $\alpha 1$ subunit involved in strychnine binding (Asp148, Tyr161, Lys200, and Tyr202) and agonist discrimination (Ile111, Phe159, Tyr161, and Thr204) (14, 84, 105). Substitution of Phe159 with Tyr, which occurs in the $GABA_A$ $\alpha 1$ subunit, confer GABA-responsiveness to the glycine receptor (84).

Spasmodic mice have a motor disorder resembling strychnine poisoning that results from mutation of Ala52 to Ser in the α_1 subunit and reduced glycine binding affinity (83). These results identifiy a sequence region that contributes to a glycine binding site, homologous to a region forming a cholinergic site on the TAChR (see above).

THE M2 DOMAIN OF THE GLYCINE RECEPTOR SUBUNITS IS IMPLICATED IN FORMATION OF THE ION CHANNEL With the use of chimeras of α and β subunits, a region of the β subunit within M2 (amino acids 278–295) was found to confer resistance to picrotoxin, which behaves as a noncompetitive antagonist of α subunit homomeric receptors (14). Startle disease, or hypereklexia, is caused by mutations of Arg271 in the glycine receptor $\alpha 1$ subunit on the extracellular end of M2 (54, 72). Mutations of Arg271 result in reduced sensitivity to glycine and conversion of the agonists β-alanine and taurine into competitive antagonists (73). These results implicate the M2 segment in lining the chloride channel.

Glycine receptors mediate agonist-responsive Cl^- ion conductance, in contrast to the cation-selective permeability of the AChR. The negatively charged residues that are clustered near the transmembrane regions of the AChR subunits (see above) are replaced by positively charged residues in the glycine receptor subunits, which may be related to differences in ion selectivity.

Synthetic peptides that correspond to the M2 segment of the glycine receptor α subunit form ion channels in phospholipid bilayers (53, 75). By replacing the amino- and carboxyl-terminal Arg with Glu residues, peptides formed cation-selective, rather than anion-selective, channels (53, 75).

THE CYTOPLASMIC DOMAIN BETWEEN M3 AND M4 OF THE GLYCINE RECEPTORS Ser391, which is in the putative cytoplasmic domain, near the M4 of the α subunit, is phosphorylated (81). This finding indicates that at least part of the region between M3 and M4 is cytoplasmic. Several mAbs are available that bind to specific regions of the aminoterminal and cytoplasmic domains of the glycine receptor α subunit. These mAbs would be useful tools for topologic mapping studies (86).

Distinct Transmembrane Topology of the Glutamate Receptor Subunits

Ionotropic glutamate receptors play a major role in rapid, excitatory neurotransmission in the central nervous system. The following subunit subfamilies are recognized, on the basis of sequence homology, agonist binding and channel activation: (*a*) α-amino-3-hydroxy-5-methyl isoxazole-4-propionic acid (AMPA) receptors (GluR1–GluR4, also designated GluA–GluD); (*b*) kainate receptor (GluR5–GluR7, KA-1, KA-2); and (*c*) N-methyl-D-aspartate (NMDA) receptors (NMR1 or ζ1, NR2A–NR2D or ϵ1–ϵ4). Alternative splicing generates a high level of combinatorial diversity for channel assembly (44, 112). Glutamate receptors have been expressed as functional homomeric or heteromeric complexes in a stoichiometry that has not been elucidated yet (112).

By analogy to the AChR, the $GABA_A$ and glycine receptors, models for the transmembrane topology of the glutamate receptor subunits were proposed on the basis of hydropathy analysis (44, 112). These models predicted four transmembrane domains (M1–M4), preceded by an amino extracellular terminal region, a cytoplasmic loop between M3 and M4, and an extracellular carboxyl terminus. Studies of N-glycosylation, phosphorylation, and epitope mapping of different glutamate receptor subunits indicate clearly that this model is incorrect.

THE AMINOTERMINAL REGION, AND THE REGION BETWEEN M3 AND M4, MAY BE EXTRACELLULAR The aminoterminal region preceding M1 may be very long in the glutamate receptors (~400–500 residues) (44 and references therein). Its extracellular disposition is supported by the results of both Ab-based and nonimmunologic studies. These studies also suggest an unexpected extracellular disposition of the segment between M3 and M4.

Functional N-glycosylation sites were found in the aminoterminal region and in the region between M3 and M4 of the Glu1, Glu3 and Glu6 subunits, which suggests these regions are extracellular (9, 45, 78, 92). Antibodies against a synthetic peptide that corresponds to residues 253–267 of the GluR1 subunit were used to demonstrate that the aminoterminal region is extracellular (65).

Site-directed mutagenesis studies indicated that the aminoterminal domain of AMPA/kainate receptors (89, 101) and NMDA receptors (52) are important in agonist binding. The agonist-binding site of the GluR3 subunit was defined with the use of Abs. Rabbits immunized with the sequence region 245–457 became seizure prone, and the Abs activated glutamate receptors in cortical neurons (94).

Mutational analysis of the glycine-binding site of the NMDA receptor NR1 subunit indicated that several aromatic residues of the aminoterminal domain that surrounds a putative extracellular disulfide loop between Cys402 and Cys418 were important for glycine/glutamate coactivation (Phe390, Tyr392, Phe466) (52). A second distinct site was shown to lie between M3 and M4 (Val666 and Ser669). The agonist binding sites of the AMPA/kainate receptor GluR6 subunit were mapped to corresponding sequence regions (89).

THE CARBOXYL TERMINUS IS INTRACELLULAR Lack of functional N-glycosylation sites on the carboxyl terminus suggested that it might be intracellular (9, 45, 78, 92). This possibility was supported by the presence of functional phosphorylation sites on this region in the AMPA/kainate and NMDA receptors (79). The intracellular locations of the carboxyl termini of GluR3 and GluR1 have been demonstrated directly with the use of an epitope protection assay (9) and an antipeptide Ab staining of permealized neurons (65).

M1, M3 AND M4 ARE TRANSMEMBRANE, BUT M2 IS NOT Mutations in the M2 region indicated that a single amino acid, either an Arg or Gln, controls the ion transport properties of glutamate receptors (44, 112). The transmembrane disposition of M1, M2, and M3 was analyzed by deleting each segment and determining the effects on N-glycosylation (45). Whereas M1 and M3 spanned the membrane, M2 did not. Epitope protection of a reporter protein demonstrated that M3 of the GluR3 subunit is a transmembrane segment (9). The conclusions from these studies is that M2 is a re-entrant loop rather than a transmembrane segment. This structural feature, called a pore loop, is common to other ion channels (62).

Like the AChR, the $GABA_A$, and the glycine receptors, glutamate receptors seem to have four domains that interact with the membrane. However, one of the glutamate receptor domains, M2, does not span the membrane. As in the AChR, the M2 segment of the glutamate receptor is important for ion channel function. Similarly, the aminoterminal region is extracellular and contributes to the forming of the agonist/antagonists binding sites. The main structural differences between AChRs and glutamate receptors is that AChRs have an intracellular domain between M3 and M4 involved in modulation of receptor function through phosphorylation, whereas in the glutamate receptors, the segment between M3 and M4 seems to be extracellular. An intracellular region that can be phosphorylated and might be involved in modulation of glutamate receptor function is found on the carboxyl terminus, after M4 (79).

ACKNOWLEDGMENTS

This work was supported by grant NS23919 from the National Institute for Neurological and Communicative Disorders and Stroke, the program project grants DA05695 and DA08131 from the National Institute for Drug Abuse, and a grant from the Council for Tobacco Research (to BMC-F).

Literature Cited

1. Akabas MH, Kaufman K, Archdeacon P, Karlin A. 1994. Identification of acetylcholine receptor channel-lining residues in the entire M2 segment of the α subunit. *Neuron* 13:919–27
2. Akabas MH, Stauffer DA, Xu M, Karlin A. 1992. Acetylcholine receptor channel structure probed in cysteine-substitution mutants. *Science* 258:307–10
3. Albuquerque EX, Pereira EFR, Castro NG, Alkondon M. 1995. Neuronal nicotinic receptors: function, modulation and structure. *Neuroscience* 7:91–101
4. Anand R, Bason L, Saedi MS, Ger-

zanich V, Peng X, Lindstrom J. 1993. Reporter epitopes: a novel approach to examine transmembrane topology of integral membrane proteins applied to the alpha 1 subunit of the nicotinic acetylcholine receptor. *Biochemistry* 32:9975–84
5. Anderson DJ, Walter P, Blobel G. 1982. Signal recognition protein is required for the integration of acetylcholine receptor δ subunit, a transmembrane glycoprotein, into the endoplasmic reticulum membrane. *J. Cell Biol.* 93:501–6
6. Appel JR, Pinilla C, Niman H, Houghten R. 1990. Elucidation of discontinuous linear determinants in peptides. *J. Immunol.* 144:976–83
7. Bahouth SW, Wang H-Y, Malbon CC. 1991. Immunological approaches for probing receptor structure and function. *Trends Biochem. Sci.* 12:338–43
8. Barlow DJ, Edwards MS, Thornton JM. 1986. Continuous and discontinuous protein antigenic determinants. *Nature* 322:747–48
9. Bennett JA, Dingledine R. 1995. Topology profile for a glutamate receptor: three transmembrane domains and a channel-lining reentrant membrane loop. *Neuron* 14:373–84
10. Beroukhim R, Unwin N. 1995. Three-dimensional location of the main immunogenic region of the acetylcholine receptor. *Neuron* 15:1–20
11. Bertrand D, Devillers-Thiery A, Revah F, Galzi J-L, Hussy N, et al. 1992. Unconventional pharmacology of a neuronal nicotinic receptor mutated in the channel domain. *Proc. Natl. Acad. Sci. USA* 89:1261–65
12. Bertrand D, Galzi JL, Devillers-Thiery A, Bertrand S, Changeux JP. 1993. Mutations at two distinct sites within the channel domain M2 after calcium permeability of neuronal $\alpha 7$ nicotinic receptor. *Proc. Natl. Acad. Sci. USA* 90:6971–75
13. Betz H. 1990. Homology and analogy in transmembrane receptor design: lessons from synaptic membrane proteins. *Biochemistry* 29: 3591–99
14. Betz H. 1992. Structure and function of inhibitory glycine receptors. *Q. Rev. Biophys.* 25:381–94
15. Blanton MP, Cohen JB. 1994. Identifying the lipid-protein interface of the *Torpedo* nicotinic acetylcholine receptor: secondary structure implications. *Biochemistry* 33:2859–72
16. Blanton MP, Raja SK, Lala AK, Cohen JB. 1995. Photolabeling *Torpedo californica* nicotinic acetylcholine receptor membranes with the hydrophobic probe [3H]diazofluorene. *Biochemistry*. Submitted
17. Brisson A, Unwin PNT. 1984. Tubular crystals of acetylcholine receptor. *J. Cell Biol.* 99:1202–11
18. Buller AL, Hastings GA, Kirkness EF, Fraser CM. 1994. Site-directed mutagensis of N-lnked glycosylation sites on the γ-aminobutyric acid type A receptor $\alpha 1$ subunit. *Mol. Pharmacol.* 46:858–65
19. Caruncho HJ, Costa E. 1994. Double-immunolabelling analysis of $GABA_A$ receptor subunits in label-fracture replicas of cultured rat cerebellar granule cells. *Recept. Channels* 2:143–53
20. Charnet P, Labarca C, Leonard R, Vogelaar NJ, Czyzk L, et al. 1990. An open channel blocker interacts with adjacent turns of α-helices in the nicotinic acetylcholine receptor. *Neuron* 4:87–95
21. Chaturvedi V, Donnelly-Roberts DL, Lentz TL. 1992. Substitution of *Torpedo* acetylcholine receptor $\alpha 1$-subunit residues with snake $\alpha 1$- and rat nerve $\alpha 3$-subunit residues in recombinant fusion proteins: effect on α-bungarotoxin binding. *Biochemistry* 31:1370–75
22. Chavez RA, Hall ZW. 1991. The transmembrane topology of the amino terminus of the α subunit of the nicotinic acetylcholine receptor. *J. Biol. Chem.* 266:15532–38
23. Chavez RA, Hall ZW. 1992. Expression of fusion proteins of the nicotinic acetylcholine receptor from mammalian muscle identifies the membrane-spanning regions in the α and δ subunits. *J. Cell Biol.* 116: 385–93
24. Cheetham JC, Raleigh DP, Griest RE, Redfield C, Dobson CM, Rees AR. 1991. Antigen mobility in the combining site of an anti-peptide antibody. *Proc. Natl. Acad. Sci. USA* 88:7968–72
25. Claudio T. 1989. Molecular genetics of acetylcholine receptor-channels. In *Frontiers in Molecular Biology*, ed. DM Glover, BD Hammes, pp. 63–142. Oxford, UK: IRL Press
26. Conti-Tronconi BM, McLane KE, Raftery MA, Grando SA, Protti MP, 1994. The nicotinic acetylcholine receptor: structure and autoimmune pathology. *Crit. Rev. Biochem. Mol. Biol.* 29:69–123

27. Corringer PJ, Galzi JL, Eisele JL, Bertrand S, Changeux J-P, Bertrand D. 1995. Identification of a new component of the agonist binding site of the nicotinic α7 homooligomeric receptor. *J. Biol. Chem.* 270:11749–52
28. Criado M, Hochschwender S, Sarin V, Fox VL, Lindstrom J. 1985. Evidence for unpredicted transmembrane domains in acetylcholine receptor subunits. *Proc. Natl. Acad. Sci. USA* 82:2004–8
29. Davies DR, Padlan EA, Sheriff S. 1990. Antibody-antigen complexes. *Annu. Rev. Biochem.* 59:439–73
30. Dennis M, Giraudat J, Kotzba-Hibert F, Goeldner M, Hirth C, et al. 1988. Amino acids of the *Torpedo marmorata* acetylcholine receptor α subunit labeled by a photoaffinity ligand for the acetylcholine binding site. *Biochemistry* 27:2346–57
31. DiPaola M, Kao PN, Karlin A. 1990. Mapping the α-subunit site photolabeled by the non-competitive inhibitor [^{3}H]quinacrine azide in the active state of the nicotinic acetylcholine receptor. *J. Biol. Chem.* 265:11017–29
32. Dougherty DA, Stauffer DA. 1990. Acetylcholine binding by a synthetic receptor: implications for biological recognition. *Science* 250:1558–60
33. Dwyer BP. 1991. Topological dispositions of lysine α380 and lysine γ486 in the acetylcholine receptor from *Torpedo californica. Biochemistry* 30:4105–12
34. Ebert B, Wafford KA, Whiting P, Krogsgaard-Larsen P, Kemp JA. 1994. Molecular pharmacology of γ-aminobutyric acid type A receptor agonists and partial agonists in oocytes injected with different α, β and γ receptor subunit combinations. *Mol. Pharmacol.* 46:957–63
35. Eisele J-L, Bertrand S, Galzi J-L, Devillers-Thiery A, Changeux J-P, Bertrand D. 1993. Chimaeric nicotinic-serotonergic receptor combines distinct ligand binding and channel specificities. *Nature* 366:479–83
36. Fairclough RH, Finer-Moore J, Love RA, Kristofferson D, Desmeules PJ, Stroud RM. 1983. Subunit organization and structure of an acetylcholine receptor. *Cold Spring Harbor Symp. Quant. Biol.* 68:9–20
37. Finer-Moore J, Stroud R. 1984. Amphipathic analysis and possible formation of the ion channel in an acetylcholine receptor. *Proc. Natl. Acad. Sci. USA* 81:155–59
38. Galzi J-L, Devillers-Thiery A, Hussy N, Bertrand S, Changeux J-P, Bertrand D. 1992. Mutations in the channel domain of a neuronal nicotinic receptor convert ion selectivity from cationic to anionic. *Nature* 359: 500–5
39. Galzi J-L, Revah F, Bessis A, Changeux J-P. 1991. Functional architecture of the nicotinic acetylcholine receptor: from electric organ to brain. *Annu. Rev. Pharmacol.* 31: 37–72
40. Ghosh P, Stroud RM. 1991. Ion channels formed by a highly charged peptide. *Biochemistry* 30:3551–57
41. Gorne-Tschelnokow U, Strecker A, Kaduk C, Naumann D, Hucho F. 1994. The transmembrane domains of the nicotinic acetylcholine receptor contain α-helical and β structures. *EMBO J.* 13:338–41
42. Hartman DS, Claudio T. 1990. Coexpression of two distinct muscle acetylcholine receptor a subunits during development. *Nature* 343:372–73
43. Helekar SA, Char D, Neff S, Patrick J. 1994. Prolyl isomerase requirement for the expression of functional homo-oligomeric ligand-gated ion channels. *Neuron* 12:179–89
44. Hollmann M, Heinemann S. 1994. Cloned glutamate receptors. *Annu. Rev. Neurosci.* 17:31–108
45. Hollmann M, Maron C, Heinemann S. 1994. N-glycosylation site tagging suggests a three transmembrane domain topology for the glutamate receptor GluR1. *Neuron* 13:1331–43
46. Imoto K, Busch C, Sakmann B, Mishina M, Konno T, et al. 1988. Rings of negatively charged amino acids determine the acetylcholine receptor channel conductance. *Nature* 335: 645–48
47. Imoto KJ, Konno T, Nakai J, Wang F, Misha M, Numa S. 1991. A ring of uncharged polar amino acids as a component of channel constriction in the nicotinic acetylcholine receptor. *FEBS Lett.* 289:193–200
48. Karlin A. 1993. Structure of nicotinic acetylcholine receptors. *Curr. Opin. Neurobiol.* 3:299–309
49. Khan ZU, Gutierrez A, DeBlas AL. 1994. The subunit composition of a $GABA_A$/benzodiazepine receptor from rat cerebellum. *J. Neurochem.* 63:371–74
50. Konno T, Busch C, Von Kitzing E, Imoto K, Wang F, et al. 1991. Rings of anionic amino acids as structural determinants of ion selectivity in the

acetylcholine receptor channel. *Proc. R. Soc. London Ser. B* 244:69–79
51. Kuhse J, Laube B, Magalei D, Betz H. 1993. Assembly of the inhibitory glycine receptor: identification of amino acid sequence motifs governing subunit stoichiometry. *Neuron* 11:1049–56
52. Kuryatov A, Laube B, Betz H, Kuhse J. 1994. Mutational analysis of the glycine-binding site of the NMDA receptor: structural similarity with bacterial amino acid-binding proteins. *Neuron* 12:1291–1300
53. Langosch D, Hartung K, Grell E, Bamberg E, Betz H. 1991. Ion channel formation by synthetic transmembrane segments of the inhibitory glycine receptor—a model study. *Biochim. Biophys. Acta* 1063:35–44
54. Langosch D, Laube B, Rundstrom N, Schmieden V, Bormann J, Betz H. 1994. Decreased agonist affinity and chloride conductance of mutant glycine receptors associated with human hereditary hyperekplexia. *EMBO J.* 13:4223–28
55. Lei S, Raftery MA, Conti-Tronconi BM. 1993. Monoclonal antibodies against synthetic sequences of the nicotinic receptor cross-react fully with the native receptor and reveal the transmembrane disposition of their epitopes. *Biochemistry* 32:91–100
56. Lentz TL, Wilson PT. 1988. Neurotoxin-binding site on the acetylcholine receptor. *Int. Rev. Neurobiol.* 29: 117–60
57. Leonard RJ, Labarca CG, Charnet P, Davidson N, Lester HA. 1988. Evidence that the M2 membrane-spanning region lines the ion channel pore of the nicotinic receptor. *Science* 242:1578–81
58. Liddington RC, Yan Y, Moulai J, Sahli R, Benjamin TL, Harrison SC. 1991. Structure of simian virus 40 at 3.8-Å resolution. *Nature* 354:278–84
59. Lindstrom J, Anand R, Peng X, Gerzanich V, Wang F, Yuebing L. 1995. Neuronal nicotinic receptor subtypes. *Ann. NY Acad. Sci.* 757:100–16
60. MacDonald RL. 1995. Ethanol, γ-aminobutyrate type A receptors, and protein kinase C phosphorylation. *Proc. Natl. Acad. Sci. USA* 92: 3633–35
61. MacDonald RL, Olsen RW. 1994. $GABA_A$ receptor channels. *Annu. Rev. Neurosci.* 17:569–602
62. MacKinnon R. 1995. Pore loops: an emerging theme in ion channel structure. *Neuron* 14:889–92
63. McLane KE, Wahlsten JL, Conti-Tronconi BM. 1993. Use of synthetic sequences of the nicotinic acetylcholine receptor to identify structural determinants of binding sites for neurotoxins and antibodies to the main immunogenic region. *Methods: Companion Methods Enzymol.* 5: 201–11
64. Mihic SJ, Whiting PJ, Klein RL, Wafford KA, Harris RA. 1994. A single amino acid of the human γ-aminobutyric acid type A receptor γ2 subunit determines benzodiazepine efficacy. *J. Biol. Chem.* 269: 32768–73
65. Molnar E, McIlhinney J, Baude A, Nusser Z, Somogyi P. 1994. Membrane topology of the GluR1 glutamate receptor subunit: epitope mapping by site-directed antipeptide antibodies. *J. Neurochem.* 63: 683–93
66. Moore CR, Yates JR III, Griffin PR, Shabanowitz J, Martino PA, et al. 1989. Proteolytic fragments of the nicotinic acetylcholine receptor identified by mass spectrometry: implications for receptor topography. *Biochemistry* 28:9184–91
67. Mulac-Jericevic B, Kurisaki J, Atassi MZ. 1987. Profile of the continuous antigenic regions on the extracellular part of the α chain of an acetylcholine receptor. *Proc. Natl. Acad. Sci. USA* 84:3633–37
68. Nayeem N, Green TP, Martin IL, Barnard EA. 1994. Quaternary structure of the native $GABA_A$ receptor determined by electron microscopic image analysis. *J. Neurochem.* 62: 815–18
69. Ohno K, Hutchinson DO, Milone M, Brengman JM, Bouzat C, et al. 1995. Congenital myasthenic syndrome caused by prolonged acetylcholine receptor channel openings due to a mutation in the M2 mutation in the ϵ subunit. *Proc. Natl. Acad. Sci. USA* 92:758–62
70. O'Leary ME, Filatov GN, White MM. 1994. Characterization of δ-tubocurarine binding site of *Torpedo* acetylcholine receptor. *Am. J. Physiol.* 266:C648–53
71. Quirk K, Gillard NP, Ragan CI, Whiting P, McKernan RM. 1994. Model of subunit composition of γ-aminobutyric acid A receptor subtypes expressed in rat cerebellum with respect to their α and γ/δ subunits. *J. Biol. Chem.* 269:16020–28
72. Rajendra S, Lynch JW, Pierce KD,

French CR, Barry PH, Schofield PR. 1994. Startle disease mutations reduce the agonist sensitivity of the human inhibitory glycine receptor. *J. Biol. Chem.* 269:18739–42
73. Rajendra S, Lynch JW, Pierce KD, French CR, Barry PH, Schofield PR. 1995. Mutation of an arginine residue in the human glycine receptor transforms β-alanine and taurine from agonists in competitive antagonists. *Neuron* 14:169–75
74. Ratnam M, Manohar, Le Nguyen D, Rivier J, Sargent PB, Lindstrom J. 1986. Transmembrane topography of nicotinic acetylcholine receptor: immunochemical tests contradict theoretical predictions based on hydrophobicity profiles. *Biochemistry* 25: 2633–43
75. Reddy GL, Iwamoto T, Tomich JM, Montal M. 1993. Synthetic peptides and four-helix bundel proteins as model systems for the pore-forming structure of channel proteins. II. transmembrane segment M2 of the brain glycine receptor is a plausible candidate for the pore-lining structure. *J. Biol. Chem.* 268:14608–15
76. Revah F, Bertrand D, Galzi J-L, Devillers-Thiery A, Mulle C, et al. 1991. Mutations in the channel domain after desensitization of a neuronal nicotinic receptor. *Nature* 353: 846–49
77. Rini JM, Schulze-Gahmen U, Wilson IA. 1992. Structural evidence for induced fit as a mechanism for antibody-antigen recognition. *Science* 255:959–65
78. Roche KW, Raymond LA, Blackstone C, Huganir RL. 1994. Transmembrane topology of the glutamate receptor subunit GluR6. *Proc. Natl. Acad. Sci. USA* 269:11679–82
79. Roche KW, Tingley WG, Huganir RL. 1994. Glutamate receptor phosphorylation and synaptic plasticity. *Curr. Opin. Neurobiol.* 4:383–88
80. Roth B, Schwendimann B, Hughes CJ, Tzartos SJ, Barkas T. 1987. A modified nicotinic acetylcholine receptor lacking the ion channel amphipatic helices. *FEBS Lett.* 221:172–78
81. Ruiz-Gomez A, Vaello M-L, Valdivieso F, Mayor F Jr. 1991. Phosphorylation of the 48-kDa subunit of the glycine receptor by protein kinase C. *J. Biol. Chem.* 266:559–66
82. Sargent PB. 1993. The diversity of neuronal nicotinic acetylcholine receptors. *Annu. Rev. Neurosci.* 16: 403–43
83. Saul B, Schmieden V, Kling C, Mulhardt C, Gass P, et al. 1994. Point mutation of glycine receptor α1 subunit in the spasmodic mouse affects agonist responses. *FEBS Lett.* 350: 71–76
84. Schmieden V, Kuhse J, Betz H. 1993. Mutation of glycine receptor subunit creates β-alanine receptor responsive to GABA. *Science* 262:256–58
85. Schroeder B, Reinhardt-Maelicke S, Schrattenholz A, McLane KE, Conti-Tronconi BM, Maelicke A. 1994. Monoclonal antibodies WF6 and FK1 define two neighboring ligand binding sites on *Torpedo* acetylcholine receptor α polypeptide. *J. Biol. Chem.* 269:10407–16
86. Schroder S, Hoch W, Beacker C-M, Grenningloh G, Betz H. 1991. Mapping of antigenic epitopes on the α1 subunit of the inhibitory glycine receptor. *Biochemistry* 30:42–47
87. Smith GB, Olsen RW. 1994. Identification of a [3H]muscimol photoaffinity substrate in the bovine γ-aminobutyric acid A receptor α subunit. *J. Biol. Chem.* 268:20380–87
88. Spangler BD. 1991. Binding to native proteins by antipeptide monoclonal antibodies. *J. Immunol.* 146:1591–95
89. Stern-Bach Y, Bettler B, Hartley M, Sheppard PO, O'Hara PJ, Heinemann S. 1994. Agonist selectivity of glutamate receptors is specified by two domains structurally related to bacterial amino acid-binding proteins. *Neuron* 13:1345–57
90. Stroud RM, McCarthy MP, Shuster M. 1990. Nicotinic acetylcholine receptor superfamily of ligand-gated ion channels. *Biochemistry* 29: 11009–23
91. Sussman JL, Harel M, Frolow F, Oefner C, Goldman A, et al. 1991. Atomic structure of acetylcholinesterase from *Torpedo californica*: a prototypic acetylcholine-binding protein. *Science* 253:872–79
92. Taverna FA, Wang L-Y, MacDonald JF, Hampson DR. 1994. A transmembrane model for an ionotropic glutamate receptor predicted on the basis of the location of asparagine-linked oligosaccharides. *J. Biol. Chem.* 269: 14159–64
93. Tomaselli GF, McLaughlin JT, Jurman ME, Hawrot E, Yellen G. 1991. Mutations affecting agonist sensitivity of the nicotinic acetylcholine receptor. *Biophys. J.* 60:721–24
94. Twyman RE, Gahring LC, Spiess J, Rogers SW. 1995. Glutamate recep-

tor antibodies activate a subset of receptors and reveal an agonist binding site. *Neuron* 14:755–62

95. Tzartos SJ, Kokla H, Walgrave S, Conti-Tronconi BM. 1988. The main immunogenic region of human muscle acetylcholine receptor is localized within residues 67–76 of the α subunit. *Proc. Natl. Acad. Sci. USA* 85: 2899–903
96. Tzartos SJ, Kouvatsou R, Tzartos E. 1995. Monoclonal antibodies as site-specific probes for the acetylcholine receptor δ subunit tyrosine and serine phosphorylation sites. *Eur. J. Biochem.* 228:463–72
97. Tzartos SJ, Loutrari HV, Tang F, Kokla A, Walgrave SL, et al. 1990. Main immunogenic region of *Torpedo* electroplax and human muscle acetylcholine receptor: localization and microheterogeneity revealed by the use of synthetic peptides. *J. Neurochem.* 54:51–61
98. Tzartos SJ, Remoundos MS. 1990. Fine localization of the major α-bungarotoxin binding site to residues α189–195 of the *Torpedo* acetylcholine receptor: residues 189, 190, and 195 are indispensable for binding. *J. Biol. Chem.* 265:21462–67
99. Tzartos, SJ, Tzartos E, Kouvatsou, R. 1995. Acetylcholine receptor tyrosine and serine phosphorylation. Monoclonal antibodies as site-specific tools for phosphorylation and channel function. *Proc. Int. Symp. Cholinergic Synapse: Structure, Function, and Regulation Site*, p. S16
100. Tzartos SJ, Valcana C, Kouvatsou R, Kokla A. 1993. The tyrosine phosphorylation site of the acetylcholine receptor β subunit is located in a highly immunogenic epitope implicated in channel function: antibody probes for β subunit phosphorylation and function. *EMBO J.* 12:5141–49
101. Uchino S, Sakimura K, Nagahari K, Mishina M. 1992. Mutations in a putative agonist binding region of the AMPA-selective glutamate receptor channel. *FEBS Lett.* 308:253–57
102. Unwin N. 1993. Nicotinic acetylcholine receptor at 9Å resolution. *J. Mol. Biol.* 229:1101–24
103. Unwin N. 1995. Acetylcholine receptor channel imaged in the open state. *Nature* 373:37–43
104. Van Regenmortel MHV. 1989. Structural and functional approaches to the study of protein antigenicity. *Immunol. Today* 10:266–72
105. Vandenberg RJ, Rajendra S, French CR, Barry PH, Scofield PR. 1993. The extracellular disulfide loop motif of the inhibitory glycine receptor does not form the agonist binding site. *Mol. Pharmacol.* 44:198–203
106. Wang T-L, Guggino WB, Cutting GR. 1994. A novel γ-aminobutyric acid receptor subunit (ρ2) cloned from human retina forms bicuculline-insensitive homooligomeric receptors in *Xenopus* oocytes. *J. Neurosci.* 14:6524–31
107. Weiland, HA, Luddens H, Seeburg PH. 1992. A single histidine in $GABA_A$ receptors is essential for benzodiazepine agonist binding. *J. Biol. Chem.* 267:1426–29
108. Wennogle LP, Changeux J-P. 1980. Transmembrane orientation of proteins present in acetylcholine receptor rich membranes from *Torpedo marmorata* studied by selective proteolysis. *Eur. J. Biochem.* 106: 381–93
109. White BH, Cohen JB. 1992. Agonist-induced changes in the structure of the acetylcholine receptor M2 regions revealed by photoincorporation of an uncharged nicotinic noncompetitive antagonist. *J. Biol. Chem.* 267: 15770–83
110. Wingrove PB, Wafford KA, Bain C, Whiting PJ. 1994. The modulatory action of loreclezole at the γ-aminobutyric acid type A receptor is determined by a single amino acid in the β2 and β3 subunit. *Proc. Natl. Acad. Sci. USA* 91:4569–73
111. Wisden W, Seeburg PH. 1992. $GABA_A$ receptor channels: from subunits to functional entities. *Curr. Opin. Neurobiol.* 2:263–69
112. Wisden W, Seeburg PH. 1993. Mammalian ionotropic glutamate receptors. *Curr. Opin. Neurobiol.* 3: 291–98

Annu. Rev. Biophys. Biomol. Struct. 1996. 25:231–258

ENGINEERING THE GRAMICIDIN CHANNEL

Roger E. Koeppe II

Department of Chemistry and Biochemistry, University of Arkansas, Fayetteville, Arkansas 72701

Olaf S. Andersen

Department of Physiology and Biophysics, Cornell University Medical College, New York, New York 10021

KEY WORDS: membrane channel, protein design, voltage dependence, protein folding, peptide conformation, nongenetic amino acids, tryptophan

ABSTRACT

The chemical design or redesign of proteins with significant biological activity presents formidable challenges. Ion channels offer advantages for such design studies because one can examine the function of single molecular entities in real time. Gramicidin channels are attractive for study because of their known structure and exceptionally well-defined function. This article focuses on amino acid sequence changes that redesign the structure or function of gramicidin channels. New, and functional, folded states have been achieved. In some cases, a single amino acid sequence can give rise to several (up to three) functional conformations. Single amino acid substitutions confer voltage-dependent channel gating. The findings provide insight into the folding of integral membrane proteins, the importance of tryptophan residues at the membrane/water interface, and the mechanism of channel gating.

CONTENTS

1056–8700/96/0610–0231$08.00

PERSPECTIVES AND OVERVIEW

Early protein chemists dreamed of redesigning the structure and function of proteins by making selected amino acid changes. With the advent of solid-phase peptide synthesis (72) and oligonucleotide-directed mutagenesis (50), it now is possible to investigate the structural and functional consequences of selected sequence changes—as a tool to understand the properties of a wide variety of proteins.

Such studies have been done at several different levels. When high-resolution structures are available for several related proteins, e.g. the serine proteases, one can obtain mechanistic information by altering residues in, or adjacent to, the active site (12, 20, 35, 41, 91, 123). As might be expected, one often cannot predict the result of the sequence alteration, even when a high-resolution structure is available. When high-resolution structures are not available, one may be able to define essential residues, and obtain mechanistic information, by examining how the replacements of conserved residues between different classes of proteins alter function (42, 43, 98). Such studies can provide insights into the molecular basis for ion selectivity (42, 43, 125), but the interpretational uncertainties become more tenuous when the sequence substitutions are done in the absence of a high-resolution structure (e.g. 61). Thus, despite the investigative power that has become available through the use of amino acid replacement technologies, few examples have emerged in which directed mutagenesis has achieved a predicted redesign of function. Further, the successful introduction of a redesigned, refolded state that is active biologically has not been reported.

A different approach to the same problem is to design a protein de novo such that it possesses a preselected structure (and function). This approach has been used to examine the structural and energetic constraints associated with the formation of coiled coils (66, 90) and to determine how the formation of a coiled coil is affected by ligand binding (38). In parallel with these studies, new classes of ion channels were designed on the basis of considerations of helix–helix interactions (65) or stacking of β-sheet–like structures (33). Synthetic ion channels also have been designed on the basis of analogies with putative membrane-spanning segments in integral membrane proteins (73).

A third approach, which we discuss in the remainder of this review, is a hybrid between the above two strategies. The known structure and well-defined functional properties of membrane-incorporated channels formed by the linear gramicidins have allowed a strategy to "redesign" or "re-engineer" both function and structure: New functional properties have been conferred on channels that remain structurally very similar to the parent channels. Further, new, and functional, folded states have been achieved through selected amino acid sequence changes. Some of these studies have used amino acids that are outside the genetic code (71, 81, 92).

WHY GRAMICIDIN CHANNELS

Traditionally, studies of protein structure and function are done on soluble proteins. Ion channels, however, offer significant advantages, because one generally can examine the function of single molecular entities in real time (e.g. 95). Moreover, electrophysiologic (single-channel) methods provide information not only about channel function but also about channel structure. This feature is shown in Figure 1, which illustrates single-channel current traces, as well as current transition amplitude and single-channel duration histograms of gramicidin channels. The channels can be characterized by their average current amplitude and duration, which are fairly unique identifiers of the molecular identity of the channel-forming molecules (cf. 9, 25, 27). The information provided by amplitude and duration histograms is equivalent to that provided by a fluorescence experiment in which the fluorophore can be identified by its emission maximum and fluorescence lifetime—except that single-channel information is accumulated one event at a time, which provides for exquisite molecular resolution. Despite these theoretical advantages, few ion channels can be used for detailed studies of structure and function, because the requisite high-resolution structures are available only for two classes of ion channels: those formed by the linear gramicidins (Figure 2; 6, 52) and the porins (18, 122).

The gramicidins offer experimental advantages because of their small size and exceptionally well-defined function. Gramicidin A (gA) forms only a single-channel type, as evidenced by the observation that more than 90% of all channel events usually fall in a narrow peak in histo-

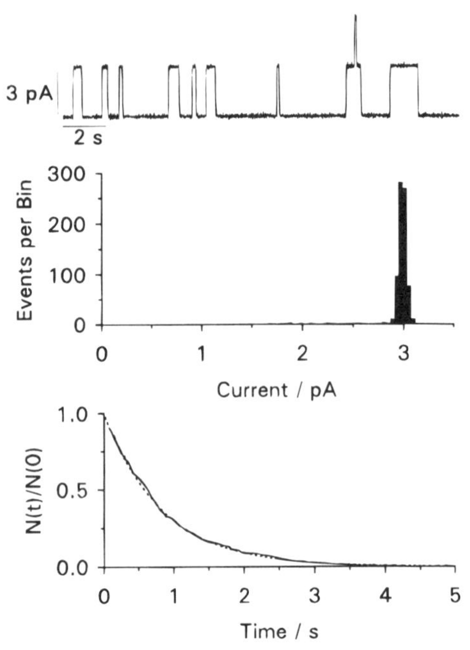

Figure 1 Single-channel current trace and characterization of channels formed by gA. (*A*) Current trace. (*B*) Current transition amplitude histogram, which shows a single predominant channel type. (*C*) Survivor distribution of channel durations (*solid line*) and a fit of a single exponential distribution to the results. Diphytanoylphosphatidylcholine/*n*-decane bilayers, 1.0 M NaCl, 200 mV. (MD Becker and OS Andersen, unpublished experiments.)

grams of current transitions (Figure 1). This functional simplicity makes gramicidin channels unique among low-molecular-weight membrane-active peptides and provides the basis for the use of gramicidin channels to understand the principles that govern the folding and function of membrane-spanning channels (and membrane proteins in general). Various amino acid substitutions can be introduced with no effect on the basic channel structure (25), which provides insight into the roles of specific residues for channel assembly and ion permeability (e.g. 9, 32, 70, 74, 78, 92). Further, gramicidin channels are particularly well suited for understanding the effects of introducing nongenetic amino acids in studies of structure and function (25, 32, 58, 74, 92). Moreover, some sequence changes endow the channels with qualitatively new functions or cause the gramicidins to take up new, functional folds. These latter observations make gramicidin channels powerful tools for understanding the molecular basis for the folding and dynamics of membrane proteins.

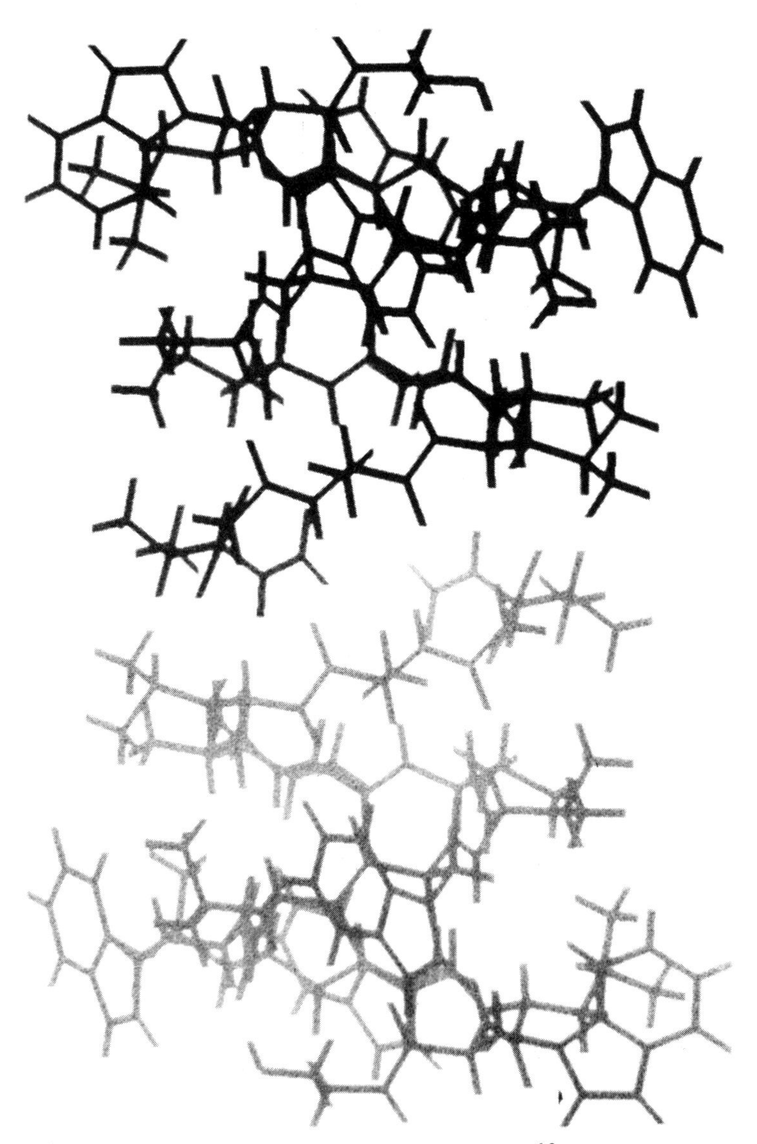

Figure 2 Wire model representation of the single-stranded $\beta^{6.3}$-helical gramicidin channel structure. (Modified from Reference 60.)

See figure in color online at http://www.annurev.org; browse to Supplementary Material and follow the links to the chapter.

Within the fields of membrane proteins (generally) and membrane channels (specifically), the gramicidins therefore are used extensively as models to study protein–lipid interactions (53), ion permeation (28, 30, 47), and structure–function relationships (3, 14, 118). Reviews of these topics are not repeated here. This article focuses on sequence changes that redesign the structure or function of gramicidin channels.

GRAMICIDIN STRUCTURE AND FUNCTION

Amino Acid Sequence and Biological Function of Gramicidin A

Gramicidin A is a hydrophobic peptide antibiotic produced by *Bacillus brevis* through enzymatic synthesis at the onset of sporulation (85); its amino acid sequence is (97)

HCO—L—Val——Gly—L—Ala—D—Leu—L—Ala—
1 2 3 4 5

—D—Val—L—Val—D—Val—L—Trp—D—Leu—L—Trp—
6 7 8 9 10 11

—D—Leu—L—Trp—D—Leu—L—Trp—$NHCH_2CH_2OH$.
12 13 14 15

In addition to the canonical gramicidins, which have a free ethanolamine-OH group, fatty acylated derivatives (at the terminal ethanolamine-O) have been discovered in natural sources (59). Similar acylated gramicidins have been synthesized chemically (117) and used to examine both local and long-range protein–lipid interactions (e.g. 57, 115). These interactions are not considered further here.

The linear gramicidins have a variety of biological activities. In the producing organism, these activities include a function during spore formation and an interaction with RNA polymerase (31). More generally, the linear gramicidins have antibiotic (23), as well as spermicidal and anti–human immunodeficiency virus, activities (13). The antibiotic activity is ultimately mediated by an increased cation permeability in biological membranes (39), and gA forms cation-selective channels in lipid bilayer membranes (45, 46, 75). The fatty acylated gramicidin derivatives have much less antibiotic activity than the conventional gramicidins have (DB Sawyer, JS Andersen, RE Koeppe, and OS Andersen, unpublished observations), which may be related to their decreased channel-forming potency (116, 124).

Folding of Gramicidin A in Bulk Solvent

The folding of a gramicidin-like molecule depends on its amino acid sequence and its environment. The alternating (L,D) amino acid sequence restricts the available conformations to β-helices (84, 108, 110) or unfolded motifs. In organic solvents, the gA structure is either an unfolded ("random coil") monomer (88) or one of several intertwined, double-stranded (DS) dimers. The intertwined dimers consist of two subunits wound around each other as β-helices in either parallel or antiparallel fashion (15, 110). Various intertwined dimer structures have been observed by vibrational spectroscopy (76, 111), two-dimensional NMR spectroscopy (1, 15), and X-ray crystallography (62, 63, 119). The intertwined arrangement of subunits has been correlated with a characteristic circular dichroism (CD) spectrum that has a negative ellipticity in the range of 210–240 nm (Figure 3B; 8, 36, 120).

Intertwined, DS gramicidin dimers sometimes are designated as being in the "pore" form (102, 118, 119). This nomenclature is unfortunate and confusing, however, because the specific meaning of pore is "a minute opening, . . . especially one by which matter passes through a membrane" (121). A pore is therefore the particular region of a "channel" through which water and ions may pass (see also Reference 44).

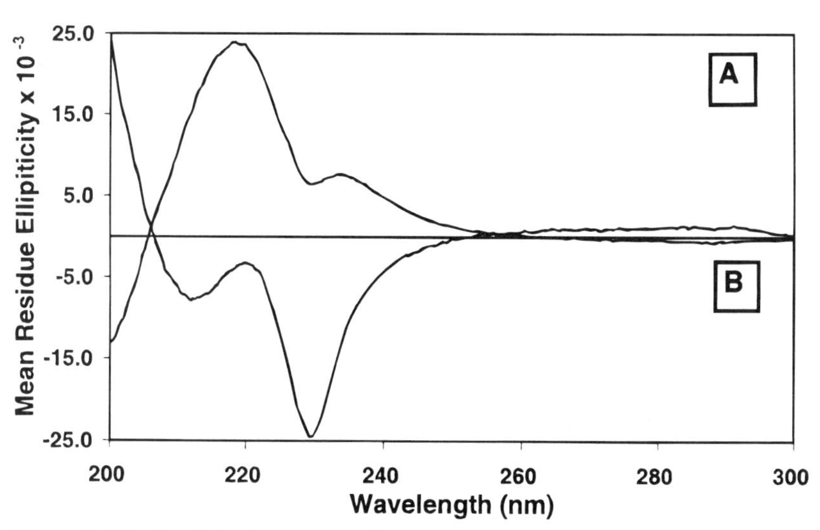

Figure 3 Circular dichroism spectra of aqueous gA dispersed in dimyristoylphosphatidylcholine (*A*), or diheptanoylphosphatidylcholine (B). In the lipid with the 14-carbon acyl chains (*A*), gA adopts a SS RH $\beta^{6.3}$-helical conformation. In the lipid with the 7-carbon acyl chains (*B*), gA adopts one or more DS conformations. (Modified from Reference 36.)

In lipid bilayers, a DS gA conformation has been observed only in membranes formed from lipids with highly unsaturated acyl chains (19, 105). These DS structures are deduced to be $\pi\pi^{5.6}$ helices (105), which appear to be impermeable to small inorganic cations and even protons (34). (Very long-lived conducting channel events that could be the functional expression of DS channels are seen occasionally in experiments with gA. These events are exceedingly rare,[1] however, and there is no detailed functional characterization.) Some sequence-substituted gramicidin analogues, however, do form conducting DS channels (see below).

Folding of Gramicidin A in Membranes

In lipid bilayer membranes, or bilayer-like environments, gA adopts a conformation that is different fundamentally from those observed in organic solvents. This novel structure is a right-handed (RH), single-stranded (SS) $\beta^{6.3}$-helical membrane-spanning dimer (5, 6, 52, 60, 108) that forms conducting channels in lipid bilayers. Correct folding into this conformation occurs only when gA is in a lipid bilayer or bilayer-like (micellar) environment. Even then, however, the folding preference depends on the dimensions of the hydrophobic core that the gramicidin dimer spans: In diacylphosphatidylcholines, a minimum acyl chain length of eight carbons is required to achieve the channel conformation (Figure 3; 36).

In RH SS $\beta^{6.3}$-helical channels (Figure 2), two SS $\beta^{6.3}$-helical monomers (''subunits'') form the membrane-spanning dimer by the transmembrane association of monomers coming from each respective monolayer (78). As proposed by Urry (108), the dimer is stabilized by six intermolecular hydrogen bonds at their formyl-NH-terminals, except that the helix sense of the $\beta^{6.3}$-helices is RH rather than left-handed (LH). The helix sense has been determined by two-dimensional NMR (5, 6), solid-state NMR (77, 103), and single-channel analysis by using appropriate reference compounds (60). The results of these different experimental approaches are all in agreement with an RH helix sense. The channel conformation is associated with a characteristic CD spectrum that has positive ellipticity in the range of 210–240 nm (Figure 3; 8, 36, 120).

---

[1]The quote attributed to JT Durkin [1986. *Biophys. J.* 49:306, Discussion Comment (see Reference 102)] is a misprint. Rather than ''one-tenth'' the fraction should be ''one-tenthousandth.''

Given the extent of the conformational transition that occurs when going from bulk organic solvent to the membrane environment, and the conformational plethora in organic solvents (1, 15, 110), one might expect that the structure of gramicidin in lipid bilayers (or micelles) would be quite sensitive to the chemical composition of the "membrane" environment. That is not the case. Somewhat surprisingly, the RH SS $\beta^{6.3}$-helical gA structure is essentially the same in sodium dodecyl sulfate micelles and dimyristoylphosphatidylcholine (DMPC) bilayers (6, 52), as evidenced from the similar CD spectra in the two environments (5, 69) and a comparison of the structures deduced from two-dimensional and solid-state NMR (52). This result shows that the channel structure is remarkably insensitive to the detailed composition of the membrane environment. The above studies, however, were all done with the use of lipids or detergents that had fully saturated acyl chains. When the gA structure is examined in bilayers that have unsaturated acyl chains (19, 105), one finds DS helical dimers in addition to SS $\beta^{6.3}$-helices.

Considering the number of conformational possibilities, it is remarkable that the alternating (L, D)-sequence of gA forms only a single type of channel. This finding raises the question, What determines the helix sense? Is it the primary amino acid sequence, the membrane environment, or a combination of these? The primary amino acid sequence is an important determinant, because the enantiomeric gramicidin, gA^- with the sequence

HCO—D—Val———Gly—D—Ala—L—Leu—D—Ala—
1 2 3 4 5

—L—Val—D—Val—L—Val—D—Trp—L—Leu—D—Trp—
6 7 8 9 10 11

—L—Leu—D—Trp—L—Leu—D—Trp—$NHCH_2CH_2OH$,
12 13 14 15

and related gA^- analogues form channels that are LH $\beta^{6.3}$-helical dimers (60). When the functional properties of channels formed by gA and gA^- are compared, all aspects of channel function are unaffected by changes in lipid chirality (83). The chirality of the membrane lipids, therefore, cannot be important; the helix sense must be determined by the amino acid sequence itself. Even though the helix sense is well established experimentally, the reason that the native gA does not adopt a LH channel conformation (108) in addition to, or instead of, the RH conformation has not been explained theoretically (see Reference 60).

Conformational energy calculations, for example, show little to choose between the RH and LH possibilities (29, 114). These calculations, however, were done on SS dimers *in vacuo*. Despite the remarkably robust and fairly invariant nature of the channel structure with respect to the precise composition of the bilayer environment, these calculations may not pertain to membrane-spanning gramicidin channels (see below). Nevertheless, the φ, ψ angles for the LH and RH SS $\beta^{6.3}$-helical channel conformations on a Ramachandran plot (Figure 4) occur relatively close to the intertwined dimer conformations, which suggests that the energetic preference for the different conformations is modest. This finding is important, because one could not recognize a conducting gA channel species experimentally if it were present at a relative concentration that was much less than 0.01 of that of the major channel type. As the channels are dimers, this concentration corresponds to a mole fraction of "misfolded" monomers of approximately 0.1, or a free energy difference between the (putative) RH and LH $\beta^{6.3}$-helical monomers that could be as small as approximately 6 kJ/mol. If misfolding were detected through the formation of hetero-

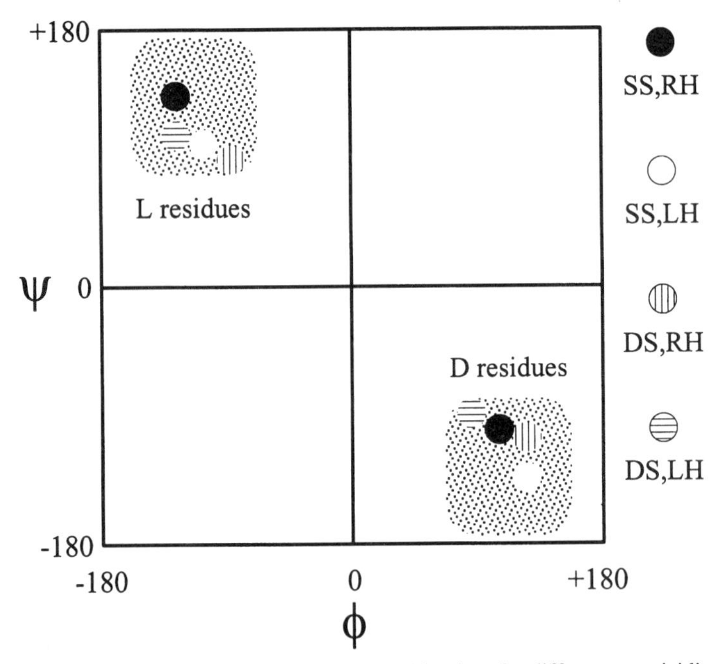

Figure 4 Ramachandran plot showing φ, ψ combinations for different gramicidin channel structures. The hatched areas show the combinations that are allowed for all α-amino acids. (Modified from Reference 2a.)

dimers, the sensitivity would be enhanced, so that one could detect a population of misfolded monomers at a mole fraction of only 0.05 or less (60).

Structural Equivalence of Single-Site Mutants in Membranes

The energetic problem identified above pertains to all attempts to redesign the structure of a protein, as one generally has a narrow energetic "window" within which several conformational states can be observed. [In a systematic study of amino acid substitutions in T4 lysozyme, Poteete et al (82, 86) found, for example, that activity would have to be decreased to approximately 3% of the wild-type activity before a substitution could be deemed deleterious.] Because gA channels are antiparallel, symmetric dimers, however, a powerful functional test is available for the structural equivalence of mutants. This test is well suited for detecting minor conformers.

The functional test is based on the observation that, within experimental accuracy, all gA dimers are channels (112) and exploits the ability of chemically distinct, but structurally similar, monomers to form heterodimers, or hybrid channels (25, 27). The test is based on five criteria (after Reference 25):

1. A substituted gramicidin that folds in only one conformation forms only a single channel type (4, 25, 99). (If multiple channel types are observed in the presence of a single gramicidin analogue, then multiple channel conformations are present; see below).
2. Chemically dissimilar subunits can form hybrid channels if they have similar backbone folding patterns (the same helix sense and complementary subunit interfaces) (25–27, 60). These heterodimeric, hybrid channels can be identified electrophysiologically by their single-channel current amplitude, average duration, or both (25, 27, 70, 113). The relative association rate constants for the formation of the heterodimeric and homodimeric channel types can be estimated from the number of channel events that underlie each peak in an amplitude histogram. The dissociation rate constants for the different channel types can be estimated as the reciprocals of their average durations. The relative concentrations can be estimated from the channel appearance rate and average duration for each channel type.
3. The activation energy ($\Delta\Delta G$) for the formation of heterodimers, relative to that for the homodimers, can be calculated from relative association rate constants:

$$\Delta\Delta G \; = -RT \ln f_H / [2(f_A f_B)^{0.5}] , \qquad 1$$

where f_A, f_B, and f_H are the appearance rates, and τ_A , τ_B, and τ_H are the average channel durations for the parent homodimeric A and B channels and the hybrid H channels, respectively (25). If Δ ΔG ∪ 0, i.e. if the transmembrane dimerization step does not depend on whether the component subunits are the same or different, the monomers and, therefore, the symmetric channels are structurally equivalent.

4. The equilibrium energy (Δ ΔG°) for the heterodimers, relative to that for the homodimers, can be calculated from relative membrane concentrations (25, 27):

$$\Delta \, \Delta G^\circ = -RT \ln f_H \tau_H \, / \, [2(f_A f_B \tau_A \tau_B)^{0.5}]. \qquad 2$$

In most cases, Δ ΔG° ∪ 0 (e.g. 25). (A Δ ΔG° 0 implies a difference in structure, dynamics, or function; see below.)

5. Finally, hybrid channel formation is a transitive property.

In addition to the functional test, one can use CD and NMR experiments to examine whether a sequence-modified gramicidin analogue forms channels that retain the basic (SS, $\beta^{6.3}$-helical) channel conformation. The structural equivalence of [Val^5-D-Ala^8]gA and gA channels and of [Ala^1-D-Ala^2]gA and gA channels have been demonstrated with the use of all these methods (54, 69).

The functional approach has a major advantage of being a "null method," i.e. one can determine whether a sequence substitution has affected the structure even when an atomic resolution structure is not available. Moreover, if the chemical modification (amino acid substitution) in question does not affect the basic structure, one can use this approach to deduce the subunit stoichiometry of a channel (16, 113). This approach also has been used to determine the subunit stoichiometry of channels formed by integral membrane proteins (17, 67).

Redesigning the Membrane Structure of Gramicidin

Within the framework of multiple environment-dependent conformations, a defined two-subunit SS RH $\beta^{6.3}$-helical conformation, and a large number of sequence-substituted analogues that retain a "structurally equivalent" channel conformation, there lingers a persistent question: How does the amino acid sequence (in concert with the membrane environment) dictate the backbone folding and subunit interactions? In this respect, it is again important that the φ , ψ peptide torsion angles for the L and D residues of LH and RH DS and SS dimers are near each other on a Ramachandran plot and within the region that is allowed for all α -amino acids (Figure 4). Thus, gramicidin channels may be useful for understanding the structural biology of conformational transitions in membranes. We first discuss the question of SS vs DS channel structures and then the question of RH vs LH SS structures.

DOUBLE-STRANDED VS SINGLE-STRANDED SUBUNITS The first hints about the molecular basis for the preference for SS $\beta^{6.3}$-helical dimers in lipid bilayers were the unexpected finding that gA crosses planar lipid bilayers quite slowly (78) and the accidental finding that a sequence-shortened gC analogue, des-Val^1-gC with a Tyr residue at position 11, forms a qualitatively new hybrid channel type in the presence of gM$^-$, a mirror-image gA analogue: [D-Phe^{9,11,13,15}]gA$^-$ (26). [The formation of conventional SS hybrid channels is prevented when the parent monomers have opposite chirality (24, 60).] These new des-Val^1-gC/gM$^-$ channels retained a cation selectivity and had an average duration that was 1000-fold longer and a relative appearance rate that was only one tenth that predicted on the basis of the durations and appearance rates of the symmetric parent channels. Importantly, similar channels were not seen with mixtures of des-Val^1-gC and gA$^-$, which highlights the significance of the amphipathic Trp residues for the conformational preference of the linear gramicidins.

The importance of the Trp residues for the stability of membrane-spanning SS $\beta^{6.3}$-helical dimers was also demonstrated by the finding that progressive Trp → Phe substitutions decrease the SS channel-forming potency of gA analogues (9). Recently (96), the importance of the Trp residues was highlighted further by the demonstration that the preferred structure of gM$^-$ in DMPC bilayers is some variety of DS dimer, which does not form conducting channels (because no conductance can be assigned to membrane-spanning DS dimers formed by gM or gM$^-$; J Girshman and OS Andersen, unpublished observations).

These results can be rationalized by considering the energetic cost of burying the amphipathic Trp residues in the membrane interior (9, 26, 49, 56, 78, 126), because the Trp indole moiety can form hydrogen bonds to polar groups at the membrane–solution interface. When gA forms an SS $\beta^{6.3}$-helix, the Trp residues will cluster at the two membrane–solution interfaces, and at least three of these residues form hydrogen bonds to water (107). This finding explains why the quite hydrophobic gA crosses lipid bilayers very slowly (78). When gA forms a DS dimer, the Trp residues are distributed rather uniformly along the exterior surface of the channel, and only one, or two, of these residues can hydrogen-bond to water. All other factors being equal, DS dimers thus are destabilized, relative to SS dimers, by the cost of burying at least two Trp residues in the membrane interior, which constitutes a major energetic destabilization of DS dimers in bilayers. [Spectroscopic investigations (19, 105) have revealed that the SS/DS conformational preference of gA in lipid bilayers varies as a function of the acyl chain unsaturation. In fully saturated bilayers no DS dimers were detected; as the unsaturation increased, however, the fraction of gA in a DS conformation increased. These results suggest that the indole– NH may interact favorably with the C=C double bonds (34).]

The importance of the Trp residues for the conformational preference of gramicidin channels is also evidenced by the influence of the arrangement of the Leu and Trp residues in the C-terminal half of the molecule on the helix "strandedness" (distribution between DS and SS dimers). If the L-Trps at positions 9, 11, 13, and 15 are replaced by L-Leu, and the D-Leus at positions 10, 12, and 14 are replaced by D-Trp, the resulting gLW

HCO—L—Val———Gly—L—Ala—D—Leu—L—Ala—
1 2 3 4 5

—D—Val—L—Val—D—Val—L—Leu—D—Trp—L—Leu—
6 7 8 9 10 11

—D—Trp—L—Leu—D—Trp—L—Leu—$NHCH_2CH_2OH$,
12 13 14 15

forms several different channel types in diphytanoylphosphatidylcholine bilayers (37, 55, 93): relatively short-lived events with durations less than approximately 0.5 s and long-lived events with average durations of approximately 60 s. The existence of several different channel types could result from inadvertent contamination, but the gLW samples are homogenous by mass spectrometry (DV Greathouse and RE Koeppe, unpublished results). The long-lived events are seen regardless of whether gLW is added to only one or to both sides of a planar bilayer, whereas the short-lived events are seen only when gLW is added to both sides of the bilayer. As standard (SS $\beta^{6.3}$-helical) gA channels are seen only when gA is added to both sides of a bilayer (78), we conclude that the long-lived events are DS gLW channels, whereas the shorter lived events are SS gLW channels. The equilibrium distribution between DS and SS dimers, therefore, is shifted in favor of the DS dimers. Surprisingly, however, the CD spectra and size-exclusion chromatography of DMPC-incorporated gLW suggest that the predominant conformation in membranes is an LH SS or unfolded monomer (37). This finding suggests that the conformational rearrangement results from a destabilization of SS channels rather than from a stabilization of the DS dimers. More detailed single-channel analysis suggests that the short-duration gLW channels are of two types: both RH and LH SS $\beta^{6.3}$-helical dimeric channels (G Saberwal and OS Andersen, unpublished observations). The DS gLW channels are antiparallel dimers, because the channels are symmetric; at present, the helix sense is not known for these DS channels.

These results show that gLW can exist in several interconverting conformations (DS; SS,RH; and SS,LH) within a lipid bilayer (Figure 5). The unfolded monomer can refold to give rise to any of the other conformations. Similar interconversions presumably are also possible for the parent gA channel, but with different relative energetics. With gA, the RH SS $\beta^{6.3}$-helical structure is so stable within a bilayer that other conformations become difficult, if not impossible, to observe. The different conformational properties of gA and gLW demonstrate the importance of the ethanolamide-CO-terminal $(\text{Trp-Leu})_n$ sequence for the conformational preference in membranes. Not only the Trp residues are important, however. When the Leu "spacer" residues are replaced by alanines, the average channel duration is decreased by approximately 70%, and the CD spectrum is altered markedly (51).

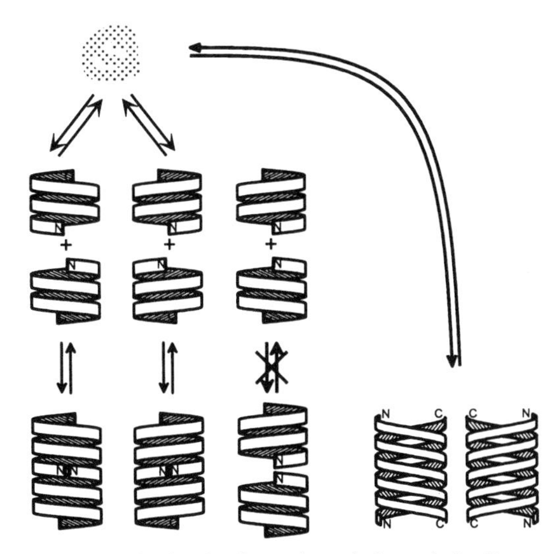

Figure 5 Folding schematic for the formation of channels by linear gramicidin analogues of the gLW family. The random coil (88) is in equilibrium with SS RH and SS LH ($\beta^{6.3}$- helical) monomers and with membrane-spanning DS channels. These different channel types can be identified by their ability to form heterodimers with reference gramicidins that form only a single-channel type and by the asymmetric or symmetric addition of the gramicidin analogue. (Modified from Reference 2a.)

Residues that are deeper in the membrane are also important for the conformation preference of gramicidin channels. Within the gLW sequence, for example, the ability to form DS dimers in lipid bilayers seems to be determined by the identity of residue 5, as demonstrated by the observation that [Val^{5}]gLW and [Val^{5},D-Ala^{8}]gLW form only SS channels, whereas gLW itself and [D-Ala^{8}]gLW form both DS and SS channels (93). In this series of analogues, therefore, a single Ala ↔ Val exchange alters the distribution among functional conformations qualitatively.

HELIX SENSE OF SINGLE-HELICAL SUBUNITS The gLW sequence framework allows LH, as well as RH, SS channels to form in the same membrane; these channels are therefore "ambidextrous" (94). This property is shared by gLW, [Val^{5}]gLW, [D-Ala^{8}]gLW, and [Val^{5},D-Ala^{8}]gLW (93). Consequently, the (Trp-Leu)$_n$ sequence also must be important for the helix sense. When the Trp residues are moved to positions of D-residues (at 10, 12, and 14), the relative energetics of LH and RH channels are placed on a more even foundation. Within this outline, one can investigate the sequence dependence of LH ↔ RH interconversions within a membrane through a combination of electrophysiologic and spectroscopic experiments and conformational energy calculations (e.g. 89).

ENGINEERING CHANNEL FUNCTION

Above we have considered the sequence determinants of conformation for gA channels and the sequence changes that lead to altered folded conformations. We now turn to the question of sequence changes that alter channel function. Can one introduce new (and interesting) functional properties simply by changing the amino acid sequence? If so, many general lessons can be inferred from studies on a system in which the basic channel structure is known. The focus here is on voltage-dependent channel "gating" (transitions between states with different conductance).

Voltage-Dependent Behavior

Ion channels exhibit several different types of voltage-dependent behavior. First, the permeability properties of the conducting (open) channel may vary as a function of potential. Second, the kinetics and equilibrium distribution between different closed and open states may vary as a function of the applied potential.

VOLTAGE-DEPENDENT PERMEABILITY PROPERTIES In symmetric membranes, two-fold symmetric channels, such as gramicidin channels that have two identical subunits, exhibit identical behavior at negative and positive membrane potentials. The permeability properties of the channels may still vary as a function of the applied potential, however, because channel-mediated ion movement results from an electrodiffusive barrier crossing, which generally gives rise to nonlinear current-voltage relationships (2, 40, 64). The extent of the nonlinearity can be altered by altering the amino acid sequence of the channels (92).

Asymmetric channels generally behave differently at positive and negative potentials (48, 100). This rectification is the behavior expected for integral membrane protein channels. For gramicidin channels, such rectification is observed for heterodimeric channels that have dipolar substitutions in which the energy profile for ion movement through the channel is asymmetric. As expected, the extent of the asymmetry increases as the position of the sequence substitution moves from the channel center toward the channel entrances (cf. 9, 32, 70, 92).

VOLTAGE-DEPENDENT GATING The general features of voltage-dependent gating have been reviewed recently (see References 10, 44, and 100). Voltage-dependent gating depends on the existence of two (or more) channel conformations and a voltage-dependent switching between them. To be detectable (and distinguishable) functionally, the states must differ in conductance (e.g. ''open'' and ''closed'').

In the absence of ''clear structural information'' (100), voltage-dependent gating usually is characterized functionally, which is largely a reflection of the ease of characterizing the dynamics of molecular function with the use of single-channel methods. There is a distinct dichotomy between the precision with which the kinetics of channel gating can be dissected and the fuzziness that characterizes most discussions of channel structure. Studies of mutations have been directed toward identifying ''gate'' domains and ''voltage sensors'' in voltage-gated sodium, potassium, and calcium channels. These studies suffer, however, from the usual problem that molecular interpretations must be considered tentative until a definitive structure has been established (61). The answers to many key questions (e.g. What is the chemical nature of the voltage sensor? What is the nature of the gate? How do the closed, open, and inactivated channel states differ in conformation? How is the necessary amount of gating charge transferred across the membrane during the closed $\leftrightarrow$ open state transitions?) therefore remain tentative or unanswered (100). Given the considerable uncertainties that persist, model systems that allow one to define possible mechanisms clearly are needed (e.g. 21).

An alternative approach is to design voltage-dependent features into simpler channels that initially do not change conformations in response to changes in the transmembrane voltage. The criteria for success in such an enterprise are whether one can actually produce chemically defined molecules that show voltage-dependent conformational transitions and, then, whether one can understand the conformational and functional behavior of these designed systems.

Voltage-Dependent Gramicidin Channels

Gramicidin channels usually have a single, well-defined conducting state with brief transitions to some low-conductance state (87, 101). The frequency of these conductance transitions can be enhanced by chemical modifications of the formyl group, which affect the stability of the junction between the monomers (101, 104, 106). Generally, however, gramicidin channels are not voltage dependent. Nevertheless, two types of amino acid alterations in gA channels have been found to confer voltage-dependent properties. The first alteration introduces a defect near the channel center through the specific deletion of a single residue from only one of the subunits, thereby yielding a mismatch between the subunits where they meet in the middle of the membrane (27). The second introduces a specific amino acid (either tyrosine or hexafluorovaline) at position 1, again on only one of the subunits near the point at which they meet in the center of the membrane. These length- and sequence-changing modifications share the common feature of breaking the twofold symmetry of the channel.

Formyl Group Modifications

In the SS $\beta^{6.3}$ conformation, the N-formyl groups of the two subunits are near each other in the center of a lipid bilayer. When the NH_2-terminal formyl group is replaced by the bulkier acetyl group, steric crowding occurs at the center of the channel and the average channel duration is reduced (106). In addition, the symmetric channels formed by N-acetyl-gA show rapid fluctuations (106) between two well-defined levels of current (101). The lower level of current is not zero but finite, which demonstrates that the fluctuations of current arise from transitions in dimers (as opposed to representing rapid monomer $\leftrightarrow$ dimer transitions).

Two gA monomers can be cross-linked to give a (symmetric) covalent dimer, e.g. with a $-CH_2-$ bridge to yield a malonyl dimer (7, 109) or with (R,R) or (S,S)-dioxolane to give two different stereospecific isomers (104). When two desformyl-gramicidin monomers are cross-linked, long-lived (>30 min) channels result (104, 109). Particularly interesting results are obtained when the linker is a dioxolane ring. The current–voltage relationship is almost linear for (S,S)-dioxolane–linked channels but strongly superlinear for the (R,R)-dimers (104), and the (R,R)-linked channels exhibit brief closures to a zero-conductance state (104). These transitions have been modeled in a study of molecular dynamics (22); the calculated block rate in the model (280/s) compares favorably with the experimental "flicker" rate of 100/s (104). There is less agreement regarding the blocked-state lifetime (9 ns in the computer model, as opposed to 100 μs experimentally). The combined experimental and theoretical results in this system demonstrate an early application toward molecular understanding of conformational transitions (gating) in membrane-spanning channels and suggest future promise.

Length Changes That Introduce Asymmetry and Voltage Dependence

The repeating motif in a $\beta^{6.3}$-helix is an L,D-dipeptide unit—not a single amino acid. If a single amino acid is added or deleted between the formyl group and the N-terminal of one subunit of a gA channel, a resulting defect occurs in the channel (27). A missing residue causes a decrease in the single-channel conductance and a destabilization of approximately 10 kJ/mol for hybrid channels in which the subunits differ in length by $\Delta n = \pm 1$ amino acid (27).

When $\Delta n = \pm 1$ and the formyl-NH-terminal residue on the lengthened or shortened subunit is Gly, then the hybrid channels that form in combination with native gA show multistate behavior with voltage-dependent transitions between two levels of conductance (27). This multistate behavior is not observed when the formyl-NH-terminal residue is D-Ala instead of Gly. [D-Ala^2 also stabilizes standard gA channels relative to Gly^2 (69).] These results suggest that both subunit asymmetry and "strain" contribute to the voltage-dependent behavior. Moreover, the structural flexibility that Gly introduces in the vicinity of the "defect" appears to be important for the obligatory conformational switch seen with voltage-dependent gating.

Sequence Changes That Introduce Asymmetry and Voltage Dependence

Two different amino acid substitutions at position 1 impart qualitatively new functions to hybrid gramicidin channels. Heterodimers usually have properties that are intermediate to those of the two symmetric channels (25). When either tyrosine (70) or hexafluorovaline (92) are substituted at position 1 on only one subunit, however, the hybrid channel conductance is lower than that of either corresponding symmetric channel; the hybrid channel average duration is likewise lower than that of either of the corresponding symmetric channels (25, 70, 92). These observations parallel those made for heterodimers that miss a residue at the subunit interface (27; see previous section).

The parallel between F_6Val^1, Tyr^1 and "missing residue" hybrid channels extends to voltage-dependent gating. In combination with either $[Val^1]gA$, $[Ala^1]gA$ or $[Gly^1]gA$, $[F_6Val^1]gA$ forms hybrid channels that show voltage-dependent transitions between (at least) two stable conductance states (79–81). The gating behavior depends on the presence of the F_6Val^1 residue in only one monomer. Changes in the identity of the position 1 residue (Val, Ala or Gly) in the *trans* monomer, however, alter the quantitative description of the voltage dependence; it appears, therefore, that both subunits are involved in the voltage-dependent conformational change from a low (closed) to a high conducting state (80).

Figure 6 shows a gating curve for hybrid channels formed by $[F_6Val^1]gA$ and (native) $[Val^1]gA$, with either Cs^+ or H^+ as the permeant ion. The voltage-dependent gating behavior does not depend on the nature of the permeant ion.

The voltage sensor in $[F_6Val^1]gA/[Val^1]gA$ heterodimeric channels most likely is the F_6Val^1 side chain that could undergo voltage-dependent rotations between different rotameric states (Figure 7). To account for the magnitude of the voltage dependence, however, a rotation of the side-chain dipole would have to trigger more extensive conformational adjustments in both subunits (81). A two-state behavior of the sensor, therefore, could lead to multistate channel behavior. Not surprisingly, the $[F_6Val^1]gA/[Val^1]gA$ results cannot be fitted by a simple two-state model (81), which raises the more general question of the origin the apparent gating charge movements that are required to account for the voltage sensitivity of membrane channels (100). $[F_6Val^1]gA/[Val^1]gA$ hybrid channels, for example, do not possess formal fixed charges, and maximal movements of the gating sensor in these channels (e.g. Figure 7) can account for only a small fraction of the apparent gating "charge" (81).

Similar voltage-dependent conformational transitions have been observed for hybrid channels between endo-Tyr^{0a}-Gly^{0b}-gA and endo-Ala^{0a}-D-Ala^{0b}-gA (LL Providence, GL Mattice, OS Andersen, and RE Koeppe, unpublished observations). These 17-residue analogues are attractive, because the voltage dependence results from the introduction

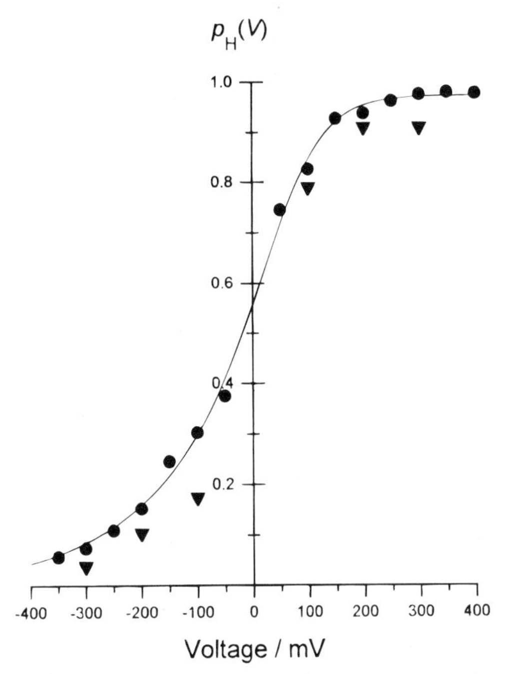

Figure 6 Voltage-dependent single-channel gating of heterodimeric $[F_6Val^1]gA/[Val^1]gA$ channels with either Cs^+ (*filled circle*) or H^+ (*filled inverted triangle*) as the permeant ion. The electric potential is referenced to the aqueous solution facing the $[F_6Val^1]gA$ half of the channel. (Reprinted from Reference 81, with permission.)

of the genetic amino acid Tyr, rather than F_6Val. Moreover, the average duration of channels formed by 17-residue analogues is more than tenfold longer than for corresponding 15-residue analogues (11, 68), which allows for a more facile functional characterization of the (relatively destabilized and voltage-dependent) hybrid channels. Future work will address the mechanism of voltage-dependent channel gating induced by both tyrosine and hexafluorovaline, through combinations of sequence modifications and spectroscopic, electrophysiologic, and molecular dynamics studies; both systems offer promise for understanding of the molecular basis for channel gating.

SUMMARY AND CONCLUSIONS

Changes in the amino acid sequence of gA have been used to introduce new structural and functional properties into the membrane-spanning channels formed by these molecules. The helix sense and backbone

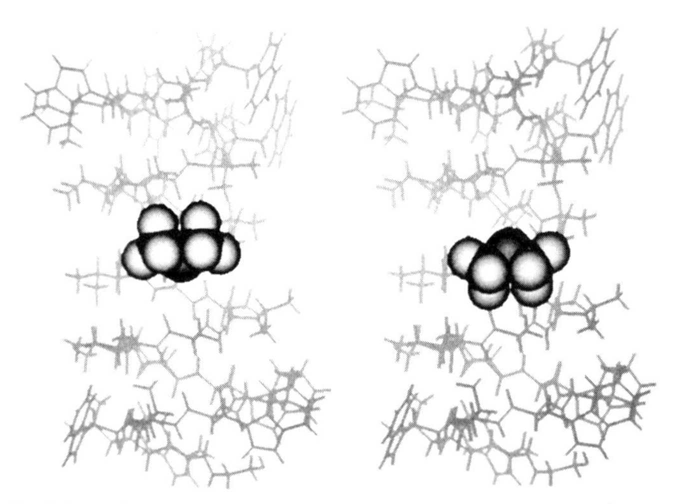

Figure 7 Schematic two-state model for the gating sensor in $[F_6Val^1]gA/[Val^1]gA$ channels. A wire model similar to that in Figure 2 is shown for a gramicidin channel, with a single hexafluorovaline side-chain dipole in two possible orientations with respect to an external transmembrane potential.

See figure in color online at http://www.annurev.org; browse to Supplementary Material and follow the links to the chapter.

interactions (subunit intertwining) can be changed while the basic channel functions are retained. In some cases, a single amino acid sequence can give three functional conformations: LH and RH SS channels and DS channels. Selected single amino acid substitutions, which break the channel symmetry and introduce a side-chain dipole, confer voltage-dependent channel gating. These findings provide general insights into the folding of membrane proteins, the importance of interfacial tryptophan residues, and the mechanism of channel gating.

ACKNOWLEDGMENTS

We are grateful for financial support for research in our laboratories from the National Institutes of Health (grants GM34968 and GM21342). We thank our colleagues for stimulating discussions, particularly those who have contributed to work in progress that is summarized here: Kenneth Blount, Jeffrey Girshman, Denise Greathouse, Anthony Jude, Nam Le, Joseph Lundquist, Gwendolyn Mattice, Lyndon Providence, and Gayatri Saberwal.

Literature Cited

1. Abdul-Manan Z, Hinton JF. 1994. Conformational states of gramicidin A along the pathway to the formation of channels in model membranes determined by 2D NMR and circular dichroism spectroscopy. *Biochemistry* 33:6773–83
2. Andersen OS. 1989. Kinetics of ion movement mediated by carriers and channels. *Methods Enzymol.* 171: 62–112
2a. Andersen OS, Greathouse DV, Koeppe RE II, Saberwal G. 1996. Gramicidin channels as tools to understand membrane protein folding. *Indian J. Biochem. Biophys.* In press
3. Andersen OS, Koeppe R II. 1992. Molecular determinants of channel function. *Physiol. Rev.* 72:S89-S158
4. Andersen OS, Koeppe RE II, Durkin JT, Mazet J-L. 1987. Structure-function studies on linear gramicidins: site-specific modifications in a membrane channel. In *Ion Transport Through Membranes*, ed. K Yagi, B Pullman, pp. 295–314. Tokyo: Academic
5. Arsen'ev AS, Barsukov IL, Bystrov VF, Lomize AL, Ovchinnikov YA. 1985. Proton NMR study of gramicidin A transmembrane ion channel. Head-to-head right-handed, single-stranded helixes. *FEBS Lett.* 186: 168–74
6. Arsen'ev AS, Lomize AL, Barsukov IL, Bystrov VF. 1986. Gramicidin A transmembrane ion-channel. Three-dimensional structure reconstruction based on NMR spectroscopy and energy refinement. *Biol. Membr.* 3: 1077–104
7. Bamberg E, Janko K. 1977. The action of a carbon suboxide dimerized gramicidin A on lipid bilayer membranes. *Biochim. Biophys. Acta* 465: 486–99
8. Bañó MC, Braco L, Abad C. 1989. HPLC study on the 'history' dependence of gramicidin A conformation in phospholipid model membranes. *FEBS Lett.* 250:67–71
9. Becker MD, Greathouse DV, Koeppe RE II, Andersen OS. 1991. Amino acid sequence modulation of gramicidin channel function: effects of tryptophan-to-phenylalanine substitutions on the single-channel conductance and duration. *Biochemistry* 30: 8830–39
10. Bezanilla F, Stefani E. 1994. Voltage-dependent gating of ionic channels. *Annu. Rev. Biophys. Biomol. Struct.* 23:819–46
11. Blount KF, Mattice GL, Greathouse DV, Taylor MJ, Koeppe RE II. 1995. Structural studies of tyrosine-1 gramicidins. *Biophys. J.* 68:A152 (Abstr.)
12. Bone R, Silen JL, Agard DA. 1989. Structural plasticity broadens the specificity of an engineered protease. *Nature* 339:191–95
13. Bourinbaiar AS, Krasinski K, Borkowsky W. 1994. Anti-HIV effect of gramicidin in vitro: potential for spermicide use. *Life Sci.* 54:PL5-PL9
14. Busath DD. 1993. The use of physical methods in determining gramicidin channel structure and function. *Annu. Rev. Physiol.* 55:473–501
15. Bystrov VF, Arsen'ev AS. 1988. Diversity of the gramicidin A spatial structure: two-dimensional proton NMR study in solution. *Tetrahedron* 44:925–40
16. Cifu A, Koeppe RE II, Andersen OS. 1992. On the supramolecular structure of gramicidin channels. The elementary conducting unit is a dimer. *Biophys. J.* 61:189–203
17. Cooper E, Sabine C, Ballivet M. 1991. Pentametric structure and subunit stoichiometry of a neuronal nicotinic acetylcholine receptor. *Nature* 350:235–38
18. Cowan SW, Schirmer T, Rummel G, Steiert M, Ghosh R. 1992. Crystal structures explain functional properties of two *E. coli* porins. *Nature* 358: 727–33
19. Cox KJ, Ho C, Lombardi JV, Stubbs CD. 1992. Gramicidin conformational studies with mixed-chain unsaturated phospholipid bilayer systems. *Biochemistry* 31:1112–17
20. Craik CS, Largman C, Fletcher T, Roczniak S, Barr PJ, et al. 1985. Redesigning trypsin: alteration of substrate specificity. *Science* 228: 291–97
21. Cramer WA, Heymann JB, Schendel SL, Deriy BN, Cohen FS, et al. 1995. Structure-function of the channel-forming colicins. *Annu. Rev. Biophys. Biomol. Struct.* 24:611–41
22. Crouzy S, Woolf TB, Roux B. 1994. A molecular dynamics study of gating in dioxolane-linked gramicidin A channels. *Biophys. J.* 67:1370–86
23. Dubos RJ, Hotchkiss RD. 1941. The production of bactericidal substances

by aerobic sporulating bacilli. *J. Exp. Med.* 73:629–40
24. Durkin JT, Andersen OS, Blout ER, Heitz F, Koeppe RE II, Trudelle Y. 1986. Structural information from functional measurements. Single-channel studies on gramicidin analogs. *Biophys. J.* 49:118–21
25. Durkin JT, Koeppe RE II, Andersen OS. 1990. Energetics of gramicidin hybrid channel formation as a test for structural equivalence. Side-chain substitutions in the native sequence. *J. Mol. Biol.* 211:221–34
26. Durkin JT, Providence LL, Koeppe RE II, Andersen OS. 1992. Formation of non-$\beta^{6.3}$-helical gramicidin channels between sequence-substituted gramicidin analogues. *Biophys. J.* 62:145–59
27. Durkin JT, Providence LL, Koeppe RE II, Andersen OS. 1993. Energetics of heterodimer formation among gramicidin analogues with an NH_2-terminal addition or deletion: consequences of missing a residue at the join in the channel. *J. Mol. Biol.* 231: 1102–21
28. Eisenman G, Sandblom JP. 1983. Energy barriers in ionic channels: data for gramicidin A interpreted using a single-file (''3B4S'') model having 3 barriers separating 4 sites. *Stud. Phys. Theor. Chem.* 24:329–48
29. Etchebest C, Pullman A. 1988. The gramicidin A channel: left versus right-handed helix. In *Transport through Membranes: Carriers, Channels, and Pumps*, ed. A Pullman, J Jortner, B Pullman, pp. 167–85. Dordrecht: Kluwer Academic
30. Finkelstein A, Andersen OS. 1981. The gramicidin A channel: a review of its permeability characteristics with special reference to the single-file aspect of transport. *J. Membr. Biol.* 59:155–71
31. Fischer R, Blumenthal T. 1982. An interaction between gramicidin and the σ-subunit of RNA polymerase. *Proc. Natl. Acad. Sci. USA* 79: 1045–48
32. Fonseca V, Daumas P, Ranjalahy-Rasoloarijao L, Heitz F, Lazaro R, et al. 1992. Gramicidin channels that have no tryptophan residues. *Biochemistry* 31:5340–50
33. Ghadiri MR, Granja JR, Buehler LK. 1994. Artificial transmembrane ion channels from self-assembling peptide nanotubes. *Nature* 369:301–4
34. Girshman J, Andersen OS, Greathouse DV, Koeppe RE II. 1996. Gramicidin channels in phospholipid bilayers having unsaturated acyl chains. *Biophys. J.* In press
35. Graf L, Craik CS, Patthy A, Roczniak S, Fletterick RJ, Rutter WJ. 1987. Selective alteration of substrate specificity by replacement of aspartic acid-189 with lysine in the binding pocket of trypsin. *Biochemistry* 26: 2616–23
36. Greathouse DV, Hinton JF, Kim KS, Koeppe RE II. 1994. Gramicidin A/ short-chain phospholipid dispersions: chain length dependence of gramicidin conformation and lipid organization. *Biochemistry* 33: 4291–99
37. Greathouse DV, Le N, Koeppe RE II, Hinton J, Saberwal G, et al. 1995. Multiple conformations for gramicidin LW. *Biophys. J.* 68:A151 (Abstr.)
38. Handel TM, Williams SA, DeGrado WF. 1993. Metal ion-dependent modulation of the dynamics of a designed protein. *Science* 261:879–85
39. Harold FM, Baarda JR. 1967. Gramicidin, valinomycin, and cation permeability of *Streptococcus faecalis. J. Bacteriol.* 94:53–60
40. Haydon DA, Hladky SB. 1972. Ion transport acrsss thin lipid membranes. Critical discussion of mechanisms in selected systems. *Q. Rev. Biophys.* 5:187–282
41. Hedstrom L, Szilagyi L, Rutter WJ. 1992. Converting trypsin to chymotrypsin: the role of surface loops. *Science* 255:1249–53
42. Heginbotham L, Abramson T, MacKinnon R. 1992. A functional connection between the pores of distantly related ion channels as revealed by mutant K+ channels. *Science* 258:1152–55
43. Heinemann SH, Terlau H, Stühmer W, Imoto K, Numa S. 1992. Calcium channel characteristics conferred on the sodium channel by single mutations. *Nature* 356:441–43
44. Hille B. 1992. *Ionic Channels of Excitable Membranes.* Sunderland, MA: Sinauer Associates. 607 pp.
45. Hladky SB, Haydon DA. 1970. Discreteness of conductance change in bimolecular lipid membranes in the presence of certain antibiotics. *Nature* 225:451–53
46. Hladky SB, Haydon DA. 1972. Ion transfer across lipid membranes in the presence of gramicidin A. I. Unit conductance channel. *Biochim. Biophys. Acta* 274:294–312

47. Hladky SB, Haydon DA. 1984. Ion movements in gramicidin channels. *Curr. Top. Membr. Transp.* 21: 327–72
48. Hodgkin AL, Huxley AF. 1952. A quantitative description of membrane current and its application to conduction and excitation in nerve. *J. Physiol. (London)* 117:500–44
49. Hu W, Lee KC, Cross TA. 1993. Tryptohpans in membrane proteins: indole ring orientations and functional implications in the gramicidin channel. *Biochemistry* 32:7035–47
50. Itakura K, Rossi JJ, Wallace RB. 1984. Synthesis and use of synthetic oligonucleotides. *Annu. Rev. Biochem.* 53:323–56
51. Jude AR, Greathouse DV, Koeppe RE II, Basu S, Providence LL, Andersen OS. 1995. Effects of β-branched amino acids at positions 10, 12, and 14 of gramicidin channels. *Biophys. J.* 68:A152 (Abstr.)
52. Ketchem RR, Hu W, Cross TA. 1993. High-resolution of gramicidin A in a lipid bilayer by solid-state NMR. *Science* 261:1457–60
53. Killian JA, Taylor MJ, Koeppe RE II. 1992. Orientation of the valine-1 side chain of the gramicidin transmembrane channel and implications for channel functioning. A ^{2}H NMR study. *Biochemistry* 31:11283–90
54. Koeppe RE II, Greathouse DV, Jude A, Saberwal G, Providence LL, Andersen OS. 1994. Helix sense of gramicidin channels as a "nonlocal" function of the primary sequence. *J. Biol. Chem.* 269:12567–76
55. Koeppe RE II, Greathouse DV, Providence LL, Andersen OS. 1991. [L-Leu9-D-Trp10-L-Leu11-D-Trp12-L-Leu13-D-Trp14-L-Leu15]-gramicidin forms both single- and double-helical channels. *Biophys. J.* 59: A319 (Abstr.)
56. Koeppe RE II, Killian JA, Greathouse DV. 1994. Orientations of the tryptophan 9 and 11 side chains of the gramicidin channel based on deuterium NMR spectroscopy. *Biophys. J.* 66:14–24
57. Koeppe RE II, Killian JA, Vogt TCB, De Kruijff B, Taylor MJ, et al. 1995. Palmitoylation-induced conformational changes of specific side chains in the gramicidin transmembrane channel. *Biochemistry* 34:9299–306
58. Koeppe RE II, Mazet J-L, Andersen OS. 1990. Distinction between dipolar and inductive effects in modulating the conductance of gramicidin channels. *Biochemistry* 29:512–20
59. Koeppe RE II, Paczkowski JA, Whaley WL. 1985. Gramicidin K, a new linear channel-forming gramicidin from Bacillus brevis. *Biochemistry* 24:2822–26
60. Koeppe RE II, Providence LL, Greathouse DV, Heitz F, Trudelle Y, et al. 1992. On the helix sense of gramicidin A single channels. *Proteins* 12:49–62
61. Kuo D, Weidner J, Griffin P, Shah SK, Knight WB. 1994. Determination of the kinetic parameters of *Escherichia coli* leader peptidase activity using a continuous assay: The pH dependence and time-dependent inhibition by β-lactams are consistent with a novel serine protease mechanism. *Biochemistry* 33: 8347–54
62. Langs DA. 1988. Three-dimensional structure at 0.86 Å of the uncomplexed form of the transmembrane ion channel peptide gramicidin A. *Science* 241:188–91
63. Langs DA, Smith GD, Courseille C, Précigoux G, Hospital M. 1991. Monoclinic uncomplexed double-stranded, antiparallel, left-handed β 5.6-helix structure of gramicidin A: alternate patterns of helical association and deformation. *Proc. Natl. Acad. Sci. USA* 88:5345–49
64. Läuger P, Neumcke B. 1973. Theoretical analysis of ion conductance in lipid bilayer membranes. In *Membranes: 2. Lipid Bilayers and Antibiotics*, ed. G Eisenman, pp. 1–59. New York: Dekker
65. Lear JD, Wasserman ZR, DeGrado WF. 1988. Synthetic amphiphilic peptide models for protein ion channels. *Science* 240:1177–81
66. Lovejoy B, Choe S, Cascio D, McRorie DK, DeGrado WF, Eisenberg D. 1993. Crystal structure of a synthetic triple-stranded α-helical bundle. *Science* 259:1288–93
67. MacKinnon R. 1991. Determination of the subunit stoichiometry of the voltage-activated potassium channel. *Nature* 350:232–35
68. Mattice GL. 1994. *The effect of chain length and second residue chirality on gramicidin A structure and function: characterization of thirteen, fifteen and seventeen residue gramicidins with L- or D-alanine at the second residue*. PhD thesis. Univ. Arkansas
69. Mattice GL, Koeppe RE II, Provi-

dence LL, Andersen OS. 1995. Stabilizing effect of D-alanine-2 in gramicidin channels. *Biochemistry* 34: 6827–37
70. Mazet J-L, Andersen OS, Koeppe RE II. 1984. Single-channel studies on linear gramicidins with altered amino acid sequences. A comparison of phenylalanine, tryptophan, and tyrosine substitutions at positions 1 and 11. *Biophys. J.* 45:263–76
71. Mendel D, Cornish VW, Schultz PG. 1995. Site-directed mutagenesis with an expanded genetic code. *Annu. Rev. Biophys. Biomol. Struct.* 24: 435–62
72. Merrifield RB. 1986. Solid phase synthesis. *Science* 232:341–47
73. Montal M. 1995. Design of molecular function: channels of communication. *Annu. Rev. Biophys. Biomol. Struct.* 24:31–57
74. Morrow JS, Veatch WR, Stryer L. 1979. Transmembrane channel activity of gramicidin A analogues: effects of modification and deletion of the amino-terminal residue. *J. Mol. Biol.* 132:733–38
75. Myers VB, Haydon DA. 1972. Ion transfer across lipid membranes in the presence of gramicidin. II. The ion selectivity. *Biochim. Biophys. Acta* 274:313–22
76. Naik VM, Krimm S. 1986. Vibrational analysis of the structure of gramicidin A. II. Vibrational spectra. *Biophys. J.* 49:1147–54
77. Nicholson LK, Cross TA. 1989. Gramicidin cation channel: an experimental determination of the right-handed helix sense and verification of β-type hydrogen bonding. *Biochemistry* 28:9379–85
78. O'Connell AM, Koeppe RE II, Andersen OS. 1990. Kinetics of gramicidin channel formation in lipid bilayers: transmembrane monomer association. *Science* 250:1256–59
79. Oiki S, Koeppe RE II, Andersen OS. 1992. A dipolar amino acid substitution induces voltage-dependent transitions between two stable conductance states in gramicidin channels. *Biophys J.* 62:28–30
80. Oiki S, Koeppe RE II, Andersen OS. 1994. Asymmetric gramicidin channels: heterodimeric channels with a single F6-Val-1 residue. *Biophys. J.* 66:1823–32
81. Oiki S, Koeppe RE II, Andersen OS. 1995. Voltage-dependent gating of an asymmetric gramicidin channel. *Proc. Natl. Acad. Sci. USA* 92: 2121–25
82. Poteete AR, Rennell D, Bouvier SE. 1992. Functional significance of conserved amino acid residues. *Proteins* 13:38–40
83. Providence LL, Andersen OS, Greathouse DV, Koeppe RE II, Bittman R. 1996. Gramicidin channel function does not depend on phospholipid chirality. *Biochemistry*. 34: 16404–11
84. Ramachandran GN, Chandrasekaran R. 1972. Studies on dipeptide conformation and on peptides with sequences of alternating L and D residues with special reference to antibiotic and ion transport peptides. In *Progress in Peptide Research*, ed. S Lande, pp. 195–215. New York: Gordon & Breach
85. Ramachandran LK. 1975. The gramicidins. *Biochem. Rev.* 46:1–17
86. Rennell D, Bouvier SE, Hardy LW, Poteete AR. 1991. Systematic mutation of bacteriophage T4 lysozyme. *J. Mol. Biol.* 222:67–88
87. Ring A. 1986. Brief closures of gramicidin A channels in lipid bilayer membranes. *Biochim. Biophys. Acta* 856:646–53
88. Roux B, Brueschweiler R, Ernst RR. 1990. The structure of gramicidin A in dimethylsulfoxide/acetone. *Eur. J. Biochem.* 194:57–60
89. Roux B, Karplus M. 1994. Molecular dynamics simulations of the gramicidin channel. *Annu. Rev. Biophys. Biomol. Struct.* 23:731–61
90. Rozzelle JE Jr, Tropsha A, Erickson BW. 1994. Rational design of a three-heptad coiled-coil protein and comparison by molecular dynamics simulation with the GCN4 coiled coil: presence of interior three-center hydrogen bonds. *Protein Sci.* 3: 345–55
91. Russell AJ, Fersht AR. 1987. Rational modification of enzyme catalysis by engineering surface charge. *Nature* 328:496–500
92. Russell EWB, Weiss LB, Navetta FI, Koeppe RE II, Andersen OS. 1986. Single-channel studies on linear gramicidins with altered amino acid side chains. Effects of altering the polarity of the side chain at position 1 in gramicidin A. *Biophys. J.* 49:673–86
93. Saberwal G, Andersen OS, Greathouse DV, Koeppe RE II. 1995. Sequence determinants of gramicidin channel folding. *Biophys. J.* 68:A151 (Abstr.)

94. Saberwal G, Greathouse D, Koeppe RE II, Andersen OS. 1994. Ambidextrous gramicidin channels. *Biophys. J.* 66:A219 (Abstr.)
95. Sakmann B, Neher E. 1983. *Single-Channel Recording*. New York: Plenum. 503 pp.
96. Salom D, Bañó MC, Braco L, Abad C. 1995. HPLC demonstrates that an all Trp → Phe replacement in gramicidin A results in a conformational rearrangement from β-helical monomer to double-stranded dimer in model membranes. *Biochem. Biophys. Res. Commun.* 209:466–73
97. Sarges R, Witkop B. 1965. Gramicidin A. V. The structure of valine- and isoleucine-gramicidin A. *J. Am. Chem. Soc.* 87:2011–20
98. Sasaki J, Brown LS, Chon YS, Kandori H, Maeda A, et al. 1995. Conversion of bacteriorhodopsin into a chloride ion pump. *Science* 269:73–75
99. Sawyer DB, Koeppe RE II, Andersen OS. 1989. Induction of conductance heterogeneity in gramicidin channels. *Biochemistry* 28:6571–83
100. Sigworth FJ. 1994. Voltage gating of ion channels. *Q. Rev. Biophys.* 27: 1–40
101. Sigworth FJ, Shenkel S. 1988. Rapid gating events and current fluctuations in gramicidin A channels. *Curr. Top. Membr. Transp.* 33:113–30
102. Smart OS, Goodfellow JM, Wallace BA. 1993. The pore dimensions of gramicidin A. *Biophys. J.* 65: 2455–60
103. Smith R, Thomas DE, Separovic F, Atkins AR, Cornell BA. 1989. Determination of the structure of a membrane-incorporated ion channel. Solid-state nuclear magnetic resonance studies of gramicidin A. *Biophys. J.* 56:307–14
104. Stankovic CJ, Heinemann SH, Delfino JM, Sigworth FJ, Schreiber SL. 1989. Transmembrane channels based on tartaric acid-gramicidin A hybrids. *Science* 244:813–17
105. Sychev SV, Barsukov LI, Ivanov VT. 1993. The double ππ 5.6 helix of gramicidin A predominates in unsaturated lipid membranes. *Eur. Biophys. J.* 22:279–88
106. Szabo G, Urry DWG. 1979. N-acetyl gramicidin: single-channel properties and implications for channel structure. *Science* 203:55–57
107. Takeuchi H, Nemoto Y, Harada I. 1990. Environments and conformations of tryptophan side chains of gramicidin A in phospholipid bilayers studied by Raman spectroscopy. *Biochemistry* 29:1572–79
108. Urry DW. 1971. The gramicidin A transmembrane channel: a proposed π(L,D) helix. *Proc. Natl. Acad. Sci. USA* 68:672–76
109. Urry DW, Goodall MC, Glickson JD, Mayers DF. 1971. The gramicidin A transmembrane channel: characteristics of head-to-head dimerized π(L,D) helices. *Proc. Natl. Acad. Sci. USA* 68:1907–11
110. Veatch WR, Blout ER. 1974. The aggregation of gramicidin A in solution. *Biochemistry* 13:5257–64
111. Veatch WR, Fossel ET, Blout ER. 1974. The conformation of gramicidin A. *Biochemistry* 13:5249–56
112. Veatch WR, Mathies R, Eisenberg M, Stryer L. 1975. Simultaneous fluorescence and conductance studies of planar bilayer membranes containing a highly active and fluorescent analog of gramicidin A. *J. Mol. Biol.* 99: 75–92
113. Veatch WR, Stryer L. 1977. The dimeric nature of the gramicidin A transmembrane channel: conductance and fluorescence energy transfer studies of hybrid channels. *J. Mol. Biol.* 113:89–102
114. Venkatachalam CM, Urry DW. 1983. Theoretical conformational analysis of the gramicidin A transmembrane channel. I. Helix sense and energetics of head-to-head dimerization. *J. Comput. Chem.* 4: 461–69
115. Vogt TCB, Killian JA, De Kruijff B. 1994. Structure and dynamics of the acyl chain of a transmembrane polypeptide. *Biochemistry* 33:2063–70
116. Vogt TCB, Killian JA, De Kruijff B, Andersen OS. 1992. Influence of acylation on the channel characteristics of gramicidin A. *Biochemistry* 31:7320–24
117. Vogt TCB, Killian JA, Demel RA, De Kruijff B. 1991. Synthesis of acylated gramicidins and the influence of acylation on the interfacial properties and conformational behavior of gramicidin A. *Biochim. Biophys. Acta* 1069:157–64
118. Wallace BA. 1990. Gramicidin channels and pores. *Annu. Rev. Biophys. Biophys. Chem.* 19:127–57
119. Wallace BA, Ravikumar K. 1988. The gramicidin pore: crystal structure of a cesium complex. *Science* 241:182–87
120. Wallace BA, Veatch WR, Blout ER. 1981. Conformation of gramicidin A

in phospholipid vesicles: circular dichroism studies of effects of ion binding, chemical modification, and lipid structure. *Biochemistry* 20:5754–60
121. *Webster's Seventh New Collegiate Dictionary*. 1967. ed. PB Gove, p. 661. Springfield, MA: Merriam
122. Weiss M, Abele U, Weckesser J, Welte W, Schiltz E, Schulz GE. 1991. Molecular architecture and electrostatic properties of a bacterial porin. *Science* 254:1627–30
123. Wells JA, Powers DB, Bott RR, Graycar TP, Estell DA. 1987. Desinging substrate specificity by protein engineering of electrostatic interactions. *Proc. Natl. Acad. Sci. USA* 84:1219–23
124. Williams LP, Narcessian EJ, Andersen OS, Waller GR, Taylor MJ, et al. 1992. Molecular and channel-forming characteristics of gramicidin K's: A family of narutally occurring acylated gramicidins. *Biochemistry* 31: 7311–19
125. Yang J, Ellinor PT, Sather WA, Zhang JF, Tsien RW. 1993. Molecular determinants of Ca2+ selectivity and ion permeation in L-type Ca2+ channels. *Nature* 366:158–61
126. Zhang Z, Pascal SM, Cross TA. 1992. A conformational rearrangement in gramicidin A: from a double-stranded left-handed to a single-stranded right-handed helix. *Biochemistry* 31:8822–28

Annu. Rev. Biophys. Biomol. Struct. 1996. 25:259–86

ELECTRON PARAMAGNETIC RESONANCE AND NUCLEAR MAGNETIC RESONANCE STUDIES OF CLASS I RIBONUCLEOTIDE REDUCTASE

A. Gräslund and M. Sahlin

Department of Biophysics and Department of Molecular Biology, Stockholm University, S-106 91 Stockholm, Sweden

KEY WORDS: ribonucleotide reductase, tyrosyl radical, diiron-oxygen protein

ABSTRACT

Ribonucleotide reductase catalyses the reduction of ribonucleotides to the corresponding deoxyribonucleotides needed for DNA synthesis. This review describes recent studies on the iron/tyrosyl free radical site in the R2 protein of iron-containing (class I) ribonucleotide reductases. The active enzyme is composed of two homodimeric proteins, R1 and R2. Active protein R2 contains a diiron-oxygen site and a neighboring free radical on a tyrosyl residue per polypeptide chain. The properties of the different redox states of the diiron center in protein R2 are discussed, as well as the formation of the iron/radical site and its possible involvement in long range electron transfer from the substrate binding site in protein R1. The EPR properties of oxidized neutral tyrosyl free radicals are described, and also of tryptophan free radicals found in studies of a mutant of the R2 protein, which lacks the tyrosyl radical site. NMR studies on protein R2 include observations of paramagnetically shifted resonances. Structural NMR studies have been performed on its highly mobile C-terminal domain as well as the corresponding oligopeptide which interacts with protein R1.

1056–8700/96/0610–0259$08.00

CONTENTS

INTRODUCTION

Ribonucleotide reduction is necessary for the biosynthesis of the deoxyribonucleotides needed for DNA synthesis. The production of deoxyribonucleotides from the corresponding ribonucleotides is regulated strictly throughout the cell cycle, and the biosynthesis of deoxyribonucleotides can be regarded as a bottleneck for cell proliferation. The enzyme that catalyzes the reduction of ribonucleotides is ribonucleotide reductase.

Different classes of ribonucleotide reductase have been characterized. Certain organisms may carry the genes for more than one class, but it seems that only one class is active for a given set of conditions. An example is *Escherichia coli*, which carries the genes for two classes of enzymes. One class is used under aerobic growth conditions, and the other is used under anaerobic growth conditions. At least three different classes of ribonucleotide reductases have been described (52), all of which use metals and free radical chemistry for the seemingly simple reaction to reduce the 2′-hydroxyl group in the ribose ring. This review focuses on the iron-containing class I ribonucleotide reductases (RDRs) [EC 1.17.4.1], which have a tyrosyl free radical in its active state. The substrates are the nucleoside diphosphates. RDR is found, for example, in mammalian cells, as well as in aerobically growing *E. coli*, and is coded for by certain viruses. Several excellent reviews have been written recently on this class of enzyme (21, 61, 63). The class II enzyme, found in *Lactobacillus leichmannii*, requires coenzyme B12 (with cobalt) for its function and acts at the nucleoside triphosphate level (52, 63). The class III enzyme is found in anaerobically growing *E. coli* and bacteriophage T4 and contains an iron sulphur center and a glycyl free radical when activated (52). A fourth type of enzyme that contains manganese has been described in *Brevibacterium ammoniagenes* (52, 63); whether this enzyme should be considered a separate class or a subclass of class I has been debated.

The common feature for the ribonucleotide reductases appears to be the involvement of a free radical in catalysis, which is localized either in the protein or in a cofactor and is created in a redox reaction involving the metal. The postulated role of the free radical is to abstract a hydrogen from the 3′ position of substrate as an initial modification of the substrate (63). An interesting aspect of this reaction in the class I enzyme is that the unpaired electron needed is stored as a stable free radical formed by oxidation of a tyrosine residue at a large distance from the active site. The postulated catalytically active thiyl radical at an active site cysteine has to be formed by long-range electron transfer and appears transiently only in the course of the catalytic reaction (61, 63).

The class I RDR is a 1:1 complex of proteins R1 and R2, each of which is a homodimer. The three-dimensional structures of each protein in the *E. coli* enzyme have been determined by X-ray crystallography (42, 43, 64). The crystal structure of protein R2 was determined for the protein without radical. Model building has also suggested a structure for the R1:R2 complex (64). Protein R1 (2×761 residues) contains the active site and the sites for the allosteric regulation. Protein R2 (2×375 residues) contains sites for two diferric clusters and variable amounts of a stable free radical localized on a tyrosyl residue (Y122) approximately 5 Å from the nearest iron ion. The diiron–radical site is buried inside protein R2 approximately 10 Å from the nearest protein surface. In the model built complex, there is a distance of approximately 35 Å between the active site in protein R1 and the iron–radical site in protein R2 (64). A chain of conserved hydrogen bonded amino acids connects the two sites (Figure 1). In protein R2, Fe1 is part of this chain, including H118, D237, W48 (42, 43), and, probably, E350 and Y356 (61, 64). The latter two residues are part of the C-terminal domain of protein R2, which was not observed in the X-ray structure of the protein. Only a small number of amino acid residues are totally conserved among the known species: per polypeptide, less than 20 in protein R2 and less than 40 in protein R1. The locations of the conserved amino acids, however, suggest similar functions and, perhaps, three-dimensional structures for the RDRs from different species. In protein R2, the conserved amino acids include the iron ligands, some residues surrounding the conserved Y122, and participants in the postulated long-range electron transfer chain.

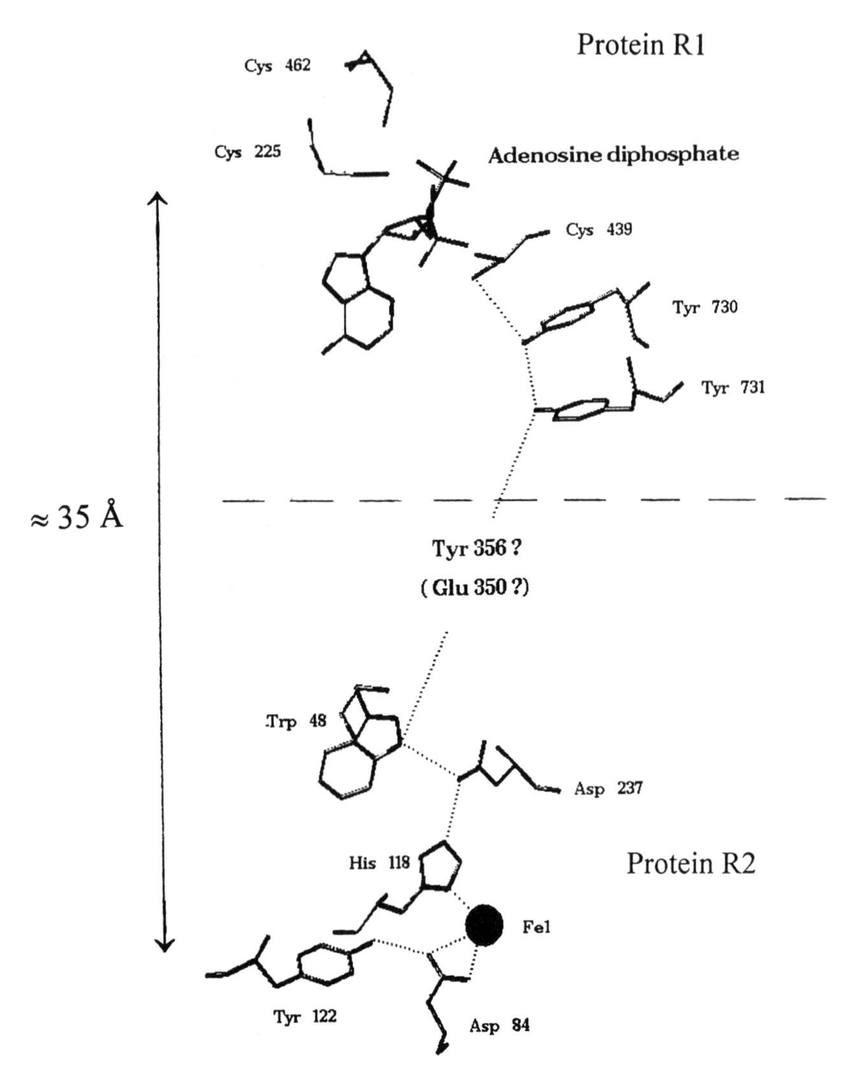

Figure 1 Conserved residues participating in the proposed hydrogen bonded long range electron transfer chain between substrate (adenine diphosphate) site in protein R1 and the tyrosyl radical in protein R2 of *E. coli* ribonucleotide reductase (42, 43, 61, 64). The dotted lines represent hydrogen bonds in the crystal structures of the proteins. The dashed line represents the boundary between proteins R1 and R2. Tyr356 and Glu350 are not visible in the crystal structure of protein R2 because of dynamic disorder but may become ordered by interaction with R1 (38). (Adapted from Reference 60a.)

THE IRON-RADICAL SITE IN PROTEIN R2

Redox States

Figure 2*a* shows the iron–radical site in *E. coli* protein R2 (43). The two high-spin ferric ions are connected by one μ-oxo and one μ-carboxylate bridge and are surrounded by carboxylate and histidine ligands. The iron pair is coupled antiferromagnetically and has a diamagnetic ground state. Y122 is the tyrosyl residue that can harbor the free radical (34). Electron-nuclear double resonance (ENDOR) studies have shown that the stable radical is of the neutral phenoxy type (9), formed by loss of one electron and one proton. Various spectroscopic techniques have been used to characterize the iron–radical site, its formation, and its different redox states.

Figure 2*b* summarizes some observed redox states of protein R2 for *E. coli* or mouse RDR (21, 39, 45). Active R2 contains the antiferromagnetically coupled electron paramagnetic resonance (EPR) silent diferric site and the tyrosyl radical that gives rise to the observable EPR spectrum. One electron reduction of the tyrosyl radical results in the met state of protein R2 (metR2). Further, one electron reduction gives a mixed valent state, easily stabilized at room temperature in mouse RDR (6). *E. coli* RDR is different in this respect, as we discuss below. Further reduction leads to the fully reduced diferrous state, which is stable under anaerobic conditions. Reduced R2 reacts spontaneously with molecular oxygen to reconstitute active R2. A shortcut between the met and active states exists, by treatment of metR2 with hydrogen peroxide (21, 58). The apo state of protein R2 (apoR2) is devoid of iron and radical.

Formation of the Iron–Radical Site

The formation of the iron–radical site in apoR2 is a complex reaction between ferrous iron and molecular oxygen in the protein environment. Several intermediates have been observed and characterized in the *E. coli* wild type and point mutant proteins (10–12, 50, 56). The in vitro reaction of one polypeptide chain in wild type R2 may be summarized as:

$$3Fe^{2+} + \text{P-tyr-OH} + O_2 + H^+ \rightarrow \text{Fe(III)-}O^{2-}\text{-Fe(III)-P-tyr-O}^{\cdot} + H_2O + Fe^{3+}. \qquad \text{Reaction 1}$$

Here, P-tyr-OH designates the protein with the normal Y122. The third ferrous iron, which may be replaced by another reductant, supplies an extra electron needed to form water (45). It is not clear whether this iron has a specific binding site in the protein.

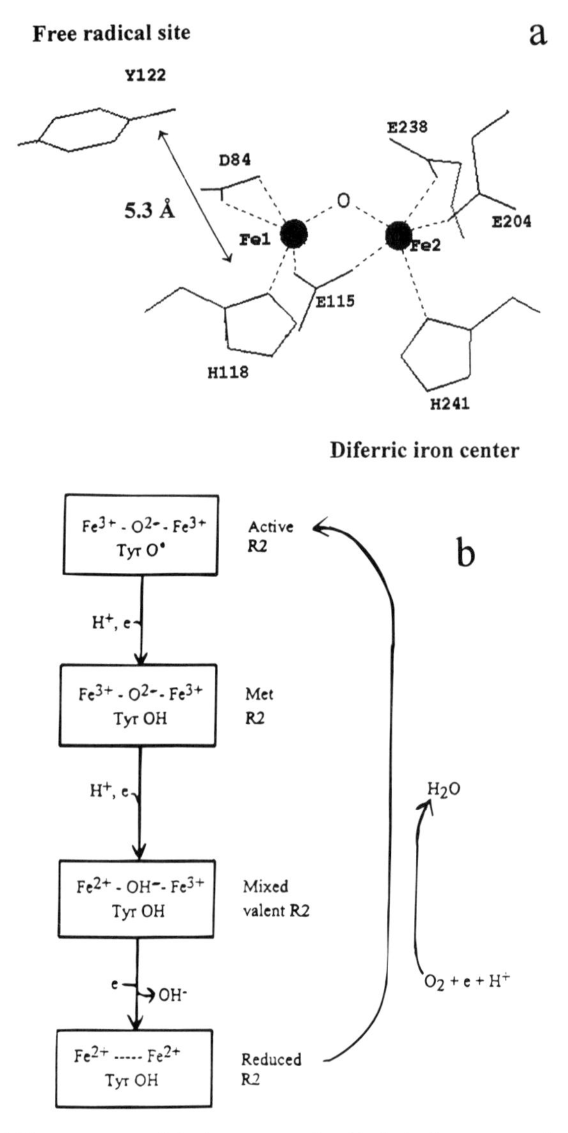

Figure 2 (*a*) The structure of the iron–tyrosyl radical site in met protein R2 of *E. coli* RDR, with the iron ligands indicated (42, 43). (*b*) Characterized redox states of the iron–free radical site of protein R2.

The time course of the reconstitution reaction and the intermediate states of the iron–radical site in the wild type protein R2 have been studied by static and kinetic optical, EPR, and Mössbauer spectroscopic techniques (10–12, 50). The time scale is subsecond or second at 5°C and depends on the stoichiometry of available ferrous iron or other reducing agents (whether substoichiometric or in excess). The observed and characterized intermediates of the iron–radical site include, in a time sequence:

1. diferrous iron,
2. diferric peroxide,
3. a so-called diferric radical, **X** + protein bound free radical, **Z**˙ (possibly on tryptophan, W48), and
4. diferric μ-oxo bridged/Y122 tyrosyl radical, which is the end product.

Stage 3, here described as two separated free radicals, might also be described as a diferryl state without free radicals or a ferryl/ferric state with one free radical, Z˙. It contains a highly oxidizing species, which can be considered formally similar to compound Q in the methane mono-oxygenase (MMOH) reaction cycle (5). The results of a detailed Mössbauer study of this state, however, have not supported the presence of any typical ferryl iron ion in **X** (50). The radical component of **X** may be a hydroxyl radical (cf 5). The reactions between stages 3 and 4 are determined by the availability of iron and other reducing agents. In case of limiting iron (typically $\leq$ 2Fe/R2), Y122˙ was observed as an early product at the expense of **Z**˙, and a slower reduction of **X** led to formation of the μ-oxo bridged diferric site (10, 12). Under conditions of excess ferrous iron (typically $\geq$ 5Fe/R2), **Z**˙ was directly reduced and **X** alone yielded the end product (stage 4) (11).

Long-Range Electron Transfer and Enzyme Reaction

When the active site in RDR protein R1 binds substrate and the enzymatic reaction is about to begin, a long-range electron transfer is postulated to take place. This transfer results in the formation of an immediate active-site radical (suggested localization at *E. coli* C439), which is an apparent prerequisite for catalytic activity. The catalytic reaction has been suggested to include the following steps (40, 41, 61, 63, 64): First, the C439 thiyl radical abstracts the 3′-hydrogen from the substrate. The 2′-hydroxyl group then protonates with a proton from the thiols of C225/462. Water is split from the substrate, and a cation radical intermediate is formed. The 2′-cation radical site adds an electron and a hydrogen from the thiols of C225/462. Finally, the 3′-hydrogen returns to its initial site, and the product is formed. Presumably, the C439 radical is reformed transiently during the last step, whereupon the long-range electron transfer occurs in reverse and the stable Y122 radical in protein R2 reappears. Because of the long distance of transfer of the unpaired electron and the presence of a hydrogen bonded chain of conserved amino acids between the active site and the iron–radical site (Figure 1), one possible interpretation is that the long-range electron transfer is aided by concomitant switching of protons in the hydrogen bonds.

Early on, it was noted that the magnitude of the $g = 2$ EPR signal of *E. coli* protein R2 was correlated directly with enzyme activity as probed in an enzymatic assay (18). This finding is strictly true only for wild type protein and is not valid for point mutants along the proposed electron transfer chain (53; B.O. Persson, M. Karlsson, I. Climent, J. Ling, J. Sanders-Loehr, et al, *J. Biol. Inorg. Chem.*, in press). Quantitations of the tyrosyl radical EPR spectra from different RDR species have yielded maximally approximately 1–1.2 radicals per protein for R2 from *E. coli*, up to 1.6 for mouse, and only approximately 0.3 for herpes simplex type 1 (HSV1). Careful quantitations for the three species at various temperatures between 4 and 77 K have given the same results. The possibility, therefore, that the temperature-dependent magnetic interaction between the iron site and radical should lead to incorrect EPR quantitations seems unlikely (A. Gräslund, unpublished data).

Larsson & Sjöberg (34) have shown that the *E. coli* Y122F mutant was devoid of EPR signal and enzyme activity. This observation was actually the first direct evidence that Y122 carries the free radical in the wild type protein. For mouse R2, the correlation between radical content and enzyme activity is less obvious, because iron is lost from the R2 protein easily (44) and the iron–radical site appears to be reconstituted continuously during assay conditions. Henriksen et al (28) observed that there was a small but significant activity in the mouse R2 mutant Y177F (corresponding to *E. coli* Y122). A suggested explanation was that because transient free radicals should be formed during the ongoing reconstitution in mouse R2 (cf studies of *E. coli* Y122F), these radicals might replace the missing Y177 radical as originator of the free radical chemistry (28, 56).

ELECTRON PARAMAGNETIC RESONANCE STUDIES

Free Radicals

TYROSYL RADICALS The X-band EPR spectrum of the first stable tyrosyl radical in a biological system was observed in *E. coli* RDR approximately 20 years ago (18, 62). The assignment was possible because of specific deuterium isotope labeling of the overexpressed protein R2 in *E. coli* bacteria. The EPR spectrum was best observed at low temperatures and was shown to arise from an oxidized form of tyrosine, a tyrosyl radical. The spin density distribution is shown in Figure 3 (*ring*) (C.W. Hoganson, M. Sahlin, B.-M. Sjöberg, G.T. Babcock, *J. Am. Chem. Soc.*, in press).

A major fraction of the spin density is localized on C1 neighboring the C_β position, and the dominant part of the hyperfine coupling arises from one of the β protons. These protons are in a locked position relative to the tyrosyl ring, and the magnitude of their hyperfine coupling is determined by their dihedral angles.

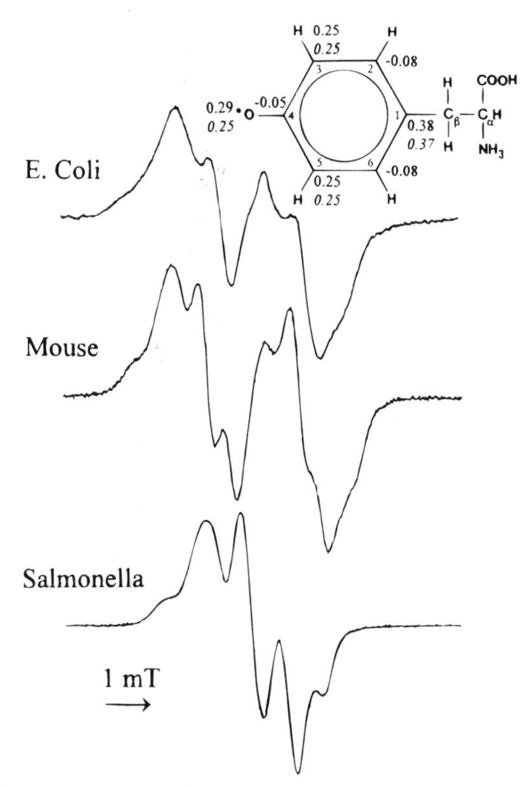

Figure 3 X-band EPR spectra of the tyrosyl radical in protein R2 from (*top*) *E. coli*, recorded at 30 K; (*middle*) mouse, recorded at 20 K; (*bottom*) R2F from *S. typhimurium*, recorded at 20 K. (*Ring*) The spin density distribution in the tyrosyl free radical in *E. coli* R2 (C.W. Hoganson, M. Sahlin, B.-M. Sjöberg, and G.T. Babcock, *J. Am. Chem. Soc.*, in press) and in *S. typhimurium* R2F (italics) from spectra simulations and comparison with the photosystem II tyrosyl radical (4a).

Figure 3 shows the EPR spectra of active proteins R2 from *E. coli* and mouse. They are quite similar and are dominated by one large β hyperfine coupling, which gives rise to the major doublet splitting. This finding indicates that the geometries of the two radicals are similar and that the same is true for several other species of protein R2, e.g. of HSV or human origin. The relatively small differences in the shapes between spectra of this type from different species may be explained by small differences in the β dihedral angles, caused by the different protein environments (9, 26).

A new study on the *E. coli* R2 protein has shown that its spin density distribution is almost the same as that of the Y_D radical of photosystem II (PSII) (C.W. Hoganson, M. Sahlin, B.-M. Sjöberg, G.T. Babcock, *J. Am. Chem. Soc.*, in press). Taken together, these results suggest that all oxidized neutral tyrosyl radicals have similar spin density distributions. The deviations around the C_4-O bond depend on the hydrogen bonding status of the phenolic oxygen.

An extreme case of an RDR with a rather different spectral shape is shown in Figure 3 (*bottom*). The spectrum was observed from the R2F protein of *Salmonella typhimurium.* This bacterium contains an active class I RDR but also has a second class RDR (class Ib), which is normally not expressed (31). The EPR spectrum of the class Ib protein R2F (Figure 3*b*, *bottom*) is strikingly similar to that observed for the tyrosyl radical Y_D of PSII (29, 31). The PSII tyrosyl radical is characterized by considerably different dihedral angles for the β protons (4a, 29). This is the reason for a narrower EPR spectrum without the dominating hyperfine doublet. The different dihedral angles, which should also be characteristic of the *Salmonella* radical, resemble those of neutral model tyrosine radicals and may indicate a "relaxed" state of a tyrosyl radical (4a). From this viewpoint, most RDR tyrosyl radicals should be in a more constrained state, perhaps reflecting a perturbed conformation caused by local charge and steric environmental factors.

The EPR spectra of the tyrosyl radicals of RDR are generally difficult to saturate by microwave power, compared with isolated free radicals. This difficulty is caused by magnetic interaction with the iron center, which has a significant population of the excited paramagnetic states toward higher temperatures. Electron paramagnetic resonance spectra recorded in solution at room temperatures are broadened severely by the interaction (26, 57). Electron paramagnetic resonance saturation recovery studies at variable temperature have been used to examine the interaction, which has exchange as well as dipolar components and varies significantly between proteins from different species (22). These studies also yield the magnitudes of the antiferromagnetic coupling constant that characterizes the diferric pair.

Recently, it has been possible to investigate the RDR radicals by EPR at extremely high magnetic fields (24), where the anisotropic *g*-values can be resolved. The magnitude of the *g* anisotropy may be related to the presence or absence of a hydrogen bond to the tyrosyl radical oxygen. It has been suggested that the relatively modest *g* anisotropy of the PSII radical correlates with the presence of a hydrogen bond (29), whereas the larger *g* anisotropy of the RDR tyrosyl radical is in agreement with the absence of such a bond (24). In a 245 GHz EPR study, the *S. typhimurium* R2F radical also showed a large *g* value anisotropy like *E. coli* RDR, which indicated the absence of an H bond in that case as well. In this respect, the *Salmonella* radical is different from that of PSII (4a).

In RDR, the hydrogen bonding status of the tyrosine ring that harbors the stable radical may have a functional significance. In the wild type *E. coli* metR2, the crystal structure (42, 43) suggests a weak hydrogen bond between the tyrosyl hydroxyl group and one oxygen of the carboxylate of D84, a bidentate ligand to Fe1 (Figure 1*a*). The Y122 radical has no proton and no hydrogen bond (9). In anticipation of a crystal structure of the active protein R2, the fate of this proton is unknown. Presumably, the hydrogen bonding network of the iron ligands is changed when the radical is formed. One guess is that the lost proton may be transferred to D84, which might become monodentately bound to Fe1 when the stable radical exists at Y122. Carboxylate metal ligands may change depending on the redox state of the metal site [so called carboxylate shifts (cf 1, 7, 49)]. For example, the reduced R2 most likely has D84 as a monodentate ligand (1, 7).

TRYPTOPHAN RADICALS Recently, two tryptophan radicals were identified by deuterium isotope labeling when the *E. coli* protein R2 mutant Y122F was reconstituted according to Reaction 1 (56; details below). One EPR spectrum had a doublet hyperfine structure (Figure 4*a*) from coupling to one of the β-methylene protons with a large fraction of the spin density at C-3 of the tryptophan. The second tryptophan radical EPR spectrum (Figure 4*b*) appeared as a quartet because of hyperfine coupling to both of the β-metylene protons. For both of these radicals it is possible to calculate the dihedral angles of the β-protons and use the crystal structure to predict assignment to a likely tryptophan residue. The radicals with EPR spectra shown in Figure 4*a* and the quartet indicated in Figure 4*b* represent the first relatively stable tryptophan

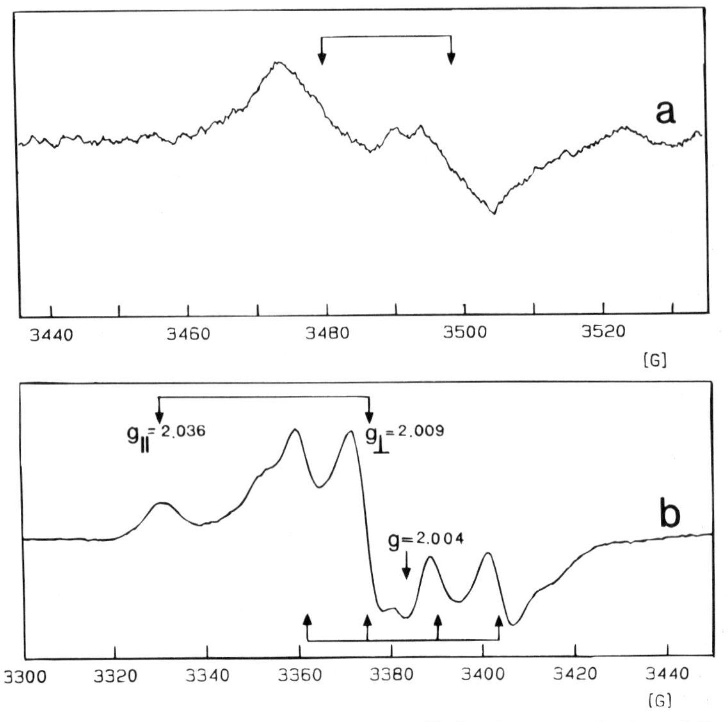

Figure 4 X-band EPR spectra of tryptophan radicals after reconstitution of *E. coli* apoprotein R2 mutant 122F (50μM) with 4Fe^{2+}/R2. (*a*) Spectrum recorded at room temperature in the time window 1–12 min after a stopped flow shot. The apoY122F was grown in a nondeuterated medium. (*b*) Spectrum recorded at 77 K after freeze-quenching the reaction 6 s after addition of Fe^{2+}. The apoY122F was grown in a medium that contained indole-deuterated tryptophan. The four arrows centered around $g = 2.004$ indicate the tryptophan quartet spectrum. Also indicated is the axial component (see text). (Results from 56.)

radicals (on the second–minute time scale) to display normal free radical spectra in a protein.

Figure 4*b* also includes the EPR spectrum of a third, quite different, tentative tryptophan radical observed transiently during reconstitution of the Y122F R2 mutant. The axial EPR line shape with $g = 2.036$ and $g = 2.009$ is very similar to that identified by selective deuteration and EPR and ENDOR studies as a tryptophan radical that interacts weakly with an oxyferryl $(Fe = 0)^{2+}$, $S = 1$ species in the heme iron protein cytochrome c peroxidase (CCP) (30). There is no direct evidence, despite the almost coinciding EPR line shapes, that this species is a tryptophan radical in Y122F R2. The line shape of the axial

Y122F R2 EPR spectrum with its *g*-value anisotropy is also consistent with a peroxy-type free radical. In another system, the self-peroxidation of metmyoglobin was shown recently to lead to formation of a tryptophan centered peroxyl radical, formed by reaction of a tryptohan radical with molecular oxygen (30a).

CYSTEINE RADICALS A thiyl free radical at an active site cysteine has been postulated to be the immediate actor on the substrate in ribonucleotide reduction by a class I enzyme. So far, no direct observation has been reported on this transient free radical. However, experiments in which suicidal substrate analogues 2′-deoxy,2′-azido-nucleotides were used instead of normal substrate in an enzymatically active mixture with either the *E. coli* or the mouse enzyme have shown the slow appearance of a new type of free radical, which might be a thiyl radical analogue with a prolonged lifetime (8, 16a, 59). The new radical appears concomitantly with the loss of the normal tyrosyl radical. The hyperfine coupling pattern of the new radical displays a major triplet splitting, because of a nitrogen originating from the azide, and a doublet splitting. A likely explanation is that this new radical is localized on a cysteine at the active site, most likely C225 or possibly C439. The new radical should be formed in a reaction between a thiyl radical and HN_3, whereby N_2 is liberated. The result would be:

$$C_\alpha\text{-}C_\beta H_2\text{-}S^{\cdot}\text{-}NH.$$

If this interpretation is correct, a significant part of the spin density should reside on the S, in agreement with a *g*-value of approximately 2.008 (8). The hyperfine doublet splitting should arise from coupling to one of the β protons of cysteine, apparently in a locked configuration so that only one of them is in a favorable position for hyperfine coupling. The induced free radical can be considered formally as a sulfenylimino type of radical. The imino proton, however, does not contribute an observable hyperfine coupling, as might be expected. A possible explanation is that the radical has lost the proton in a reaction with oxygen. This hypothesis might also explain the nitroxide-like *g*-value anisotropy of the spectrum.

A concluding remark on all free radicals hitherto observed on different amino acids in RDR is that they are formed by an oxidative process and have significant spin densities at atoms neighboring C_β and, therefore, exhibit hyperfine couplings to one or both of the β protons. The $C_\beta H_2$ group is in a locked position relative to the sidechain at both low temperatures and at room temperature for the Y122 radical and the azido nucleotide induced transient radical (26). This may be a general feature of a relatively stable amino acid–based free radical held in a protein environment.

The Iron Center

The iron center of RDR (Figure 2*a*) is similar to other diiron-oxygen centers in proteins with different functions (5). One example is hemerythrin, an oxygen carrier protein of lower invertebrates, which has a μ-oxo bridge between the iron ions and many spectroscopic properties similar to those of active protein R2. Other proteins in this class have been shown to have a hydroxobridge, for instance, the hydroxylase component of MMOH and purple acid phosphatases.

As outlined in Figure 2*b* the diiron center of RDR can exist in several redox states. The diferric state is invisible by EPR but is well characterized by its near UV absorption bands. Mild chemical reduction methods with dithionite and redox mediators were used in several attempts to produce the mixed valent state of *E. coli* protein R2 (A Gräslund and B-M Sjöberg, unpublished). These methods, however, proved unsuccessful, and only special methods, such as hydrazine incubation in solution or low-temperature reduction by ionizing radiation at 77 K, have been found to induce the mixed valent state of *E. coli* protein R2 (15, 23, 27).

In contrast, mild chemical reduction was found to induce a large fraction of the mixed valent state for mouse protein R2 (6). One reason for the different behavior of the iron sites of *E. coli* and mouse proteins R2 may lie in their redox potentials. The direct two-electron transfer potential for *E. coli* protein R2 has been determined to be -115 mV at 4°C (60), whereas recent studies on mouse R2 suggest that the redox potentials may be close to those of MMOH, i.e. $+76$ mV for the first electron and $+21$ mV for the second electron at 4°C (48). The mixed valent state of mouse R2 does not react with molecular oxygen (6). Another reason for the different redox behaviors may be different accessibilities to the iron/radical sites (31a).

The fully reduced state of protein R2 can be obtained by the two-electron reduction of the diferric state or by the addition of ferrous iron to the apo form of the protein under anaerobic conditions. Crystal structures have been solved for the fully reduced state of the mutant S211A (1) and for Mn^{2+} containing *E. coli* protein R2 (7), which should mimic the diferrous state. These structures show only amino acid carboxylates as bridging ligands between the iron ions and no μ-oxobridge.

THE MIXED VALENT STATE Figure 5 (*top*) shows an EPR spectrum recorded at 4 K of the mixed valent state of mouse protein R2. The rhombic $g < 2$ EPR signal with g-values of 1.92, 1.73, and 1.60 is typical for a mixed valent iron center with $S = \frac{1}{2}$. Quantitatively, the signal corresponds to approximately 20% of the dinuclear iron sites (6). From the temperature dependence of the microwave saturation the coupling constant J for Heisenberg exchange ($H = 2JS_1 S_2$) could be estimated as $J = -7.5 \text{ cm}^{-1}$. The weak antiferromagnetic coupling suggests that the oxo bridge protonates in the mixed valent state and becomes a hydroxo bridge. In addition, the spectrum of Figure 5*a* also shows a weak $g = 4.3$ signal from nonspecifically bound ferric iron in a rhombic environment and another weak signal at a very low mag-

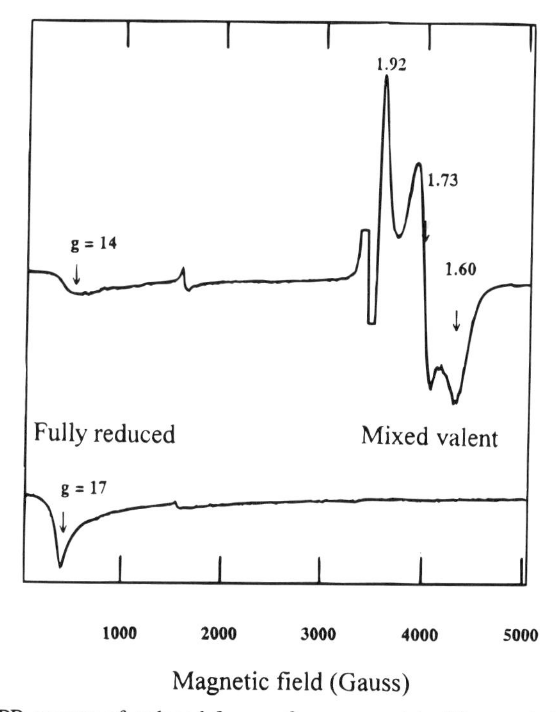

Figure 5 EPR spectra of reduced forms of mouse protein R2 recorded at 4 K. (*top*) Perpendicular mode EPR. Mixed valent rhombic signal with g values 1.92, 1.73, and 1.60, obtained after incubation with the redox mediator phenazine metho-sulfate (PMS) and dithionite; (*bottom*) Parallel mode EPR. Fully reduced iron site signal with g value 17, obtained after incubation with the redox mediator methyl viologen and dithionite. (Adapted from Reference 7a.)

netic field around $g = 14$, which arises from a small fraction of the fully reduced state (discussed below).

Weak, transient, mixed valent EPR signals were observed during reduction of *E. coli* protein R2 with nitrogen-containing compounds, such as diimide or hydrazine (23). X-irradiation (27) and γ-irradiation (15) in a frozen protein solution at 77 K were also used successfully to produce a mixed valent state of *E. coli* protein R2. The mobile electrons produced by ionizing radiation in the frozen matrix (containing a large fraction of glycerol for a high yield) are efficient for one-electron reduction of metalloproteins (15). The primary products observed at 77 K after this low-temperature reduction treatment are expected to keep the conformation of the oxidized state. Only after suitable annealing, the coordination environment of the metal site should relax to a new equilibrium conformation relevant for the new electronic state of the metal site, possibly through one or more intermediate states.

The mixed valent state of *E. coli* metR2 was obtained after low-temperature reduction at 77 K. It showed an almost axial EPR spectrum with $g_\perp = 1.94$ and $g_\parallel = 1.82$, which is indicative of an $S = \frac{1}{2}$ state (15). The EPR signal was best seen at low temperatures, but could be observed also at 130 K. Quantitatively, the $S = \frac{1}{2}$ signal corresponded to approximately 30% of the iron sites. Annealing of the sample to 142 K produced small spectral changes, whereas annealing to 165 K gave rise to a completely changed EPR signal observable only at temperatures less than 35 K with g values of 5.4, 6.7, and 14.8. This signal indicates a ferromagnetically coupled $S = \frac{9}{2}$ state (15). A similar switch from antiferromagnetic to ferromagnetic coupling mode has been observed for deoxyhemerythrin on azide complex formation and was ascribed to protonation of the hydroxo bridge to form an aquo bridge (51). The $S = \frac{9}{2}$ signal disappears after annealing at temperatures greater than 230 K, in agreement with the failures to observe a stable mixed valent state of *E. coli* R2 at higher temperatures.

The low-temperature reduction method also has been applied recently to active *E. coli* R2. In that case, the primary products give rise to two somewhat different EPR spectral components (R. Davydov, M. Sahlin, S. Kuprin, A. Gräslund, and A. Ehrenberg, *Biochemistry*, in press). The simplest explanation is that one component corresponds to the met state and the other one to the active state of the iron–radical site. Low-temperature reduction of mouse R2 yields a primary $S = \frac{1}{2}$ state, which undergoes only minor changes on annealing. The relaxed mixed valent state, which has EPR spectrum similar to that obtained with mild chemical reduction (6), is stable at room temperature (R

Davydov, R Ingemarson, L Thelander, A Ehrenberg, and A Gräslund, unpublished data).

THE FULLY REDUCED STATE The fully reduced state of *E. coli* protein R2 was obtained by treatment with dithionite in the presence of methyl viologen ($E_m = -430$ mV) as redox mediator (55). An indirect proof of the presence of the fully reduced state is that treatment with air immediately results in formation of the active diferric site with accompanying tyrosyl radical. Direct spectroscopic evidence of the presence of this state can be obtained from observation of paramagnetically shifted ^{1}H-NMR resonances from the iron ligands (55) (Figure 6), or from observation of EPR signals from the iron ions at very low magnetic fields (Figure 5, *bottom*). Azide binding to the diferrous site in *E. coli* R2 was studied by parallel mode EPR, which greatly enhances the signals from integer spin systems (19). It was suggested that binding of azide converted the weakly antiferromagnetically coupled iron sites to ferromagnetically coupled sites, possibly with an aquo bridge.

Figure 5 (*bottom*) shows an EPR spectrum of the fully reduced state

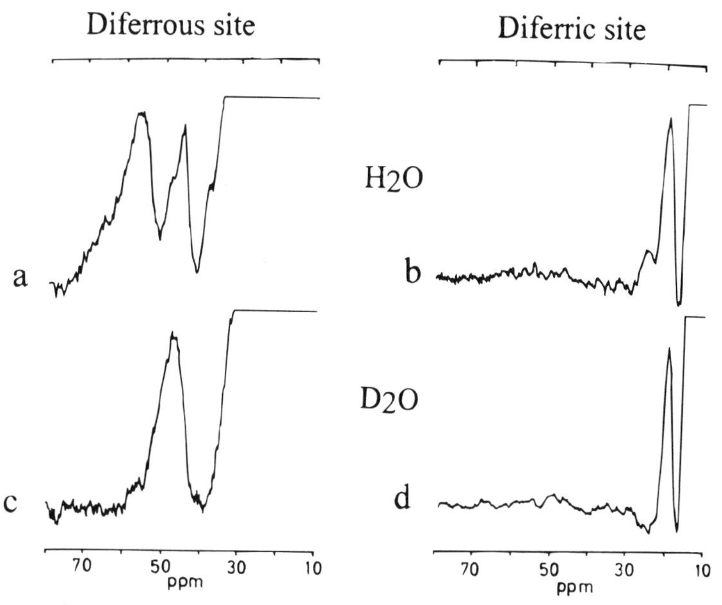

Figure 6 ^{1}H-NMR (400 MHz) spectra of *E. coli* protein R2 (~1.5 mM protein in 25 mM phosphate buffer, pH 7.5) in different redox states. (*a*) Reduced R2 at 37°C; (*b*) active R2 at 37°C; (*c*) reduced R2 in deuterated buffer at 37°C; and (*d*) active R2 in deuterated buffer at 30°C. (Adapted from Reference 55.)

of mouse protein R2, which displays a prominent, low-field signal, with an observed g value of approximately 14. This signal is seen even better in parallel mode EPR, because it arises from an integer spin system (7a). Adding azide (0.2 M) or glycerol (20%) to the sample enhanced the EPR signal intensity greatly. As in *E. coli* R2, azide probably binds to the diferrous site. Magnetic susceptibility measurements at several temperatures showed that an azide-containing mouse R2 sample had a weakly ferromagnetically coupled ($J = +0.3\ cm^{-1}$) $S = 4$ diiron site, in qualitative agreement with the observations on *E. coli* R2. In contrast, the iron ions in the absence of azide were essentially noncoupled in mouse R2.

Transient Free Radicals in Mutant R2 Proteins

Y122F MUTANT IN *E. COLI* R2 Transient tryptophan radicals have been observed on reconstitution of the *E. coli* mutant Y122F (Reaction 1). Because the mutant Y122F lacks the oxidizable Y122, the reconstitution reaction leads to formation of free radicals on other amino acid residues in the protein. In this reaction, EPR with either stopped flow or freeze quench techniques has been used to observe six transient paramagnetic species (56). One of the transients arises from a mixed valent Fe(II)/Fe(III) center observed in the time window 6 s to a few minutes.

In the time window before and during formation of the dinuclear iron center (up to 2 s) the following EPR active species were observed (56): 1. A $g = 2.001$ singlet EPR spectrum observed at 77 K after freeze quenching about 0.3 s after mixing. This singlet is similar to that observed in wild type R2 reconstitution and assigned to a diferric radical species (10, 12). This species corresponds to **X** in reconstitution stage 3 (discussed above). 2. At room temperature, a second singlet with an unknown origin was observed with the use of stopped flow EPR. The signal remained unchanged in the double mutant Y122F/Δ30C (lacking the 30 C-terminal amino acids). This finding excludes the possibility that the EPR singlet should be assigned to Y356, a participant of the postulated electron transfer pathway. With optical spectroscopy, a transient absorbance at 410 nm was observed in the same time window as the room temperature EPR singlet. With the use of the double mutants Y122F/Δ30C and Y122F/Y356A, the light-absorbing species was identified as a tyrosyl radical at position 356, as was also suggested by Bollinger et al (10).

After complete formation of the iron center, and in the time window greater than 6 s, a complex EPR spectrum was observed at low temperatures (Figure 4*b*). Two components could be resolved in the evaluation, on the basis of their different power saturation and lifetimes (56). With the use of indole-deuterated tryptophan, the hyperfine quartet was assigned to a tryptophan radical. The most likely candidate for this tryptophan radical, judging from the saturation behavior and structure of protein R2, is Trp111, which is at a distance of 4 Å from Fe2 (cf Figure 2*a*). Trp111 is H bonded to Glu204, which in turn is H-bonded to Fe2. The second component displayed an axial line shape, also described in the section on tryptophan radicals (Figure 4*b*). The origin was suggested to be a W48 radical that was coupled weakly to the iron, because of its similarity with the line shape of the CCP tryptophan radical. The CCP radical is coupled to the heme iron through an H-bonded triad of H175-D235-W191, where H175 is an iron ligand (30). A similar H-bonded triad (Fe1-His118-Asp237-Trp48) was highlighted as a potential electron transfer pathway in the structure of protein R2 (42, 43).

At times later than 10 s after starting the reconstitution of Y122F R2, an EPR doublet spectrum observable at room temperature started to grow (56). This doublet had its maximum after approximately 2 min and decayed after approximately 10 min. The use of indole-deuterated tryptophan identified this radical as localized on a tryptophan. A possible candidate for this signal is Trp107, which connects to Fe2 through an H-bonded network, including a water molecule and His241. The distance from the edge of the indole ring of Trp107 to Fe2 is approximately 8 Å.

When Y122 is absent in the immediate neighborhood of the strongly oxidizing ferrous iron-oxygen binding site in protein R2, other tryptophan or tyrosine side chains may become oxidized and turned into free radicals by slow electron transfer, provided that they are connected properly to the iron center. Of the three suggested candidates for tryptophan radicals, however, only W48 is conserved. W111 is glutamine in mouse, HSV1 and human R2, whereas W107 is generally aromatic, preferably F, but Y in HSV1.

MUTANTS SURROUNDING THE IRON SITE IN *E. COLI* PROTEIN R2 Studies on the reconstitution reaction for protein R2 (Reaction 1) have been reported for mutants with changes in the hydrophobic patch surrounding Y122, namely F208Y, F212W, F212Y, and I234N. One aim was to study the importance of charge and space for stability and formation of the tyrosyl radical (47). F208, F212, and I234 are conserved residues. Among the mutants, only I234N formed a radical stable enough at 4°C to display an EPR spectrum after the protein purification procedure (~4 days). The others had half-lives in the range 30 s to 10 min. The EPR spectra were all similar to wild type R2 but had some variations in the hyperfine splitting pattern. For F208Y, the transient radical was assigned to Y122 by studies on the double mutant. For the others, the similarity in line shape and/or saturation behavior was considered a strong indication for a tyrosyl radical at Y122. The common effect from changes in the hydrophobic patch was concluded to be a destabilizing effect on the tyrosyl radical (47).

F208Y had further interesting features, because the introduced tyrosine is close enough to Fe1 to induce a drastic change in the radical chemistry on reconstitution with ferrous iron and dioxygen (2, 35, 46). Instead of forming a stable radical at Y122 and a Fe(III)-O^{2-}-Fe(III) center, hydroxylation of the Y208 residue occurs. The resulting DOPA becomes a bidentate ligand bound to Fe1, and D84 becomes a monodentate ligand. No bridging oxygen is seen in the crystal structure (compare with Figure 2).

Model Substances

Over the years, numerous model substances for dinuclear iron centers have been synthesized to give information about structural and physical properties of dinuclear iron proteins (32). Recently, a long-awaited compound, $[Fe_2O(XDK)(BIDPhE)_2(NO_3)_2]$, which can harbor a phenoxy radical in addition to a diiron center, was synthesized (25). The light absorption of the radical form of BIDPhE was reported to be like that of a typical phenoxy radical. The spectroscopic data, including magnetic susceptibility measurements, were consistent with the crystallographic data and showed one μ-oxo and two μ-carboxylate bridges.

A model that can harbor a high-valent iron intermediate has also been presented recently: $[Fe_2(\mu\text{-}O)_2(5\text{-}Me\text{-}TPA)_2](ClO_4)_3$ (16). On the basis of a variety of spectroscopic studies and crystallographic data, it was suggested that this complex had an $Fe_2(\mu-O)_2$ core. Its formal state is Fe(III)Fe(IV), with $S = 3/2$, and it showed an EPR spectrum with the g values of 4.45, 3.90, and 2.01.

NUCLEAR MAGNETIC RESONANCE STUDIES

Because of the high-molecular weights for *E. coli* proteins R1 and R2 (174 kDa and 86 kDa, respectively), ^{1}H-NMR spectra of RDR proteins are not expected to show well-resolved resonances. Paramagnetically shifted resonances have been observed in protein R2, however, and a mobile C-terminal domain of protein R2 gives well-resolved resonances. It is also possible to study protein R1:R2 interactions and the substrate binding site in the enzyme by NMR methods.

Paramagnetically Shifted Resonances

Protein R2 without iron bound, apo R2, does not show any ^1H resonances in the region 10–100 ppm, where paramagnetically shifted resonances are expected to appear (55). On addition of ferrous ions to *E. coli* apo R2 in an anaerobic H_2O containing buffer, a spectrum like that presented in Figure 6*a* is registered from the diferrous form of the protein. Broad resonances are observed at 45 and 57 ppm. If the protein is reconstituted in D_2O containing buffer, the ^1H NMR spectrum instead displays only one resonance centered at 45 ppm from a nonexchangeable iron ligand proton (Figure 6*c*). Because these lines are very broad and somewhat asymmetric, there may be more than one similar iron ligand resonance superimposed to give rise to each broad spectral line. The ratio of intensities in the 45 and 57 ppm peaks is approximately 1:1. Figure 6*b* and *d* shows the ^1H-NMR spectra obtained with active R2 in H_2O and D_2O containing buffers, respectively. The resonances have shifted to 19 ppm for the nonexchangeable protons and 24 ppm for the exchangeable protons, the latter assigned to the histidine ligands. The intensity ratio between nonexhangeable:exchangeable proton resonances has changed to 3:1, which may be indicative of a change in ligand structure in going from a diferrous to a diferric iron center. The spectrum obtained with metR2 was identical qualitatively to that of active R2.

The *J*-values for coupled systems can be estimated from the temperature dependence of the paramagnetically shifted resonances. For the shifted resonances associated with the diferrous site, a very weak antiferromagnetic coupling was estimated as $J = -5\ (\pm 5)\ cm^{-1}$. This observation is in good agreement with the conclusion from EPR studies that most diferrous sites in *E. coli* R2 are EPR silent (18) because of antiferromagnetic coupling. For the strongly ($J \simeq -100\ cm^{-1}$) antiferromagnetically coupled active R2 (22), there is no shift of the ligand resonances in the temperature range (55) as expected from theory. Nuclear magnetic resonance hence proved to be a good method to verify the presence of a reduced iron center and, perhaps, to indicate the carboxylate shifts found much later (1, 7, 49), which were observed when protein R2 changes oxidation state.

Recently, NMR studies on the paramagnetically shifted resonances in Co-substituted *E.coli* R2 were performed (20). Here, a better resolved spectrum was achieved than in the Fe(II)–containing protein because of the more efficient electronic relaxation properties of Co(II) compared with Fe(II). Several assignments were suggested, mostly for exchangeable and nonexchangeable protons on the histidines 118 and 241.

Two-Dimensional ^{1}*H-Nuclear Magnetic Resonance on Protein R2*

A one-dimensional ^{1}H-NMR spectrum of protein R2 from *E. coli* (Figure 7*a*) has several sharp peaks superimposed on the major broad protein resonances, which suggests the presence of a domain with higher mobility than the rigid protein. The likely presence of a mobile segment is also suggested by the fact that 32 C-terminal amino acids could not be resolved in the crystal structure (42, 43). A particular interest in this small protein "domain" also relates to the emerging knowledge that the C terminus of protein R2 is crucial for the specific recognition and formation of the functional R1:R2 complex (64). The C-terminal domain of protein R2 is species specific and varies both in size and in amino acid composition.

A nonamer peptide of HSV1 was shown to compete effectively with HSV1 protein R2 in binding to protein R1 (14, 17). A modified peptide was also constructed and found to have lower IC_{50} as an antiherpes agent than previously studied "unmodified" peptides (36). From a functional point of view, the C terminus carries two conserved residues, E350 and Y356 (in *E. coli* numbering), which have been suggested to be parts of the electron transfer chain between the substrate binding site in protein R1 and the iron–tyrosyl radical site in protein R2 (Figure 2).

Two-dimensional NMR spectroscopy, e.g. Nuclear Overhauser Enhancement Spectroscopy (NOESY), has been applied to demonstrate the dynamic properties and to determine the lengths and possible structures of the C terminus of protein R2 from HSV (33), mouse (37), and *E. coli* (38). At least 6, 10, and 26 amino acid residues were sequentially assigned in the C terminus of the respective proteins. A well-resolved NOESY spectrum of the *E. coli* protein R2 is shown in Figure 7*b*. The assignment was aided by the observations that (*a*) a deletion mutant of *E. coli* R2, Δ30C, lacked the crosspeaks corresponding to the 26 assigned amino acids and that (*b*) the crosspeaks of a 20-mer oligopeptide that corresponded to the 20 most C-terminal amino acids in *E. coli* protein R2 overlapped with the crosspeaks observed in the full-length protein. The crosspeaks in the NOESY spectrum disappear on complex formation with protein R1, which indicates that complex formation locks the C-terminal residues of R2 into a more rigid state. The secondary structure of the mobile C terminus should resemble a random coil in isolated protein R2. The specific interaction with protein R1 in the active enzyme complex then induces a more ordered structure. The same strategies with deletion mutants and appropriate peptides were

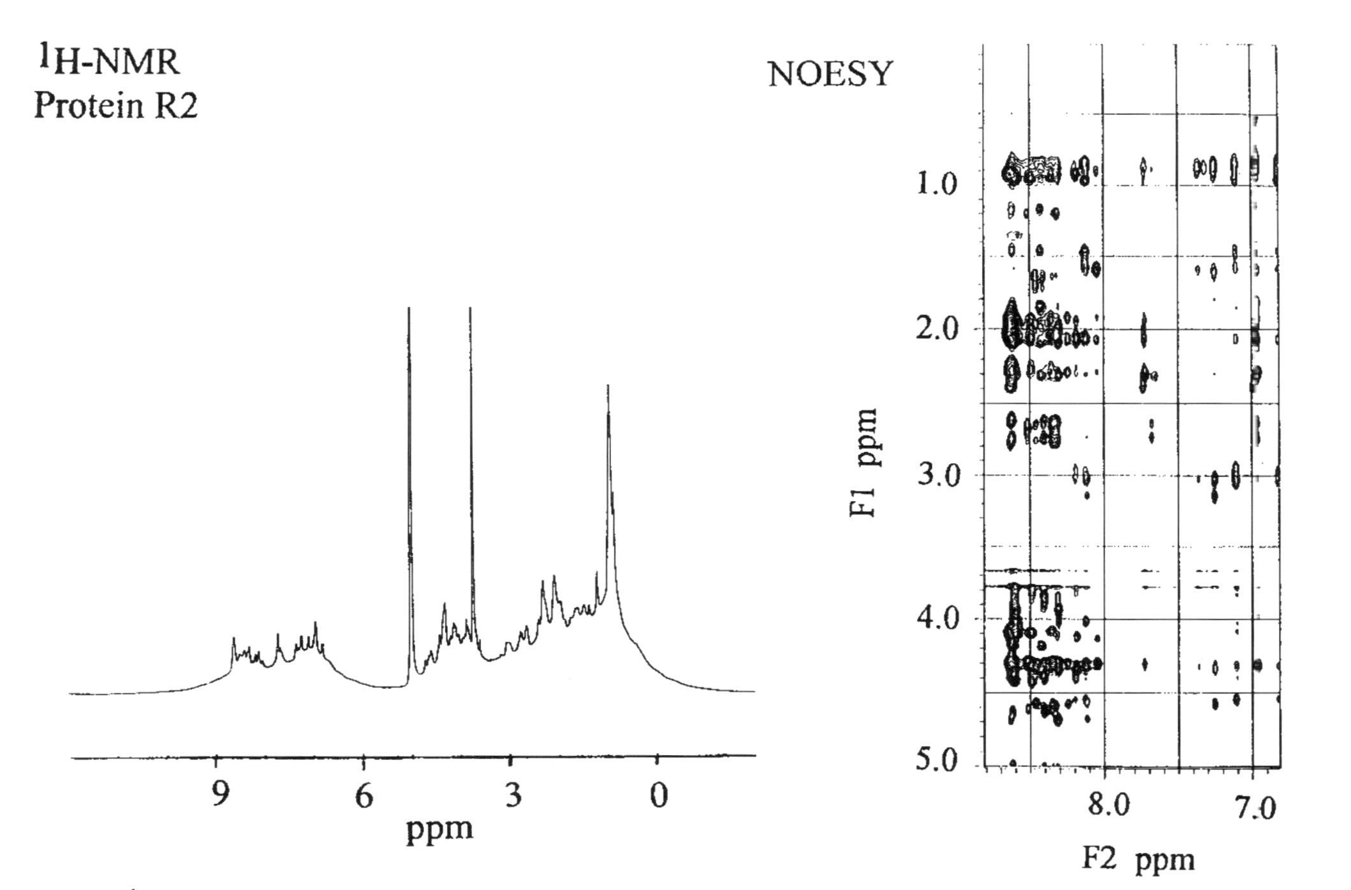

Figure 7 (*a*) ^{1}H-NMR (500 MHz) spectrum of 0.9 mM *E. coli* protein R2 in 25 mM deuterated phosphate buffer, pD 7.5 at 10°C. (*b*) Partial NOESY spectrum of protein R2 in 25 mM phosphate buffer, pH 7.5 at 6°C. (Results from Reference 38.)

also used for assignment in the HSV and mouse proteins. A difference between these two proteins and that of *E. coli* protein R2 is that the former two appear to have more flexible regions than can be accounted for by the C terminus alone.

The conformation of an octapeptide inhibitor to *E. coli* RDR in complex with protein R2 has been probed by transferred NOE spectroscopy (13). A large excess of peptide over protein R1 and rather weak peptide–protein association (rapid dissociation rate) were prerequisites for the success of the studies. The octapeptide could be sequentially assigned. The observed NOEs were interpreted to suggest that the hydrophobic sidechains in the peptide formed a cluster when interacting with protein R1 and that a nonclassic turn was induced in the backbone of the peptide. Similar studies were recently reported also for mouse RDR and its specific inhibitory 7-mer peptide (20a).

Binding of Competitive Inhibitors

Deoxyribocytidine diphosphate (dCDP) is a product of the enzymatic reaction but is also a competitive inhibitor to RDR (4). In ^{1}H-NMR experiments, it is in fast exchange, and its line broadening in the presence of protein R1 or a R1:R2 complex is a probe of its binding characteristics (4). Results on the *E. coli* enzyme showed that the available binding sites on protein R1 decreases (probably from 2 to 1) when R1 forms a complex with R2. The results were supported by a ^{19}F-NMR study of binding of 5-fluoro-deoxyribouridine-diphosphate (5-FdUDP) (54). Transferred NOE studies of bound dCDP gave information on interproton distances in the nucleotide, which suggested that dCDP had the base in anticonformation with S-type deoxyribose puckering when bound to either protein R1 or the R1:R2 complex (3).

CONCLUDING REMARKS

Ribonucleotide reductase is a fascinating enzyme that carries out a difficult reaction, which is essential for cell proliferation. The iron/oxygen and free radical chemistry that occurs inside the protein matrix is spectacular. Equally spectacular is the long-range electron transfer that seems to occur inside protein R2 during the iron–oxygen reaction, as well as between proteins R1 and R2 during the catalytic cycle. At present, the detailed mechanisms of these processes are understood only vaguely but remain as promising targets for studies spanning from molecular biology to chemistry and biophysics.

ACKNOWLEDGMENTS

We thank Anders Ehrenberg for valuable comments on the manuscript. Work on this topic in the authors' laboratories was supported by the Swedish Natural Science Research Council, the Bank of Sweden Tercentenary Foundation, the Carl Trygger Foundation, and the Swedish Cancer Society.

Literature Cited

1. Åberg A. 1993. *Ribonucleotide reductase from* Escherichia coli*: Structural aspects of protein function*. PhD thesis. Stockholm Univ. 59 pp.
2. Åberg A, Ormö M, Nordlund P, Sjöberg B-M. 1993. Autocatalytic generation of Dopa in the engineered protein R2 F208Y from *Escherichia coli* ribonucleotide reductase and crystal structure of the Dopa-208 protein. *Biochemistry* 32:9845–50
3. Allard P, Kuprin S, Ehrenberg A. 1994. Conformation of dCDP bound to protein R1 of *Escherichia coli* ribonucleotide reductase. *J. Magn. Reson.* B103:242–46
4. Allard P, Kuprin S, Shen B, Ehrenberg A. 1992. Binding of the competitive inhibitor dCDP to ribonucleoside-diphosphate reductase from *Escherichia coli* studied by ^{1}H NMR. *Eur. J. Biochem.* 208:635–42

4a. Allard P, Barra AL, Andersson KK, Schmidt PP, Atta M, Gräslund A. 1996. Characterization of a new tyrosyl free radical in *Salmonella typhimurium* ribonucleotide reductase with EPR at 9.45 and 245 GHz. *J. Am. Chem. Soc.* In press

5. Andersson KK, Gräslund A. 1995. Diiron-oxygen proteins. *Adv. Inorg. Chem.* 43:359–408
6. Atta M, Andersson KK, Ingemarsson R, Thelander L, Gräslund A. 1994. EPR studies of mixed-valent [FeII-FeIII] clusters formed in the R2 subunit of ribonucleotide reductase from mouse or herpes simplex virus: mild chemical reduction of the diferric centers. *J. Am. Chem. Soc.* 116:6429–30
7. Atta M, Nordlund P, Åberg A, Eklund H, Fontecave M. 1992. Substitution of manganese for iron in ribonucleotide reductase from *Escherichia coli*. *J. Biol. Chem.* 267:20682–88

7a. Atta M, Debaecker N, Andersson KK, Latour JM, Thelander L, Gräslund A. 1995. EPR and multi-field magnetization of reduced forms of the binuclear iron center in ribonucleotide reductase from mouse. *J. Biol. Inorg. Chem.* In press

8. Behravan G, Sen S, Rova U, Thelander L, Eckstein F, Gräslund A. 1995. Formation of a free radical of the sulfenylimine type in the mouse ribonucleotide reductase reaction with 2′-azido-2′deoxycytidine 5′diphosphate. *Biochim. Biophys. Acta.* 1264: 323–29
9. Bender CJ, Sahlin M, Babcock GT, Barry BA, Chandrashekar TK, et al. 1989. An ENDOR study of the tyrosyl free radical in ribonucleotide reductase from *Escherichia coli*. *J. Am. Chem. Soc.* 111:8076–83
10. Bollinger JM Jr, Edmondson DE, Huyhn BH, Filley J, Norton JR, Stubbe J. 1991. Mechanism of assembly of the tyrosyl radical-dinuclear iron cluster cofactor of ribonucleotide reductase. *Science* 253:292–98
11. Bollinger JM Jr, Hang Tong W, Ravi N, Hahn Huynh B, Edmondson DE, Stubbe J. 1994. Mechanism of assembly of the tyrosyl radical-diiron(III) cofactor of *E. coli* ribonucleotide reductase. 2. Kinetics of the excess Fe^{2+} reaction by optical, EPR, and Mössbauer spectroscopies. *J. Am. Chem. Soc.* 116:8015–23
12. Bollinger JM Jr, Hang Tong W, Ravi N, Hahn Huynh B, Edmondson DE, Stubbe J. 1994. Mechanism of assembly of the tyrosyl radical-diiron(III) cofactor of *E. coli* ribonucleotide re-

ductase. 3. Kinetics of the limiting Fe^{2+} reaction by optical, EPR, and Mössbauer spectroscopies. *J. Am. Chem. Soc.* 116:8024–32
13. Bushweller JH, Bartlett PA. 1991. Investigation of an octapeptide inhibitor of *Escherichia coli* ribonucleotide reductase by transferred nuclear overhauser effect spectroscopy. *Biochemistry* 30:8144–51
14. Cohen EA, Gaudreau P, Brazeau P, Langelier Y. 1986. Specific inhibition of herpesvirus ribonucleotide reductase by a nonapeptide derived from the carboxy terminus of subunit 2. *Nature* 321:441–43
15. Davydov R, Kuprin S, Gräslund A, Ehrenberg A. 1994. Electron paramagnetic resonance study of the mixed-valent diiron center in *Escherichia coli* ribonucleotide reductase produced by reduction of radical-free protein R2 at 77 K. *J. Am. Chem. Soc.* 116:11120–28
16. Dong Y, Fuji H, Hendrich MP, Leising RA, Pan G, et al. 1995. A high-valent nonheme iron intermediate. Structure and properties of $[Fe_2(\mu\text{-}O)_2(5\text{-MeTPA})_2](ClO_4)_3$. *J. Am. Chem. Soc.* 117:2778–92
16a. van der Donk WA, Stubbe J, Gerfen GJ, Bellew BF, Griffin RG. 1995. EPR investigations of the inactivation of *E. coli* ribonucleotide reductase with 2′-azido-2′-deoxyuridine 5′-diphosphate: Evidence for the involvement of the thiyl radical of C225-R1. *J. Am. Chem. Soc.* 117:8908–16
17. Dutia BM, Frame MC, Subak-Sharpe JH, Clark WN, Marsden HS. 1986. Specific inhibition of herpesvirus ribonucleotide reductase by synthetic peptides. *Nature* 321:439–41
18. Ehrenberg A, Reichard P. 1972. Electron spin resonance of the iron-containing protein B2 from ribonucleotide reductase. *J. Biol. Chem.* 247:3485–88
19. Elgren TE, Hendrich MP, Que L Jr. 1993. Azide binding to the diferrous clusters of the R2 protein of ribonucleotide reductase from *Escherichia coli.* *J. Am. Chem. Soc.* 115:9291–92
20. Elgren TE, Ming L-J, Que L Jr. 1994. Spectroscopic studies of Co(II)-reconstituted ribonucleotide reductase R2 from *Escherichia coli.* *Inorg. Chem.* 33:891–94
20a. Fisher A, Laub PB, Cooperman BS. 1995. NMR structure of an inhibitory R2 C-terminal peptide bound to mouse ribonucleotide reductase R1 subunit. *Nature Struct. Biol.* 2:951–55
21. Fontecave M, Nordlund P, Eklund H, Reichard P. 1992. The redox centers of ribonucleotide reductase of *Escherichia coli.* *Adv. Enzymol.* 65:147–83
22. Galli C, Atta M, Andersson KK, Gräslund A, Brudvig GW. 1995. Variations of the diferric exchange coupling in the R2 subunit of ribonucleotide reductase from four species as determined by saturation-recovery EPR spectroscopy. *J. Am. Chem. Soc.* 117:740–46
23. Gerez C, Fontecave M. 1992. Reduction of small subunits of *Escherichia coli* ribonucleotide reductase by hydrazines and hydroxylamine. *Biochemistry* 31:780–86
24. Gerfen GJ, Bellew BF, Un S, Bollinger JM Jr, Stubbe J, et al. 1993. High frequency (139.5 GHz) EPR spectroscopy of the tyrosyl radical in *Escherichia coli* ribonucleotide reductase. *J. Am. Chem. Soc.* 115:6420–21
25. Goldberg DP, Koulougliotis D, Brudvig GW, Lippard SJ. 1995. A (μ-Oxo) bis (μ-carboxylato)diiron (III) complex with a tethered phenoxyl radical as a model for the active site of the protein R2 of ribonucleotide reductase. *J. Am. Chem. Soc.* 117:3134–44
26. Gräslund A, Sahlin M, Sjöberg B-M. 1985. The tyrosyl free radical in ribonucleotide reductase. *Environ. Health Perspect.* 64:139–49
27. Hendrich MP, Elgren TE, Que L Jr. 1991. A mixed valence form of the iron cluster in the B2 protein of ribonucleotide reductase from *Escherichia coli.* *Biochem. Biophys. Res. Commun.* 176:705–10
28. Henriksen M, Cooperman BS, Salem JS, Li L-S, Rubin H. 1994. The stable tyrosyl radical in mouse ribonucleotide reductase is not essential for enzymatic activity. *J. Am. Chem. Soc.* 116:9773–74
29. Hoganson K, Babcock GT. 1992. Protein-tyrosyl radical interactions in photosystem II studied by electron spin resonance and electron nuclear resonance spectroscopy: comparison with ribonucleotide reductase and in vitro tyrosine. *Biochemistry* 31:11874–80
30. Houseman ALP, Doan PE, Goodin DB, Hoffman BM. 1993. Comprehensive explanation of the anomalous EPR spectra of wild-type and mutant cytochrome c peroxidase compound ES. *Biochemistry* 32:4430–43
30a. Gunther MR, Kelman DJ, Corbett

JT, Mason RP. 1995. Self-peroxidation of metmyoglobin results in formation of an oxygen-reactive tryptophan-centered radical. *J. Biol. Chem.* 270: 16075–81

31. Jordan A, Pontis E, Atta M, Krook M, Gibert I, et al. 1994. A second class I ribonucleotide reductase in Enterobacteriaceae: characterization of the *Salmonella typhimurium* enzyme. *Proc. Natl. Acad. Sci. USA* 91:12892–96

31a. Kjöller Larsen I, Sjöberg BM, Thelander L. 1982. Characterization of the active site of ribonucleotide reductase of *Escherichia coli*, bacteriophage T4 and mammalian cells by inhibition studies with hydroxyurea analogues. *Eur. J. Biochem.* 125:75–81

32. Kurtz DM Jr. 1990. Oxo- and hydroxybridged diiron complexes: a chemical perspective on a biological unit. *Chem. Rev.* 90:585–606

33. Laplante SR, Aubry N, Liuzzi M, Thelander L, Ingemarson R, Moss N. 1994. The critical C-terminus of the small subunit of the herpes simplex virus ribonucleotide reductase is mobile and conformationally similar to C-terminal peptides. *Int. J. Peptide Protein Res.* 44:549–55

34. Larsson Å, Sjöberg B-M. 1986. Identification of the stable free radical tyrosine radical in ribonucleotide reductase. *EMBO J.* 5:2037–40

35. Ling J, Sahlin M, Sjöberg B-M, Loehr TM, Sanders-Loehr J. 1994. Dioxygen is the source of the μ-oxo bridge in iron ribonucleotide reductase. *J. Biol. Chem.* 269:5595–601

36. Liuzzi M, Déziel R, Moss N, Beaulieu P, Bonneau A-M, et al. 1994. A potent peptido-mimetic inhibitor of HSV ribonucleotide reductase with antiviral activity in vivo. *Nature* 372:695–98

37. Lycksell P, Ingmarson R, Davies R, Gräslund A, Thelander L. 1994. ^{1}H NMR studies of mouse ribonucleotide reductase: The R2 protein carboxyl-terminal tail, essential for subunit interaction, is highly flexible but becomes rigid in the presence of protein R1. *Biochemistry* 33:2838–42

38. Lycksell P, Sahlin M. 1995. Demonstration of segmental mobility in the functionally essential carboxy terminal part of ribonucleotide reductase from *Escherichia coli*. *FEBS Lett.* 368: 441–44

39. Mann GJ, Gräslund A, Ochai E-I, Ingmarson R, Thelander L. 1991. Purification and characterization of recombinant mouse and herpes simplex virus ribonucleotide reductase R2 subunit. *Biochemistry* 30:1939–47

40. Mao SS, Holler TP, Yu GX, Bollinger JM, Booker S, et al. 1992. A model for the role of multiple cysteine residues involved in ribonucleotide reductase: amazing and still confusing. *Biochemistry* 31:9733–43

41. Mao SS, Yu GX, Chalfoun D, Stubbe J. 1992. Characterization of C439SR1, a mutant of *Escherichia coli* ribonucleotide diphosphate reductase: evidence that C439 is a residue essential for nucleotide reduction and C439SR1 is a protein possesing novel thioredoxin-like activity. *Biochemistry* 31: 9752–59

42. Nordlund P, Eklund H. 1993. Structure and function of the *Escherichia coli* ribonucleotide reductase protein R2. *J. Mol. Biol.* 232:123–64

43. Nordlund P, Sjöberg B-M, Eklund H. 1990. Three-dimensional structure of the free radical protein of ribonucleotide reductase. *Nature* 345:593–98

44. Nyholm S, Mann GJ, Johansson AG, Bergeron RJ, Gräslund A, Thelander L. 1993. Role of ribonucleotide reductase in inhibition of mammalian cell growth by potent iron chelators. *J. Biol. Chem.* 268:26200–5

45. Ochai EI, Mann GJ, Gräslund A, Thelander L. 1990. Tyrosyl free radical formation in the small subunit of mouse ribonucleotide reductase. *J. Biol. Chem.* 265:15758–61

46. Ormö M, de Maré F, Regnström K, Åberg A, Sahlin M, et al. 1992. Engineering of the iron site in ribonucleotide reductase to a selfhydroxylating monooxygenase. *J. Biol. Chem.* 267: 8711–14

47. Ormö M, Regnström K, Wang Z, Que L Jr, Sahlin M, Sjöberg B-M. 1995. Residues important for radical stability in ribonucleotide reductase from *Escherichia coli*. *J. Biol. Chem.* 270: 6570–76

48. Paulsen KE, Liu Y, Fox GF, Lipscomb JD, Münck E, Stankovich MT. 1994. Oxidation-reduction potentials of the methane monooxygenase hydroxylase components from *Methylosinus trichosporium* OB3b. *Biochemistry* 33: 713–22

49. Rardin LR, William BT, Lippard SJ. 1991. Monodentate carboxylate complexes and the carboxylate shift: implications for polymetalloprotein structure and function. *N. J. Chem.* 15: 417–30

50. Ravi N, Bollinger JM, Hahn Huynh B, Edmondson DE, Stubbe J. 1994.

Mechanism of assembly of the tyrosyl radical-diiron(III) cofactor of *E. coli* ribonucleotide reductase. 1. Mössbauer characterization of the diferric radical precursor. *J. Am. Chem. Soc.* 116:8007–14
51. Reem RC, Solomon EI. 1987. Spectroscopic studies of the binuclear ferrous active site of deoxyhemerythrin: coordination number and probable bridging ligands for the native and ligand bound forms. *J. Am. Chem. Soc.* 109: 1216–26
52. Reichard P. 1993. From RNA to DNA, why so many ribonucleotide reductases? *Science* 260:1773–77
53. Rova U, Goodtzova K, Ingmarson R, Behravan G, Gräslund A, Thelander L. 1995. Evidence by site-directed mutagenesis supports long-range electron transfer in mouse ribonucleotide reductase. *Biochemistry* 34:4267–75
54. Roy B, Decout J-L, Béguin C, Fontecave M, Allard P, et al. 1995. NMR studies of binding of 5-FdUDP and dCDP to ribonucleoside-diphosphate reductase from *Escherichia coli. Biochim. Biophys. Acta* 1247:284–92
55. Sahlin M, Gräslund A, Petersson L, Ehrenberg A, Sjöberg B-M. 1989. Reduced forms of the iron-containing small subunit of ribonucleotide reductase from *Escherichia coli. Biochemistry* 28:2618–25
56. Sahlin M, Lassmann G, Pötsch S, Sjöberg B-M, Gräslund A. 1995. Transient free radicals in iron/oxygen reconstitution of the mutant protein R2 Y122F. Possible participants in electron transfer chains in ribonucleotide reductase. *J. Biol. Chem.* 270: 12361–72
57. Sahlin M, Petersson L, Gräslund A, Ehrenberg A, Sjöberg B-M, Thelander L. 1987. Magnetic interaction between the tyrosyl free radical and the antiferromagnetically coupled iron center in ribonucleotide reductase. *Biochemistry* 26:5541–48
58. Sahlin M, Sjöberg B-M, Backes G, Loehr T, Sanders-Loehr J. 1990. Activation of the iron-containing B2 protein of ribonucleotide reductase by hydrogen peroxide. *Biochem. Biophys. Res. Commun.* 167:813–18
59. Salowe S, Bollinger JM, Ator M Jr, Stubbe J, McCracken J et al. 1993. Alternative model for mechanism-based inhibition of *E. coli* ribonucleotide reductase by 2′-azido–2′-deoxyuridine 5′-diphosphate. *Biochemistry* 32: 12749–60
60. Silva KE, Elgren TE, Que L Jr, Stankovich MT. 1995. Electron transfer properties of the R2 protein or ribonucleotide reductase from *E. coli. Biochemistry* 34:14093–14103
60a. Sjöberg BM. 1994. The ribonucleotide jigsaw puzzle: a large piece falls into place. *Structure* 2:793–96
61. Sjöberg B-M. 1995. Structure of ribonucleotide reductase from *Escherichia coli.* In *Nucleic Acids and Molecular Biology*, ed. F Eckstein, DMJ Lilley, 9:192–221. Berlin/Heidelberg: Springer-Verlag
62. Sjöberg B-M, Reichard P, Gräslund A, Ehrenberg A. 1978. The tyrosine free radical in ribonucleotide reductase from *Escherichia coli. J. Biol. Chem.* 253:6863–65
63. Stubbe J. 1990. Ribonucleotide reductases: amazing and confusing. *J. Biol. Chem.* 265:5329–32
64. Uhlin U, Eklund H. 1994. Structure of ribonucleotide reductase protein R1. *Nature* 370:533–39

Annu. Rev. Biophys. Biomol. Struct. 1996. 25:287–314

ACTIVATING MUTATIONS OF RHODOPSIN AND OTHER G PROTEIN–COUPLED RECEPTORS

Vikram R. Rao and Daniel D. Oprian

Graduate Department of Biochemistry and the Volen Center for Complex Systems, Brandeis University, Waltham, Massachusetts 02254

KEY WORDS: constitutive activity, seven-helix receptor, Schiff base, visual pigments

ABSTRACT

Rhodopsin, the visual pigment of rod photoreceptors cells, is a member of the large family of G protein-coupled receptors. Rhodopsin is composed of two parts: a polypeptide chain called opsin and an 11-*cis*-retinal chromophore covalently bound to the protein by means of a protonated Schiff base linkage to Lys296 located in the seventh transmembrane segment of the protein. Several mutations have been described that constitutively activate the apoprotein opsin. These mutations appear to activate the protein by a common mechanism of action. They disrupt a salt-bridge between Lys296 and the couterion Glu113 that helps constrain the protein to an inactive conformation. Four of the mutations have been shown to cause two different diseases of the retina, retinitis pigmentosa and congenital night blindness. Recently, several other human diseases have been shown to be caused by constitutively activating mutations of G protein-coupled receptors.

CONTENTS

1056–8700/96/0610–0287$08.00

PERSPECTIVES AND OVERVIEW

The G protein–coupled receptors form a large family of eucaroytic signal transduction proteins (24, 25, 71, 103a). The receptors consist of seven α-helical transmembrane domains, which are thought to form a bundle in the membrane and serve as receptors for neurotransmitters, hormones, and sensory stimuli. Binding of an agonist ligand to the receptor activates the protein, which in turn activates a heterotrimeric GTP-binding regulatory protein or G protein. The G proteins activate effector proteins, such as adenylyl cyclase, phospholipases, phosphodiesterases, and ion channels (44), which in turn regulate the level of second messengers in the cell.

Activating mutations have been identified recently in rhodopsin and several other G protein–coupled receptors. These mutations in rhodopsin are associated with two different diseases of the retina: retinitis pigmentosa and congenital night blindness. This article focuses on recent work on the mechanism by which these mutations activate the protein, possible mechanisms for the disease, and potential therapies suggested by an understanding of the mechanism. The mutations in rhodopsin are considered in the broader context of constitutively active mutants of other G protein–coupled receptors.

RHODOPSIN STRUCTURE AND FUNCTION

Rhodopsin is the visual pigment of retinal rod photoreceptor cells that mediates visual transduction under conditions of dim light intensity in vertebrates. Approximately 10^9 molecules of rhodopsin per cell account for a concentration of roughly 3 mM in the highly specialized rod outer segment. The high concentration of rhodopsin is responsible, in part, for the exquisite sensitivity of these cells, which can detect a single photon of light (11–13, 27).

Rhodopsin consists of two components: an apoprotein opsin and an 11-*cis*-retinal chromophore (110). Opsin is an integral membrane protein and a prototypic member of the large family of G protein–coupled receptors. It is composed of 348 amino acid residues organized into seven highly hydrophobic segments. These segments, which are arranged in the form of a helical bundle, traverse the membrane bilayer and orient the protein with its amino terminus on the external (intradiscal) side of the membrane and carboxy-terminus on the cytoplasmic side (Figure 1).

The 11-*cis*-retinal chromophore is bound covalently to the protein by means of a protonated Schiff base linkage to the ϵ-amino group of Lys-296, which is located in the seventh transmembrane segment of the protein (17). In the dark, rhodopsin has an absorption maximum at 500 nm. On exposure to light, the chromophore isomerizes to the all-

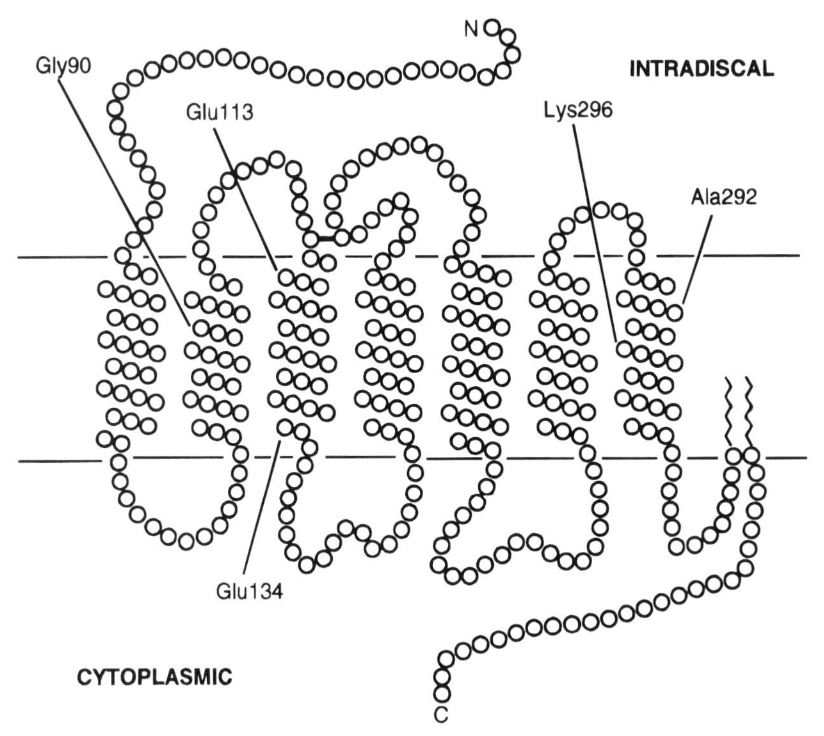

Figure 1 Primary structure of rhodopsin in the rod cell membrane showing transmembrane helices. Residues that activate the protein constitutively are indicated. The cytoplasmic face of the protein interacts with the G protein transducin. Black zig-zag lines indicate palmitoyl groups.

trans form, thereby inducing a series of conformational alterations in the protein. These alterations result ultimately in the spectral intermediate metarhodopsin II (MII) with absorption maximum at 380 nm, shifted some 120 nm to the blue from that of rhodopsin (64). Metarhodopsin II is the only intermediate that can activate the G protein transducin and is therefore equated with the activated state of rhodopsin (31, 53).

Activation of transducin by photolysed rhodopsin initiates an enzymic cascade that results in closure of cation channels in the rod cell membrane. The activity is transient and decays rapidly under in vivo conditions because of the action of two other proteins in the rod outer segment: rhodopsin kinase and arrestin (72, 73a). Rhodopsin kinase is a member of the GRK family of Ser/Thr kinases, which are responsible for phosphorylation of the activated forms of G protein–coupled receptors. Rhodopsin kinase phosphorylates MII at multiple sites on the carboxy-terminal tail of the protein (48, 69, 72, 73). The phosphorylated protein then binds arrestin in a reaction that inhibits the interaction between rhodopsin and transducin competitively and brings the activation reaction to an immediate halt (73a, 94, 113).

Once the activity of MII is quenched by phosphorylation and the binding of arrestin, the dark state of rhodopsin, with 11-*cis*-retinal bound to the protein, is regenerated by a series of reactions that involve hydrolysis of the Schiff base linkage, dissociation of all-*trans*-retinal from the protein, rebinding of 11-*cis*-retinal, dissociation of arrestin, and, finally, dephosphorylation by a rod cell specific phosphatase 2A (72). The newly regenerated rhodopsin is now poised for the absorption of a second photon of light to initiate the enzymic cascade once again.

The Retinal Schiff Base Counterion

The pKa of the retinal Schiff base nitrogen in rhodopsin is highly perturbed. Typically, the pKa of a retinylidene Schiff base complex free in solution is approximately 7. In contrast, the Schiff base nitrogen in rhodopsin remains protonated until approximately pH 11, at which point the protein denatures. This finding suggests that the pKa in the native pigment is at least 11 or 12 and may be much higher. Indeed, Steinberg et al (102) have estimated that the pKa of the Schiff base nitrogen in rhodopsin is at least 16, because the nitrogen remains protonated at pH 11 when the protein is reconstituted with 20,20,20-trifluoro-9-*cis*-retinal instead of 11-*cis*-retinal. The pKa of a Schiff base complex of the trifluoro analogue free in solution is 1.8, shifted by the highly electron withdrawing fluorine atoms some 5 units more acid than that of a typical Schiff base of 11-*cis*-retinal. Thus, the pKa of the Schiff base nitrogen in rhodopsin appears to be displaced by at least 9 pH units from its expected pKa free in solution.

The highly perturbed Schiff base pKa in rhodopsin has long been postulated to result from the close juxtaposition of a negatively charged carboxylate counterion in the protein. On the basis of mutagenesis studies, the Schiff base counterion was identified as Glu113 in the third transmembrane segment of the protein (68, 89, 116). Changing Glu113 to Gln results in a dramatic change in the pKa of the Schiff base: The absorption maximum of the mutant isolated at pH 7.4 is 380 nm, which is typical of an unprotonated retinylidene Schiff base, and is shifted some 120 nm to the UV from that of the wild-type pigment. The Schiff base nitrogen of the Glu113Gln mutant can be titrated cleanly and reversibly with a pKa of approximately 6, shifted at least 10 pH units from native rhodopsin and much closer to that expected of a Schiff base free in solution. Under these conditions, the positively charged nitrogen is stabilized in the mutant protein by interaction with a chloride ion recruited from bulk solution (68, 90). The interpretation of the pH-dependent spectral changes of the Glu113Gln mutant and the identification of Glu113 as the Schiff base counterion have been supported by resonance Raman microprobe spectroscopy studies (61).

The detailed structure of the retinal binding pocket in rhodopsin has been the subject of considerable interest. On the basis of two-photon absorption spectroscopy of rhodopsin with a locked 11-*cis*-retinal chromophore, Birge et al (15) concluded that the binding pocket was electrostatically neutral and, therefore, contained only one carboxylate sidechain from the protein. This conclusion is in agreement with the mutagenesis studies described above (68, 89, 116). Further, there appears to be a consensus in the field that the carboxylate of Glu113 is located close to carbon atom 12 of the retinal chromophore and makes contact with the Schiff base nitrogen by hydrogen bonding through one or more intermediary water molecules (42, 46). This conclusion has been supported by two very different approaches: one from Nakanishi and Honig and coworkers (57), who used various dihydroretinal derivatives of the chromophore, and the other from Griffin and Smith and others (46, 101), who used C13 solid state NMR to determine the relative chemical shift of individual carbon nuclei of the retinal chromophore in rhodopsin.

The Covalent Bond to the Chromophore

Rhodopsin and other visual pigments are distinguished as a subgroup among the G protein–coupled receptors, because they bind the retinal ligand covalently. All the other receptors bind their ligands by strictly noncovalent bonds. This finding makes sense, because the other receptors must respond rapidly to changes in the concentration of neurotransmitters and hormones in the external milieu, and it would be disadvantageous to use covalent bonding in this process. The fact that the retinal chromophore is bound covalently suggested that there might be a fundamental difference between the ligand-binding domain of rhodopsin and the other G protein–coupled receptors. On the other hand, rhodopsin is highly similar to the other members of this family, especially the biogenic amine receptors (e.g. the serotonin, dopamine, histamine, and adrenergic receptors), and it seems unlikely that these proteins experienced such divergent evolutionary paths. For example, all the biogenic amine receptors have a conserved Asp counterion in the third transmembrane segment of the protein that interacts with the positively charged ammonium ion of the ligand (71). This counterion is four residues removed, roughly one helical turn, from the position of the Glu113 counterion in rhodopsin. Significantly, the Glu113 counterion in rhodopsin can be moved to exactly the same position as the ammonium ion counterion in the biogenic amine receptors and still retain the native function of the protein (118, 120). This finding strongly supports the concept that there is not a fundamental difference in the ligand-binding domain of these receptors. But what, then, is the function of the covalent bond in rhodopsin?

To test whether the covalent bond to the chromophore was essential for the native function of rhodopsin, the active site Lys296 residue was changed to Gly. As expected, 11-*cis*-retinal did not bind to this mutant. However, the mutant could be reconstituted with a Schiff base complex of 11-*cis*-retinal and n-propylamine to form a pigment with near wild-type spectral properties (117). Therefore the covalent bond is not essential for reconstitution of the protein with chromophore. More significantly, the Lys296Gly mutant reconstituted with the n-propylamine Schiff base of 11-*cis*-retinal displayed wild-type specific activity for the light-dependent activation of transducin (87, 117). Thus, the covalent bond is also not essential for activation of the transducin.

Cohen et al (19) found that noncovalent binding of the chromophore was mediated, in part, by the Glu113 counterion, because the double mutation Lys296Gly/Glu113Gln, in which the counterion is neutralized, fails to bind the Schiff base complexes. This finding is analogous to the situation in β-adrenergic receptor, where mutation of the ammonium counterion abolishes the binding of both agonist and antagonists to the receptor (104).

If the covalent bond between the chromophore and the protein is not necessary for the activation of rhodopsin, what is its purpose? There are probably two roles served by the covalent bond, both of which are consequences of the greater binding energy inherent in a covalent linkage and the resulting increased occupancy of the receptor by retinal. The first role, to ensure the efficient detection of an arriving photon of light, is rather obvious. Less obvious, however, is that high occupancy of the receptor by retinal will also decrease photoreceptor noise that arises from opsin. This phenomenon is discussed in the next section.

CONSTITUTIVE ACTIVATION OF OPSIN

A surprising and unexpected result was found when the Lys296Gly mutant was assayed for its ability to activate transducin in the absence of the retinal chromophore, i.e. when the apoprotein, or opsin, form was assayed. Unlike wild-type opsin, which does not activate transducin (see below), the Lys296Gly mutant activates transducin with a specific activity that is essentially identical to that observed for photoactivation of the wild-type protein (87). This light-independent or constitutive activity of the Lys296Gly mutant is inhibited in the dark by the n-propylamine Schiff base complex of 11-*cis* retinal. On exposure to light, this chromophore-bound, inactive mutant pigment again activates transducin (87). These results demonstrate clearly that mutation of Lys296 constitutively activates opsin.

The Salt Bridge

In all, 12 different amino acids have been substituted for the active-site Lys of position 296 (19, 20). Of these, only two amino acids have been observed not to result in constitutive activation of the apoprotein: the naturally occurring Lys, and Arg. All others display varying degrees of constitutive activity, which suggests that the positive charge on Lys296 is important for maintaining the inactive state of opsin. The positive charge on the nitrogen of Lys296 interacts electrostatically with the negative charge of Glu113 when the chromophore is bound to the protein. Perhaps, then, the positive charge on the nitrogen also interacts electrostatically with Glu113 in the apoprotein, and this salt-bridge maintains, in part, the protein in an inactive state (Figure 2). In complete accord with this simple salt-bridge model for activation and inactivation of opsin, mutation of Glu113 to Gln also activates the protein constitutively (19, 87). Addition of 11-*cis*-retinal suppresses this constitutive activity in the dark; on exposure to light, the inactivated, retinal-bound Glu113Gln mutant becomes active once again.

The interpretation of the fact that the Lys296Arg mutant is inactive

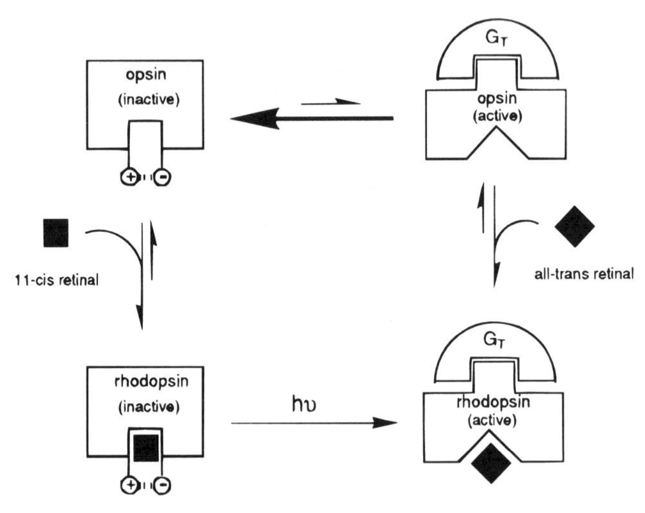

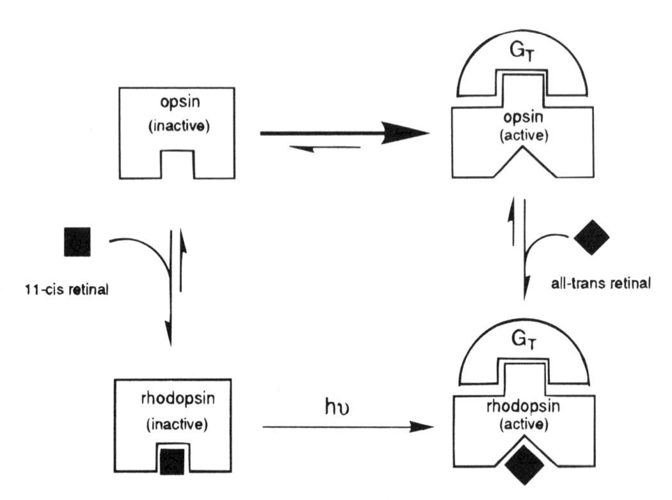

Figure 2 Equilibrium model of the active and inactive states of the apoprotein, opsin, and the holoprotein, rhodopsin. The Lys296–Glu113 salt bridge is indicated with the positive sign that denotes Lys296 and the negative sign that denotes Glu113. Note that in wild-type opsin the equilibrium favors the inactive state, whereas in the constitutively active form of opsin the equilibrium is displaced toward the active state. G_T is transducin.

is not straightforward. The Lys296Arg mutant cannot bind chromophore and cannot be assayed for light-dependent activity. It is impossible, therefore, to tell whether the absence of activity is caused by a positive charge at 296 or by denaturation of the protein. If, however, a second mutation is introduced into the Lys296Arg mutant that changes the Glu113 counterion to Gln, the resulting double-mutant opsin becomes active constitutively, thus showing essentially the same activity as the Glu113Gln single mutant and adding further support to the salt-bridge model (20). Both a positive charge at 296 and a negative charge at position 113 are necessary, therefore, to keep the opsin inactive. We speculate that breaking the salt-bridge leads to activation of the protein because it allows helix 3 to move relative to helix 7 and this movement is required for activation of rhodopsin. Electron paramagnetic resonance (EPR) spin-labeling experiments of Hubbell, Khorana and coworkers (36, 37, 82) have documented just such movement at the cytoplasmic surface of helix 3 on photoactivation of rhodopsin (see below).

Two-State Model for Activation

The results from mutation experiments for activation and inactivation of the protein may be accommodated into a simple two-state model (see Figure 2), where the protein exists in two different conformations: one inactive and one active. The inactive state binds 11-*cis*-retinal, is stabilized by a salt-bridge between Glu113 and Lys296, and does not activate transducin. The active state binds preferentially the all-*trans* form of retinal. It also binds to and activates transducin and does not have a salt-bridge between Glu113 and Lys296. According to this model, mutation of either Lys296 or Glu113 to a neutral amino acid will break the salt-bridge and promote the transition to the activated state, whereas binding of 11-*cis*-retinal will favor the inactive state of the protein. One prediction of the model is that 11-*cis*-retinal will stabilize the inactive state of rhodopsin, even in the absence of the salt-bridge. This stabilization is exactly what has been observed for the Glu113Gln counterion mutant; when complexed with 11-*cis*-retinal, the mutant is completely inactive in the dark, even at high pH where the Schiff base nitrogen is not protonated (87, 89). The 11-*cis*-retinal chromophore functions, therefore, as an antagonist or, more properly, as an inverse agonist in this system.

Another prediction of this simple two-state model is that wild-type opsin is not completely inactive. The equilibrium between the inactive and active states favors the inactive state heavily, but there must be a finite amount of the active state present that displays some small amount of activity in the absence of the retinal chromophore. The salt bridge contributes an estimated 2–3 kcal/mol to the stability of the inactive state (20), which suggests that, under appropriate assay conditions, the activity of opsin should be observed in vitro. In accord with this prediction, both recombinant opsin expressed in tissue culture cells (20) and native bovine opsin (109) have been shown to activate transducin with low but significant activity. In both systems, 11-*cis* retinal inhibited the reactions in the dark.

The fact that wild-type opsin has some ability to activate transducin suggests the possibility that opsin itself might be responsible for setting the threshold of light detection in rod photoreceptor cells. Electrophysiologic experiments in isolated rod outer segments and psychophysical experiments have been interpreted to indicate that the level of internal noise in the eye that sets the limit for visual detection is the result of thermal isomerization of the 11-*cis*-retinal chromophore in rhodopsin (1, 8–10, 12). The rate of spontaneous thermal isomerization of retinal in rhodopsin derived from these experiments is thought to be approximately 0.02/s/rod cell (12). However, the activity observed for opsin in vitro suggests another possible source for this internal noise. Retinal is thought to exchange between rhodopsins in the dark at a rate of approximately 30 molecules/s/rod cell (23). If we use this rate as the rate of dissociation of the retinal chromophore from opsin and estimate that one opsin of every 1000 is active (19, 20), we expect active opsin to arise at a rate of approximately 0.03/s/rod cell. This rate is similar to the spontaneous noise rate attributed to thermal isomerization of the chromophore. It is possible, therefore, that noise generated from active opsin molecules could be responsible for limiting the level of light detection in the eye. Indeed, experiments on isolated amphibian rods and cones which have been exposed to bright light suggest that the persistent desensitization of these photoreceptors may be mediated by the activity of free opsin (21a, 49a). The fact that 11-*cis*-retinal inhibits this activity suggests that high occupancy of the retinal binding pocket ensured by the covalent bond to the chromophore might also be important in decreasing activity of the native opsin molecules in a rod photoreceptor cell.

An underlying assumption in this analysis is that an activated opsin has the same lifetime as a photoactivated rhodopsin. This assumption is unlikely to be the case for the isolated proteins *in vitro*, where the photoactivated intermediate MII has a lifetime of many minutes. It is much more likely to hold true within a rod photoreceptor cell, however, where the lifetime of the activated state is less than 1 s and is thought to be limited by reaction with rhodopsin kinase rather than the intrinsic rate of decay of the activated intermediate.

Effect of Size at Position 296 and Charge at Position 134

Further work on constitutively active mutants has shown that at least two ionizable groups other than Glu113 and Lys296 regulate the transition to the active state such that activation has at least two requirements: breaking of the salt-bridge and uptake of protons from the solvent (19). Activation of transducin by opsin and rhodopsin, therefore, is pH dependent. When the transducin activation rates for various constitutively active mutants are plotted as a function of pH, they give a characteristic bell-shaped rate profile that is defined by two ionization constants, pKa_1 and pKa_2 (19). Although the value of pKa_1 does not vary with the identity of amino acid at position 296, the value of pKa_2 does (19, 20). Analysis of pKa_2 values for various mutants found that, in addition to charge, the size of the side chain at position 296 also has an effect on the ability of opsin to activate transducin (20). In general, large side chains at this position favored the inactive state and small side chains favored the active state suggesting that van der Waals contacts with the Lys296 side chain are lost on activation of the protein. Remarkably, the correlation between stabilization energy and side chain size (20 cal/A^2) was similar to that found for cavity-creating mutants in T4 lysozyme (32). From an analysis of the effect of side chain size and charge at position 296, the salt bridge stabilizes the inactive conformation of opsin by an estimated 2–3 kcal/mol (20). This value is approximately the same as that found for the His31–Asp70 salt-bridge–mediated thermal stabilization of T4 lysozyme (4). The salt bridge between Glu113 and the retinal Schiff base in rhodopsin, however, is likely to be much stronger, given the shift of 10 units in pKa of the Schiff base upon neutralization of Glu113.

In addition to the ability of mutations at positions 113 and 296 to constitutively activate opsin, changing Glu134 at the cytoplasmic surface of transmembrane segment 3 to Gln also activates the protein constitutively (20). Glu134 is of interest because it is one of the most highly conserved residues among all G protein-coupled receptors; with only a handful of exceptions an Asp or a Glu is always found at this position. The constitutive activity of Glu134Gln is weak but easily detectable, whereas mutant Glu134Asp is not constitutively active, suggesting that neutralization of the carboxylate side chain is important for activation of transducin (20). Neutralization of the Glu134 carboxylate appears to be important also for photoactivation of rhodopsin as the Glu134Gln mutant is more active than is the wild-type protein in that reaction (5, 34, 89).

ACTIVATION OF OPSIN AS A CAUSE OF RETINAL DISEASE

Retinitis Pigmentosa

The constitutively activating mutations are of interest not only because they give us insight into the mechanism of activation and inactivation of the protein, but also because they are found in two different human retinal diseases: autosomal dominant retinitis pigmentosa (ADRP) and congenital night blindness. Autosomal dominant retinitis pigmentosa is a severe degenerative disease of the retina. Approximately 30% of all cases of ADRP are linked to rhodopsin mutations, and approximately 70 different mutations have been identified. Several mechanisms for the etiology of the disease have been proposed, including inability of the mutant rhodopsin to fold properly (51, 108), inability to activate transducin (66), and inability of the protein to be transported properly in the cell (21, 107).

Two different families have been identified with mutations of Lys296 that cause ADRP: one in Great Britain, with a Glu at position 296 (52), and the other in the United States, with a Met at position 296 (105). The mutations are intriguing because they render the protein incapable of binding the retinal chromophore. Despite the inability to bind retinal, both mutants appear to fold properly when expressed in transfected COS cells (84, 87). In addition, they are both active constitutively, as is expected from neutralization of the charge at position 296 (84, 87). It is not clear, however, how a constitutively active rhodopsin would cause degeneration of rod photoreceptor cells. In fact, recent experiments with a transgenic mouse model of the Lys296Glu mutation suggest that, under in vivo conditions, the protein is not constitutively active at all (60). Rather, the mutant protein is phosphorylated quantitatively, presumably by rhodopsin kinase, and bound in inactive complexes with the inhibitory protein arrestin (60). The mutation likely causes the disease by forcing the protein into an active conformation that accumulates in inactive complexes in the cell. Perhaps the binding of arrestin to the phosphorylated opsin is preventing the binding of other important proteins, such as those necessary for protein transport, as is thought to be the case for some of the other mutations (21, 106–108). This overabundance of malprocessed protein in the rod cell could trigger apoptotic pathways, which could lead to retinal degeneration (79). The results of the transgenic mouse study appear to be at odds with those of Robinson et al (86), which show that the constitutively active mutants are not constitutively phosphorylated by rhodopsin kinase. Rim & Oprian (84), however, have found that the constitutively active mutants are phosphorylated constitutively by rhodopsin kinase. Reasons for the different observations are far from clear at this time, and a full resolution of this issue will require additional work.

Congenital Night Blindness

Recently, two mutations in rhodopsin have been shown to cause congenital night blindness: Ala292Glu (28) and Gly90Asp (100). Patients who have night blindness cannot adapt to darkness and have difficulty seeing under dim light conditions. Unlike patients who have ADRP, however, those who have congenital night blindness do not have progressive retinal degeneration (14).

The mutation of Ala292 to Asp was observed in a single individual who showed a complete inability to adapt to dark conditions or to sense dim light but showed no retinal degeneration in 16 years (28). Polymerase chain reaction amplification and DNA sequence analysis of the rhodopsin gene locus of this patient showed that he was heterozygous for a mutation of Ala292 to Glu in the seventh transmembrane segment of the protein. In vitro characterization of the Ala292Glu mutant showed that it was expressed in transfected COS cells in good yield, it bound the 11-*cis*-retinal chromophore to give a near wild-type absorption spectrum, and it had essentially wild-type specific activity for the light-dependent activation of transducin (28). When assayed in the absence of chromophore, however, the Ala292Glu mutant opsin activated transducin constitutively, just as did mutations at position 296 and 113. The nature and location of the mutation immediately suggested a molecular mechanism for activation of the protein. Ala292 is located four amino acids and approximately one helical turn from the active site Lys296. The mutation is a change of a neutral Ala residue to a negatively charged Glu. The change therefore places a negatively charged carboxylate in close proximity to Lys296. The newly introduced charge can compete with Glu113 for the positively charged nitrogen on Lys296 and, by a competitive mechanism, break the salt-bridge between Lys296 and Glu113 and activate the protein. The Ala292Glu mutant is less active than either Lys296Gly or Glu113Gln, which is consistent with the idea that the salt-bridge is broken by a competitive mechanism in Ala292Glu.

The Gly90Asp mutation was found in a large Michigan kindred with autosomal dominant congenital night blindness (100). These patients had symptoms identical to those of the Ala292Glu individual, i.e. they could not adapt to darkness and showed little retinal degeneration. Sequencing of DNA revealed that all the afflicted individuals carried the heterozygous mutation Gly90Asp in the second transmembrane helix of rhodopsin. In vitro characterization of the Gly90Asp mutant showed it to be highly similar to the Ala292Glu mutant. Gly90Asp was expressed in transfected COS cells in good yield, it bound the 11-*cis*-retinal chromophore to give a somewhat blue-shifted absorption spectrum, and it had essentially wild-type specific activity for the light-dependent activation of transducin (51, 80). When assayed in the absence of chromophore, however, the Gly90Asp mutant also activated transducin constitutively, just like Ala292Glu opsin (80).

The fact that the Gly90Asp mutant was so similar to the Ala292Glu mutant suggested that the molecular mechanism for activation of the protein was also similar, i.e. competitive breaking of the Lys-296–Glu113 salt-bridge. Certainly, the nature of the mutation suggested that this mechanism could be a possibility: a change of a neutral Gly residue to a negatively charged Asp. Unlike Ala292Glu, where the ectopic carboxylate is located one helical turn from Lys296, there was no direct evidence that position 90 was close to either Lys296 or Glu113 and hence in a position to disrupt the salt-bridge. Proof that 90 was close to 296 came from the double mutant Gly90Asp/Glu113Gln, in which an Asp carboxylate at position 90 was shown to substitute for the loss of the Schiff base counterion at position 113 (80). These data clearly show that an Asp at position 90 can furnish an alternative counterion to the Schiff base in the absence of Glu113; thus, the Gly at position 90 is close to the Glu113–Lys296 salt bridge. This result is in good agreement with the projection structure of rhodopsin (92) and recent three-dimensional models of G-protein–coupled receptors (7, 25). The proximity of helix 2 and helix 7 has also been shown recently in the gonadotropin-releasing hormone receptor (115) and the serotonin 5-HT_{2A} receptor (96).

It appears, therefore, that Gly90Asp and Ala292Glu activate opsin constitutively by the same molecular mechanism of disrupting the Glu-113–Lys296 salt-bridge competitively. Although an Asp at position 90 can furnish a counterion for the Schiff base in the Gly90Asp/Glu113Gln mutant rhodopsin, it does not substitute for Glu113 as far as constitutive activity of the protein is concerned. The Gly90Asp/Glu113Gln double mutant is as constitutively active as is the Glu113Gln single mutant (VR Rao and DD Oprian, unpublished results). This finding is exactly as expected from the salt-bridge model. The formation of the salt-bridge between helices 2 and 7 cannot substitute for the one broken between 3 and 7. A salt-bridge between helices 2 and 7 still allows helix 3 to move relative to 7.

The fact that the Ala292Glu and Gly90Asp mutations cause constitutive activation of opsin provides a satisfying explanation for the cause of disease in patients who have night blindness. Simply put, the rod photoreceptor cells adapt to the apparent internal light signal generated by the mutant opsins and desensitize until the cell no longer responds to external stimuli. Both mutant opsins bind 11-*cis*-retinal efficiently and should be predominantly in the inactive state under physiologic conditions in the dark. At any given time, however, a small fraction of the protein will lose retinal from the binding pocket as a result of a finite binding constant and thermal dissociation of the chromophore. We expect opsin to arise spontaneously, therefore, by dissociation of 11-*cis*-retinal from the binding pocket. We also expect that this rate would correspond roughly to the rate of chromophore exchange in the dark, estimated to be approximately 30/s/rod cell (23). Both wild-type and mutant rhodopsin will lose retinal by this mechanism, but the low level of activity associated with the wild-type opsin is of little concern in this context. In contrast, the rate of 30/s/rod cell for a constitutively active mutant would be expected to desensitize the cell significantly. For example, careful psychophysical tests of the Gly90Asp patients in the dark indicate that their rod cells are 3.5 log units desensitized (100). This value would be expected from background light that causes 10 rhodopsin photoisomerizations/s/rod cell (100).

Interestingly, a recently discovered mutation in the β subunit of photoreceptor cGMP phosphodiesterase (PDE), the effector protein immediately downstream of transducin in the visual cascade, is also associated with autosomal dominant congenital night blindness (40). The mutation lies close to the inhibitory γ-subunit binding site on the protein and might disrupt the ability of the γ subunit to bind to and inhibit PDE. Consequently, the PDE would be activated constitutively, and the molecular mechanism for the disease would be similar to that of the rhodopsin Ala292Glu and Gly90Asp mutants.

There is an apparent paradox in this discussion. If mutations at Lys296, Ala292, and Gly90 all result in constitutive activation of the protein, how can two very different diseases arise from the mutations. Mutations of Lys296 cause a devastating degenerative disease of the retina, whereas mutation of Ala292 and Gly90 cause the comparatively benign condition of night blindness. The reason, simply, is that the retinitis pigmentosa mutations at Lys296 cannot bind to 11-*cis* retinal. In wild-type rhodopsin, rebinding of 11-*cis*-retinal by photoactivated rhodopsin inactivates the protein, thus releasing arrestin and allowing dephosphorylation by phosphatase 2A. The Lys296 mutants accumulate in inactive complexes with arrestin (60), because 11-*cis*-retinal cannot reverse this reaction. In contrast, the constitutively active Gly90Asp and Ala292Glu opsins are also phosphorylated by rhodopsin kinase and bound in inactive complexes with arrestin, but these reactions, like those with the wild-type protein, can be reversed when the proteins are regenerated with 11-*cis*-retinal.

Inhibition of Constitutive Activity by Modified Retinals

If the proposed mechanisms of disease by the constitutively active receptors are correct, then it should be possible to stop the constitutive activity of these mutants and, hence, the disease by designing active site–directed reagents that function as inverse agonists. It is clear that 11-*cis*-retinal and its analogues inhibit constitutive activity in the mutants, and these are obvious candidates for modification (87, 117). The inverse agonists would have to possess certain characteristics to be effective: They must not be activated by light; they must not inactivate wild-type rhodopsin or, more importantly, the cone cell visual pigments; and they must inhibit the constitutive activity of the mutants irreversibly. To this end, several amine derivatives of 9-*cis* and 11-*cis* retinal were synthesized (45). These analogues inhibited the constitutive activity of Lys296Gly opsin irreversibly and did not inactivate the wild-type protein. Unfortunately, none of these analogues inhibited the activity of the retinitis pigmentosa mutants Lys296Glu or Lys296Met because of unfavorable steric interactions within the retinal binding pocket, as Glu and Met have much larger side chains than Gly. To solve this problem, a retinylamine analogue that is shorter by one carbon than naturally occurring retinal was synthesized (T Yang and DD Oprian, unpublished results). This analogue inactivates both of the ADRP mutants, Lys296Glu and Lys296Met, in the dark and in the light effectively and does not bind to or inactivate the wild-type protein. Additionally, this analogue also prevents constitutive phosphorylation of the mutant opsins by rhodopsin kinase and the subsequent binding by arrestin. Therefore, preliminary data with this compound are very encouraging, and further studies will test the efficacy of this analogue in treating retinal degeneration in a Lys296Glu transgenic mouse.

PHOTOACTIVATION OF RHODOPSIN

There are several lines of evidence suggesting that activation of opsin by mutation and activation of rhodopsin by light are similar and that, consequently, the active conformation in the two cases might be the same. Primarily, it appears that breaking of the salt bridge between Glu113 and Lys296 is relevant to both mechanisms of activation.

When rhodopsin is exposed to light, the isomerization of retinal from 11-cis to all-*trans* occurs in less than 200 fs (95). The protein then goes through a series of spectrally defined intermediate states that are in temperature and pH dependent equilibrium with each other. At temperatures near −10°C metarhodopsin I (MI) is observed. This intermediate has an absorption maximum of 480 nm, which indicates that the retinal Schiff base is protonated (64). At higher temperatures, the metarhodopsin II (MII) intermediate, with an absorption maximum of 380 nm indicative of an unprotonated Schiff base, predominates (6, 26, 41, 64, 74, 110). The Schiff base in MII has an extremely low pK and is not titratable. This observation suggests that the Schiff base has moved into an hydrophobic environment where protonation of the nitrogen is unfavorable (19, 20, 49).

Metarhodopsin II is the only intermediate capable of activating transducin (31, 53). The fact that MII has an unprotonated Schiff base suggests that loss of the proton, and hence the positive charge on the nitrogen, may be required for activation of rhodopsin. To show this, Rando and coworkers (41, 63, 97) prepared a chemically modified form of rhodopsin in which Lys296 had a methyl group attached to the nitrogen. This modified protein could bind 11-*cis*-retinal to form a positively charged Schiff base linkage, but the Schiff base could not lose the charge on the nitrogen. This modified protein could not activate transducin upon exposure to light.

The fact that the methylated rhodopsin was unable to activate transducin is consistent with the mechanism proposed for activation of opsin. As described earlier, the salt-bridge between Lys296 and Glu113 helps to keep opsin in the inactive state and breaking of this salt-bridge activates the protein. In the methylated rhodopsin, the Schiff base remains positively charged after photoisomerization of the chromophore and thus continues to interact electrostatically with Glu113. The salt-bridge is intact and the protein is unable to activate transducin. In native rhodopsin, deprotonation of the Schiff base in MII is accompanied by loss of the salt-bridge and this promotes activation of the protein.

Recently, several mutants of rhodopsin have been reported which appear to violate the principle that deprotonation of the Schiff base is obligate for activation of the protein. When exposed to light at room temperature these mutants exhibit "slow bleaching" behavior; that is, the retinal Schiff base does not immediately deprotonate upon photoactivation as it does in the wild-type protein (28, 51, 67, 80, 112, 120). Instead the mutants form a "MI-like" intermediate with absorption maximum at approximately 480 nm, indicative of a protonated Schiff base. Based on these spectral data it was assumed that the mutations perturbed the MI-MII equilibrium and slowed the transition to MII (112, 120). If this were in fact the case, the mutants would not be expected to activate transducin with the same specific activity as wild-type rhodopsin. These mutants, however, did activate transducin with near wild-type specific activity (28, 80, 120) and so an alternative explanation was needed.

The mutant rhodopsins that exhibit this anomalous behavior include those associated with night blindness, Ala292Glu and Gly90Asp, as well as mutants in which the native Glu113 counterion has been moved one helical turn away to position 117, and they appear to have one thing in common—all can provide an alternative counterion to the Schiff base nitrogen (28, 80, 118, 120). Recall that the Ala292Glu and Gly90Asp opsins are constitutively activated because carboxylates at these positions are thought to form an alternative salt-bridge with Lys296 in the active conformation of the apoprotein. This observation led to the idea that the alternative counterion can stabilize the protonated Schiff base in the photoactivated state of the mutant rhodopsins (80). According to this interpretation, the observed ''MI-like'' intermediate is in an active MII-conformation of rhodopsin but with a protonated Schiff base. Transducin binds to and stabilizes this ''MI-like'' intermediate in much the same way that transducin stabilizes the MII intermediate of wild-type rhodopsin lending support to the suggestion that the intermediate is in the active conformation (119). Strong evidence that the conformation of the protein corresponds to the MII intermediate was provided by Fourier transform infrared spectroscopy showing that the infrared spectrum of the photoactivated mutant was essentially identical to photoactivated wild-type rhodopsin (35). It thus appears that these mutants do not have a perturbed MI-MII equilibrium, but instead have an active, MII conformation with a protonated Schiff base. This slow-bleaching phenomenon highlights the danger of ascribing biological activity to photo-intermediates simply on the basis of spectral characteristics.

Recently, Hubbell and Khorana and coworkers have combined Cys-scanning mutagenesis of rhodopsin with electron spin resonance (EPR) spectroscopy of spin-labeled mutants in elegant experiments to document movements in the protein upon photoactivation. In early studies, only a few Cys residues were examined, but these covered all of the cytoplasmic domains of the protein and the predominant movement was localized to the second cytoplasmic loop connecting transmembrane segments 3 and 4 (36, 37). Subsequently, each residue in this loop was replaced systematically with Cys (82). EPR spectra of the spin-labeled mutants showed movement at both ends of the loop in the region bordering the cytoplasmic surface of transmembrane segments 3 and 4 (82). The observation of movement at the top of helix 3 upon photoactivation of rhodopsin is consistent with and complementary to the conclusion from mutagenesis studies with constitutively active mutants that a salt bridge is broken between Glu113 and Lys296 upon activation of the protein.

The second cytoplasmic loop is not the only region of the protein involved in the activation of transducin. It is now well established from mutagenesis studies that the third loop is also critical for the binding and activation of transducin by rhodopsin (38, 39, 47, 55, 99). In addition, the α-subunit of transducin has been shown to be cross-linked to position 240 in the third cytoplasmic loop through a photo-activatable reagent (83). Transducin could be cross-linked only to MII and only to position 240. This finding clearly demonstrates that the transducin α subunit binds to and contacts the third cytoplasmic loop in the photo-activated form of rhodopsin. The third cytoplasmic loop is critical for the activation of other G protein-coupled receptors as has been recently reviewed (103a), and is also a mutational hot-spot for consititutively activating mutations as will be discussed in the next section.

OTHER CONSTITUTIVELY ACTIVE G PROTEIN–COUPLED RECEPTORS

The first constitutively active mutants of G protein-coupled receptors were discovered by Lefkowitz and coworkers working with the adrenergic receptors. Substitution of seven amino acids from the C-terminal end of the third cytoplasmic loop of the β_2-adrenergic receptor into the corresponding region of the α_1-adrenergic receptor activated the α-receptor constitutively and increased affinity for agonists 100-fold (22). Further dissection of this region of the protein showed that mutations at Lys290 and Ala293 were chiefly responsible for these effects. In fact, single mutations of Ala293 to each of the other 19 amino acids all resulted in constitutive activation of the protein (54). Since the mutants differed in activity and no correlation of activity with chemical properties of the side chains was evident, the mutations presumably activate the protein by sterically disrupting packing interactions important for stabilization of the inactive state of the receptor. As expected, antagonists (or more properly, inverse agonists) inhibited the activity of these receptors and shifted the equilibrium back toward the inactive state (54). The reciprocal mutation replacing residues of the third cytoplasmic loop of the β_2-adrenergic receptor with amino acids from the α_{1B}-receptor activated the β-receptor and increased affinity of the receptor for agonists (77, 91). Comparable mutations in the α_2-adrenergic receptor behaved similarly (81). Importantly, mutation of the adrenergic receptors resulted not only in constitutive activation of the G protein, but also in constitutive phosphorylation of the receptor by β-adrenergic receptor kinase (77) adding further strength to the argument that the constitutively active state of the mutant G protein-coupled receptors mimics the agonist activated state of the wild-type receptors.

In addition to the adrenergic receptors, constitutively activating mutations have been found in the third cytoplasmic loop of the yeast **a**-factor receptor (16). The mutations in this G protein-coupled pheromone receptor increase both the level of basal activity and agonist sensitivity. In most other receptors, activating mutations express a dominant phenotype. Curiously, those of the yeast **a**-factor are recessive. The reason for this anomalous behavior is unknown.

The transforming ability of G protein-coupled receptors was first realized with the report that continuous application of serotonin to cultured fibroblasts ectopically expressing the 5-hydroxytryptamine$_{1C}$ receptor induced the formation of transformed foci in these cells (50). Later, expression of constitutively active α_{1B}-adrenergic receptors in Rat-1 and NIH 3T3 cells was shown to result in agonist-independent formation of transformed foci, and cells derived from these foci induced tumors when injected into nude mice (3). These observations suggested that activating mutations in G protein-coupled receptors might be responsible for tumor formation among humans, and within a short time such mutations were identified in the thyrotropin receptor isolated from patients with hyperfunctioning thyroid adenomas (75).

Somatic mutations of the thyrotropin receptor were identified at five independent positions in the C-terminal domain of the third cytoplasmic loop and in the cytoplasmic border of the sixth helix (56, 75, 76, 78, 88). Heterologous expression of these receptors in cultured cells caused thyroid-stimulating hormone–independent activation of adenylate cyclase (75, 76, 88). The individual mutations were Asp619 to Gly; Ala623 to Ile, Val, or Ser; Thr632 to Ile; Phe631 to Cys; and Asp633 to Glu or Tyr (56, 75, 76, 78, 88). Remarkably, the mutations at positions 619 and 623 were in exactly the same positions as those found to activate the adrenergic receptors (59).

Two germline mutations of the thyrotropin receptor, Val509Ala and Cys672Tyr, constitutively activate adenylate cyclase and cause nonautoimmune autosomal dominant hyperthyroidism (29). Interestingly, position 509 is in the third transmembrane helix of the thyrotropin receptor and corresponds to a position in rhodopsin that is three α-helical turns from the Glu113 Schiff base counterion. Position 672 is in the seventh transmembrane helix of the thyrotropin receptor and corresponds to a position in rhodopsin that is one helical turn from Lys296. Perhaps a movement of helix 3 relative to 7 is also important for activation of the thyrotropin receptor. It is striking that no mutation investigated so far of the thyrotropin receptor, which activates adenylate cyclase constitutively, can also activate inositol phosphate production constitutively, although activation of inositol phosphate production is observed normally when the receptor is stimulated with agonist (56).

In addition to the constitutively activating mutations of the thyrotropin receptor associated with disease described above, truncation of the C-terminal domain of the TRH receptor also activates that protein constitutively. When assayed in oocytes or tissue culture cells a receptor missing its last 59 amino acids showed a basal activity that was elevated by two-fold over that of the wild-type receptor (65). Elevated basal activity has also been observed in membranes isolated from cultured cells transfected with a similar mutation of the β-adrenergic receptor (74a).

Constitutively active G protein-coupled receptors are also found in other human diseases. For example, several mutations in the fifth and sixth helices, as well as the third cytoplasmic loop of the luteinizing hormone (LH) receptor, have been shown to cause familial male precocious puberty (58, 98). The LH-receptor controls levels of testosterone through regulation of G_s and its effector, adenylate cyclase. Familial male precocious puberty is characterized by elevated levels of testosterone produced independently of LH stimulation and the appearance of secondary sex characteristics by 3–4 years of age. The most commonly occurring activating mutation is Asp578Gly in the middle of the sixth transmembrane segment of the protein (98). Less common mutations include Asp564Gly in the third cytoplasmic loop, Cys581Arg in the sixth helix, Ile542Leu in the fifth helix, and a second mutation at position 578 where the Asp is changed to a Tyr (58). Similar to Asp578, the amino acids Ile542, Asp564, and Cys581 are conserved in the glycoprotein hormone receptors, which implies that these residues are important in the structure and function of the protein. It is of interest that a constitutively active mutant of the α subunit of G_s has also been shown to lead to precocious puberty in males (62).

Another disease causing mutation is found in a type of short-limbed dwarfism known as Jansen-type metaphyseal chondrodysplasia where mutation of His223 to Arg in the C-terminal domain of the first cytoplasmic loop of the parathyroid hormone-parathyroid hormone-related peptide (PTH-PTHrP) receptor was found to activate adenylate cyclase constitutively and inhibit production of inositol phosphate (93). The His223Arg mutation in the PTP-PTHrP receptor noted here and the Gly90Asp mutation in rhodopsin discussed previously appear to highlight another mutational hot-spot in the G protein-coupled receptors centered around the second transmembrane segment and the first cytoplasmic loop.

Constitutively activating mutations in the second transmembrane segment of the melanocyte-stimulating hormone (MSH) receptor affect coat color in mice (85). The tobacco darkening (E^{tob}) and sombre (E^{so} and $E^{so\text{-}3J}$) alleles encode constitutively active mutations of the receptor, which causes dark coats in the absence of MSH stimulation (85). The E^{tob} gene contained the mutation Ser69Leu, which is located in the C-terminal end of the first cytoplasmic loop and produces an MSH receptor with a higher basal activity, a twofold higher affinity for MSH, and a greater ability to stimulate adenylate cyclase on exposure to MSH than the wild-type receptor (85). The E^{so} and $E^{so\text{-}3J}$ alleles contain the mutations Leu98Pro and Glu92Lys, respectively, which are located in the second transmembrane helix close to the extracellular side of the membrane. Although the E^{so} mutant was not assayed in vitro for constitutive activity, the $E^{so\text{-}3J}$ (Glu92Pro) protein was constitutively active and not responsive to hormone (85).

SUMMARY

Activating mutations have been identified in many different G protein–coupled receptors. The mutations are of interest not only because they give us insight into the molecular mechanism of activation and inactivation of these proteins, but also because they are found to cause a variety of human diseases, including retinitis pigmentosa, night blindness, hyperfunctioning thyroid adenomas, precocious puberty, and a type of short-limbed dwarfism. Studies of these mutations in rhodopsin have illuminated a salt-bridge mechanism for activation and inactivation of the protein. In turn, an understanding of this mechanism has suggested a possible therapeutic approach for rescue of retinitis pigmentosa that arises from one of the mutations through the design of specific antagonists or inverse agonists. Continued study of other receptors undoubtedly will elucidate similar insights and the development of potential therapies.

Literature Cited

1. Aho AC, Donner K, Hyden C, Larsen LO, Reuter T. 1988. Low retinal noise in animals with low body temperature allows high visual sensitivity. *Nature* 334:348–50
2. Deleted in proof.
3. Allen LF, Lefkowitz RJ, Caron MG, Cotecchia S. 1991. G-protein-coupled receptor genes as protooncogenes: constitutively activating mutation of the α- adrenergic receptor enhances mitogenesis and tumorigenicity. *Proc. Natl. Acad. Sci. USA* 88: 11354–58
4. Anderson DE, Becktel WJ, Dahlquist FW. 1990. pH-Induced denaturation of protein: a single salt bridge contributes 3–5 kcal/mol to the free energy of folding of T4 lysozyme. *Biochemistry* 29:2403–8
5. Arnis S, Fahmy K, Hoffman KP, Sakmar TP. 1994. A conserved carboxylic acid group mediates light-dependent proton uptake and signalling by rhodopsin. *J. Biol. Chem.* 269: 23879–81
6. Bagley KA, Balogh-Nair V, Croteau AA, Dollinger G, Ebrey TG, et al. 1985. Fourier-transform infrared difference spectroscopy of rhodopsin and its photoproducts at low temperature. *Biochemistry* 24:6055–71
7. Baldwin JM. 1993. The probable arrangement of the helices in G protein-coupled receptors. *EMBO J.* 12: 1693–703
8. Barlow HB. 1956. Retinal noise and absolute threshold. *J. Opt. Soc. Am.* 46:634–39
9. Barlow HB. 1988. The thermal limit to seeing. *Nature* 334: 296–97
10. Barlow RB, Birge RR, Kaplan E, Tallent JR. 1993. On the molecular origin of photoreceptor noise. *Nature* 366:64–66
11. Baylor DA, Lamb RD, Yau K-W. 1979. Responses of retinal rods to single photons. *J. Physiol.* 288: 613–34
12. Baylor DA, Matthews G, Yau K-W. 1980. Two components of electrical dark noise in toad retinal rod outer segments. *J. Physiol.* 309:591–621
13. Baylor DA, Nunn BJ, Schnapf JL. 1984. The photocurrent noise and spectral sensitivity of rods of the monkey *Macaca fascicularis*. *J. Physiol.* 357:575–607
14. Berson EL. 1993. Retinitis pigmentosa: the Friedenwald lecture. *Invest. Ophthalmol. Vis. Sci.* 34: 1659–76
15. Birge RR, Murray LP, Pierce BM, Akita H, Balogh-Nair V, et al. 1985. Two-photon spectroscopy of locked-11-*cis*-rhodopsin: evidence for a protonated Schiff base in a neutral protein binding site. *Proc. Natl. Acad. Sci. USA* 82:4117–21
16. Boone C, Davis NG, Sprague GF Jr. 1993. Mutations that alter the third cytoplasmic loop of the **a**-factor receptor lead to a constitutive and hypersensitive phenotype. *Proc. Natl. Acad. Sci. USA* 90:9921–25
17. Bownds D. 1967. Site of attachment of retinal in rhodopsin. *Nature* 216: 1178–81
18. Deleted in proof.
19. Cohen GB, Oprian DD, Robinson PR. 1992. Mechanism of activation and inactivation of opsin: role of Glu113 and Lys296. *Biochemistry* 31:12592–601
20. Cohen GB, Yang T, Robinson PR, Oprian DD. 1993. Constitutive activation of opsin: influence of charge at position 134 and size at position 296. *Biochemistry* 32:6111–15
21. Colley NJ, Cassil JA, Baker EK, Zuker CS. 1995. Defective intracellular transport is the molecular basis of rhodopsin-dependent retinal degeneration. *Proc. Natl. Acad. Sci. USA* 92:3070–74

21a. Corson DW, Cornwall MC, MacNichol EF, Jin J, Johnson R, et. al. 1990. Sensitization of bleached rod photoreceptors by 11 *cis*-locked analogues of retinal. *Proc. Natl. Acad. Sci. USA* 87:6823–27

22. Cotecchia S, Exum S, Caron MG, Lefkowitz RJ. 1990. Regions of the α-adrenergic receptor involved in coupling to phosphatidylinositol hydrolysis and enhanced sensitivity of biological function. *Proc. Natl. Acad. Sci. USA* 87:2896–900
23. Defoe DM, Bok D. 1983. Rhodopsin chromophore exchanges among opsin molecules in the dark. *Invest. Ophthalmol. Vis. Sci.* 24:1211–26
24. Dohlman HG, Thorner J, Caron MG, Lefkowitz RJ. 1991. Model systems for the study of seven-transmembrane-segment receptors. *Annu. Rev. Biochem.* 60:653–88
25. Donnelly D, Findlay JBC. 1994. Seven-helix receptors: structure and modelling. *Curr. Opin. Struct. Biol.* 4:582–89
26. Doukas AG, Aton B, Callender RH, Ebrey TG. 1978. Resonance raman

studies of bovine metarhodopsin I and metarhodopsin II. *Biochemistry* 17:2430–35
27. Dratz EA, Hargrave PA. 1983. The structure of rhodopsin and the rod outer segment disk membrane. *Trends Biochem. Sci.* 8:128–31
28. Dryja TP, Berson EL, Rao VR, Oprian DD. 1993. Heterozygous missense mutation in the rhodopsin gene as a cause of congenital stationary night blindness. *Nature Genet.* 4: 280–83
29. Duprez L, Parma J, Van Sande J, Allgeier A, Leclere J, et al. 1994. Germline mutations in the thyrotropin receptor gene cause non-autoimmune autosomal dominant hyperthyroidism. *Nature Genet.* 7:396–401
30. Deleted in proof.
31. Emeis D, Kuhn H, Reichert J, Hoffman KP. 1982. Complex formation between metarhodopsin II and GTP-binding protein in bovine photoreceptor membranes leads to a shift of the photoproduct equilibrium. *FEBS Lett.* 143:29–34
32. Erikkson AE, Baase WA, Zheng X-J, Heinz DW, Blaker M, et al. 1992. Response of protein structure to cavity-creating mutations and its relation to the hydrophobic effect. *Science* 255:178–83
33. Fahmy K, Jager F, Beck M, Zvyaga TA, Sakmar TP, Siebert F. 1993. Protonation states of membrane-embedded carboxylic acid groups in rhodopsin and metarhodopsin II: a Fourier-transform infrared spectroscopy study of site-directed mutants. *Proc. Natl. Acad. Sci. USA* 90: 10206–10
34. Fahmy K, Sakmar TP. 1993. Regulation of the rhodopsin-transducin interaction by a highly conserved carboxylic acid group. *Biochemistry* 32: 7229–36
35. Fahmy K, Siebert F, Sakmar TP. 1994. A mutant rhodopsin photoproduct with a protonated Schiff base displays an active-state conformation: a Fourier-transform infrared spectroscopy study. *Biochemistry* 33: 13700–5
36. Farahbakhsh ZT, Hideg K, Hubbell WL. 1993. Photoactivated conformational changes in rhodopsin: a time-resolved spin label study. *Science* 262:1416–19
37. Farahbakhsh ZT, Ridge KD, Khorana HG, Hubbell WL. 1995. Mapping light-dependent structural changes in the cytoplasmic loop connecting helices C and D in rhodopsin: a site-directed spin labelling study. *Biochemistry* 34:8812–19
38. Franke RR, Konig B, Sakmar TP, Khorana HG, Hofmann KP. 1990. Rhodopsin mutants that bind but fail to activate transducin. *Science* 250: 123–25
39. Franke RR, Sakmar TP, Graham RM, Khorana HG. 1992. Structure and function in rhodopsin: studies of the interaction between the rhodopsin cytoplasmic domain and transducin. *J. Biol. Chem.* 267:14767–74
40. Gal A, Orth U, Baehr W, Schwinger E, Rosenberg T. 1994. Heterozygous missense mutation in the rod cGMP phosphodiesterase β-subunit gene in autosomal dominant stationary night blindness. *Nature Genet.* 7:64–68
41. Ganter UM, Longstaff C, Pajares MA, Rando RR, Siebert F. 1991. Fourier transform infrared studies of active-site-methylated rhodopsin: implications for chromophore-protein interaction, transducin activation, and the reaction pathway. *Biophys. J.* 59:640–44
42. Gat Y, Sheves M. 1993.A mechanism for controlling the pKa of the retinal protonated Schiff base in retinal proteins. A study with model compounds. *J. Am. Chem. Soc.* 115: 3772–73
43. Deleted in proof.
44. Gilman AG. 1987. G-proteins: transducers of receptor generated signals. *Annu. Rev. Biochem.* 56:615–49
45. Govardhan CP, Oprian DD. 1994. Active site-directed inactivation of constitutively active mutants of rhodopsin. *J. Biol. Chem.* 269:6524–27
46. Han M, Smith SO. 1995. NMR constraints on the location of the retinal chromophore in rhodopsin and bathorhodopsin. *Biochemistry* 34: 1425–32
47. Hargrave PA, Hamm HE, Hofmann KP. 1993. Interaction of rhodopsin with the G-protein, transducin. *BioEssays* 15:43–50
48. Inglese J, Freedman NJ, Koch WJ, Lefkowitz RJ. 1993. Strucure and mechanism of the G protein coupled receptor kinases. *J. Biol. Chem.* 268: 23735–38
49. Jager F, Fahmy K, Sakmar TP, Siebert F. 1994. Identification of glutamic acid 113 as the schiff base proton acceptor in the metarhodopsin II photointermediate of rhodopsin. *Biochemistry* 33:10878–82
49a. Jin J, Crouch RK, Corson DW, Katz

BM, MacNichol EF, Cornwall MC. 1993. Non-covalent occupancy of the retinal-binding pocket of opsin diminishes bleaching adaptation of retinal cones. *Neuron* 11:513–22
50. Julius D, Livelli TJ, Jessell TM, Axel R. 1989. Ectopic expression of the serotonin 1c receptor and the triggering of malignant transformation. *Science* 244:1057–62
51. Kaushal S, Khorana HG. 1994. Structure and function in rhodopsin. 7. Point mutations associated with autosomal dominant retinitis pigmentosa. *Biochemistry* 33:6121–28
52. Keen TJ, Inglehearn CF, Lester DH, Bashir R, Jay M, et al. 1991. Autosomal dominant retinitis pigmentosa: four new mutations in rhodopsin, one of them in the retinal binding site. *Genomics* 11:199–205
53. Kibelbek J, Mitchell DC, Beach JM, Litmann BJ. 1991. Functional equivalence of metarhodopsin II and the G activating form of photolyzed bovine rhodopsin. *Biochemistry* 30:6761–68
54. Kjelsberg MA, Cotecchia S, Ostrowski J, Caron MG, Lefkowitz RJ. 1992. Constitutive activation of the α-adrenergic receptor by all amino acid substitutions at a single site. *J. Biol. Chem.* 267:1430–33
55. Konig B, Arendt A, McDowell JH, Kahlert M, Hargrave PA, Hofmann KP. 1989. Three cytoplasmic loops of rhodopsin interact with transducin. *Proc. Natl. Acad. Sci. USA* 86: 6878–82
56. Kosugi S, Shenker A, Mori T. 1994. Constitutive activation of cyclic AMP but not phosphatidylinositol signalling caused by four mutations in the sixth transmembrane helix of the human thyrotropin receptor. *FEBS Lett.* 356:291–94
57. Koutalos Y, Ebrey TG, Tsuda M, Odashima K, Lien T, et al. 1989. Regeneration of bovine and octopus opsins in situ with retinal and artificial retinals. *Biochemistry* 28:2732–39
58. Laue L, Chan W-Y, Hseuh AJ, Kudo M, Hsu SY, et al. 1995. Genetic heterogeneity of constitutively activating mutations of the human luteinizing hormone receptor in familial male-limited precocious puberty. *Proc. Natl. Acad Sci. USA* 92: 1906–10
59. Lefkowitz RJ. 1993. Turned on to ill effect. *Nature* 365:603–4
60. Li T, Franson WK, Gordon JW, Berson EL, Dryja TP. 1995. Constitutive activation of phototransduction by K296E opsin is not the cause of photoreceptor degeneration. *Proc. Natl. Acad. Sci. USA* 92:3551–55
61. Lin SW, Sakmar TP, Franke RR, Khorana HG, Matthies RA. 1992. Resonance Raman microprobe spectroscopy of rhodopsin mutants: effect of substitutions in the third transmembrane helix. *Biochemistry* 31: 5105–11
62. Liri T, Herzmark P, Nakamoto J, van Dop C, Bourne HR. 1994. Rapid GDP release from G in patients with gain and loss of endocrine function. *Nature* 371:164–68
63. Longstaff C, Calhoon RD, Rando RR. 1986. Deprotonation of the Schiff base of rhodopsin is obligate in the activation of the G protein. *Proc. Natl. Acad. Sci. USA* 83: 4209–13
64. Matthews RG, Hubbard R, Brown PK, Wald G. 1963. Tautomeric forms of metarhodopsin. *J. Gen. Physiol.* 47:215–40
65. Matus-Leibovitch N, Nussenzveig DR, Gershengorn MC, Oron Y. 1995. Truncation of the thyrotropin-releasing hormone receptor carboxy tail causes constitutive activity and leads to impaired responsiveness in *Xenopus* oocytes and AtT20 cells. *J. Biol. Chem.* 270:1041–47
66. Min KC, Zvyaga TA, Cypess AM, Sakmar TP. 1993. Characterization of mutant rhodopsin responsible for autosomal dominant retinitis pigmentosa: mutations on the cytoplasmic surface affect transducin activation. *J. Biol. Chem.* 268:9400–4
67. Nakayama TA, Khorana HG. 1991. Mapping of the amino acids in membrane-embedded helices that interact with the retinal chromophore in bovine rhodopsin. *J. Biol Chem.* 266: 4269–75
68. Nathans J. 1990. Determinants of visual pigment absorbance: identification of the retinylidenes Schiff's base counterion in bovine rhodopsin. *Biochemistry* 29:9746–52
69. Ohguro H, Palczewski K, Ericsson LH, Walsh KA, Johnson RS. 1993. Sequential phosphorylation of rhodopsin at multiple sites. *Biochemistry* 32:5718–24
70. Deleted in proof.
71. Oprian DD. 1992. The ligand-binding domain of rhodopsin and other G protein-linked receptors. *J. Bioenerg. Biomembr.* 24:211–17
72. Palczewski K, Benovic JL. 1991. G-

protein coupled receptor kinases. *Trends Biochem. Sci.* 16:387–91
73. Palczewski K, Buczylko J, Kaplan MW, Polans AS, Crabb JW. 1991. Mechanism of rhodopsin kinase activation. *J. Biol. Chem.* 266:12949–55
73a. Palczewski K, Rispoli G, Detwiler PB. 1992. The influence of arrestin (48K protein) and rhodopsin kinase on visual transduction. *Neuron* 8: 117–26
74. Palings I, Pardoen A, Van den Berg E, Winkel C, Lugtenberg J, Mathies RA. 1987. Assignment of fingerprint vibrations in the resonance raman spectra of rhodopsin, isorhodopsin, and bathorhodopsin: implications for the chromophore structure and environment. *Biochemistry* 24:2544–56
74a. Parker EM, Ross EM. 1991. Truncation of the extended carboxyl-terminal domain increases the expression and regulatory activity of the avian β-adrenergic receptor. J. Biol. Chem. 266: 9987–96
75. Parma J, Duprez L, Van Sande J, Cochaux P, Gervy C, et al. 1993. Somatic mutations in the thyrotropin receptor gene cause hyperfunctioning thyroid adenomas. *Nature* 365: 649–51
76. Paschke R, Tonacchera M, Van Sande J, Parma J, Vassart G. 1994. Identification and functional characterization of two new somatic mutations causing constitutive activation of the thyrotropin receptor in hyperfunctioning autonomous adenomas of the thyroid. *J. Clin. Endocrinol. Metab.* 79:1785–89
77. Pei G, Samama, Lohse M, Wang M, Codina J, Lefkowitz RJ. 1994. A constitutively active mutant β-adrenergic receptor is constitutively desensitized and phosphorylated. *Proc. Natl. Acad. Sci. USA* 91:2699–702
78. Porcellini A, Ciullo I, Laviola L, Amabile G, Fenzi G, Avvedimento VE. 1994. Novel mutations of thyrotropin receptor gene in thyroid hyperfunctioning adenomas. Rapid identification by fine needle aspiration biopsy. *J. Clin. Endocrinol. Metab.* 79:657–61
79. Portera-Cailliau C, Sung C-H, Nathans J, Adler R. 1994. Apoptotic photoreceptor cell death in mouse models of retinitis pigmentosa. *Proc. Natl. Acad. Sci. USA* 91:974–78
80. Rao VR, Cohen GB, Oprian DD. 1994. Rhodopsin mutation G90D and a molecular mechanism for congenital night blindness. *Nature* 367: 639–42
81. Ren Q, Kurose H, Lefkowitz RJ, Cotecchia S. 1993. Constitutively active mutants of the α-adrenergic receptor. *J. Biol. Chem.* 268:16483–87
82. Resek JF, Farahbakhsh ZT, Hubbell WL, Khorana HG. 1993. Formation or the meta II photointermediate is accompanied by conformational changes in the cytoplasmic surface of rhodopsin. *Biochemistry* 32:12025–32
83. Resek JF, Farrens D, Khorana HG. 1994. Structure and function in rhodopsin: Covalent crosslinking of the rhodopsin (metarhodopsin II)-transducin complex-the rhodopsin cytoplasmic face links to the transducin α subunit. *Proc. Natl. Acad. Sci. USA* 91:7643–47
84. Rim J, Oprian DD. 1995. Constitutive activation of opsin: interaction of mutants with rhodopsin kinase and arrestin. *Biochemistry* 34:11938–45
85. Robbins LS, Nadeau JH, Johnson KR, Kelly MA, Roselli-Rehfuss L, et al. 1993. Pigmentation phenotypes of variant extension locus alleles result from point mutations that alter MSH receptor function. *Cell* 72:827–34
86. Robinson PR, Buczylko J, Ohguro H, Palczewski K. 1994. Opsins with mutations at the site of chromophore attachment constitutively activate transducin but are not phosphorylated by rhodopsin kinase. *Proc. Natl. Acad. Sci. USA* 91:5411–15
87. Robinson PR, Cohen GB, Zhukovsky EA, Oprian DD. 1992. Constitutively active mutants of rhodopsin. *Neuron* 9:719–25
88. Russo D, Arturi F, Wicker R, Chazenbalk GD, Schlumberger M, et al. 1995. Genetic alterations in thyroid hyperfunctioning adenomas. *J. Clin. Endocrinol. Metab.* 80:1347–51
89. Sakmar TP, Franke RR, Khorana HG. 1989. Glutamic acid-113 serves as the retinylidene Schiff base counterion in bovine rhodopsin. *Proc. Natl. Acad. Sci. USA* 86:8309–13
90. Sakmar TP, Franke RR, Khorana HG. 1991. The role of the retinylidene Schiff base counterion in determining wavelength absorbance and Schiff base pK . *Proc. Natl. Acad. Sci. USA* 88:3079–83
91. Samama P, Cotecchia S, Costa T, Lefkowitz RJ. 1993. A mutation-induced activated state of the β-adrenergic receptor. *J. Biol. Chem.* 268: 4625–36

92. Schertler GFX, Cilla C, Henderson R. 1993. Projection structure of rhodopsin. *Nature* 362:770–72
93. Schipani E, Kurse K, Juppner H. 1995. A constitutively active mutant PTH-PTHrP receptor in Jansen-type metaphyseal chondrodysplasia. *Science* 268:98–100
94. Schleicher A, Kuhn H, Hofmann KP. 1989. Kinetics, binding constant, and activation energy of the 48 kDa protein-rhodopsin complex by extra-metarhodopsin II. *Biochemistry* 28: 1770–75
95. Schoenlein RW, Peteanu LA, Mathies Ra, Shank CV. 1991. The first step in vision: femtosecond isomerization of rhodopsin. *Science* 254: 412–15
96. Sealfon SC, Chi L, Ebersole BJ, Radic V, Zhang D, et al. 1995. Related contribution of specific helix 2 and 7 residues to conformational activation of the serotonin 5-HT receptor. *J. Biol. Chem.* 270:16683–88
97. Seckler B, Rando RR. 1989. Schiff-base deprotonation is mandatory for light-dependent rhodopsin phosphorylation. *Biochem. J.* 264:489–93
98. Shenker A, Laue L, Kosugi S, Merendino J Jr, Minegishi T, Cutler GB Jr. 1993. A constitutively activating mutation of the luteinizing hormone receptor in familial male precocious puberty. *Nature* 365:652–54
99. Shi W, Osawa S, Dickerson CD, Weiss ER. 1995. Rhodopsin mutants discriminate sites important for the activation of rhodospin kinase and transducin. *J. Biol. Chem.* 270: 2112–19
100. Sieving PA, Richards JE, Naarendorp F, Bingham EL, Scott K, Alpern M. 1995. Dark-light: model for nightblindness from the human rhodopsin Gly-90→Asp mutation. *Proc Natl Acad Sci USA*. 92:880–84
101. Smith SO, Palings I, Miley M, Courtin J, de Groot H, et al. 1990. Solid-state NMR studies of the mechanism of the opsin shift in the visual pigment rhodopsin. *Biochemistry* 29: 8158–64
102. Steinberg G, Ottolenghi M, Sheves M. 1993. pK of the protonated Schiff base of bovine rhodopsin: a study with artificial pigments. *Biophys. J.* 64:1499–502
103. Deleted in proof.
103a. Strader CD, Fong TM, Tota MR, Underwood D. 1994. Structure and function of G protein-coupled receptors. *Annu. Rev. Biochem.* 63:101–32
104. Strader CD, Sigal IS, Register RB, Candelore MR, Ronds E, Dixon RAF. 1987. Identification of residues required for ligand binding to the β-adrenergic receptor. *Proc. Natl. Acad. Sci. USA.* 84:4384–88
105. Sullivan JM, Scott KM, Falls HF, Richards JH, Sieving PA. 1993. A novel rhodopsin mutation at the retinal binding site (Lys296-Met) in ADRP. *Invest. Ophthalmol. Vis. Sci.* 34:1149
106. Sung C-H, Davenport CM, Nathans J, 1993. Rhodopsin mutations responsible for autosomal dominant retinitis pigmentosa. *J. Biol. Chem.* 268:26645–49
107. Sung C-H, Makino C, Baylor D, Nathans J. 1994. A rhodopsin gene mutation responsible for autosomal dominant retinitis pigmentosa results in a protein that is defective in localization to the photoreceptor outer segment. *J. Neurosci.* 14:5818–33
108. Sung C-H, Schneider BG, Agarwal N, Papermaster DS, Nathans J. 1991. Functional heterogeneity of mutant rhodopsins responsible for retinitis pigmentosa. *Proc. Natl. Acad. Sci. USA* 88:8840–44
109. Surya A, Foster KW, Knox BE. 1995. Transducin activation by the bovine opsin apoprotein. *J. Biol. Chem.* 270:5024–31
110. Wald G. 1968. Molecular basis of visual excitation. *Science* 162: 230–39
111. Deleted in proof.
112. Weitz CJ, Nathans J. 1993. Rhodopsin activation: effects of the metarhodopsin I-metarhodopsin II equilibrium of neutralization or introduction of charged amino acids within putative transmembrane segments. *Biochemistry* 32:14176–82
113. Wilden U, Hall SW, Kuhn H. 1986. Phosphodiesterase activation by photoexcited rhodopsin is quenched when rhodopsin is phosphorylated and binds the intrinsic 48-kDa protein of rod outer segments. *Proc. Natl. Acad. Sci USA* 83:1174–78
114. Deleted in proof.
115. Zhou W, Flanagan C, Ballesteros JA, Konvicka K, Davidson JS, et al. 1994. A reciprocal mutation supports helix 2 and helix 7 proximity in the gonadotropin-releasing hormone receptor. *Mol. Pharmacol.* 45:165–70
116. Zhukovsky EA, Oprian DD. 1989. Effect of carboxylic acid side chains on the absorption maximum of visual pigments. *Science* 246:928–30

117. Zhukovsky EA, Robinson PR, Oprian DD. 1991. Transducin activation by rhodopsin without a covalent bond to the 11-*cis*-retinal chromophore. *Science* 251:558–60
118. Zhukovsky EA, Robinson PR, Oprian DD. 1992. Changing the location of the Schiff base counterion in rhodopsin. *Biochemistry* 31:10400–5
119. Zvyaga TA, Fahmy K, Sakmar TP. 1994. Characterization of rhodopsin-transducin interaction: a mutant rhodopsin photoproduct with a protonated Schiff base activates transducin. *Biochemistry* 33:9753–61
120. Zvyaga TA, Min KC, Beck M, Sakmar TP. 1993. Movement of the retinylidene Schiff base counterion in rhodopsin by one helix turn reverses the pH dependence of the metarhodopsin I to metarhodopsin II transition. *J. Biol. Chem.* 268:4661–67 (Corr: *J. Biol. Chem.* 1994. 269: 13056)

Annu. Rev. Biophys. Biomol. Struct. 1996. 25:315–42

COMPUTATIONAL STUDIES OF PROTEIN FOLDING

Richard A. Friesner

Department of Chemistry, Columbia University, New York, NY 10027

John R. Gunn

Department of Chemistry, University of Montreal, C. P. 6128, Succursale A, Montreal, Quebec, Canada H3C 3J7

KEY WORDS: protein structure, computational chemistry, molecular modeling, myoglobin

ABSTRACT

This review describes computational approaches to the determination of protein structure from sequence. The emphasis is on reduced protein models that are sufficiently accurate to represent protein structure at low resolution, yet are computationally efficient enough to allow the extensive search of phase space required to locate the global minimum from an unfolded state. A discussion of both potential functions and algorithmic simulation strategies for such models are presented, along with a number of specific models that have been developed and successfully applied to proteins as large as myoglobin. The results indicate that significant progress is being made in understanding the requirements for computational prediction of protein structure.

CONTENTS

1056–8700/96/0610–0315$08.00

INTRODUCTION

The determination of protein structure from amino acid sequence via computational methodology is a major goal of theoretical molecular biology. The determination of structure by experimental means, such as NMR spectroscopy and X-ray crystallography, cannot be expected to keep pace with the explosion of protein sequence information provided by DNA sequencing techniques. In any case, such techniques are costly in terms of both equipment and human effort. With the cost/performance of computational hardware being reduced by a factor of roughly 2 every 18 months, and with improvements in software and algorithm development providing an even greater enhancement of computational efficiency, the goal of computation-based structure determination will be reached eventually.

However, there are still formidable problems, both conceptual and numerical, in making this objective a reality. These problems involve both the issue of the appropriate representation of the protein (most critically, the potential functions that assign energies to any given protein structure) and the algorithmic issue of finding the native structure, starting from an unfolded chain, in a finite amount of computation time. These two central areas are, of course, intertwined: If computational expense were irrelevant, one could simply solve the Schrodinger equation for each protein configuration (averaging over all positions of the solvent molecules), thereby guaranteeing accurate evaluation of the potential energy. The coupled problem is then to have the total computation time required to obtain the predicted structure, which is roughly the cost of evaluating the energy of each structure multiplied by the number of such evaluations, be tractable with current computing technology. The development of new potential functions and the investigation of improved methods for the minimization of functions (to reduce the number of evaluations of the potential function that are necessary) are, therefore, two key areas in computational studies of protein folding.

During the past 5 years, a wide range of approaches to the computational protein folding problem has emerged from many different research groups. At one extreme, simple cubic lattice models have been used to investigate fundamental features of heteropolymer folding that are more or less independent of the details of protein structure (15, 19, 20, 32, 61, 62, 65). At the other extreme, a small number of atomic-level molecular dynamics simulations of protein unfolding, including explicit water molecules, have been carried out (5, 17, 70). In between, several different "intermediate" or "reduced" protein models have been developed and parametrized. The variety of computational techniques that have been applied to these models is equally diverse, ranging from standard versions of Monte Carlo molecular dynamics to novel approaches like genetic algorithms (71) and the diffusion equation method of Scheraga and co-workers (44, 64).

Useful theoretical insights into the folding process, many of which have been discussed in recent reviews (10, 19), have emerged from all these approaches, as well as from analytically based theory. In this paper, however, we focus primarily on the practical problem of the determination of structure from sequence by computational means. Further, we emphasize the thermodynamics of the problem as opposed to kinetics, i.e. the detailed pathway taken by the protein to fold in nature. Although some insights into kinetics arise naturally in the course of generating ensembles of structures during the minimization process, care must be exercised in drawing conclusions about real time events from such simulations. Finally, we restrict our discussion to the problem of prediction of novel folds that are not represented in the current protein database. Although homology modeling (63) and inverse folding (6) are extremely important, these areas assume the existence of an approximate three-dimensional template structure, and the methods relevant to this problem differ substantially from those needed to span a wide range of phase space to search for novel folds.

The paper is organized as follows. In the next section, we give a general overview of the folding problem from a physical and computational viewpoint. We then discuss the models and algorithms that we, and others, have used to generate folded structures. After this discussion, we present results from these calculations and attempt to assess the degree of progress that has been made by following various lines of attack. Finally, we identify what we believe to be the key issues in making future progress and how these issues are likely to be addressed successfully.

OVERVIEW OF THE PROTEIN FOLDING PROBLEM

Qualitative Physical Features of Protein Folding

The two most striking features of native protein structures in water are (*a*) the overwhelming tendency of charged or polar side chains to appear on the "exterior" of the protein (i.e. exposed to solvent) and hydrophobic side chains to be located in the "interior" of the protein (protected from solvent), thereby forming a unique compact globular structure for a given sequence; and (*b*) the formation of secondary structure segments, α-helices and β-sheets, so as to satisfy backbone hydrogen bonds. In a naturally occurring globular protein, the pattern of hydrophobic and hydrophilic residues is such that the formation of a micellar structure, with the hydrophilic groups exposed to water and the hydrophobic groups buried in the interior, is possible. A central feature of proteins, as compared with other polymers or colloidal structures, is that this micellar formation must be compatible with the two secondary structure classes. If the backbone groups inside the protein cannot make a sufficient number of internal hydrogen bonds, the protection of hydrophobic side chains from solvent will not compensate for the removal of the backbone carbonyl and amino moieties from aqueous solution.

Simple topological arguments would then suggest that a relatively large fraction of the protein (~40–70%) must be in either an α-helix or a β-sheet. Further, each individual secondary structure region cannot be too long, or the protein would lack the ability to fold into a compact globular structure. The "loop" regions joining the secondary structure elements cannot be too long, either, or the backbone residues in the interior of this region would be have difficulty forming hydrogen bonds efficiently, as suggested above. The picture that then emerges is one of rigid, rodlike segments (ignoring for the moment the twists and deformations of α-helices and β-strands) joined by flexible loops, with each segment of length 3–30 residues. This description characterizes the vast majority of proteins whose structures are available in the protein data bank.

Both theoretical and experimental results have shown that secondary structure elements in proteins are marginally stable; indeed, helical and β-sheet segments in proteins rarely retain their form when removed from the protein environment and studied as small peptides (2, 49). Protein folding, therefore, is a highly cooperative phenomenon in which the final secondary structure pattern and the tertiary architecture are determined by global optimization of the hydrophobic effect and backbone hydrogen bonding. The native structure is a result of a delicate balance of energetic terms; the overall thermodynamic stability of a native protein at room temperature is many orders of magnitude smaller than the total energy of the system. These features make the computational determination of protein structure extremely demanding with regard to the accuracy of the potential function and the magnitude of the conformational searching problem.

Computational Strategies for Prediction of Protein Structure

In this section, we assume that the native structure of a protein corresponds to the global free energy minimum of a physically accurate potential function (this assumption may not be valid universally for all proteins but is likely to be true for a large number). Further, as stated above, we are not concerned here with the kinetics of in vivo protein folding, e.g. the issue of the timing of the onset of secondary structure formation as compared with collapse into a compact tertiary structure. The problem is then reduced to construction of a potential function and algorithms for minimization of that potential function. The first critical issue, therefore, in devising a strategy for protein structure prediction is to evaluate the suitability of the different types of theoretical protein models for this task.

We begin by eliminating several types of models as unsuitable on the grounds of either accuracy or computational tractability. First, models that use explicit solvent representations are too expensive to allow determination of the global energy minimum, starting from an unfolded state, with any degree of efficiency. This problem arises not only because of the substantial CPU cost involved in evaluating the solvent–solvent and solvent–protein interactions but also because of the enormous amount of averaging over solvent configurations that is required to determine the free energy of any compact protein structure. The only real hope would be to run energy-conserving molecular dynamics and obtain formation of the native state on the same timescale as is observed for the in vivo system. At present, the discrepancy between the timescale for practicable molecular dynamics runs (nanoseconds) and the timescale for protein folding (milliseconds at best) is a minimum of 6 orders of magnitude; even parallel computers will not bridge this gap any time in the near future.

At the other end of the spectrum, we believe that oversimplified models, e.g. simple cubic lattice models with nearest neighbor interactions, are inadequate to obtain predictive results for protein structures at sufficient resolution to be of biological interest. The lack of both proper geometrical structure of the peptide bond and the absence of a reasonably accurate representation of secondary structure are serious flaws that are difficult to see how to overcome in the potential function. There are a number of results in the literature, in which cubic lattice models have yielded coarse representations of native-like geometries, that are ranked in the top several thousand structures out of those surveyed (15, 16, 32). It is unclear, however, how useful such results would be if the native structure were not known a priori, and virtually all such studies have been carried out on quite small proteins. The question of whether simplified models can provide qualitative insight into folding thermodynamics or kinetics via universal features that are transferable to more accurate representations, as has been suggested quite plausibly by several researchers (10, 19, 61, 65), is a more controversial subject, and we do not discuss it here. Even if this were the case, however, this in no way implies that detailed a priori structural predictions can be made with such models.

If these judgments are accepted, three important types of models remain. The first of these is a detailed, atomic-level model of the protein coupled with a continuum model for the solvation free energy. There is a long history of the development of molecular modeling potential functions for proteins based on the physical chemistry of intermolecular interactions, and a significant number of protein force fields [CHARMM (7), ECEPP (60), AMBER (72), OPLS (69)] are available. Considerable progress has been made in the development of continuum solvent models during the past decade, and several alternatives (21, 66), the most accurate of which are based on numerical solution of the Poisson-Boltzmann equation (24, 34), are currently available. Although issues regarding accuracy of both parts of the potential remain, and the computational expense of evaluating the energy with such a potential is far from trivial, these models have the advantages of being improvable systematically, as more accurate physical-chemical potential functions are developed, and of providing atomic-level detail for the structure and energetics without the necessity of explicitly averaging over ensembles of solvent molecules.

The second type is a reduced model that can represent the geometry of the peptide bond and the various secondary structure elements realistically but treats side chains and intermolecular forces in an approximate manner, thus reducing substantially the computational effort required to evaluate the potential function. In this case, potential functions must be constructed heuristically, utilizing statistical information from the protein database as well as physical-chemical energetics. Such models are probably limited to 4–6 Å resolution; however, one can hope to start from a structure of this type and then use a detailed atomic-level model to refine the resolution further. This approach has been pioneered by Skolnick and co-workers (38–43, 55, 56), who began with simple lattice models but, with the use of increasingly complex lattices, have emerged ultimately with rather fine-grained representation of the torsion angle phase space of the protein, along with an accurate geometric mapping from the lattice to three-dimensional space. The alternative approach to a fine-grained lattice representation is to use off-lattice models with an accurate treatment of the backbone structure. In practice, it is doubtful that there is much difference between these two types of models, from the point of view of either computational efficiency or achievable structural accuracy. The crucial issues for either lattice or off-lattice models of this type lie in the details of the potential functions.

The third type of modeling is unique to the protein folding problem: prediction of secondary structure from the protein sequence. These methods typically are based exclusively on statistical analysis of the database and employ a variety of mathematical procedures, ranging from simple secondary structure propensities (23) to sophisticated pattern recognition techniques like neural networks (59) or segment-based approaches (54), to predict the secondary structural state of each residue. At present, accuracy is on the order of 70% for the most successful methods. There are difficulties in going beyond this value, partly because of the lack of tertiary contact information inherent in such procedures. This is an active area of research, however, and new methods are being developed and tested continually. For example, Benner and co-workers (3, 4) recently used heuristic, as opposed to statistical, methods to achieve interesting results in blind test predictions.

These three types of models in turn suggest three distinct strategies for computational prediction of protein structure. First, one could simply minimize the energy of the atomic-level model with continuum solvation. In essence, Scheraga and co-workers have pursued this strategy for the past 30 years. Putting aside questions concerning the accuracy of the potentials, the minimization of an atomic-level model is formidably difficult: Not only is the cost of evaluation of the potential per structure high, but the use of atomic-level detail leads to a very "rough" energy landscape, where it is easy to become trapped in a high-energy local minimum. Recent work by Scheraga's group, however, has led to the development of a novel minimization approach, based on solution of a diffusion equation (44), which appears to have the potential to circumvent these problems. Although only small systems have as yet been minimized successfully, there is real hope that an approach along these lines will render direct location of the global minimum tractable at least for small proteins, albeit at substantial computational expense.

The second strategy is to carry out reduced-model simulations to determine secondary and tertiary structure simultaneously. Skolnick and co-workers have pursued this approach most vigorously. They recently achieved some impressive results for small proteins, which are described in more detail below (42,43). The key problems here are construction of a potential capable of reliably discriminating secondary structure formation and whether the correct structures will be formed on a reasonable simulation timescale. Subsequent mapping of the final structure onto an atomic-level model is relatively straightforward.

The third strategy is to carry out reduced-model simulations with fixed secondary structure, obtained from either secondary structure prediction algorithms or from experimental data, such as NMR data. Cohen et al (14) initiated this approach many years ago, and Cohen and co-workers (57) and our research groups (28, 51, 52) have pursued it during the past several years. The approach is based on the idea that once secondary structure is specified, the phase space for the folding problem is reduced qualitatively simplified, thus allowing a relatively rapid determination of the correct tertiary packing. Recent analytical work by Wolynes and co-workers (47) suggests that a dramatic reduction of phase space when secondary structure is specified will, in fact, allow tractability in the conformational searching problem to be achieved. As described below, considerable success has been achieved by this approach, even for relatively large, complex structures, such as myoglobin.

Despite the improved prospects for direct minimization of atomic-level models described above, it is our belief that the development of reliable potentials and simulation strategies for reduced models is essential for solving the general computational protein folding problem, particularly if one wants to treat the full range of protein sizes and topologies. A typical reduced-model evaluation of a protein potential function costs 100–1000 times less than a detailed atomic-level model, and this advantage will be decisive whatever minimization algorithm is used. The issue of how best to incorporate secondary structure into reduced models is far less clear-cut at present. It is likely that some combination of the second and third strategies described above ultimately will be optimal for this purpose.

REDUCED MODELS, POTENTIAL FUNCTIONS, AND ALGORITHMS FOR PREDICTION OF PROTEIN STRUCTURE

We now focus primarily on reduced-model and algorithm development and results. In addition, we discuss briefly issues that arise in building atomic-level models from reduced templates.

Models

The distinguishing feature of any reduced model of proteins is the goal of decreasing the number of degrees of freedom to render the problem more manageable from a computational point of view. In addition to reducing the dimensionality of the problem, these models can also incorporate a coarse graining of the representation of the molecule. This feature reduces the level of detail in the model, as well as the discretization of the space of possible conformations the model can adopt, which limits the sampling required for optimization. The use of simple lattice models provides the most direct form of discretization, but often at the expense of being restricted to model polymers rather than real proteins. For a cubic lattice in which each site corresponds to an amino acid residue, the structural prediction for a real sequence is limited to roughly 7–8 Å rms for relatively small (50–75 residues) molecules (15, 16). One novel approach involves determining the topology of the molecule by exhaustive enumeration of paths on a diamond lattice but with a flexible assignment of residues to the lattice sites (32). This approach allows determination of the most favorable contacts and generation of a more realistic distance map.

An alternate method has shown, however, that the flexibility of the backbone can be approximated reasonably by using an expanded set of basis vectors (and therefore possible nearest-neighbor bonds), along with a number of lattice sites, to represent the volume of each residue (25). The results do not differ qualitatively from dynamics with continuous degrees of freedom (40). The asymmetric torsional potential (caused by the chirality of the C_α centers) can be modeled with many-body interactions, which can give rise to secondary structure (16, 38, 41) that would not appear otherwise (39).

Similar models have also been introduced by using the representation of the backbone and side chains by the C_α and C_β atoms, respectively, but with continuous spatial coordinates (26, 56, 67, 68). Other methods that use a finite set of discrete values for the internal degrees of freedom (typically backbone dihedral angles) combine the advantage of a discretized conformational space with unrestricted atomic coordinates (11, 28, 35, 51). The set of values can be selected from a small number of representative local conformations (11, 51) that can reproduce global structure (58) or from a uniform grid in the dihedral angle space (28, 35). This type of discretization has also been extended to side chains, where the possible conformations of each amino acid are described by a small number of rotamer states (1, 35, 42).

An important question arises concerning the formation of secondary structure in the context of a reduced model, because the specific interactions that determine the detailed features of secondary structural motifs are difficult to represent with a coarse-grained model. This problem can be overcome by including specific potential interactions designed to stabilize the secondary structure (38, 42, 56, 67, 68) or simply by holding it fixed a priori during the course of the minimization (12, 28, 51). This approach represents a simplification of the folding problem but, of course, begs the question of knowing the secondary structure in advance. If other methods are available to assign the secondary structure, however, this approach effectively becomes a very simple way to represent the local interactions that stabilize the structure, which leads not only to a reduction in the number of degrees of freedom but also to a qualitative simplification in the potential function. The large-scale features of the tertiary structure, therefore, can be sampled more efficiently.

With the secondary structure held in place during the simulation, there is an enormous computational advantage to using a coarse-grained model based on the secondary-structure motifs, such as the "cylinder-and-sphere" (CS) model that we have used successfully. At that level of resolution, however, there is a question regarding the predictive value of the resulting structures. We have therefore implemented a strategy designed to take advantage of this coarse-graining while retaining the resolution of the residue-based (RB) reduced models described above (28). Our hierarchical model is based on the nested minimization of the two levels of representation, while a strict one-to-one correspondence between the two is maintained. The CS model is used as a crude screen of the trial conformations; only those structures left at the end of a CS minimization cycle are passed on as trial conformations at the RB level. The use of multiple levels of resolution has also been introduced in the context of lattice models (42), with a crossover from one model to another as the simulation progresses.

In the hierarchical model, the CS-level representation consists of the endpoints of the cylinder axes, which define the secondary structural elements, and the endpoints of the connecting loops. These points define the reduced chain, which can be simulated by changing the corresponding internal coordinates. The RB-level representation consists of the series of backbone dihedral angles that determine the positions of the C_α and C_β atoms. Because the pitch of the corresponding helix for a given set of ϕ-ψ dihedral angles is calculated easily, the specification of the axis and starting point completely determines the position of all backbone atoms within the helix. The key to the hierarchical method lies in maintaining a similar identity for the flexible loop regions. This process is done by using a precalculated list of loops where each entry lists both the backbone dihedral angles and the corresponding "loop geometry," which specifies the internal coordinates in the CS representation. A rapid screening of structures can be carried out, therefore, with the use of only the CS model, but the complete set of RB coordinates can be generated at any time to evaluate the RB-level potential function. The use of both models throughout the simulation leads to a qualitative improvement of the results relative to a CS-level minimization followed by fitting to a RB representation.

Potential Functions

In the previous section, we described briefly the geometrical representations of some reduced models. For each model of the structure, however, there must also be a corresponding model of the force-field, and this problem is much more difficult. In addition, from a computational point of view, the evaluation of the potential is the dominant cost compared with the manipulation of the coordinates. Reduced potential functions generally have two roles: First, they must represent the true potential at a comparable level of coarse graining as the model. Second, they must compensate for structural features omitted from explicit consideration. The former generally includes average packing forces and the effects of solvent and hydrophobicity, whereas the latter tries to mimic specific, short-range forces, such as backbone steric interactions and hydrogen bonding.

The backbone torsional potential can be included explicitly as a function of ϕ and ψ (67, 68), indirectly as C_α pseudo-angle and pseudo-dihedral potentials (42, 56), or implicitly in using a biased selection of trial moves (1, 28, 35, 51, 68). Hydrogen bonding can be included with specific multibody interactions designed to identify backbone conformations characteristic of hydrogen-bonded structures (35, 41). It also can be combined with nonlocal, orientation-dependent interactions to stabilize secondary structure further (38, 42). Another approach, which we are currently pursuing as a generalization of using fixed secondary structure, consists of selecting secondary structural elements, including β-strand pairings, from a predetermined list of possible choices. In this approach, the effects of short-range interactions can be included implicitly in the choice of trial structures, rather than as a complicated set of terms in the potential function.

The coarse-grained, nonlocal potential poses the greatest challenge, however, because that is what determines the overall folded topology. It is precisely the subtle cooperativity of the entire sequence, which cannot be built from known elements (excluding the case of homology modeling), that makes the protein-folding problem so difficult in the first place. In the case of globular proteins, to predict the structure correctly, in the sense of identifying the native contacts in some discretized sense, the energy of those contacts needs to be known to within a few percent (9). In other words, if the error in the contact energy is too great, the structure with the native set of contacts is unlikely to be a minimum of the potential. With a level of coarse graining that includes solvent effects and side-chain conformational averaging, as well as being discretized in space, this is a formidable challenge.

Not surprisingly, there are several reduced-contact potentials currently in use in the literature (8, 27, 32, 48, 50, 73). These potentials are generally in the form of discrete values that correspond to a square well potential (in some cases with several "steps") or fit to a simple functional form (13). Because of the complexity of the interactions that are represented, these potentials are determined empirically from the distribution of known structures in the crystallographic database. This approach leads to a compromise between the level of resolution desired and the statistical reliability of each parameter. The pairwise interactions can be supplemented with single-residue "context" potentials (27, 42), which consider the surface exposure of a residue. Solvation can also be represented by a single size–dependent "external pressure" term in the potential (28, 35). Results obtained with a few of these potentials are discussed further in the next section.

For the hierarchical model, a further coarse graining was developed to calculate directly the interactions between entire secondary structure elements (28). This approach was done by taking a long-range potential with a smooth distance dependence and expanding it around the center–center vector between cylinders (13). The second term in the expansion corresponds to a "hydrophobic dipole" interaction, which depends on the distribution of hydrophobic and hydrophilic residues within each helix. Although this approach gives a credible approximation to the solvation energy of helices, it fails to represent site-specific contact energies and needs to be supplemented with a specific "pairing" potential to model β-sheet formation reasonably (52). The resolution of any potential at this level of detail is limited but is nonetheless useful in screening out unlikely conformations.

Improved resolution within the context of a residue–residue potential can be achieved by incorporating orientation-dependent or many-body interactions (38, 42). This approach considers the fact that the contact potential between two side chains depends on the relative positions and orientations of the backbone segments, as well as on the position of the side-chain centers. This feature, which is just a renormalization of the averaged-over internal degrees of freedom of the side chains (and solvent), is the rationale behind our recently developed multidimensional contact potential (52). In this potential, the interaction between two residues depends simultaneously on all four of the interresidue C_α and C_β distances. This allows for a significant improvement over a sum of pairwise interactions of the same atoms, because the correlation among the indices contains information on how the residues "fit together." This component is important for representing the excluded volume without explicit side chains. Recent results suggest that the problems of sparse statistics can be overcome partially by fitting the potential to a simple functional form as a means of filling the gaps in the data.

Simulation Algorithms

Several methods have been used to determine the global minima of the various models discussed so far. The predominant technique is that of Monte Carlo simulated annealing (15, 16, 35, 38, 41, 42, 56); however, molecular dynamics (26) and Langevin dynamics (12, 40) have been used as well. In addition, noncanonical sampling methods have been suggested to improve the barrier-crossing efficiency (29–31). Alternatively, we have introduced a systematic variation of the potential function itself (parameter annealing), in the context of Monte Carlo annealing, as a means of smoothing out the potential barriers in the early stages of the simulation (28). This approach is most significant for the excluded volume, where a relaxation of the hard-core constraints speeds up the development of compact structures substantially. The resulting atomic overlaps are then annealed out slowly as the simulation progresses. The same idea is also applied to the coefficients of each term in the potential to generate a crossover from long-range to short-range interactions.

The use of a biased distribution of trial moves, mentioned above in the context of the potential function, has also been used as a means of improving the efficiency of the conformational sampling (1). This idea has been extended (67, 68) to include the selection of segments of up to several residues from a biased distribution or from a library of fragments from the database. The precalculation of a list of multiresidue trial moves has also been used in the context of a lattice model as a computational shortcut (38, 41, 42) and, as mentioned above, forms the basis of the hierarchical model (28). In addition to simply specifying the correspondence between the CS and RB models, the loop lists also can be screened according to the loop geometries to bias the distribution toward those loops already present in an ensemble of structures. The acceptance rate of the trial moves can be improved significantly, therefore, by generating moves that are, on average, smaller displacements from the existing structures. It is highly nontrivial to generate a loop of several residues with endpoints in a desired relative position, and the use of a precalculated list makes the resulting segments available for a large number of subsequent trial moves.

The substitution of larger and larger segments of the structure as trial moves leads naturally to the concept of genetic algorithms (especially because the system in question has a well-known genetic code!). Generalized algorithms that involve cross-breeding and incorporate ''offspring'' structures into the ''parent'' population have been established for simple lattice models (36, 71), and a simple implementation has been used with the hierarchical model (28). In this method, the molecule was divided into three parts: the N-terminal end, which was taken from one structure in the ensemble; the C-terminal end, which was taken from another structure; and a loop taken from the list to connect them. In this way, a large number of hybrids (3200 per parent) were generated, and those with the lowest energy at the CS level (1 per parent) were then used as additional trial moves at the RB level. The genetic algorithm led to a qualitative improvement in the results, because it can generate new conformations that are different from any of the existing structures by more than one simultaneous loop substitution. In addition, the crossover effectively allows the other structures to be used as a reservoir of trial loops, thereby allowing low-energy fragments of the structure to replicate throughout the ensemble.

The success of the genetic algorithm is also coupled to the hierarchical structure of the trial moves. The genetic algorithm allows for the effective sampling at a coarse-grained level of a ''gene pool'' of loop fragments, which themselves are already constructed from smaller fragments and screened independently. In this way, trial structures are built from an evolving library of structural elements, whereas the random mutations are carried out independently within the individual fragments.

Reconstructing Atomic-Level Detail

If the results of reduced-model calculations are to provide detailed structural predictions, an atomic-level structure, based on the reduced representation, must be generated. This step will naturally depend on the simplicity of the model, but the important requirement is that it be local, i.e. that it can be carried out systematically without qualitatively changing the global structure. If not, one is faced with yet another global optimization problem, and the utility of the initial coarse-grained structure becomes questionable.

From a starting point consisting of (some or all) C_α coordinates, there are two methods in use to generate a complete set of backbone coordinates. The first method, which involves fitting to a library of fragments from the X-ray database, has proved highly accurate in several examples (33, 46). An alternative method makes use of geometric criteria to place the C_β atoms and then deduces the values of ϕ and ψ

by assuming a planar peptide geometry (55). This method has also shown excellent performance in tests of known structures, thereby confirming that resolution of the C_α positions is sufficient to construct a (nearly) unique all-atom backbone.

The addition of side chains (beyond the C_β) is more challenging, as the backbone trace no longer provides a constraint. The method of template fitting, however, can also be applied to generate complete side chains (46), albeit with reduced accuracy compared with that of the backbone. Other methods make use of discrete "rotamer states," which correspond to the observed values of χ angles of each amino acid (53). The problem is then determining the state of each side chain simultaneously. Although the problem is coupled, because of the excluded volume, it is still local in that a change in the state of any side chain affects only its neighbors and does not alter the overall structure. This problem has been addressed with the use of an iterative mean-field technique (37) and Monte Carlo simulated annealing (22, 33). In addition, pairwise interactions can be used to eliminate possible rotamers rigorously from inclusion in the global minimum, thereby reducing the combinatorial problem drastically (18). Good results can be obtained with these techniques by using only a simple Lennard-Jones potential function (33, 37, 45).

The success of using rotamer states has some implications for the construction of reduced models: There is a certain flexibility in choosing the point at which the reduced model ends and the "reconstruction" begins. An additional degree of freedom representing the rotamer state can be included in the coarse-grained description of a residue conformation. This description considers more of the local constrains around each C_α center both in the potential function (42) and the choice of trial moves (1). Alternatively, the side-chain conformations can be considered as another level of the hierarchical structure, in which new side chains are selected from an appropriately evolving list. The important principle is the continuity of refinement, i.e. successive levels of detail can be built on simpler representations, and the knowledge of the folded structure is passed on intact to a more detailed calculation. Reduced models (or series of models) for which this holds thus offer the promise of bridging the gap between effective sampling of conformations and accurate structural resolution.

RESULTS

Recent Progress

There has been a substantial amount of work carried out with simple lattice models of proteins in the past few years, but relatively little work has been aimed at the predictive folding of real sequences. Some recent results in this area have shown only moderate success (15, 16, 32), with typical rms deviations from the native structures on the order of 7–8 Å for small proteins in the range of 50–70 residues. Although the simplicity of the models makes them easy to work with, the lack of sufficient flexibility will likely hinder their extension to more complex proteins.

The more sophisticated lattice models of Skolnick and co-workers have fared much better, largely because of a much more detailed representation of secondary structure. Simple examples of both four-helix bundles (38, 43) and β-sheets (31) have been assembled from random initial conditions. The 46-residue protein crambin has also been folded recently to less than 4 Å rms deviation from its native structure (43). The authors state, however, that results with larger, more complex proteins have shown mixed results, with secondary structural elements generally being stable, but with limited success at reproducing the overall folded topology. It is not yet known whether further refinements to the potential function and simulation algorithm will be able to surmount this barrier. The off-lattice reduced model of Sun (68) has also shown mixed results. This model can reproduce segments of detailed structure but fails at finding the globally correct conformation. Results for small proteins typically show larger errors, including, for example, a deviation on the order of 8 Å for crambin.

On the other hand, work to date with larger proteins has proceeded generally with simplified representations of the secondary structure elements, with the emphasis on overcoming the combinatorial problem of packing them into a compact structure. In a recent example (12), a model of myoglobin (with 153 residues and eight helices) that used rigid helices connected by flexible loops was folded to roughly 6 Å rms deviation, as determined by the helical C_α positions, with a clear resemblance to the native topology at the cylinder-and-spring level of resolution. This result shows the efficacy of coarse-grained models in determining the overall features of the tertiary structure; however, it lacks a clear path to atomic resolution.

The Hierarchical Model

In this section, we discuss simulations for 5 proteins, the X-ray crystallographic structures of which we have obtained from the Protein Data Bank (PDB) [ref PDB]. These proteins are: myoglobin (1MBO); L7/L12 50 S ribosomal protein, C-terminal domain (1CTF); 434 repressor, amino terminal domain (1R69); myohemerythrin (2MHR); cytochrome b562 (B256). In what follows, we use PDB identifications (in parenthesis above) interchangeably with the full name of the protein in question.

The initial tests of the hierarchical model were concerned with the question of whether rigid secondary structure elements could be packed together into the correct tertiary structure. Simulations were carried out (51) for proteins that form four-helix bundles with the use of a residue-based reduced model and a simple long-range potential function (13). The trial moves were restricted to changes in the backbone dihedral angles of the loop residues, and the conformation was minimized with a simple Monte Carlo quenching algorithm. Results on the order of 4 Å rms were obtained for cytochrome b-562 (B526) (44a) and myohemerythrin (2MHR) (65a) (shown in Figure 1), each of which folded into

Figure 1 Overlap of the X-ray structure of myohemerythrin (65a) (*dark*) and one of the low-energy structures produced by our algorithm (*light*). The C_α rms deviation between the two structures is 4.12 Å.

essentially the correct overall structure. Despite the simplicity of the structures in this case, the significance of this result is highlighted by the fact that the corresponding native X-ray crystallographic structures differ by 7.5 Å rms. In addition, experiments in which randomized or monomeric sequences were used failed to generate comparable structures. Most importantly, the results showed a clear difference in energy between correctly folded and misfolded structures. The native-like folds always had a lower level of energy, and the crystal structure itself had the lowest level of all.

Despite the ease with which the above results were obtained, positive results for more complex structures required considerable development of the minimization algorithm. This led to the hierarchical combination of the CS and RB models (28), as introduced above. In particular, the use of the genetic algorithm and the annealing of the excluded volume constraints had an enormous influence on the speed with which compact structures were generated. It also became clear that a large ensemble of structures was necessary to generate results that were reproducible in a statistical sense. This observation underscores the magnitude of the increase in complexity associated with a mere doubling of the sequence length.

A final distribution of the 256 lowest energy structure from a typical myoglobin simulation (~48 h of run time on a 16 node partition of a CM-5 massively parallel supercomputer) is shown in Figure 2. These structures represent the final results of examining more than 10^{10} structures in the course of the simulations. The distribution was analyzed by calculating the rms deviation between all pairs of structures and grouping those within 5 Å into distinct clusters. The results thus fall into only two different folded topologies, one of which is clearly native-like. This result is remarkable in that the native structure itself (being more than 5 Å from any of the calculated structures) was not used in any way to identify the clusters. This finding suggests that there is indeed a native-like ''attractor'' in the potential surface, because the simulation is producing variations on a common theme rather than a uniform distribution of wrong structures. The results are even more striking in the case of the misfolded structures, given that the number of ways of generating such structures grows extremely rapidly with distance from the native, and yet again only one distinct topology is observed. We note that the energy in Figure 2 and in subsequent figures is in arbitrary units, as the potential function is not scaled to physical interaction energies.

The low-energy structures from each cluster are shown in Figures 3 (for the low-rms, superimposed on the native) and 4 (for the high-rms).

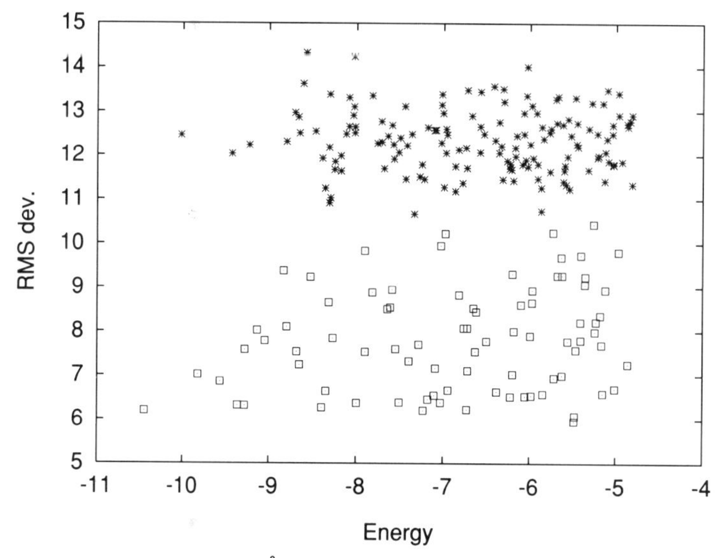

Figure 2 The energy vs rms in Å deviation distribution of two clusters of low-energy structures from a minimization of myoglobin.

The 6-Å (low-rms) structure closely resembles the native in the packing of the helices and has the same overall topology (52a). In fact, most of the rms error comes from the loop regions, with the helical residues differing by only 4 Å from the native. Also striking is that the misfolded structure shares many of the same features, including a densely packed hydrophobic core and a solvated flexible loop region. The entire C-terminal half of the molecule is virtually the same. Without knowing the correct structure, it would be very difficult to choose among these possibilities by any intuitive means, thus demonstrating the ability of the simulation to generate ''intelligently'' misfolded structures that are viable tests of the potential function.

Comparable results were also obtained for the C-terminal fragment of L7/L12 ribosomal protein, which is a mixed α/β protein with 66 residues (52). The low-energy structure obtained from the simulation is superimposed on the native in Figure 5 (46a). This structure shows the correct overall topology but considerable error in the structure of the β-sheet. This error results from the complete neglect of hydrogen bonding and electrostatic interactions in the model, which cause the β-strands to bunch together without any specific structure (although a simple strand–strand angular term was included in the potential in an attempt to compensate for this). In the case of helical proteins, most

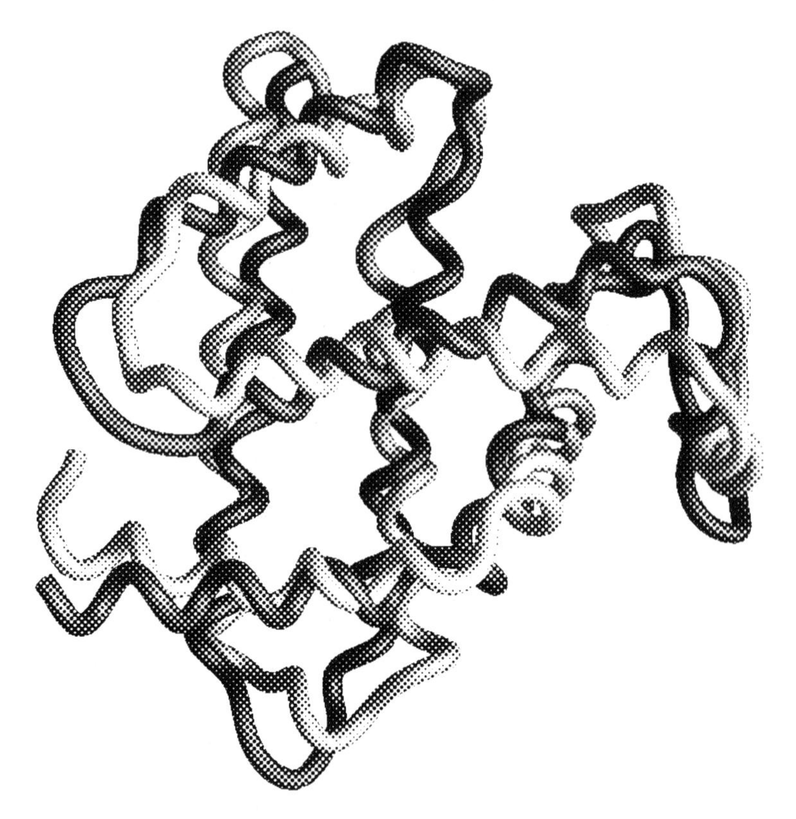

Figure 3 α-Carbon backbone worm structures for the native (52a) (*light*) and calculated (*dark*) conformations of myoglobin shown in the optimal superposition with 6.2 Å rms deviation from the nature structure.

of the backbone hydrogen bonding is inside the helices and, therefore, is well accounted for by holding the secondary structure fixed. That approximation clearly breaks down, however, in the case of β-sheet formation.

Evaluation of Potential Functions

The next example studied was the small α-helical protein 1R69 (the N-terminal domain of the phage 434 repressor protein) (50a), for which the distribution of structures is shown in Figure 6 (44a). In this case, there is no native-like cluster of structures, and the crystal structure itself is nowhere near the minimum in energy. This result, as well as similar problems with larger β-sheet proteins, prompted us to look more

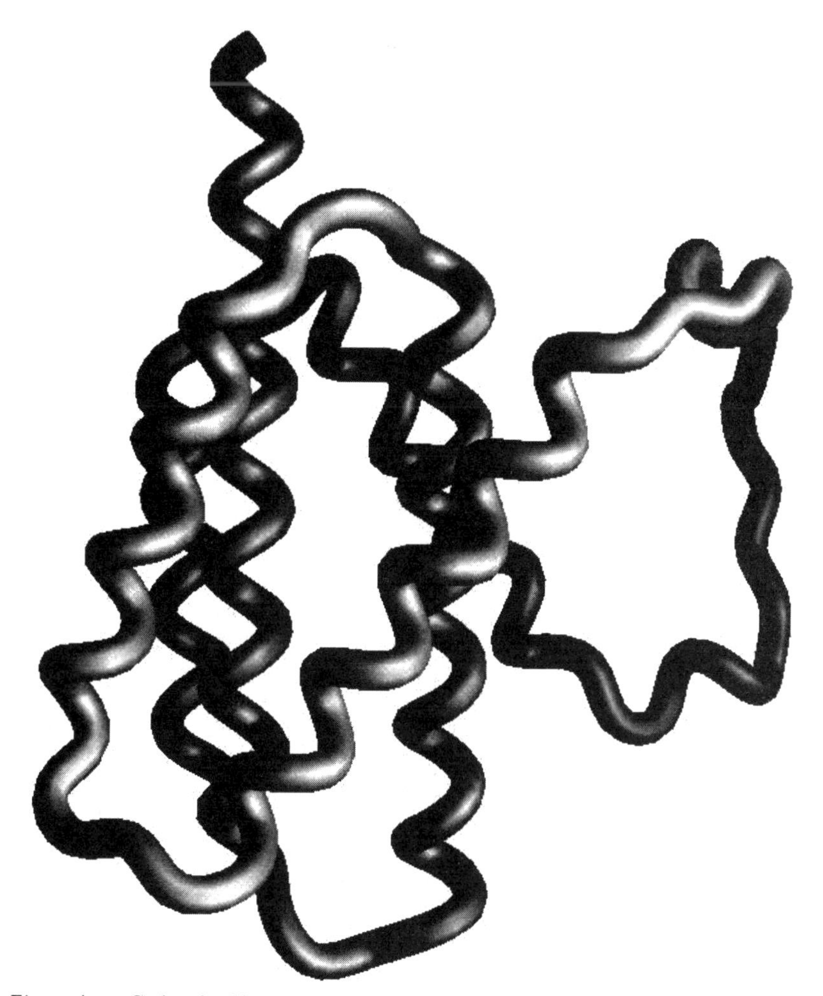

Figure 4 α-Carbon backbone worm structure of a misfolded conformation of myoglobin with 12.5 Å rms deviation.

deeply at the characteristics of the potential function. This study was done by taking selected structures from the hierarchical simulations and comparing them in the context of an all-atom model with an atomic-level continuum solvent potential function, in this case the AMBER protein model of Kollman and co-workers (71a) and the generalized Born model of Still and co-workers (66).

Side chains were added using the RSA algorithm (22) followed by simulated annealing of the structure to find a local minimum of the

Figure 5 Superimposition of C_α worms for the native (*light*) and calculated (*dark*) structures of 1CTF (46a). The rms deviation between the two structures is 5.0 Å.

all-atom potential. The results of repeated trials (52) indicate that this can be done reliably to give structures that differ by typically 1–2 Å from the reduced-model coordinates, comparable to the change observed in directly minimizing the crystal structure with the same potential. This confirms that the reduced model can provide a reasonable starting point for all-atom refinement. The resulting distributions (albeit with only a handful of structures) show that the all-atom potential can distinguish the native from the calculated structures for myoglobin (1MBO) (52a) and L7/L12 50 S ribosomal protein, C-terminal domain

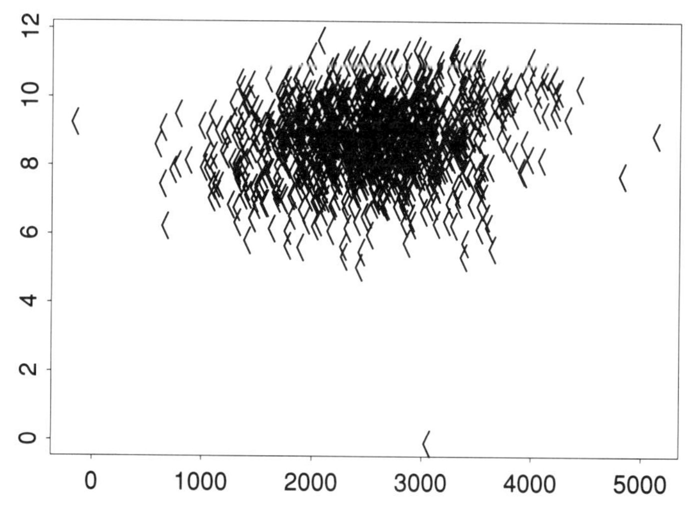

Figure 6 Distribution of reduced-model structure plotted with rms deviation vs total energy for 1R69 (50a). The energy of the crystal structure is shown with rms = 0.

(ICTF) (46a). In the case of 1R69, some structures have a slightly lower energy than the native; however, this still represents a substantial improvement over the reduced-model potential. On the other hand, the all-atom potential still did not show any significant difference between the energies of the two competing topologies for myoglobin. Further analysis of the different contributions to the energy shows that all structures have roughly the same total electrostatic energy (Coulombic plus solvation) but that the native has a much lower level of van der Waals packing energy. Further, the native structure generally has a lower internal electrostatic energy, whereas the calculated structures have lower solvation energies. This suggests that the reduced-model potential does a good job of predicting the overall solvation energy (consistent with the idea of hydrophobic–hydrophilic segregation) but fails to account for the details of the internal structure, such as the close packing of side chains and close-range electrostatic interactions. Although the native structure can accommodate these interactions simultaneously, the calculated structures have more unfavorable contacts or, alternatively, must relax the all-atom structure at the expense of overall compactness.

CONCLUSIONS

A central conclusion of the work described above is that the problem of determining tertiary structure once secondary structure is specified, although nontrivial from the point of view of both algorithms and potential functions, is tractable with current computing technology. The success of relatively simple potential functions for a significant number of proteins, including several with rather complicated structures, suggests that a refined potential of similar structure, in conjunction with some of the algorithmic improvements outlined above (particularly for complex β-sheet topologies), will be able to generate tertiary structures robustly in the 4–6 Å range. Such a potential can be developed by optimizing parameters and functional form against an increasingly large database of misfolded structures. This procedure must be carried out via an iterative cycle to optimize a potential with the current X-ray database, add the new low energy misfolded structures, and repeat until reasonable results are obtained for a substantial number of proteins in the database (the "training set") with the use of a single potential function. The resulting potential can then be tested in its ability to determine the structure of proteins it was not optimized for, i.e. those proteins in the "test set." The ultimate test, of course, is the prediction of new structures. We are currently pursuing this strategy utilizing the orientation-dependent reduced potential mentioned above, and preliminary results indicate substantial improvement over the simplified potential functions that we have used in the past.

Determination of secondary structure from first principles is considerably more difficult. Progress is being made, however, with novel simulation approaches (e.g. the work of Skolnick's group) and heuristic prediction methods (e.g. the work of Benner's group). These methods can perhaps be synthesized with the secondary structural packing algorithms; indeed, Skolnick already freezes in secondary structure components, once the simulation has identified their location as an algorithmic device. It is certainly straightforward to allow secondary structural elements to fluctuate in length during a simulation. A more difficult issue is whether one can obtain a reasonable approximation to the native structure for large proteins if the initial guess for the secondary structure contains a major flaw. Alternatively, one could try several different initial guesses for the secondary structural pattern and see which one produces a better final structure. These approaches will likely require significantly more computation time, but this should be available with the next generation of computer hardware, particularly massively parallel machines such as the IBM SP2.

An alternative approach, which we are pursuing in our laboratories, is to use experimental data in new ways in combination with reduced-model simulations to determine low-resolution structures. The most obvious step along these lines is to extract secondary structure from

NMR data; this approach is considerably easier than obtaining sufficient long-range Nuclear Overhauser Effect (NOE) constraints to determine a high-resolution tertiary structure. With the use of a small subset of NMR constraints (available from the initial experiments required to assign resonances), we have improved the resolution of the structures described above by approximately 2 Å, as well as fold even more complicated proteins such as lysozyme to satisfactory resolution. This effort, however, is only a beginning along this path. Once it is established that determination of secondary structure plus a very small number of additional distance constraints is sufficient to determine structure, there will undoubtedly be clever new experimental techniques, not only NMR but other types of spectroscopy as well, designed to reach this goal with as little effort as possible. The reduction of effort in structure determination, although not as spectacular a goal as de novo folding, may still prove to be extremely valuable in a practical sense.

Finally, methods are still needed to take a 4–6 Å rms structure and reduce the error to 1–2 Å rms if the structures are to be used in such applications as rational drug design. This process must necessarily involve atomic-level models with improved accuracy as compared with current potential functions, which cannot yet achieve this level of resolution, at least in the tests that have been carried out to date. Although the paths toward progress are clear—better solvation models, hydrogen bonding energies, and torsional potentials—the precision needed to obtain high resolution is not yet apparent. Nevertheless, significant advances in the next 5 years can be expected toward this goal.

ACKNOWLEDGMENTS

This work was supported in part by grants from the National Institutes of Health Institute of General Medicine (GM-52018) and by the National Institutes of Health Division of Research Resources (P41-RR-06892) (RAF) and by a grant from the Canadian government (JRG). We thank Barry Honig and his research group for generously sharing their knowledge of all aspects of protein modeling with us over the past several years and Peter Shenkin and Clark Still for providing access to the Macromodel molecular modeling program.

Literature Cited

1. Abagyan R, Totrov M. 1994. Biased probability Monte Carlo conformational searches and electrostatic calculations for peptides and proteins. *J. Mol. Biol.* 235:983–1002
2. Baldwin RL. 1995. Alpha-helix formation by peptides of defined sequence. *Biophys. Chem.* 1–2:127–35
3. Benner SA, Gerloff D. 1991. Patterns of divergence in homologous proteins as indicators of secondary and tertiary structure: a prediction of the structure of the catalytic domain of protein kinases. *Adv. Enzyme Regul.* 31:121–81
4. Benner SA, Gerloff DL, Jenny TF. 1994. Predicting protein crystal structures. *Science* 265:1642–44

4a. Bernstein FC, Koetzle TF, Williams GJB, Meyer EF, Brice MD, Rodgers JR, Kennard O, Shimanouchi T, Tasumi M. 1977. The Protein Data Bank: a computer-based archival file for macromolecular structures. *J. Mol. Biol.* 112:535–542

5. Boczko EM, Brooks CL. 1995. First-principles calculation of the folding free energy of a three-helix bundle protein. *Science* 222:393–96
6. Bowie JU, Eisenberg D. 1993. Inverted protein structure prediction. *Curr. Opin. Struct. Biol.* 3:437–44
7. Brooks CR, Bruccoleri R, Olafson B, States D, Swaminathan S, Karplus M. 1983. CHARMM: a program for macromolecular energy, minimizations and dynamics calculations. *J. Comput. Chem.* 4:187
8. Bryant SH, Lawrence CE. 1993. An empirical energy function for threading protein sequence through the folding motif. *Proteins* 16:92–112
9. Bryngelson JD. 1994. When is a potential accurate enough for structure prediction? Theory and application to a random heteropolymer model of protein folding. *J. Chem. Phys.* 100:6038–45
10. Bryngelson JD, Onuchic JN, Socci ND, Wolynes PG. 1995. Funnels, pathways, and the energy landscape of protein folding—a synthesis. *Proteins* 3:167–95
11. Buturovic LJ, Smith TF, Vajda S. 1994. Finite-state and reduced-parameter representations of protein backbone conformations. *J. Comput. Chem.* 15:300–12
12. Callaway DJE. 1994. Solvent-induced organization: a physical model of folding myoglobin. *Proteins* 20:124–38
13. Casari G, Sippl MJ. 1992. Structure-derived hydrophobic potential: hydrophobic potential derived from X-ray structures of globular proteins is able to identify native folds. *J. Mol. Biol.* 224:725–32
14. Cohen FE, Richmond TJ, Richards FM. 1980. Protein folding: evaluation of some simple rules for the assembly of helices into tertiary structures with myoglobin as an example. *J. Mol. Biol.* 137:9–22
15. Covell DG. 1992. Folding protein α-carbon chains into compact forms by Monte Carlo methods. *Proteins* 14:409–20
16. Covell DG. 1994. Lattice model simulations of polypeptide chain folding. *J. Mol. Biol.* 235:1032–43
17. Daggett V, Levitt M. 1994. Protein folding and unfolding dynamics. *Curr. Opin. Struct. Biol.* 2:291–95
18. Desmet J, De Maeyer M, Hazes B, Lasters I. 1992. The dead-end elimination theorem and its use in protein side-chain positioning. *Nature* 356:539–42
19. Dill KA, Bromberg S, Yue KZ, Fiebig KM. 1995. Principles of protein folding—a perspective from simple exact models. *Protein Sci.* 4:561–602
20. Dinner A, Sali A, Karplus M, Shakhnovich E. 1994. Phase diagram of a model protein derived by exhaustive enumeration of the conformations. *J. Chem. Phys.* 2:1444–51
21. Eisenberg D, Wesson M, Yamahshita M. 1989. Interpretation of protein folding and binding with atomic solvation parameters. *Chem. Scr.* 29a:217–21
22. Farid H, Shenkin PS, Greene J, Fetroe J. 1992. Prediction of side chain conformations in protein core and loops from rotamer libraries. *Biophys. J.* 61:A350 (Abstr.)
23. Fasman G. 1989. In *The Development of the Prediction of Protein Structure*, ed. G Fasman, pp. 193–316. New York: Plenum
24. Gilson MK, Davis ME, Luty BA, McCammon JA. 1993. Computation of electrostatic forces on solvated molecules using the Poisson-Boltzmann equation. *J. Phys. Chem.* 97:3591–3600
25. Godzik A, Kolinski A, Skolnick J. 1993. Lattice representations of globular proteins: How good are they? *J. Comput. Chem.* 14:1194–1202
26. Goldstein RA, Luthey-Schulten ZA, Wolynes PG. 1992. Optimal protein-folding codes from spin-glass theory.

Proc. Natl. Acad. Sci. USA 89: 4918–22
27. Goldstein RA, Luthey-Schulten ZA, Wolynes PG. 1992. Protein tertiary structure recognition using optimized Hamiltonians with local interactions. *Proc. Natl. Acad. Sci. USA* 89: 9029–33
28. Gunn JR, Monge A, Friesner RA. 1994. Hierarchical algorithm for computer modeling of protein tertiary structure: folding of myoglobin to 6.2Å resolution. *J. Phys. Chem.* 98: 702–11
29. Hansmann UHE, Okamoto Y. 1993. Prediction of peptide conformation by multicanonical algorithm: new approach to the multiple-minimia problem. *J. Comput. Chem.* 14:1333–38
30. Hansmann UHE, Okamoto Y. 1994. Comparative study of multicanonical and simulated annealing algorithms in the protein folding problem. *Physica A* 212:415–37
31. Hao MH, Scheraga HA. 1994. Monte Carlo simulation of a first-order transition for protein folding. *J. Phys. Chem.* 98:4940–48
32. Hinds DA, Levitt M. 1992. A lattice model for protein structure prediction at low resolution. *Proc. Natl. Acad. Sci. USA* 89:2536–40
33. Holm L, Sander C. 1991. Database algorithm for generating protein backbone and side-chain coordinates from a C_α trace: application to model building and detection of coordinate errors. *J. Mol. Biol.* 218:183–94
34. Honig B, Sharp K, Yang A. 1993. Macroscopic models of aqueous solutions—biological and chemical applications. *J. Phys. Chem.* 97:1101–9
35. Hunt NG, Gregoret LM, Cohen FE. 1994. The origins of protein secondary structure: effects of packing density and hydrogen bonding studied by a fast conformational search. *J. Mol. Biol.* 241:214–25
36. Judson RS. 1992. Teaching polymers to fold. *J. Phys. Chem.* 96:10102–4
37. Koehl P, Delarue M. 1994. Application of a self-consistent mean field theory to predict protein side-chains conformation and estimate their conformational entropy. *J. Mol. Biol.* 239: 249–75
38. Kolinski A, Godzik A, Skolnick J. 1993. A general method for the prediction of the three-dimensional structure and folding pathway of globular proteins: application to designed helical proteins. *J. Chem. Phys.* 98:7420–33
39. Kolinski A, Milik M, Skolnick J. 1991. Static and dynamic properties of a new lattice model of polypeptide chains. *J. Chem. Phys.* 94:3978–85
40. Kolinski A, Skolnick J. 1991. Comparison of lattice Monte Carlo dynamics and Brownian dynamics folding pathways of α-helical hairpins. *Chem. Phys.* 158:199–219
41. Kolinski A, Skolnick J. 1992. Discretized model of proteins. I. Monte Carlo study of cooperativity in homopolypeptides. *J. Chem. Phys.* 97:9412–26
42. Kolinski A, Skolnick J. 1994. Monte Carlo simulations of protein folding. I. Lattice model and interaction scheme. *Proteins* 18:338–52
43. Kolinski A, Skolnick J. 1994. Monte Carlo simulations of protein folding. II. Application to protein A, ROP, and crambin. *Proteins* 18:353–66
44. Kostrowicki J, Scheraga HA. 1992. Application of the diffusion equation method for global optimization to oligopeptides. *J. Phys. Chem.* 18: 7442–49
44a. Lederer F, Glatigny A, Bethge P, Bellamy HD, Mathews FS. 1981. Improvement of the 2.5 Å resolution model of cytochrome b562 by redetermining the primary structure and using molecular graphics. *J. Mol. Biol.* 148: 427
45. Lee C, Subbiah S. 1991. Prediction of protein side-chain conformation by packing optimization. *J. Mol. Biol.* 217:373–88
46. Levitt M. 1992. Accurate modeling of protein conformation by automatic segment matching. *J. Mol. Biol.* 226: 507–33
46a. Liejonmarch M, Liljas A. 1987. Structure of the C-terminal domain of the ribosomal protein L7/L12 from *Escherichia coli* at 1.73 Å. *J. Mol. Biol.* 195:555
47. Luthey-Schulten Z, Ramirez BE, Wolynes PG. 1995. Helix-coil, liquid crystal, and spin glass transitions of a collapsed heteropolymer. *J. Phys. Chem.* 7:2177–85
48. Maiorov VN, Crippen GM. 1992. Contact potential that recognizes the correct folding of globular proteins. *J. Mol. Biol.* 227:876–88
49. Merutka G, Morikis D, Bruschweller R, Wright PE. 1993. NMR evidence for multiple conformations in a highly helical model peptide. *Biochemistry* 48:13089–97
50. Miyazawa S, Jernigan RL. 1985. Estimation of effective interresidue contact energies form protein crystal

structures: quasi-chemical approximation. *Macromolecules* 18:534–52
50a. Mondragon A, Subbiah S, Almo S, Drottar M, Harrison SC. 1989. Structure of the amino terminal domain of phage 434 repressor at 2.0 Å resolution. *J. Mol. Biol.* 205:189
51. Monge A, Friesner RA, Honig B. 1994. An algorithm to generate low-resolution protein tertiary structures from knowledge of secondary structure. *Proc. Natl. Acad. Sci. USA* 91: 5027–29
52. Monge A, Lathrop EJP, Gunn JR, Shenkin PS, Friesner RA. 1995. Computer modeling of protein folding: conformational and energetic analysis of reduced and detailed protein models. *J. Mol. Biol.* 247:995–1012
52a. Phillips, SEV. 1980. Structure and refinement of oxymyoglobin at 1.62 Å resolution. *J. Mol. Biol.* 142:531
53. Ponder JW, Richards FM. 1987. Tertiary templates for proteins. Use of packing criteria in the enumeration of allowed sequences for different structural classes. *J. Mol. Biol.* 193:775–91
54. Presnell SR, Cohen BI, Cohen FE. 1992. A segment based approach to secondary structure prediction. *Biochemistry* 31:983–93
55. Rey A, Skolnick J. 1992. Efficient algorithm for the reconstruction of a protein backbone from the α-carbon coordinates. *J. Comput. Chem.* 13:443–56
56. Rey A, Skolnick J. 1994. Computer simulation of the folding of coiled coils. *J. Chem. Phys.* 100:2267–76
57. Ring CS, Cohen FE. 1993. Modeling protein structures: construction and their applications. *FASEB J.* 7:783–90
58. Rooman MJ, Kocher JPA, Wodak SJ. 1991. Prediction of protein backbone conformation based on seven structure assignments: influence of local interactions. *J. Mol. Biol.* 221:961–79
59. Rost B, Sander C. 1994. Combining evolutionary information and neural networks to predict protein secondary structure. *Proteins* 19:55–72
60. Roterman I, Lambert M, Gibson K, Scheraga HA. 1989. A comparison of the CHARMM, AMBER, and ECEPP potentials for peptides. II. ϕ-χ Maps for N-acetylalanine N′-methylamide: comparisons, contrasts, and simple experimental tests. *J. Biomol. Struct. Dyn.* 7:421–53
61. Sali A, Shakhnovich E, Karplus M. 1994. How does a protein fold? *Nature* 477:248–51
62. Sali A, Shakhnovich E, Karplus M. 1994. Kinetics of protein folding—a lattice model study of the requirements for folding to the native state. *J. Mol Biol.* 5:1614–36
63. Sander C, Schneider R. 1991. Database of homology-derived protein structures and the structural meaning of sequence alignment. *Proteins* 9: 56–68
64. Scheraga HA. 1994. Toward a solution of the multiple-minima problem in protein folding. *J. Protein Chem.* 5: 468–69
65. Shakhnovich EI. 1994. Proteins with selected sequences fold into unique native conformation. *Phys. Rev. Lett.* 24:3907–10
65a. Sheriff S, Hendrickson W, Smith JL. 1987. Structure of myohemerythrin in the azidomet state at 1.71/1.3 Å resolution. *J. Mol. Biol.* 197:273
66. Still WC, Tempczyk A, Hawley R, Hendrickson T. Semianalytical treatment of solvation for molecular mechanics and dynamics. *J. Am. Chem. Soc.* 112:6117–29
67. Sun S. 1993. Reduced representation model of protein structure prediction: statistical potential and genetic algorithms. *Protein Sci.* 2:762–85
68. Sun S. 1995. Reduced representation approach to protein tertiary structure prediction: statistical potential and simulated annealing. *J. Theor. Biol.* 172:13–32
69. Tirado-Rives J, Jorgensen WL. 1990. Molecular dynamics of proteins with the OPLS functions. Simulation of the third domain of silver pheasant ovomucoid in water. *J. Am. Chem. Soc.* 112:2773–79
70. Tirado-Rives J, Jorgensen WL. 1993. Molecular dynamics simulations of the unfolding of apomyoglobin in water. *Biochemistry* 16:4175–84
71. Unger R, Moult J. 1993. Genetic algorithms for protein folding simulations. *J. Mol. Biol.* 231:75–81
71a. Weiner S, Kollman P, Case D, Singh U, Ghiao C, Alagona G, Profets S, Weiner P. 1984. A new force field for molecular mechanical simulation of nucleic acids and proteins. *J. Am. Chem. Soc.* 106:765–784
72. Weiner S, Kollman P, Nguyen D, Case D. 1986. An all atom force field for simulations of proteins and nucleic acids. *J. Comput. Chem.* 7:230
73. Wilson C, Doniach S. 1989. A computer model to dynamically simulate protein folding: studies with crambin. *Proteins* 6:193–209

Annu. Rev. Biophys. Biomol. Struct. 1996. 25:343–65

PROTEIN FUNCTION IN THE CRYSTAL

Andrea Mozzarelli and Gian Luigi Rossi

Institute of Biochemical Sciences, University of Parma, 43100 Parma, Italy

KEY WORDS: catalysis in enzyme crystals; biological activity in the crystalline state; structure-function relationships; single-crystal microspectrophotometry; polarized absorption spectroscopy; X-ray crystallography of active intermediates

ABSTRACT

Protein crystals contain wide solvent-filled channels that allow for traffic of metabolites and intramolecular motility. Ligand binding, catalysis and allosteric regulation occur in the crystalline environment but intermolecular interactions may hinder function-associated transitions and alter activity with respect to solution. Lattice constraints have, however, provided the opportunity to isolate and characterize conformational states that are poorly populated in solution. New methods are being developed to initiate reactions rapidly and synchronously throughout the crystal and to monitor their time course. A model consistent with kinetics in the crystal is necessary to interpret the results of time-resolved macromolecular crystallography.

1056–8700/96/0610–0343$08.00

PERSPECTIVES AND OVERVIEW

The structure of proteins sustains ligand binding and catalytic activity in a variety of milieus that substitute for their biological environment. In suitable solvents, proteins associate to form highly ordered, three-dimensional arrays whereby their spatial coordinates can be determined to atomic resolution by X-ray crystallography. The wide, solvent-filled channels, which occupy a large portion of the crystal volume, allow for considerable breathing of the protein molecules and provide a route toward the active sites for substrates and other reagents that diffuse from the surrounding medium. Access of ligands to the active site is hindered sterically by residues from a neighboring molecule only in a few cases. One would therefore expect proteins within the crystal lattice to feature most of their characteristic ligand-induced fits and, thus, unravel the molecular basis of events as complex as enzyme catalysis. However, the specific intermolecular interactions, which account for formation and stability of the lattice in the mother liquor, might select a subset of conformational states among those accessible to the protein in solution and exert adverse effects on conformational flexibility, thus distorting expression of function. In other cases, ligand-induced conformational changes may break the preexisting contacts and lead to crystal disorder and cracking.

The diverse responses of proteins to the call for action in their crystalline microenvironment demand a detailed characterization before function in solution can be correlated with structure in the crystal. Numerous techniques have been used to compare the overall or local conformation and motility of a protein in the two physical states. Biological activity has been demonstrated and assayed either by measuring the conversion of substrates to products in the solvent phase of a crystalline suspension or by using a microspectrophotometer to monitor formation and breakdown of chromophoric enzyme–substrate complexes within single crystals. Bioactivity studies are of paramount relevance in the planning and interpretation of the results of classical crystallographic experiments. These studies are even more important, however, in the perspective of time-resolved diffraction studies, in which reactions are initiated by synchronized exposure of crystalline proteins to substrates or other reagents. A microspectrophotometer on line with the X-ray apparatus provides the means to trail chemistry while catching structure.

Makinen & Fink (72) have summarized and discussed early work on bioactivity of proteins in the crystalline state in a definitive review that encompasses a thorough description of cryoenzymologic techniques as a means to trap transient intermediates in the crystalline state. A brief report on recent studies appeared in 1992 (107). The reviews of Moffat (82) and Hajdu & Andersson (39) and the Proceedings of a 1992 Royal Society discussion meeting (20) are excellent accounts of the rapidly evolving state of the art in fast crystallography and time-resolved macromolecular structures.

THE CRYSTAL ENVIRONMENT

Packing of water-soluble proteins in a crystal lattice involves specific intermolecular interactions which are favored by the precipitating agents present in the crystallization medium, usually salts, organic molecules, or poly(ethylene) glycol. A large fraction, 40–70%, of the crystal volume is occupied by liquid channels that allow for diffusion of relatively large solutes (72 and references therein). In a limit case, molecules of cytochrome c, 30 Å in diameter, have been found to diffuse through the 100-Å wide channels of crystals of tetrameric flavocytochrome b_2 from baker's yeast and to form a reversible active complex (134).

The available solvent grants macromolecules a considerable degree of conformational flexibility (3, 34). This flexibility is demonstrated, for example, by the lack of crystallographic structure for part of the sequence of some proteins, by the temperature dependence of the crystallographic B factors (98), and by extensive structural heterogeneity in some crystals (122).

Various techniques enable us to compare protein motility in the crystal and in solution over a wide range of time scales. Recent two-dimensional NMR spectroscopic studies on hen egg white lysozyme (90) and on bovine pancreatic trypsin inhibitor (32) have shown, at the resolution of individual hydrogen atoms, similarities and differences in their hydrogen isotope exchange behavior in solution and in the crystal. The most dramatic differences between the two physical states are not found in intermolecular contact regions, where they arise, but in the interior of proteins, as a result of long-range interactions that selectively damp local motions.

Myoglobin has been investigated thoroughly to probe lattice-imposed restrictions on its motility. Long-range intramolecular interactions between tryptophanyl residues and the heme have been studied by picosecond time-resolved fluorescence and found to be unaffected by crystallization (142). Resonance Raman spectroscopic studies have provided evidence that the equilibrium distribution of interconverting conformational states of carbonmonoxy myoglobin is controlled kinetically by different energy barriers in the crystal and in solution (144) and that CO motion within the protein matrix is slowed down in the crystallized state (49, 143). An early study had shown that a small fraction of metmyoglobin or deoxymyoglobin affects the equilibrium distribution of carbonmonoxy myoglobin conformers in the crystal (73). With the use of single-crystal infrared linear dichroism, CO has been found to bind to the iron of the heme almost in a linear geometry (50), in agreement with the results of photoselection experiments in solution (67) but in contrast with a bent geometry detected by X-ray crystallography (63). These studies, and similar ones conducted on other proteins (31, 99, 114, 136), indicate that lattice interactions affect, to a different and unpredictable extent, various features of protein motility.

Intermolecular contact regions, although less specific and extended than interacting surfaces within oligomers, may select a particular conformer among those present in dilute solutions. In fact, structural differences have been detected with the use of high-resolution X-ray crystallography in different packings of a few proteins (19, 44, 61, 62, 100). The alternative conformations of bovine pancreatic trypsin inhibitor have been shown to correspond to low energy states, which are reproducible by molecular dynamic simulations (62). In the case of dimeric aspartate aminotransferase, subunits have been found in either of two conformational states, which correspond to one or another catalytic state of the enzyme (44). Further, this enzyme (2, 27) and dimeric malate dehydrogenase (6) exhibit structural and functional symmetry in solution but lose this symmetry in some crystal packings as a consequence of environmental asymmetry.

The stability and resistance to physical stress of protein crystals depend on the nature and concentration of the precipitating agents. High molecular weight poly(ethylene) glycol stabilizes lattices, even at subzero temperatures. A systematic procedure to replace salts with poly (ethylene) glycol and a co-solute has been described for rabbit muscle phosphoglucomutase and suggested to be of general applicability (102). The dynamic situation created by progressive desalting in the presence of substrates allows the formation of catalytically active complexes in unfractured crystals. Such complexes could neither be obtained in the presence of salts, which inhibit substrate binding, nor after complete removal of salts, as lattice-based binding cooperativity causes crystal cracking (103).

FUNCTION IN THE CRYSTAL ENVIRONMENT

Bioactivity in the crystal, first described by pioneering studies in the early 1960s, has been exploited to validate structure-based mechanisms, to characterize a unique structure in a defined environment, and to obtain stable derivatives for X-ray crystallography. Since the advent of synchrotron radiation and the revival of the Laue methods, the new field of time-resolved protein crystallography has created the expectation of bringing high-resolution structural studies on the time scale of physiologic events (38, 40). Experiments designed to trigger rapidly and monitor catalysis in the crystalline state represent an essential stage in the planning of the crystallographic work.

A Criterion to Compare Structure in the Crystal and in Solution

Functional studies of proteins in the crystalline state were undertaken initially to determine whether the molecular models that emerge from high-resolution crystallographic studies represent the species that is active in aqueous solution or an artifact of lattice interactions. In some cases, the active site structure could not be reconciled with the mechanism of action derived from chemical and kinetic studies of the protein in solution. Bioactivity provided a stringent, albeit indirect, criterion to compare protein structure and dynamics in the two physical states. Richards (104) and Rupley (111) summarized the original thoughts and work. Most of these studies involved binding and activity assays on microcrystalline suspensions. The provisional conclusion was that equilibrium binding constants are usually similar in solution and in the crystal, whereas rate constants may be greatly different because of lattice-induced restrictions to conformational changes associated with catalysis.

The Property of a Uniquely Defined Structure

Proteins in solution exist as an equilibrium mixture of conformers that may differ by even large displacements of one region with respect to another, including transitions from open to closed conformation, hinge bending motions, and quaternary rearrangements. Binding of substrate, or the ensuing catalytic reactions, provides the energy to stabilize one conformer over the others. In the crystal, packing interactions are an alternative source of stabilization energy.

An interesting example is provided by hexokinase crystals grown in the presence of glucose and adenosine diphosphate (ADP), a competitive inhibitor of adenosine triphosphate (ATP) that induces the catalytically active state of the enzyme (141). On release of ADP, lattice interactions prevent the enzyme from collapsing to the inactive conformation of the binary complex. The stabilized enzyme can bind MgATP and, thus, catalyze the phosphoryl transfer reaction. Further, the affinity for glucose of enzyme crystallized in its presence in the closed conformation is two to three orders of magnitude higher than in solution, where a fraction of the binding energy is expended to induce this conformation. For the same reason, chicken heart mitochondrial aspartate aminotransferase crystallized in the closed form exhibits a higher affinity for oxo-acid substrates than does enzyme in solution (75). The functional asymmetry exhibited by the same dimeric enzyme in triclinic crystals has no counterpart in solution as it reflects the prevention of the substrate-dependent conformational change by the microenvironment of one of the two subunits (59, 60). Another interesting example of the role of lattice interactions in the stabilization of an elusive conformation is offered by crystals of human hemoglobin grown in the presence of poly(ethylene) glycol, in which the molecules remain in the T quaternary state in the presence of oxygen (9). A modest fraction of oxidized hemes prevents crystal cracking even in the presence of saturating oxygen, probably because methemoglobin molecules have a lesser tendency to convert to the R state than do oxygenated molecules. It has been possible, therefore, to characterize the oxygen-binding properties (88, 105, 109) of a T-state hemoglobin of known structure (9, 65, 66).

The structures of three complexes among proteins involved in electron transfer reactions have been determined to high resolution (11, 12, 91). The question arises, however, whether intermolecular recognition sites are the same in these crystals and in the physiologic milieu. Our studies on the binary complex between methylamine dehydrogenase and amicyanin from *Paracoccus denitrificans* and on the ternary complex that contains cytochrome c_{551i} demonstrate catalytic activity and electron transfer, supporting the view that the observed structures are useful models to identify the intermolecular pathway.

A Means to Obtain Derivatives for X-Ray Crystallography

Stable derivatives for X-ray diffraction studies can be prepared either by crystallizing the complex or by soaking crystals of the parent protein in a medium that contains the suitable reagent. The success of the latter method depends on accessibility of the active site and retention of bioactivity and, further, on compatibility of the new structure with the preexisting lattice. For example, the active site on the catalytic domain of a thermophilic endocellulase is open, and the enzyme fragment can be crystallized with bound cellobiose; however, this inhibitor cannot gain access to the site in a soaking experiment (124). Exposure of deoxyhemoglobin crystals, grown in the presence of salts, to oxygen leads to severe cracking, likely because of the onset of the quaternary transition; therefore, R-state oxyhemoglobin has been crystallized directly from solution (96). Conversely, oxygenation of deoxyhemoglobin crystals grown in poly(ethylene) glycol solutions results in the formation of T-state oxyhemoglobin (9, 88). Remarkably, CTP-ligated aspartate carbamoyltransferase in the T state can be converted progressively, on soaking with L-aspartate and phosphate, to the R state, which is usually obtained by crystallization in the presence of N-(phosphonoacetyl)-L-aspartate (35). The possible coexistence of the T and R states of an oligomeric molecule within a single crystal, and the role of lattice interactions in the control of the relative amounts, are confirmed by structural studies on L-lactate dehydrogenase from *Bifidobacterium longum* (52). Crystals that contain an equimolar mixture of the two states resulted from a solution in which the R state largely prevailed over the T state. This situation offers the unique opportunity to determine the structures of both quaternary forms within the same lattice.

Protein activity in the crystal has been exploited to prepare both reversible complexes and dead-end compounds. Stabilization of transient intermediates has been achieved by reducing temperature, maintaining thc crystal pH outside the range of protein activity, depriving the system of a substrate necessary for the progress of reaction, and crystallizing mutant enzymes with an altered kinetic profile.

Photorelease of a cinnamoyl group, which inhibits the active site of γ chymotrypsin covalently, has been used to switch on activity simultaneously throughout the crystal and thus initiate reaction with the mechanism-based inhibitor 3-benzyl-6-chloro-2-pyrone. The structures of the ''caged,'' nascent-free and pyrone inhibited enzyme have been collected by Laue X-ray crystallography (127–129). Although transition state analogues usually have been prepared by co-crystallization (68), formation in situ has been obtained by letting crystalline glycogen phosphorylase b catalyze the synthesis of heptulose-2-phosphate from heptenitol and phosphate (24, 55, 77). Structural snapshots along the catalytic pathway of a class μ glutathione transferase have been obtained by soaking crystals of the native enzyme with a transition-state analogue and with the product of a nucleophilic aromatic substitution reaction (53, 54). Similarly, by exploiting the effects of pH and temperature on the rates of the elementary steps in the conversion of 1,2-dichloroethane to 2-chloroethanol and chloride, catalyzed by haloalkane dehalogenase from *Xanthobacter autotrophicus*, the structures of the enzyme with reversibly bound substrate, of the alkylated intermediate, and, after its hydrolysis, of the complex with chloride have been isolated and, with the use of classic crystallographic methods, determined (138). The use of the Laue method of white-beam crystallography has allowed the visualization, with a temporal resolution of a few minutes, of the process of activation of p-guanidinobenzoyl-trypsin after a pH jump of three units. As an early event, a novel water molecule entered the active site and achieved a position suitable for nucleophilic attack of the acyl bond (121).

The techniques of crystallographic cryoenzymology (23) have permitted the determination of the structure of a specific acylenzyme intermediate in the hydrolysis of N-carbobenzoxy-L-alanine p-nitrophenyl ester, catalyzed by porcine pancreatic elastase, by forming it at −26°C and collecting diffraction data at −55°C (21). Below this temperature, a glasslike transition in the dynamic properties of proteins occurs with concomitant loss of the ability to bind or release ligands, as shown by a crystallographic study on ribonuclease A (101). Timed soaking of *Staphylococcus aureus* β-lactamase crystals with clavulanate, followed by flash-cooling to −120°C, has allowed the trapping of two acylenzyme intermediates, *cis* enamine and decarboxylated *trans* enamine, which represent two different stages of inhibitor degradation (10). By soaking with penicillin G a deacylation-defective mutant of β-lactamase from a penicillin-resistant strain of *Escherichia coli*, it has become possible to isolate and determine the structure of a stable natural acylenzyme at room temperature (130).

The status of active intermediate for the crystalline ternary complex, formed by glutaminyl-tRNA-synthetase, tRNA2Gln, and ATP, which is stable and suitable for diffraction studies, has been demonstrated by the hydrolysis of ATP and release of pyrophosphate after addition of glutamine (95).

Guanosine triphosphatase (GTP-ase) activity, retained in the crystalline state by the oncogene product Ha-Ras p21 protein, has been the premise to collect time-resolved structural data after initiation of a single transient reaction by flash photolysis of caged GTP (116, 118). This technique of reaction initiation is also being used in time-resolved studies on phosphorylase b (24, 37). Experiments on these enzymes are examples of the use of photolabile caged compounds as precursors of substrates to be made available synchronously and uniformly throughout a crystal (18, 76).

In the case of enzymes that catalyze fully reversible reactions, as D-xylose isomerase (17) or aspartate aminotransferase (2, 26, 75, 81), stable mixtures of true catalytic intermediates suitable for crystallographic analysis have been accumulated by soaking crystals with substrates and products.

Two sequential intermediates in the catalytic pathway of the nicotinamide adenine dinucleotide phosphate (NADP)-dependent isocitrate dehydrogenase from *E. coli* have been analyzed by Laue crystallography. Each intermediate has been trapped under steady-state conditions in the crystal of a mutant in which one key catalytic residue has been substituted to change the limiting step of reaction and drastically reduce its rate. High occupancy of catalytic sites in the crystal has been achieved and maintained by continuous flow of saturating substrates (7). In this study, enzyme activity in the crystal has been investigated quantitatively by steady-state kinetics and compared with the activity in solution.

ACTIVITY ASSAYED WITHOUT THE CRYSTAL

For several decades, the cornerstone in the quantitative description of enzyme activity has been steady-state kinetics. In principle, steady-state kinetic parameters can be determined for enzymes in different physical states and are straightforward reporters of binding and catalytic efficiency. As realized by early students of enzyme function in the crystal, however, diffusion of substrates through the crystal channels is restricted, whereas turnover within the surface layers occurs rapidly. Consequently, it is difficult to establish how many sites actually participate in catalysis. Numerous studies have addressed the diffusion problem (22, 45, 72, 140, 145, and references therein). A critical crystal thickness has been defined, below which diffusion restrictions are not appreciable. This critical thickness, in reactions with natural substrates, generally has been found to be of the order of a few microns. In the case of an extremely efficient catalyst, e.g. Δ^5–3-ketosteroid isomerase (140), however, it has been estimated to be of the same order of the unit cell, whereas for slow enzymes and poor substrates or at subzero temperatures, it can approach the value typical of a crystal suitable for X-ray studies.

To abolish undesired crystal dissolution in the course of activity studies, microcrystals sometimes need to be stabilized by cross-linking reagents. A preliminary characterization of microcrystals by electron microscopy is usually required to verify that their habits conform to those of the crystals used for X-ray crystallography and, therefore, that their functional properties are representative of those of the larger samples.

Micron-sized crystals are not handled as individual entities but as microcrystalline suspensions, the equivalent of dialysis chambers (140), that sequester enzyme molecules but allow free access to small ligands. Activity assays are based on the detection of time-dependent changes in the concentration of substrates or products in the suspending medium, without the crystal.

Phosphorylases a and b have been the objects of some of the most detailed and thought-provoking studies of structure and activity of an enzyme in the crystalline state. In an early work, both forms were found to be active catalytically in the direction of saccharide synthesis. The maximum activity of cross-linked microcrystalline suspensions was reduced, however, with respect to solution by a factor of 10–50 for phosphorylase b and a factor of 50–100 for phosphorylase a (56). Because the K_m values were unaffected by crystallization, the loss of catalytic activity was attributed tentatively to a reduced conformational flexibility rather than to an altered conformation. An interesting finding was the partial retention of homotropic cooperativity by crystalline phosphorylase a under assay conditions in which glucose-1-P was kept constant while the acceptor maltoheptaose concentration varied. The Hill coefficient for this reaction was $n = 1.17$ in solution and $n = 1.08$ in the crystal. Recent studies on the activity of microcrystalline suspensions of R-state tetrameric glycogen phosphorylase b have demonstrated that the reduction in activity is caused more by association of dimers to form tetramers than by crystallization itself (64).

A very detailed steady-state kinetic analysis of cross-linked microcrystalline carboxypeptidase A was undertaken to find functional differences with respect to solution that could explain the inconsistency of the observed active site structure with the proposed catalytic mechanism (125).

The consequences of asymmetric packing on the function of a symmetric oligomer are emphasized by a kinetic investigation, in which suspensions of triclinic microcrystals of chicken heart mitochondrial aspartate aminotransferase were used. The kinetic behavior reflected the presence of two equivalent populations of active sites, with distinct Michaelis and catalytic constants (59, 60). Similarly, kinetic studies on cross-linked microcrystalline suspensions of pig heart cytoplasmic malate dehydrogenase revealed a dramatic decrease and different pH dependence of enzyme activity with respect to solution, as well as a biphasic rate of alkylation (145). The results of these studies are of general importance, as they underline how sensitive the conformational and functional equivalence of identical subunits in oligomeric proteins is to environmental conditions.

Aldolase in a monoclinic crystal form retains essentially full catalytic activity (131, 133). The minor differences with respect to solution apparently are caused by a somewhat slower dissociation of products. The crystalline conformer, however, has the unique property of being inactivated by phosphate (132).

Microcrystalline suspensions of the $\alpha_2\beta_2$ complex of tryptophan synthase from *Salmonella typhimurium* have been studied with the use of the standard assay procedures to measure the α, β, $\alpha\beta$, and serine deaminase reactions (1). These studies demonstrated that both the α and β sites are accessible to the respective substrates and are catalytically competent. In addition, they indicated the limited extent to which packing interactions exert restrictions on intersubunit communication within this finely tuned oligomeric enzyme.

The possibility of nonproductive binding of substrate has been considered to explain steady-state data obtained for crystalline suspensions of D-alanyl-D-alanine peptidase from *Streptomyces* R61 (58). Similarly, alternative binding modes have been suggested to explain the hydrolysis pattern of hexa-N-acetylglucosamine by crystalline turkey egg white lysozyme, as established by thin layer chromatography of products released in the crystal surrounding medium (46). This work is of remarkable interest, as it creates the premises for a time-resolved X-ray crystallographic study, with the use of substrate diffusion or a modest pH jump as a means to initiate the slow hydrolytic reaction of a radiolabeled substrate-analogue that leads to a unique product.

ACTIVITY MONITORED WITHIN THE CRYSTAL BY MICROSPECTROPHOTOMETRY

Many of the spectroscopic techniques that allow the investigation of structural, dynamic, and functional properties of proteins in solution can also be applied in solid-state studies. Among these techniques, polarized absorption microspectrophotometry and linear dichroism have been used extensively to investigate the electronic structure and the orientation of chromophores within crystals (15, 16, 25, 33, 43, 69–71, 74, 81, 92, 139). Single-crystal microspectrophotometric studies of protein function have included the identification of the absorbing species (4, 5, 13, 27–29, 74, 79, 80, 93, 94, 119, 120, 134) and the determination of equilibrium binding and kinetic parameters (8, 47, 57, 75, 78, 84–89, 105, 106, 108–110, 113, 115, 126, 137).

Absorption of plane polarized light with the electric vector parallel to the principal optical directions of the crystal obeys the Beer-Lambert law. Spectra can be analyzed and resolved into individual components by use of lognormal distribution curves (81), singular value decomposition (42), and linear combination of reference spectra (88, 105). By normalizing the polarized absorptions along principal optical directions to a common crystal thickness, the isotropic spectrum can be calculated.

Microspectrophotometric measurements provide information essential in the planning and in the interpretation of X-ray crystallographic studies. Microspectrophotometers have been designed to be used on-line with the X-ray apparatus (14, 36) to monitor reactions in the course of time-resolved diffraction experiments (24, 30, 37, 89, 135).

Single Transient Kinetics

Absorption microspectrophotometry allows us to monitor the time course of accumulation or disappearance of metastable catalytic intermediates on soaking a protein crystal with an appropriate reagent. With natural substrates and at room temperature, the observed rates are usually limited by diffusion. The use of cryoenzymology, the choice of an appropriate pH outside the range of optimal enzyme activity, the selection of a less active mutant, and the use of slow-reacting substrate-analogues provide the means to measure diffusion-independent rates. Photoactivation of caged substrates preaccumulated in the crystal channels may open the possibility of investigating fast reactions.

An early example of a slow-reacting substrate-analogue being used to compare the rates of individual catalytic steps and the cooperative behavior of an oligomeric enzyme in the crystal and in solution is the characterization of the reaction between crystalline D-glyceraldehyde-3-phosphate dehydrogenase from lobster and pig muscle and the chromophoric acylating reagent β-(2-furyl)acryloyl phosphate (4, 84, 137). In solution, this compound labels specifically two of the four chemically equivalent sites of enzyme from various sources, including yeast but, surprisingly, not *Bacillus stearothermophilus*. The kinetics of three elementary steps of reaction were investigated by exploiting the large spectral changes associated with each of them: formation of a metastable β-(2-furyl)acryloyl enzyme intermediate, activation of the acyl bond toward nucleophilic attack, and arsenate-assisted hydrolysis of the activated acyl bond. Lattice forces did not alter the intersubunit interactions that control the kinetic mechanism of acylation leading to half of the sites reactivity, nor did they prevent the conversion of the inert to the active conformational state of acylated subunits caused by nicotinamide adenin dinucleotide (NAD) binding. On the other side, the rate of active site alkylation by iodoacetamide was the same for the four subunits, as in solution (41).

As an extension of the early work on crystalline indoleacryloyl-α-chymotrypsin (108), the deacylation kinetics of indoleacryloyl- and furylacryloyl-γ-chymotrypsin were monitored over the full pH range of catalytic interest (78). Diffusion-limited acylation was obtained by soaking crystals of the native enzyme with the acylimidazole substrates in media of low pH, in which the resulting acylenzymes were stable for days. After completion of acylation and removal of excess substrate, deacylation was initiated, as in the case of α-chymotrypsin, by a pH jump obtained by replacing the initial crystal suspending medium with another of similar composition but higher pH. The time required by this procedure, several seconds to a few minutes, was short compared with the duration of the ensuing reaction. The maximal rates were the same as in solution, whereas the pK_a of the catalytic system was shifted by 0.9 units. This finding might reflect a reported structural anomaly of the active site of γ-chymotrypsin, with respect to other serine proteases; a distance between O^{γ} of Ser 195 and $N^{\epsilon}2$ of His 57, which apparently is too long for hydrogen bond formation, was observed. In the case of α-lytic protease, solid-state NMR studies have explained contradictory results of solution NMR and X-ray crystallographic studies on the occurrence of a hydrogen bond between Ser 195 and His 57; in the crystal, histidine titrates with a pK_a nearly one unit higher than in solution (123).

In a recent study, the time course of the reaction of cytochrome c peroxidase with hydrogen peroxide to generate a doubly oxidized enzyme intermediate was monitored with the use of a portable diode array microspectrophotometer (30). These measurements have provided the basis for the interpretation of the diffraction data concomitantly collected by Laue techniques. The photochemical activation of a caged substrate has been monitored in the course of the crystallographic experiment with the same microspectrophotometric equipment (37). Photorelease of phosphate from highly concentrated 3,5-dinitrophenyl phosphate within crystals of glycogen phosphorylase b appears to be restricted to the outermost layers of the crystal, where most of the exciting light is absorbed. Reaction initiation is limited, therefore, by diffusion of substrate to the interior of the crystal.

Reversible Ligand Binding

In a typical binding experiment, individual crystals are suspended in media that contain different concentrations of ligand. As long as the number of ligand molecules is large with respect to the number of binding sites within the crystal, and the free ligand concentration can be assumed to be the same in the outside medium and within the crystal channels, binding is governed simply by the dissociation constant of the crystalline protein (86). For example, assuming a protein concentration of 10 mM, the number of sites in a crystal of linear dimensions

100 μm × 100 μm × 100 μm, would be only one tenth the number of ligand molecules contained in 100 μL of a 1 μM solution. Thus, if the dissociation constant is in the micromolar range, the ligand will concentrate within the crystal to bind to an appreciable fraction of the active sites, although its concentration in the surrounding medium remains nearly constant.

A recent interesting example of equilibrium binding measurements is the determination of the oxygen saturation curve for crystalline human hemoglobin A (88, 105), hemoglobin Rothschild (106), and des-(Arg141α) hemoglobin (57), crystallized in the T state from poly(ethylene) glycol. In the presence of oxygen, the protein remains in the same quaternary state with intact salt bridges (9, 65, 66). Single-crystal polarized absorption spectra have been recorded as a function of oxygen pressure and analyzed as a linear combination of the spectra of deoxyhemoglobin, oxyhemoglobin, and methemoglobin. In the case of hemoglobin A, the saturation function is hyperbolic, as predicted, in the absence of the allosteric transition, by the Monod, Wyman, and Changeux model (83). In agreement with the stereochemical mechanism proposed by Perutz (97), the Bohr effect is absent. Hemoglobin Rothschild (Trp37βArg) and des(Arg141α) hemoglobin crystals bind oxygen with an affinity 10 and 15 times higher, respectively, than does hemoglobin A. Because the intermolecular interactions are essentially the same for the three hemoglobins, these findings indicate that lattice forces do not prevent the expression of molecular differences.

From the different projections of the α and β hemes of hemoglobin A on two optical directions of the crystal, it has been possible to separate the contribution of the α and β hemes to oxygen binding (105). In agreement with current studies on crystals of metal hybrids, $\alpha(Ni^{2+})_2\beta(Fe^{2+})_2$ and $\alpha(Fe^{2+})_2\beta(Ni^{2+})_2$, where Ni^{2+} does not bind oxygen, the affinity values for α and β hemes have been found to differ by a factor of less than four.

Similar oxygen binding measurements carried out on crystals of the homodimeric hemoglobin I from *Scapharca inaequivalvis*, grown in the deoxygenated state, show that this protein retains the positive cooperativity exhibited in solution (85). This finding indicates that the conformational changes that support the cooperative behavior are not affected by lattice interactions and, therefore, can be observed by crystallographic methods.

Equilibrium Distribution of Catalytic Intermediates

The pyridoxal 5′-phosphate–dependent enzyme aspartate aminotransferase was one of the first enzymes to be investigated by polarized absorption microspectrophotometry (27, 69, 70, 80, 86, 110). The transamination reaction in solution is fully reversible, and interconverting catalytic intermediates can be accumulated in proportions dependent on the relative concentrations of substrates and products in the suspending medium. Crystals of cytosolic and mitochondrial chicken and pig heart enzymes soaked with the amino- and oxo-acid substrates and analogues exhibit the absorption spectra characteristic of the catalytic complexes in solution. For the pig heart mitochondrial isoenzyme, the dissociation constants of the four and five carbon pairs of amino- and oxo-acid substrates have been determined and found to be very close to the corresponding constants in solution. The pseudo–first order rates of transamination of the slowly reacting substrates alanine and pyruvate are also similar (86, 110). After the crystallographic observation that crystal packing prevents the open-to-closed conformation transition of one of the two subunits, a spectroscopic study demonstrated that the cytosolic isoenzyme from pig heart binds natural substrates mostly at one site, which retains the affinity of the soluble state (47). Polarized light absorption spectra confirm the occurrence of coenzyme rotation during transamination (2, 69, 70, 80, 139), in general agreement with a proposed reaction model (51). A detailed study of orthorhombic crystals of the pig heart cytosolic isoenzyme (81) has enabled researchers to assign principal molecular directions of the transition dipole moment within the plane of the pyridoxal 5′-phosphate ring and to measure the coenzyme tilt on binding of 2-methylaspartate to one of the two subunits. The empty subunit maintains the spectrum of the high pH form of the native enzyme even at pH 5.8.

Malashkevich et al (75) combined X-ray diffraction and polarized absorption studies of true interconverting catalytic intermediates of chicken heart mitochondrial aspartate aminotransferase to prove that the closed form of the enzyme, stabilized by lattice interactions, has a four orders of magnitude higher affinity for oxo-acid substrates than does enzyme in solution or in crystals in which substrate binding induces the open-to-closed conformational transition.

Polarized absorption studies of the tryptophan synthase $\alpha_2\beta_2$ complex from *S. typhimurium*, the structure of which is known (48), have shown that, on reaction of serine with the β-site–bound pyridoxal 5′-phosphate, the predominant intermediates are the external aldimine and the α-aminoacrylate Schiff base, as in solution. Their equilibrium distribution is controlled allosterically by α-subunit ligands and modulated by pH and monovalent cations (87, 93).

Relaxation Kinetics

T jump has been used as a convenient means to perturb equilibria in solution rapidly. The method cannot find immediate application in the study of crystalline samples because of the difficulties inherent to heat pulse delivery as well as to heat and ligands diffusion (82). A pioneering study, however, reports the measurement, by single-crystal polarized absorption spectroscopy, of the association equilibrium and association kinetics between thiocyanate and methemoglobin (126). Two relaxations have been detected, in the millisecond and second ranges, that describe differential ligand binding to the α and β hemes.

Photoreceptor protein crystals are an ideal system for time-resolved studies, as a light pulse can initiate a response synchronously and uniformly throughout the sample. A time-resolved X-ray diffraction study of bacteriorhodopsin in purple membranes of *Halobacterium halobium* has demonstrated that its tertiary structure changes during the photocycle. Three-dimensional crystals of bacteriorhodopsin, grown in a needle-shaped pseudohexagonal form, have been used for a detailed polarized absorption microspectrophotometric study in the dark- and light-adapted state (113). The most stable intermediate in the cycle, the M_{410} species, formed approximately 10 times more rapidly and decayed approximately 15 times more slowly in the crystal than in the purple membrane and, therefore, could be accumulated to reach a high proportion at room temperature with continuous illumination from a xenon-lamp. Further enrichment of the intermediate was achieved by using crystals of the bacteriorhodopsin mutant Asp96Asn. Some of these spectroscopic studies have been extended to a better ordered orthorhombic crystal form of the protein (112).

The approach toward a time-resolved crystallographic study on the time scale of less than 1 s is documented by a single-crystal spectroscopic study of the photoactive yellow protein, which is a low-molecular-weight photoreceptor, isolated from *Ectothiorhodopsin halophila*, the structure of which has been refined to high resolution (89). On initiation of the reversible photocycle by laser excitation, the kinetics of relaxation from the photostationary state was monitored and found to be described by two exponentials, rather than one as in solution. This finding has been attributed to the lattice-induced presence of two metastable bleached intermediates. Because a model consistent with kinetics is necessary to extract the underlying structural intermediates from time-resolved X-ray diffraction measurements, the results of this exciting study stress once more the impossibility of relying exclusively on the protein behavior in solution.

The recent extension of X-ray crystallography of proteins to liquid helium temperatures has enabled two groups to investigate the time-resolved structure of photolyzed carbonmonoxy myoglobin (117, 135).

In this context, single-crystal microspectrophotometry has been used to obtain spectral data from which the kinetics of ligand rebinding could be calculated (14, 135).

CONCLUSIONS

The number of structures which are determined by X-ray crystallography is rapidly increasing and includes highly organized complexes and short-lived reaction intermediates. Their biological relevance must be assessed because lattice interactions perturb conformational equilibria, as it has been most clearly demonstrated by the templatetype binding behavior of proteins that undergo ligand-dependent transitions in solution. Laue diffraction methods permit the collection of high resolution data on the time scale of protein function in the physiologic milieu, but catalysis in the crystal may be affected by kinetic barriers that retard function-associated conformational changes. The difficulties inherent to initiation of fast reactions within a crystal have not yet been overcome, except in the case of photoreceptor proteins. A full description of protein action in the crystal that correlates structural changes with the progress of catalysis is still a challenge for the future.

Literature Cited

1. Ahmed SA, Hyde CC, Thomas G, Miles EW. 1987. Microcrystals of tryptophan synthase $\alpha_2\beta_2$ complex from *Salmonella typhimurium* are catalytically active. *Biochemistry* 260:5492–98
2. Arnone A, Rogers PH, Hyde CC, Makinen MW, Feldhaus R, et al. 1984. Crystallographic and chemical studies on cytosolic aspartate aminotransferase. In *Chemical and Biological Aspects of Vitamin B6 Catalysis*, Pt. B, pp. 171–93. New York: Liss
3. Bennet WS, Huber R. 1984. Structural and functional aspects of domain motions in proteins. *CRC Crit. Rev. Biochem.* 15:291–384
4. Berni R, Mozzarelli A, Pellacani L, Rossi GL. 1977. Catalytic and regulatory properties of D-glyceraldehyde-3-phosphate dehydrogenase in the crystal. *J. Mol. Biol.* 110:405–15
5. Bignetti E, Rossi GL, Zeppezauer E. 1979. Microspectrophotometric measurements on single crystals of coenzyme containing complexes of horse liver alcohol dehydrogenase. *FEBS Lett.* 100:17–22
6. Birktoft JJ, Banaszak LJ. 1983. The presence of a histidine-aspartic acid pair in the active site of 2-hydroxyacid dehydrogenases. X-ray refinement of cytoplasmic malate dehydrogenase. *J. Biol. Chem.* 258:472–82
7. Bolduc JM, Dyer DH, Scott WG, Singer P, Sweet RM, et al. 1995. Mutagenesis and Laue structures of enzymes intermediates: isocitrate dehydrogenase. *Science* 268:1312–18
8. Bolognesi M, Cannillo E, Ascenzi P, Giacometti GM, Merli A, Brunori M. 1982. Reactivity of ferric *Aplysia* and sperm whale myoglobins towards imidazole. X-ray and binding study. *J. Mol. Biol.* 158:305–15
9. Brzozowski A, Derewenda Z, Dod-

son E, Dodson G, Grabowski M, et al. 1984. Bonding of molecular oxygen to T state human haemoglobin. *Nature* 307:74–76

10. Chen CCH, Herzberg O. 1992. Inhibition of β-lactamase by clavulanate. Trapped intermediates in cryocrystallographic studies. *J. Mol. Biol.* 224: 1103–13
11. Chen L, Durley RCE, Mathews FS, Davidson VL. 1994. Structure of an electron transfer complex: methylamine dehydrogenase, amicyanin and cytochrome c551i. *Science* 264: 86–90
12. Chen L, Durley R, Poliks BJ, Hamada K, Chen Z, et al. 1992. Crystal structure of an electron-transfer complex between methylamine dehydrogenase and amicyanin. *Biochemistry* 31:4959–64
13. Chen L, Mathews FS, Davidson VL, Tegoni M, Rivetti C, Rossi GL. 1993. Preliminary crystal structure studies of a ternary electron transfer complex between a quinoprotein, a blue copper protein, and a c-type cytochrome. *Protein Sci.* 2:147–54
14. Chen Y, Srajer V, Ng K, LeGrand A, Moffat K. 1994. Optical monitoring of protein crystals in time-resolved x-ray experiments: microspectrophotometer design and performance. *Rev. Sci. Instrum.* 65:1506–11
15. Clark PA, Jansonius JN, Mehler EL. 1993. Semiempirical calculations of the electronic absorption spectrum of mitochondrial aspartate aminotransferase. *J. Am. Chem. Soc.* 115: 1894–902
16. Clark PA, Jansonius JN, Mehler EL. 1993. Semiempirical calculations of the electronic absorption spectrum of mitochondrial aspartate aminotransferase. The 2-methylaspartate complex. *J. Am. Chem. Soc.* 115: 9789–93
17. Collyer CA, Blow DM. 1990. Observations of reaction intermediates and the mechanism of aldose-ketose interconversion by D-xylose isomerase. *Proc. Natl. Acad. Sci. USA* 87: 1362–66
18. Corrie JET, Katayama Y, Reid GP, Anson M, Trentham DR. 1992. The development and application of photosensitive caged compounds to aid time-resolved structure determination of macromolecules. *Philos. Trans. R. Soc. London Ser.* A 340: 233–44
19. Crosio MP, Janin J, Jullien M. 1992. Crystal packing in six crystal forms of pancreatic ribonuclease. *J. Mol. Biol.* 228:243–51
20. Cruickshank DWJ, Helliwell JR, Johnson LN, eds. 1992. Time-resolved macromolecular crystallography. *Philos. Trans. R. Soc. London Ser. A* 340:167–334
21. Ding X, Rasmussen BF, Petsko GA, Ringe D. 1994. Direct structural observation of an acyl-enzyme intermediate in the hydrolysis of an ester substrate by elastase. *Biochemistry* 33: 9285–93
22. Doscher MS, Richards FM. 1963. The activity of an enzyme in the crystalline state: ribonuclease S. *J. Biol. Chem.* 238:2399–406
23. Douzou P, Petsko GA. 1984. Proteins at work: ''stop-action'' pictures at subzero temperatures. *Adv. Protein Chem.* 36:245–361
24. Duke EMH, Wakatsuki S, Hadfield A, Johnson LN. 1994. Laue and monochromatic diffraction studies on catalysis in phosphorylase b crystals. *Protein Sci.* 3:1178–96
25. Eaton WA, Hofrichter J. 1981. Polarized absorption and linear dichroism spectroscopy of hemoglobin. *Methods Enzymol.* 76:175–261
26. Eichele G, Ford GC, Glor M, Jansonius JN, Mavrides C, Christen P. 1979. The three-dimensional structure of mitochondrial aspartate aminotransferase at 4.5 Å resolution. *J. Mol. Biol.* 133:161–80
27. Eichele G, Karabelnik D, Halonbrenner R, Jansonius NJ, Christen P. 1978. Catalytic activity in crystals of mitochondrial aspartate aminotransferase as detected by microspectrophotometry. *J. Biol. Chem.* 253: 5239–42
28. Fotinou C, Kokkinidis M, Fritzsch G, Haase W, Michel H, Ghanotakis DF. 1993. Characterization of a photosystem II and its three-dimensional crystals. *Photosynth. Res.* 37:41–48
29. Frank HA, Aldema ML, Violette CA, Parot PH. 1991. Low temperature polarized absorption microspectroscopy of single crystals of the reaction center from *Rhodobacter sphaeroides* wild type strain 2.4.1. *Photochem. Photobiol.* 54:151–55
30. Fülöp V, Phizackerley RP, Soltis SM, Clifton IJ, Wakatsuki S, et al. 1994. Laue diffraction study on the structure of cytochrome c peroxidase compound-I. *Structure* 2:201–8
31. Gabellieri E, Strambini GB, Gualtieri P. 1988. Tryptophan phosphorescence and the conformation of liver

alcohol dehydrogenase in solution and in the crystalline state. *Biophys. Chem.* 30:61–67

32. Gallagher W, Tao F, Woodward C. 1992. Comparison of hydrogen exchange rates for bovine pancreatic trypsin inhibitor in crystals and in solution. *Biochemistry* 31:4673–80
33. Gay RR, Solomon EI. 1978. Polarized single crystal spectroscopic studies of oxyhemerytrin. *J. Am. Chem. Soc.* 100:1972–73
34. Gerstein M, Lesk AM, Chothia C. 1994. Structural mechanisms for domain movements in proteins. *Biochemistry* 33:6739–49
35. Gouaux JE, Lipscomb WN. 1989. Structural transitions in crystals of native aspartate carbamoyltransferase. *Proc. Natl. Acad. Sci. USA* 86: 845–48
36. Hadfield A, Hajdu J. 1993. A fast and portable microspectrophotometer for protein crystallography. *J. Appl. Crystallogr.* 26:839–42
37. Hadfield A, Hajdu J. 1994. On the photochemical release of phosphate from 3,5-dinitrophenyl phosphate in a protein crystal. *J. Mol. Biol.* 236: 995–1000
38. Hajdu J, Acharya KR, Stuart DI, Badford D, Johnson LN. 1988. Catalysis in enzyme crystals. *Trends Biochem. Sci.* 13:104–9
39. Hajdu J, Andersson I. 1993. Fast crystallography and time-resolved structures. *Annu. Rev. Biophys. Biomol. Struct.* 22:467–98
40. Hajdu J, Johnson LN. 1990. Progress with Laue diffraction studies on protein and virus crystals. *Biochemistry* 29:1669–78
41. Halasz P, Polgar L. 1982. Lack of asymmetry in the active sites of tetrameric D-glyceraldehyde-3-phosphate dehydrogenase during alkylation in the crystalline state. *FEBS Lett.* 143: 93–95
42. Henry ER, Hofrichter J. 1992. Singular value decomposition: application to the analysis of experimental data. *Methods Enzymol.* 210:129–92
43. Hofrichter J, Eaton WA. 1976. Linear dichroism of biological chromophores. *Annu. Rev. Biophys. Bioeng.* 5:511–60
44. Hohenester E, Jansonius JN. 1994. Crystalline mitochondrial aspartate aminotransferase exists in only two conformations. *J. Mol. Biol.* 236: 963–68
45. Hoogenstraaten W, Sluyterman AÆ. 1969. The activity of papain in the crystalline state. *Biochim. Biophys. Acta* 171:284–287 (Appendix)
46. Howell PL, Warren C, Amatayakul-Chantler S, Petsko GA, Hajdu J. 1992. Activity of crystalline turkey egg white lysozyme. *Proteins* 12: 91–99
47. Hubert E, Martinez-Carrion M. 1982. Equilibrium kinetics of substrate-enzyme interactions in single crystals of cytoplasmic aspartate aminotransferase. *J. Protein Chem.* 1:163–75
48. Hyde CC, Ahmed SA, Padlan EA, Miles EW, Davies DR. 1988. Three-dimensional structure of the tryptophan synthase $\alpha_2\beta_2$ multienzyme complex from *Salmonella typhimurium*. *J. Biol. Chem.* 263:17857–71
49. Iben IET, Keszthelyi L, Ringe D, Varo G. 1989. Charge motion in MbCO crystals after flash photolysis. *Biophys. J.* 56:459–63
50. Ivanov D, Sage JT, Kleim M, Powell JR, Asher SA, Champion PM. 1994. Determination of CO orientation in myoglobin by single-crystal infrared linear dichroism. *J. Am. Chem. Soc.* 116:4139–40
51. Ivanov VI, Karpeisky MY. 1969. Dynamic three-dimensional model for enzymic transamination. *Adv. Enzymol.* 32:21–53
52. Iwata S, Kamata K, Yoshida S, Minowa T, Ohta T. 1994. T and R states in the crystals of bacterial L-lactate dehydrogenase reveal the mechanism for allosteric control. *Nature Struct. Biol.* 1:176–85
53. Ji XH, Armstrong RN, Gilliland GL. 1993. Snapshots along the reaction coordinate of an S_NAr reaction catalyzed by glutathione transferase. *Biochemistry* 32:12949–54
54. Ji XH, Johnson WW, Sesay MA, Dickert L, Prasad SM, et al. 1994. Structure and function of the xenobiotic substrate binding site of a glutathione S-transferase as revealed by X-ray crystallographic analysis of product complexes with the diastereomers of 9-(S-glutathionyl)-10-hydroxy-9,10-dihydrophenanthrene. *Biochemistry* 33:1043–52
55. Johnson LN, Acharya KR, Jordan MD, McLaughlin PJ. 1990. Refined crystal structure of the phosphorylase-heptulose 2-phosphate-oligosaccharide-AMP complex. *J. Mol. Biol.* 211:645–61
56. Kasvinsky PJ, Madsen NB. 1976. Activity of glycogen phosphorylase in the crystalline state. *J. Biol. Chem.* 251:6852–59

57. Kavanaugh JS, Chafin, DR, Arnone A, Mozzarelli A, Rivetti C, et al. 1995. Structure and oxygen affinity of crystalline des(arg 141α) human hemoglobin A in the T state. *J. Mol Biol.* 248:136–50
58. Kelly JA, Waley SG, Adam M, Frère JM. 1992. Crystalline enzyme kinetics: activity of the *Streptomyces* R61 D-alanyl-D-alanine peptidase. *Biochim. Biophys. Acta* 1119:256–60
59. Kirsten H, Christen P. 1983. Catalytic activity of non-crosslinked microcrystals of aspartate aminotransferase in poly(ethylene glycol). *Biochem J.* 211:427–34
60. Kirsten H, Gehring H, Christen P. 1983. Crystalline aspartate aminotransferase: lattice-induced functional asymmetry of the two subunits. *Proc. Natl. Acad. Sci. USA* 80: 1807–10
61. Kishan KVR, Zeelen JPH, Noble MEM, Borchert TV, Wierenga RK. 1994. Comparison of the structures and the crystal contacts of trypanosomal triosephosphate isomerase in four different crystal forms. *Protein Sci.* 3:779–87
62. Kossiakoff AA, Randal M, Guenot J, Eigenbrot C. 1992. Variability of conformations at crystal contacts in BPTI represent true low-energy structures: correspondence among lattice packing and molecular dynamics structures. *Proteins* 14:65–74
63. Kuriyan J, Wilz S, Karplus M, Petsko GA. 1986. X-ray structure and refinement of carbon-monoxy (Fe II)-myoglobin at 1.5 Å resolution. *J. Mol. Biol.* 192:133–54
64. Leonidas DD, Oikonomakos NG, Papageorgiou AC, Sotiroudis TG. 1992. Kinetic properties of tetrameric glycogen phosphorylase *b* in solution and in the crystalline state. *Protein Sci.* 1:1123–32
65. Liddington R, Derewenda Z, Dodson E, Hubbard R, Dodson G. 1992 High resolution crystal structure and comparisons of T-state deoxyhaemoglobin and two liganded T-state haemoglobins: T(α-oxy)haemoglobin and T(met)haemoglobin. *J. Mol. Biol.* 228:551–79
66. Liddington R, Derewenda Z, Dodson G, Harris D. 1988. Stucture of the liganded T state of haemoglobin identifies the origin of cooperative oxygen binding. *Nature* 331:725–28
67. Lim M, Jackson TA, Anfinrud PA. 1995. Binding of CO to myoglobin from a heme pocket docking site to form a nearly linear Fe-C-O. *Science* 269:962–66
68. Lolis E, Petsko GA. 1990. Transition-state analogues in protein crystallography: probes of the structural source of enzyme catalysis. *Annu. Rev. Biochem.* 59:597–630
69. Makarov VL, Kochkina VM, Torchinsky YM. 1980 Polarized light absorption spectra of single crystals of aspartate transaminase from chicken heart cytosol. *FEBS Lett.* 114:79–82
70. Makarov VL, Kochkina VM, Torchinsky YM. 1981. Reorientations of coenzyme in aspartate transaminase studied on single crystals of the enzyme by polarized-light spectrophotometry. *Biochim. Biophys. Acta* 659:219–28
71. Makinen MW, Churg AK. 1983. Structural and analytical aspects of the electronic spectra of hemeproteins. *Phys. Bioinorg. Chem. Ser. Iron-Porphyrins* 1 (Pt. 1):141–235
72. Makinen MW, Fink AL. 1977. Reactivity and cryoenzymology of enzymes in the crystalline state. *Annu. Rev. Biophys. Bioeng.* 6:301–43
73. Makinen MW, Houtchens RA, Caughey WS. 1979. Structure of carboxymyoglobin in crystals and in solution. *Proc. Natl. Acad. Sci. USA* 76: 6042–46
74. Makinen MW, Schichman SA, Hill SC, Gray HB. 1983. Heme-heme orientation and electron transfer kinetic behaviour of multi-site oxidation-reduction enzymes. *Science* 222: 929–31
75. Malashkevich VN, Toney MD, Jansonius JN. 1993. Crystal structures of true enzymatic reaction intermediates: aspartate and glutamate ketimines in aspartate aminotransferase. *Biochemistry* 32:13451–62
76. McCray JA, Trentham DR. 1989. Properties and uses of photoreactive caged compounds. *Annu. Rev. Biophys. Biophys. Chem.* 18:239–70
77. McLaughlin PJ, Stuart DJ, Klein HW, Oikonomakos NG, Johnson LN. 1984. Substrate-cofactor interactions for glycogen phosphorylase b: a binding study in the crystal with heptenitol and heptulose 2-phosphate. *Biochemistry* 23:5862–73
78. Merli A, Rossi GL. 1986. Deacylation kinetics of γ-chymotrypsin in solution and in the crystal. *FEBS Lett.* 199:179–81
79. Merli A, Rossi GL, Bolognesi M, Gatti G, Morpurgo L, Finazzi-Agrò

A. 1988. Single crystal absorption spectra of ascorbate oxidase from green zucchini squash. *FEBS Lett.* 231:89–94

80. Metzler CM, Metzler DE, Martin DS, Newman R, Arnone A, Rogers P. 1978. Crystalline enzyme-substrate complexes of aspartate aminotransferase. *J. Biol. Chem.* 253:5251–54
81. Metzler CM, Mitra J, Metzler DE, Makinen MW, Hyde CC, et al. 1988. Correlation of polarized absorption spectroscopic and X-ray diffraction studies of crystalline cytosolic aspartate aminotransferase of pig hearts. *J. Mol. Biol.* 203:197–220
82. Moffat K. 1989. Time-resolved macromolecular crystallography. *Annu. Rev. Biophys. Biophys. Chem.* 18: 309–32
83. Monod J, Wyman J, Changeux J. 1965. On the nature of allosteric transitions: a plausible model. *J. Mol. Biol.* 12:88–118
84. Mozzarelli A, Berni R, Rossi GL, Vas M, Bartha F, Keleti T. 1982. Protein isomerization in the NAD^+ dependent activation of β-(2-furyl)acryloyl-glyceraldehyde-3-phosphate deydrogenase in the crystal. *J. Biol. Chem.* 257:6739–44
85. Mozzarelli A, Bettati S, Rivetti C, Rossi GL, Colotti G, Chiancone E. 1996. Cooperative oxygen binding to *Scapharca inaequivalvis* hemoglobin in the crystal. *J. Biol. Chem.* 271: 3627–32
86. Mozzarelli A, Ottonello S, Rossi GL, Fasella P. 1979. Catalytic activity of aspartate aminotransferase in the crystal. Equilibrium and kinetic analysis. *Eur. J. Biochem.* 98:173–79
87. Mozzarelli A, Peracchi A, Rossi GL, Ahmed SA, Miles EW. 1989. Microspectrophotometric studies on single crystals of the tryptophan synthase $\alpha_2\beta_2$ complex demonstrate formation of enzyme-substrate intermediates. *J. Biol. Chem.* 264:15774–80
88. Mozzarelli A, Rivetti C, Rossi GL, Henry ER, Eaton WA. 1991. Crystals of haemoglobin with the T quaternary structure bind oxygen noncooperatively with no Bohr effect. *Nature* 351:416–19
89. Ng K, Getzoff ED, Moffat K. 1995. Optical studies of a bacterial photoreceptor protein, photoactive yellow protein, in single crystals. *Biochemistry* 34:879–90
90. Pedersen TG, Sigurskjold BW, Andersen KV, Kjaer M, Poulsen FM, et al. 1991. A nuclear magnetic resonance study of the hydrogen-exchange behaviour of lysozyme in crystals and solution. *J. Mol. Biol.* 218:413–26
91. Pelletier H, Kraut J. 1992. Crystal structure of a complex between electron transfer partners, cytochrome c peroxidase and cytochrome c. *Science* 258:1748–55
92. Penfield KW, Gay RR, Himmelwright RS, Eickman NC, Norris VA, et al. 1981. Spectroscopic studies on plastocyanin single crystals: a detailed electronic structure determination of the blue copper active site. *J. Am. Chem. Soc.* 103:4382–88
93. Peracchi A, Mozzarelli A, Rossi GL. 1995. Monovalent cations affect dynamic and functional properties of the tryptophan synthase $\alpha_2\beta_2$ complex. *Biochemistry* 34:9459–65
94. Peracchi A, Mozzarelli A, Rossi GL, Dominici P, Borri Voltattorni C. 1994. Single crystal polarized absorption microspectrophotometry of aromatic L-amino acid decarboxylase. *Protein Pept. Lett.* 1:98–105
95. Perona JJ, Rould MA, Steitz TA. 1993. Structural basis for transfer RNA aminoacylation by *Escherichia coli* glutaminyl-tRNA synthetase. *Biochemistry* 32:8758–71
96. Perutz MF. 1968. Preparation od hemoglobin crystals. *J. Crystallogr. Growth* 2:54–56
97. Perutz MF. 1970. Stereochemistry of cooperative effects in haemoglobin. *Nature* 228:726–39
98. Petsko GA, Ringe D. 1984. Fluctuations in protein structure from X-ray diffraction. *Annu. Rev. Biophys. Bioeng.* 13:331–71
99. Potter WT, Houtchens RA, Caughey WS. 1985. Crystallization-induced changes in protein structure observed by infrared spectroscopy of carbon monoxide liganded to human hemoglobins A and Zurich. *J. Am. Chem. Soc.* 107:3350–52
100. Raghunathan S, Chandross RJ, Kretsinger RH, Allison TJ, Penington J, Rule GS. 1994. Crystal structure of human class mu glutathione transferase GSTM2-2. Effects of lattice packing on conformational heterogeneity. *J. Mol. Biol.* 238:815–32
101. Rasmussen BF, Stock AM, Ringe D, Petsko GA. 1992. Crystalline ribonuclease A loses function below the dynamical transition at 220 K. *Nature* 357:423–24
102. Ray WJ Jr, Bolin JT, Puvathingal JM, Minor W, Liu Y, Muchmore SW.

1991. Removal of salt from a salt-induced protein crystal without cross-linking. Preliminary examination of "desalted" crystals of phosphoglucomutase by X-ray crystallography at low temperature. *Biochemistry* 30:6866–75
103. Ray WJ Jr, Puvathingal JM, Liu Y. 1991. Formation of substrate and transition-state analogue complexes in crystals of phosphoglucomutase after removing the crystallization salt. *Biochemistry* 30:6875–85
104. Richards FM. 1963. Structure of proteins. *Annu. Rev. Biochem.* 32: 269–300
105. Rivetti C, Mozzarelli A, Rossi GL, Henry ER, Eaton WA. 1993. Oxygen binding by single crystals of hemoglobin. *Biochemistry* 32:2888–906
106. Rivetti C, Mozzarelli A, Rossi GL, Kwiatkoswski LD, Wierzba AM, Noble RW. 1993. Effect of chloride on oxygen binding to crystals of hemoglobin Rothschild (β37 trp-arg) in the T quaternary structure. *Biochemistry* 32:6411–18
107. Rossi GL. 1992. Biological activity in the crystalline state. *Curr. Opin. Struct. Biol.* 2:816–20
108. Rossi GL, Bernhard SA. 1970. Are the structure and function of an enzyme the same in aqueous solution and in the wet crystal? *J. Mol. Biol.* 49:85–91
109. Rossi GL, Mozzarelli A, Peracchi A, Rivetti, C. 1992. Time course of chemical and structural events in protein crystal measured by microspectrophotometry. *Philos. Trans. R. Soc. Lond. Ser. A* 340:191–207
110. Rossi GL, Ottonello S, Mozzarelli A, Tegoni M, Martini F, et al. 1978. Time correlated events in enzyme active site: the perspectives of single crystal microspectrophotometry. In *Protein: Structure, Function and Industrial Application,* ed. E Hofman, W Pfeil, A Aurich, pp. 249–58. Oxford, UK: Pergamon
111. Rupley JA. 1969. In *Structure and Stability of Macromolecules,* ed. SA Timasheff, GD Fasman, pp. 291–352. New York: Dekker
112. Schertler GFX, Bartunik HD, Michel H, Oesterhelt D. 1993. Orthorhombic crystal form of bacteriorhodopsin nucleated on benzamidine diffracting to 3.6 Å resolution. *J. Mol. Biol.* 234: 156–64
113. Schertler GFX, Lozier R, Michel H, Oesterhelt D. 1991. Chromophore motion during the bacteriorhodopsin photocycle: polarized absorption spectroscopy of bacteriorhodopsin and its M-state in bacteriorhodopsin crystals. *EMBO J.* 10:2353–61
114. Scheule RK, Van Wart HE, Vallee BL, Scheraga HA. 1980. Resonance Raman spectroscopy of arsanilazo-carboxypeptidase A: conformational equilibria in solution and crystal phases. *Biochemistry* 19:759–66
115. Schirch LV, Mozzarelli A, Ottonello S, Rossi GL. 1981. Microspectrophotometric measurements on single crystals of mitochondrial serine hydroxymethyltrasferase. *J. Biol. Chem.* 256:3776–80
116. Schlichting I, Almo SC, Rapp G, Wilson K, Petratos K, et al. 1990. Time-resolved X-ray crystallographic study of the conformational change in H-Ras p21 protein on GTP hydrolysis. *Nature* 345:309–15
117. Schlichting I, Berendzen J, Phillips GN, Sweet RM, 1994. Crystal structure of photolysed carbonoxy-myoglobin. *Nature* 371:808–12
118. Schlichting I, John J, Rapp G, Wittinghofer A, Pai EF, Goody RS. 1989. Biochemical and crystallographic characterization of a complex of p21 and caged GTP using flash photolysis. *Proc. Natl. Acad. Sci. USA* 86: 7687–90
119. Schreuder HA, van der Laan JM, Swarte MBA, Kalk KH, Hol WGJ, Drenth J. 1992. Crystal structure of the reduced form of p-hydroxybenzoate hydroxylase refined at 2.3 Å resolution. *Proteins* 14:178–90
120. Shinzawa-Itoh K, Yamashita H, Yoshikawa S, Fukumotu Y, Abe T, Tsukihara T. 1992. Single crystals of bovine heart cytochrome c oxidase at fully oxidized resting, fully reduced and CO-bound fully reduced states are isomorphous with each other. *J. Mol. Biol.* 228:987–90
121. Singer PT, Smalas A, Carty RP, Mangel WF, Sweet RM. 1993. The hydrolytic water molecule in trypsin, revealed by time-resolved Laue crystallography. *Science* 259:669–73
122. Smith JL, Hendrickson WA, Honzatko RB, Sheriff S. 1986. Structural heterogeneity in protein crystals. *Biochemistry* 25:5018–27
123. Smith SO, Farr-Jones S, Griffin RG, Bachovchin WW. 1989. Crystal versus solution structures of enzymes: NMR spectroscopy of a crystalline serine protease. *Science* 244:961–64
124. Spezio M, Wilson DB, Karplus PA. 1993. Crystal structure of the cata-

lytic domain of a thermophilic endocellulase. *Biochemistry* 32:9906–16
125. Spilburg CA, Bethune JL, Vallee BL. 1977. Kinetic properties of crystalline enzymes. Carboxypeptidase A. *Biochemistry* 16:1142–50
126. Steinhoff HJ, Schrader J, Schlitter J. 1992. Temperature-jump studies and polarized absorption spectroscopy of methemoglobin-thiocyanate single crystals. *Biochim. Biophys. Acta* 1121:269–78
127. Stoddard BL, Bruhnke J, Koenigs P, Porter N, Ringe D, Petsko GA. 1990. Photolysis and deacylation of inhibited chymotrypsin. *Biochemistry* 29: 8042–51
128. Stoddard BL, Bruhnke J, Porter N, Ringe D, Petsko GA. 1990. Structure and activity of two photoreversible cinnamates bound to chymotrypsin. *Biochemistry* 29:4871–79
129. Stoddard BL, Koenigs P, Porter N, Petratos K, Petsko GA, Ringe D. 1991. Observation of the light-triggered binding of pyrone to chymotrypsin by Laue X-ray crystallography. *Proc. Natl. Acad. Sci. USA* 88: 5503–7
130. Strynadka NCJ, Adachi H, Jensen SE, Johns K, Sielechi A, et al. 1992. Molecular structure of the acyl-enzyme intermediate in *β*-lactam hydrolysis at 1.7 Å resolution. *Nature* 359:700–5
131. Sygusch J, Beaudry D. 1984. Catalytic activity of rabbit skeletal muscle aldolase in the crystalline state. *J. Biol. Chem.* 259:10222–27
132. Sygusch J, Beaudry D. 1985. Phosphate ion inactivation of rabbit skeletal muscle aldolase in the crystalline state. *Biochem. Biophys. Res. Commun.* 128:417–23
133. Sygusch J, Beaudry D, Allaire M. 1987. Molecular architecture of rabbit skeletal muscle aldolase at 2.7 Å resolution. *Proc. Natl. Acad. Sci. USA*. 84:7846–50
134. Tegoni M, Mozzarelli A, Rossi GL, Labeyrie F. 1983. Complex formation and intermolecular electron transfer between flavocytochrome b_2 in the crystal and cytochrome c. *J. Biol. Chem.* 258:5424–27
135. Teng TY, Srajer V, Moffat K. 1994. Photolysis-induced structural changes in single crystals of carbonmonoxy myoglobin at 40 K. *Nature Struct. Biol.* 1:701–5
136. Tsubaki M, Shinzawa K, Yoshikawa S. 1992. Effects of crystallization on the heme-carbon monoxide moiety of bovine heart cytochrome c oxidase carbonyl. *Biophys. J.* 63: 1564–71
137. Vas M, Berni R, Mozzarelli A, Tegoni M, Rossi GL. 1979. Kinetic studies of crystalline enzymes by single crystal microspectrophotometry. Analysis of a single catalytic turnover in a D-glyceraldeyde-3-phosphate deydrogenase crystal. *J. Biol. Chem.* 254:8480–86
138. Verschueren KHG, Seljee F, Rozeboom HJ, Kalk KH, Dijkstra BW. 1993. Crystallographic analysis of the catalytic mechanism of haloalkane dehalogenase. *Nature* 363: 693–98
139. Vincent MG, Picot D, Eichele G, Jansonius JN, Kirsten H, Christen P. 1984. Linear dichroism measurements on single crystals of mitochondrial aspartate aminotransferase. In *Chemical and Biological Aspects of Vitamin B6 Catalysis*, Pt. B, pp. 233–43. New York: Liss
140. Westbrook EM, Sigler PB. 1984. Enzymatic function in crystals of Δ^5-3-ketosteroid isomerase. Catalytic activity and binding of competitive inhibitors. *J. Biol. Chem.* 259:9090–95
141. Wilkinson KD, Rose IA. 1980. Glucose exchange and catalysis by two crystalline hexokinase-glucose complexes. *J. Biol. Chem.* 255:7569–74
142. Willis KJ, Szabo AG, Krajcarski DT. 1991. Fluorescence decay kinetics of the tryptophyl residues of myoglobin single crystals. *J. Am. Chem. Soc.* 113:2000–2
143. Zhu L, Sage JT, Champion PM. 1993. Quantitative structural comparison of heme protein crystals and solutions using resonance Raman spectroscopy. *Biochemistry* 32: 11181–85
144. Zhu L, Sage JT, Rigos AA, Morikis D, Champion PM. 1992. Conformational interconversion in protein crystals. *J. Mol. Biol.* 224:207–15
145. Zimmerle CT, Alter GM. 1983. Crystallization-induced modification of cytoplasmic malate dehydrogenase structure and function. *Biochemistry* 22:6273–81

Annu. Rev. Biophys. Biomol. Struct. 1996. 25:367–94

MODELING DNA IN AQUEOUS SOLUTIONS: Theoretical and Computer Simulation Studies on the Ion Atmosphere of DNA

B. Jayaram

Department of Chemistry, Indian Institute of Technology, Hauz Khas, New Delhi 110016, India

D. L. Beveridge

Department of Chemistry and Molecular Biophysics Program, Wesleyan University, Middletown, Connecticut 06459

KEY WORDS: counterion condensation, Poisson-Boltzmann studies, structure, energetics, dynamics

ABSTRACT

This article provides a review of current theoretical and computational studies of the ion atmosphere of DNA as related to issues of both structure and function. Manning's elementary yet elegant concept of "counterion condensation" is revisited and shown to be well supported by current state-of-the-art molecular simulations. Studies of the ion atmosphere problem based on continuum electrostatics, integral equation methods, Monte Carlo simulation, molecular dynamics, and Brownian dynamics are considered. Grand canonical Monte Carlo and non-linear Poisson Boltzmann studies have recently focussed on the determination and significance of the index of non-ideality in solution known as the "preferential interaction coefficient," for which the relevant current literature is cited. The review concludes with a survey of applications to ligand binding problems involving drug-DNA and protein-DNA interactions.

1056–8700/96/0610–0367$08.00

CONTENTS

PERSPECTIVES AND OVERVIEW

A growing realization of the utility of computer simulation studies in developing an atomic view of biomolecular systems and advances made in the adaptation of numerical techniques to biomolecular problems have led to a surge in theoretical explorations of DNA structure and function in recent years. These theoretical efforts have sought to arrive at an accurate description of DNA in aqueous solutions, to present a molecular perspective on conformational transitions of DNA and sequence-specific structural modulations (fine structure) of DNA, and to understand the molecular basis of recognition in protein-DNA and drug-DNA systems. We present herein an overview of the recent theoretical and computational studies aimed at developing accurate theoretical models of DNA in aqueous solutions. We focus particularly on issues related to the structure and energetics of the counterion atmosphere of DNA. For a larger purview on the subject, see the more comprehensive and specialized reviews on condensation theory (77–79), ion atmosphere of DNA (2, 3, 113, 114), molecular electrostatic potentials (106, 107), electrostatic interactions in biomolecules (7, 30, 47, 52, 122, 123, 134), nucleic acid hydration (8, 9, 143), molecular mechanics and dynamics of DNA (14, 72, 111, 138, 141), free energy simulations on biomolecular systems (12, 68, 83), DNA supercoiling (99, 140), condensation of DNA by multivalent cations (18), and modeling DNA (15, 16, 53) and drug-DNA interactions (94).

Because the acidic phosphate groups are fully ionized, DNA at physiologic pH occurs as a polyanion (118). The presence of net equivalent number of counterions (e.g. Na^+, K^+, or Mg^{2+}) in the system ensures electroneutrality. The ions, being mobile, assume a statistical distribution around DNA, the microscopic details of which are not yet established fully. Additional ions from any added salt (e.g. NaCl, KCl, or $MgCl_2$), together with the original counterions and usually some organic ions, form the total ion atmosphere. Along with solvent water, this atmosphere constitutes the DNA environment in vitro. Both the base sequence and the aqueous ionic environment are expected to be major determinants of the structure and function of DNA.

COUNTERION CONDENSATION IN NUCLEIC ACID SYSTEMS

The counterion atmosphere of DNA neutralizes the charges of the anionic phosphates and imparts electrostatic stability to the system. DNA structure is sensitive to the composition and concentration of the ion atmosphere as well as to water activity. This sensitivity is most dramatically illustrated by the change in helix sense involved in the transition from right-handed B-DNA to left-handed Z-DNA at high salt in guanine-cytosine (GC)–rich sequences. The nature of the ion atmosphere of DNA has been the subject of considerable research attention, both experimental and theoretical, in recent years. Manning advanced an elementary theoretical model to account for the diverse macroscopic properties of DNA (78).

Manning's Theory

An important organizing principle in Manning's theory is the phenomenon of "counterion condensation" (CC) (77–79, 100): No matter how dilute the solution, a number of the counterions remain in close proximity to the DNA, thereby compensating a large percentage of the phosphate charges. These counterions, some fraction of the total, are said to be "condensed." The remaining counterions are considered to form a Debye-Huckel–type diffuse ionic cloud. Manning (78) analyzed the problem by a minimization of the phenomenologic free energy expression in the limit of infinite dilution and came to the vexingly simple conclusion that the percent condensation of monovalent mobile cations on B-form DNA was 76% and independent of salt concentration. (The net charge/phosphate in aqueous solutions of canonical B-DNA according to CC theory is -0.24 with sodium counterions and -0.12 with magnesium counterions.) The idea of CC follows simply from thermodynamic arguments. Small ion pairing decreases with dilution because of the increased potential for a large entropy of mixing, which favors dissociation. For polyions like DNA, the superposition of the electrostatic potentials of the phosphate groups on any given mobile counterion makes enthalpic effects dominant in the equilibrium. Consequently, a significant fraction of counterions remain associated with the DNA. The fraction turns out to be essentially independent of the bulk salt concentration owing to the particular nature of enthalpy-entropy compensation in nucleic acid systems. Counterion condensation, therefore, is unique in polyelectrolytes compared with simple electrolyte systems.

The detailed structure (or lack of it) assumed by the counterions within the region of condensation is a point of interest. This feature is important in such diverse areas as the development of a molecular view of salt effects on protein-DNA and drug-DNA complexation and the selection of a suitable initial configuration for the treatment of counterions in molecular dynamics simulation studies on nucleic acids. Considerable ambiguity exists with regard to the underlying structural details of the ion atmosphere. Manning is careful to distinguish condensation (i.e. ions remaining associated with the DNA) from actual site binding. Nuclear magnetic resonance (NMR) experiments have been cited in support of Manning's (78) idea that "all small cations (condensed counterions) are in a state of complete hydration and free translational and rotational mobility," i.e. delocalized and rather loosely associated with DNA. The Manning radius (i.e. the radius of the coaxial cylinder around DNA that encloses 76% of net counterionic charge per phosphate) of the counterion condensate is typically approximately 7 Å beyond the surface of the DNA (and ~17 Å from the helix axis). Manning's theory has provided an account of diverse observed properties of DNA (e.g. colligative properties, transport properties, binding equilibria, melting temperatures, and intercalative binding of drugs to DNA) (40, 79), and the agreement between experiment and CC theory is considered good. Counterion condensation theory is phenomenologic in origin, and a rigorous structural interpretation/extension is beyond its scope. However, the basic concepts of CC in aqueous solutions of DNA have been strongly confirmed by computer simulation (59), as we discuss later. Figure 1 shows the counterion distribution around a DNA oligonucleotide calculated from a recent Monte Carlo computer simulation (59). The distinction between the condensed counterions and those beyond the CC region, some 10 Å from the DNA surface, is remarkably clear.

The CC theory was adapted recently to a more realistic representation of DNA as a three-dimensional discrete charge distribution (33). The results conformed to the inferences based on experiment and the simple linear lattice model with a uniform dielectric constant (78), when dielectric saturation function was considered by means of a distance-dependent dielectric function for the interactions between phosphates on a double-helical array in a B-DNA geometry. The simpler the DNA model, the simpler the solvent description. Stated alternatively, a more realistic solute model necessitates a more complicated solvent description. Recent further extensions of Manning theory have been described in the area of structural and excluded volume effects (112a) and superhelical DNA (33a). Some new perspectives on the limitations of CC theory have been provided (102a, 108).

Related Experimental Studies

Results on counterions in nucleic acid systems from X-ray crystallographic studies are fragmentary (143). The NMR literature is extensive on cation resonance in DNA systems (17, 19, 36, 63, 97). Experiments involving ^{23}Na NMR indicate support in large measure to the condensation hypothesis (78), but the data have proved to be difficult to interpret unequivocally in terms of structure. Bleam et al (17) demonstrated a two-state approximation in which the observed invariance of the slope of the plot of NMR line widths as a function of salt concentration implies the constancy of the product of the extent of counterion association, and $(r_B - r_F)$ is the difference in relaxation rates of the bound and free sodium ions. Beyond this, there is some ambiguity in concluding that the two factors, r and $(r_B - r_F)$, are individually constant and in extracting a value for the condensed fraction. The NMR predictions of the condensed counterion fractions are in the range of 0.65 to 0.85 (17 and references therein) and, from a more recent study, approximately 0.53 (104). In both cases, however, a constancy of the condensed fraction is implied. Some uncertainty persists with regard to the character of the diffusional motion that dominates the relaxation mechanism and the extent of counterion association (25).

The DNA ion atmosphere has been the subject of numerous further experimental studies. The duplex rotation angle of DNA was found to vary systematically with cation type (97). An explanation for this finding might require some degree of site binding or at least local specificity. Multivalent cations seem to be more disposed to site binding than do monovalents (36). Recent ^{25}Mg, ^{43}Ca, and ^{59}Co NMR studies on the titration of NaDNA with multivalent cations point to the existence of multiple binding environments on B-DNA for the divalent cations (19, 145). DNA flexibility, torsional stiffness, and helical repeat were not influenced significantly by increases in the NaCl salt concentration (133). Light scattering experiments (60), however, showed a strong ionic strength dependence of the persistence length, which is interpretable in terms of Manning's theory. Effects of Na^+ ions on the persistence length and excluded volume of T7 bacteriophage DNA were interpreted in terms of the CC theory (126). Counterion dependence on the dynamic behavior of water of hydration, studied by low frequency Raman and differential scanning calorimetry, was different for different counterions (K^+, Rb^+, Sr^{2+}, Ba^{2+}) (117, 136). Experimental studies of salt effects on helix-coil transitions (19, 145) in oligonucleotide systems indicate that the thermodynamic equivalent of approximately 0.08–0.13 Na^+ ions per phosphate are released on melting, which is consistent qualitatively with the CC theory predictions. The results also imply some sequence dependence in the condensed fraction. Small angle X-ray scattering measurements (23) of the monovalent (Tl^+) and divalent (Ba^{2+}) counterion distributions, that surround 500 Å long DNA fragments in aqueous solutions, appear to be in agreement with the Poisson-Boltzmann (PB) predictions. The agreement improved if the grooves of the double helix were assumed to accommodate 10% of the counterions.

POISSON-BOLTZMANN STUDIES

A well-known theoretical approach for determining the electrostatic potentials around macromolecules is based on solutions to the PB equation. The PB equation can be solved analytically for simple geometries, such as a line of charge and a uniformly charged cylinder at zero added salt in a dielectric continuum solvent characterized by a uniform dielectric constant. Numerical solutions can be sought in other cases. Solutions to the PB equation yield potentials from which other properties, such as electrostatic free energies and small ion (counter- and coion) concentrations, ensue.

Uniform Dielectric Models

The first studies of this genre were conducted on simplified models of DNA. (For earlier literature, see 2, 45, 54.) Zimm & LeBret (151) used the PB equation to show elegantly how a rodlike polyanion, like DNA, will condense counterions naturally on increasing dilution at a level intermediate between that of a charged sheet (100% condensation, the Gouy-Chapman double layer) and that of a charged sphere (~0%). Anderson & Record (2, 3) have investigated numerous aspects of the PB treatment of the ion atmosphere vis-a-vis thermodynamic measurements and NMR spectroscopy, with a focus on the salt dependence of the fraction of the condensed counterions. Limitations of the PB theory due to neglect of the finite size of the mobile ions and spatial correla-

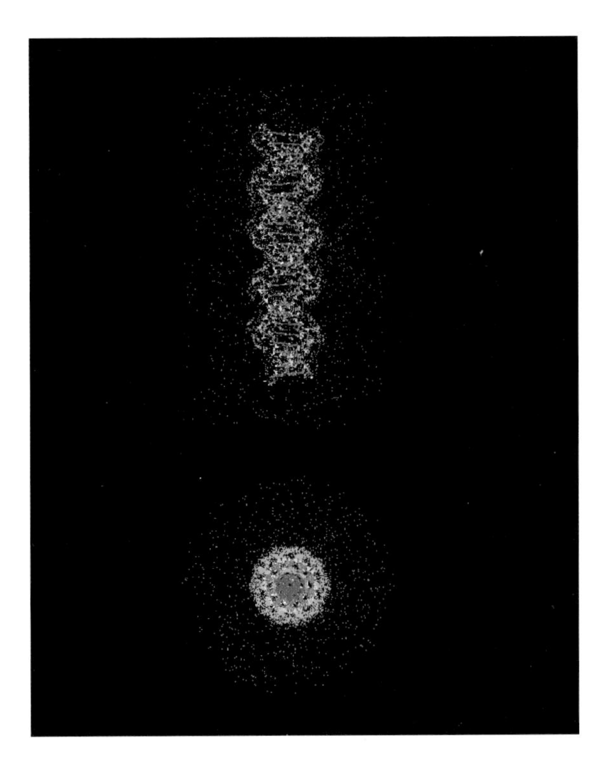

Figure 1 Counterion condensation of Na^+ cations on a DNA oliganucleotide. The calculated cation density is depicted by a superposition of 200 configurations from a 4 million step Monte Carlo Metropolis simulation on 22 mobile Na^+ ions around a fixed canonical B form of the DNA sequence. The effect of water is treated by means of a variable dielectric continuum (59). Graphic presentation prepared by M.A. Young.

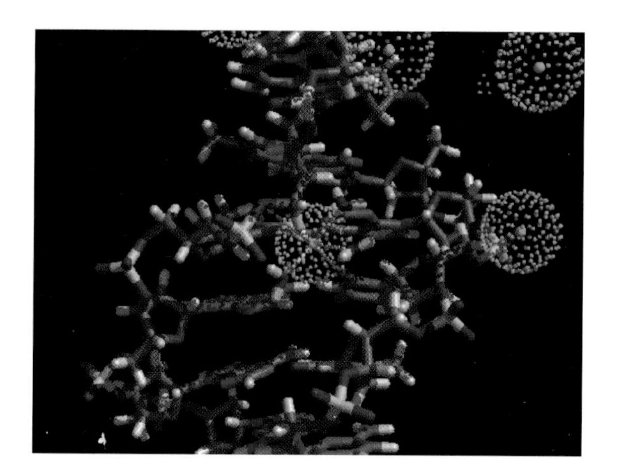

Figure 2 A snapshot from an MD simulation on the d(CGCGAATTCGCG) duplex, 4000 water molecules and 22 Na^+ counterions, depicting the intrusion of a mobile counterion into the minor groove spine of hydration (146a).

tions were characterized. In a recent analysis of the solutions of the cylindrical PB equation as applied to DNA in aqueous solutions, Rajasekaran & Jayaram (108) suggest the existence of certain spatial scales other than the "Manning radius," which are invariant to added salt. Lamm et al (71), on the basis of their PB and Monte Carlo calculations, advocate that the radius enclosing the region in which electrostatic potential is less than $-kT/e$, where e is the charge of an electron, offers a more suitable criterion than the Manning radius. A theory for the second virial coefficients of short DNA, in which interactions between charged cylinders of finite length were modeled, was developed with the use of the PB equation (128). Gueron & Demaret (44) arrived at an algebraic approximation for calculating the PB free energy of a composite cylinder and provided an explanation of the relative stabilities of the B and Z forms of DNA.

Theoretical studies of electrostatic interactions at finite salt concentrations have been extended to treat all-atom models of DNA. Klein & Pack (66) obtained electrostatic potentials from an iterative PB solution to a combination of Coulombic potentials from the fixed macromolecular charges and the distribution of mobile charges obtained from the Boltzmann equation. The results predicted significant concentration of mobile cations in the minor groove as well as along the sugar phosphate backbone, a consequence of the superposition of trans-groove anionic phosphate potentials. Hydrogen ion concentrations at the surface of the DNA, which might be of relevance to interaction of mutagenic epoxides, were estimated with the use of the PB approach (70). The pH was less by approximately 2–3 pH units compared with that of the bulk solvent. The above theoretical treatments assume that the dielectric constant ϵ is everywhere equal to 80, including inside the DNA.

Two-Dielectric Models

The potential influence of the dielectric boundary between the DNA (ϵ = 2–4?) and solvent at $\epsilon \cup 80$ was explored by several groups (58, 81, 135, 142). The Tanford-Kirkwood model was extended to arbitrary charge distributions with an overall cylindrical symmetry (54). This work suggested that a consideration of the details of the three-dimensional charge distribution, as opposed to a linear lattice description, was necessary to account for the stability of the B form of DNA relative to the Z form in aqueous solutions at low salt.

Troll et al (135) conducted a macroscopic simulation of duplex DNA represented by a clay model in an electrolyte tank. These authors found that interactions between charges on the same side of the DNA were enhanced as a consequence of a concentration of field lines by the low dielectric DNA, thereby increasing both attractive phosphate–cation and repulsive phosphate–phosphate and cation-cation interactions. They also found that shielding was increased for charges on opposite sides of DNA by the presence of the low dielectric medium. Jayaram et al (58) incorporated dielectric boundary, solvent screening, and ionic strength into finite difference solutions to the PB equation (FDPB) for an all-atom model of DNA. In this study, the sequence dependence in electrostatic potentials appears as a natural consequence of the model with no extra assumptions. The distribution of the electrostatic potential in B-DNA shows an intrinsic dissymmetry (the minor groove of adenine-thymine (AT) sequences is more negative than the minor groove of GC sequences). Deepest potentials were located in the grooves rather than in the phosphate regions. The results are in general agreement with the pioneering work of Pullman and co-workers (106, 107) on the basis of quantum mechanical studies. The numerical results also support qualitatively the electrolyte tank observations on the angle dependence of the charge–charge interactions (135). Electrostatics appears to dominate the base pairing interactions (N Aneja, B Jayaram, and B Honig, unpublished data) but contributes negligibly to stacking interactions (37). The FDPB approach provides a conceptually simple and theoretically sound basis to investigate electrostatic interactions involving nucleic acids (biomolecules in general) under physiologic conditions (52). Pack et al (103) found that the presence of a dielectric boundary altered the electrostatic potentials significantly within the grooves of DNA, as in the FDPB studies cited above. These effects appear to be confined close to the surface. Application of the finite element technique to solve the nonlinear PB equation as applied to DNA was also reported (110).

Distance-Dependent Dielectric Models

Electric fields near an ion are on the order of 1 million volts/cm and are expected to be even larger for a polyanion such as DNA. Dielectric saturation thus assumes importance in studies on DNA. Hingerty et al (50) studied the role of dielectric inhomogeneity and the influence of dielectric saturation on oligonucleotide–small ion interactions by means of a distance-dependent dielectric screening function. This method was subsequently reparameterized and used in energy minimization studies on DNA by Lohman (75) and Ramstein & Lavery (109) with the JUMNA algorithm. Mazur & Jernigan (82) reviewed the applicability of distance-dependent dielectric functions. Their analysis indicated that the local helix geometry of the base pairs was affected strongly by the contributions of the electrostatic interactions to the stacking energy. A lower dielectric constant ($\epsilon = 10$) in the layer at the first 1 Å from the surface resulted in a 70% increase in the calculated counterion density (103). Support for the inclusion of dielectric saturation in dealing with charge–charge interactions in DNA systems in the dielectric continuum solvent approach appears to be mounting. The results of Hecht et al (49), however, seem to imply that a distance-dependent dielectric function is not necessarily superior to the two-dielectric model for calculating electrostatic potentials around DNA in solution. Applications of the PB equation to cylindrical models of polyions and various other models of DNA continue to hold research attention (1, 29, 105, 125).

Numerical solution of the non-linear PB equation by finite difference methods for highly charged systems such as nucleic acids has been subject to convergence and stability problems, addressed heretofore by multigridding techniques (51b, 97a). A rapid new method for non-linear PB calculations, stable even for highly charged systems, has recently been developed based on the inexact Newton method (51a). The effect of ion size has been introduced into non-linear PB using self-consistent iteration schemes (103).

INTEGRAL EQUATION METHODS

The practice of deducing an effective two-body potential of mean force (pmf) and incorporating that in the Hamiltonian of a many-body system has a long history. Soumpasis (127) considered the phosphate backbone of DNA and the ion atmosphere as a fully dissociated 1:1 electrolyte of specified composition and thermodynamic state for the purpose of estimating the phosphate–phosphate pmf. The pmfs were calculated with the use of either the hyper-netted chain (HNC) formalism or the exponential mean spherical approximation (EXP-MSA), depending on the level of accuracy desired. This approach was shown to account quantitatively for the salt dependence of free energy differences between B and Z conformations of DNA from 0.1 to 4 M NaCl (62). The theory uses only one adjustable parameter, the distance of closest approach of an anion–cation pair (set equal to 4.9 Å for NaCl). The success of the theory is intriguingly remarkable considering that the derived pmfs between phosphate charges incorporate neither the effects caused by the presence of other atoms on DNA nor their connectivity along the backbone of DNA. The pmf approach was integrated subsequently into the AMBER molecular mechanics force field (41, 65), thereby replacing the electrostatic terms in it by the pmfs. This step enabled an investigation of the ionic distributions (64), as well as structural transitions of DNA in solution.

More recently, Hirata & Levy (51) discussed the application of a restricted interaction site model (RISM) theory developed for solvated polymers to a study of the salt effects on DNA conformation. Their studies predicted a transition from B to Z form at 3.6 M added salt in qualitative accord with experiments without any optimization of the model parameters involved. Their work further showed that the superposition approximation inherent in the pmf approach might be less severe than anticipated (in the B to Z transition) because of a cancellation of higher order correlations when the free energy difference between B and Z DNA was calculated (127).

Bacquet & Rossky (4) applied the HNC integral equation method to the system of DNA in aqueous 1:1 electrolyte solution. The HNC and CC approaches were in general qualitative accord but differed in quantitative predictions regarding the extent and concentration dependence of the CC approach. In a related study on ion distributions and competitive association in DNA/mixed salt solutions, composition near the polyion was more responsive to the changes in the bulk composition (5). The counterion size was of minor importance compared with its valence in determining the ionic distribution, except at immediate contact with the polyion. The calculated fluctuations in the electric field gradients experienced by the sodium nuclei correlated well with the observed NMR line widths.

The integral equation methods, accepting the approximations, are computationally more expedient than are molecular simulation methods for estimating free energies.

MOLECULAR SIMULATIONS

Canonical Ensemble Monte Carlo Studies

HYDRATION DNA presents a variable and sequence-dependent hydrogen bonding pattern in the major and minor grooves (119). Solvent water is expected to interact differently with the four bases—adenine, guanine, thymine, and cytosine—and show a differential, sequence-dependent, stabilization effect on the overall conformation. With the observation of the "spine of hydration" in the minor groove of B-DNA (31), the plausible role of solvent in deciding the fine structure of DNA, the implication of water-mediated hydrogen bonds in protein-DNA interactions (102), and their potential contribution to the thermodynamics of DNA-ligand binding (80, 117), there is a continued interest in the hydration of nucleic acids (8, 9, 143).

Early computer simulation studies on hydration used Monte Carlo (MC) methods and focused on nucleic acid constituents (13). MC studies were extended subsequently to hydration of oligonucleotides and in particular the Eco RI dodecamer of sequence d(CGCGAATTCGCG) (15). Subramanian et al (129) gave a theoretical account of the sequence-dependent hydration of the Eco RI dodecamer vis-a-vis X-ray and calorimetric studies. They noted that both AT and GC regions in the minor groove could support an extended network of water molecules owing to the geometric versatility of water–water hydrogen bonding. As anticipated, the penetration of water into the DNA was clearly greater for AT than for GC tracts. A recent analysis of the crystallographic data on the hydration of bases in oligonucleotide systems shows conformation-dependent differences in both geometry and extent of hydration (8, 9). (For a survey of the hydration of nucleic acid systems with a nearly complete citation to the literature, see 143.) MC simulations on water and counterions around DNA are sometimes performed as part of the equilibration procedure in molecular dynamics simulations of DNA in solution.

ION ATMOSPHERE MC or molecular dynamics (MD) calculations based on a fully explicit consideration of DNA, water, counterions, and added salt at a concentration of physiologic interest are a computationally expensive proposition, even for current day supercomputers. Studies aimed at pursuing the role of electrostatic interactions and the nature of the ion atmosphere of DNA on the basis of molecular simulation often take recourse to the primitive model. In all the simulation studies with the primitive model and variations thereof, ions are treated explicitly, and water is represented as a dielectric continuum. This is clearly a serious approximation, because the molecular nature of water as an associated liquid is neglected and its particular capacity for hydration, bonding, and solvation in different modes—hydrophilic, hydrophobic, and ionic—is denied. Also, there is no way to parametrize the heterogeneous qualities of the dielectric medium accurately from experimental data. Results of theoretical studies and simulations based on primitive models have to be interpreted in this context.

Extensive canonical MC studies on DNA counterion systems in the absence and presence of added salt have been reported by several groups. LeBret & Zimm (73) modeled DNA as a linear lattice of charges imbedded in an impenetrable cylinder and as a double helical array of charges, with mobile ions represented as hard spheres that interact with each other and with DNA via Coulombic potentials in a solvent treated as a uniform dielectric continuum. They found a striking accumulation of counterions in a layer of concentration exceeding 1 M at the surface of the polyion in agreement with conclusions from previous PB studies and CC theory. Murthy et al (93) and Mills et al (86, 88) subsequently explored ion distributions in a solvent treated as a uniform dielectric continuum around a uniformly charged cylinder (and a lattice of discrete phosphates) that represented DNA. A comparison of their Monte Carlo results with those emerging from PB and HNC studies suggested that the PB approach underestimated counterion concentrations near DNA treated as a cylinder by approximately 12–18%. The empirical Manning radius decreased with increasing salt concentration in all these studies. Conrad et al (28) incorporated effects caused by dielectric discontinuity in their interaction potentials to evaluate small-ion DNA interactions. The modifications to Coulomb's law tended to drive the ions out of the grooves, especially the major groove. They ascribed this observation to repulsions between the ions, the low permittivity of the helix, and, partly, to the focusing of field lines of phosphates at the surface of the helix caused by dielectric discontinuity.

Jayaram et al (59) performed comparative simulations of the counterion distribution around DNA on the basis of a series of models for the aqueous dielectric medium that were likely to bracket the true physical nature of the system. The results showed that, in all cases of continuum solvent, counterion concentrations near DNA (i.e. ~10 Å from the helical axis) exceeded 1 M, even in the absence of excess salt. This finding was consistent with CC theory and previous MC and PB studies on DNA-counterion systems. An analysis of the simulation results based on different dielectric models indicated that the dielectric saturation model favored increased CC relative to the Coulombic model, with DNA–counterion interactions dominating the small ion repulsions. The dielectric saturation model produced an essentially salt-independent result for the fraction of condensed counterions over an added salt concentration range of 0 to 150 mM, which is consistent with inferences based on NMR and CC theory. The results with an energetic criterion, as opposed to a geometric radial criterion, for computing the condensed fraction of counterions around DNA were essentially similar to the above results with the dielectric saturation model (B Jayaram and DL Beveridge, unpublished data).

Gordon & Goldman (42) recently reported counterion and solvent distributions around a polyion, as seen in their MC calculations conducted with 15 counterions and explicit water. Interestingly, the uniformly charged cylindrical model led to a high degree of solvent polarization, which caused the counterions to avoid regions vicinal to the polyion. When a helical necklace of charges at the surface of a cylinder was considered as a model for the polyion in conjunction with molecular solvent, the counterions formed a compact double layer near the surface. This formation resulted in an extensive screening of the polyionic charges, as found earlier in the numerous studies based on continuum solvent. Nishio (96) presented some exploratory MC simulations on the potentiometric behavior of aqueous salt-free solution of rodlike polyelectrolytes in a cylindrical cell system. A comparative study of the MC and mean field theory results on the titration behavior of linear polyelectrolytes was reported recently (116). Mills et al (89) reported MC calculations of ion distributions surrounding an oligonucleotide $d(ATATATATAT)_2$ in the B, A, and wrinkled D conformations. Significant counterion accumulation in the major groove of A-DNA and, to a lesser extent, of B-DNA was noticed. Results on D-DNA were similar to those for an impenetrable cylinder. Excluded volume effects therefore appeared to alter the ion distributions strongly, whereas the detailed charge distribution was affected minimally. Erie et al (32) extended MC methodology to generate a distribution of oligonucleotide conformations with feasible dinucleotide steps. Several loop conformations were compatible with the given experimental (spectroscopic and thermodynamic) data.

Grand Canonical Monte Carlo (GCMC) Studies

The (T,V, μ) ensemble simulations lead to direct calculation of the excess chemical potentials and excess free energies, which are not as easily accessible in the canonical Monte Carlo calculations. Almost all the GCMC studies to date on DNA use the primitive model to study the role of ion atmosphere on the bulk solution properties of DNA, and no hydration studies have been reported. Grand canonical Monte Carlo simulations with explicit solvent are slow to converge at particle densities of experimental interest. The GCMC simulations on the ion atmosphere have provided interesting information on non-idealities in aqueous solutions of DNA (55).

Vlachy & Haymet (139) used the GCMC method to obtain structural and thermodynamic data for model polyelectrolyte solutions. They treated the polyion as an impenetrable, rigid, infinitely long cylinder. The results were compared with PB and HNC integral equation studies with the use of the MSA. The authors concluded that the PB equation retained its semiquantitative utility even in the range of moderate to high (1 M) concentrations of added salt. Extensive GCMC studies on aqueous solutions of sodium salt of DNA in the presence of added simple salt were reported by Anderson, Record, and coworkers (3, 87, 98). The simulations were used to calculate mean ionic activity coefficients and ''preferential interaction coefficients,'' [similar to the

Donnan salt exclusion factor (79) and the Donnan membrane equilibrium parameter (43)] for a cell model representation of NaDNA with added salt. The role of end effects on molecular and thermodynamic properties in oligoelectrolyte solutions was characterized via the GCMC method (98 and references therein). The results indicated that an oligoion that contained 48 or more phosphates behaved as a polyion. The thermodynamics of denaturation of oligonucleotides was also investigated. The GCMC studies of Anderson, Record, and coworkers on the preferential interaction coefficient, which they idenify as a thermodynamic measure of nonideality caused by small ion–polyion interactions, provided a considerably enhanced perspective on the problem (3, 97b). A recent detailed exposition of the theory has been provided (3a). Sharp (122a) has developed the relationship between preferential interaction coefficients and PB. Valleau (137) presented an extension of the GCMC method to a flexible polyelectrolyte immersed in primitive model aqueous electrolyte solutions.

Jayaram & Beveridge (55) recently reported GCMC simulation studies on aqueous solutions of sodium chloride and sodium salt of DNA in the presence of added salt. Results for the simple electrolyte indicated that a soft-sphere potential function for the ions in a solvent treated as a dielectric continuum, supplemented with a Gurney correction term for desolvation, described the behavior of activity coefficients as a function of concentration quite well, over a concentration range of 5 to 500 mM. The results on the NaDNA system in the presence of added simple electrolyte provided an account of the contravariant behavior of nonideality of mobile ions in polyelectrolyte vs simple electrolyte solutions. Excess chemical potentials calculated from these GCMC simulations suggested that the counterions interacted more strongly with DNA at low salt than at high salt (55). Nonideality of counter and coions in sodium salts of DNA in water in the presence of added sodium chloride salt, therefore, increases with dilution and decreases with added salt contrary to the behavior in simple electrolyte solutions.

Among the limitations of the canonical and GC Monte Carlo studies, as described above, are that the internal degrees of freedom of the solute (DNA) are frozen and that the solute does not respond to the fluctuating environment of the solvent and ion atmosphere. Zhurkin et al (149) presented a Monte Carlo study on the sequence-dependent bending of DNA by sampling the internal coordinates. Gabb et al (39) reported an internal coordinate furanose model on the basis of the pseudorotational variables, phase and amplitude, thus providing a method to deal with flexible rings in the context of Monte Carlo simulations. One further desirable, albeit difficult, course of research to follow would be to perform an internal coordinate Monte Carlo on DNA alongside the counterion/solvent Monte Carlo. This study should enable probing the sequence-, solvent-, and ionic strength–dependent conformational flexibility of DNA. Of course, some methodologic improvements, particularly on the convergence and improving acceptance ratios, are desired to facilitate this line of research.

Theoretical studies of counterion atmosphere have recently been extended to branched nucleic acids (98b), conformational transitions and DNA triplexes (18a).

Molecular Dynamics Simulations

Currently, only a few molecular dynamics calculations on DNA systems that consider counterions and solvent water explicitly have been reported (14, 35, 111, 124, 131, 132). Considering, however, that the first molecular dynamics simulation on DNA was reported little more than 10 years ago (74), the popularity of this technique, with simpler models for solvent and small ions, is phenomenal. Most studies to date treat the effect of counterions implicitly, by reducing the charges on the phosphates from -1 to -0.25, -0.32, -0.34, -0.5 (14, 16, 95), i.e. fractional charge justified by CC theory, or even to zero (74)! This approach does make the DNA more stable electrostatically but obviously lacks the correct physics. Explicit inclusion of counterions has produced convergence problems in the simulations that currently are being investigated actively (111). The effect of the solvent and counterions has been incorporated implicitly via a distance-dependent dielectric function into numerous studies; the justification is in the quality of the results. Further, problems in analyzing molecular dynamics literature on DNA in aqueous solutions are twofold. No uniform method of analysis is adopted by different authors to facilitate a comparison among the different dynamic models. Root mean square deviations are too simplistic. Analysis based on Curves, Dials, and Windows (112) is helpful in providing a detailed characterization of the dynamic models of DNA. The other problem relates to comparison with experiment. We do not know yet what is absolutely correct regarding DNA structure in solution. Gross structural distortions, such as collapse of the groove structure and folding of the DNA, may be termed incorrect. Progress in the solution structure determinations via two-dimensional NMR, combined with the information collected from crystal structures, holds promise for a rigorous characterization of the different simulations on DNA, but the situation at present remains a complex one and nothing should be taken for granted. For an overview of all the molecular dynamics simulations of DNA published to date, see the reviews by Beveridge et al (14, 16, 111).

A general concern emerging from the molecular dynamics studies on DNA is that, in simulations with explicit ions and solvent molecules, the trajectories may be sensitive to the initial location of counterions. Overall, a better agreement with crystal structure has been obtained so far by starting with a scaled phosphate charge model. The truncation of potentials is also a particularly important issue in nucleic acid systems. A switching function is applied around the cutoff to feather the potentials smoothly off to zero, so that calculation of forces remains well conditioned. Truncation effects can be rather serious for a polyionic system such as DNA. In addition, if the range of the switching function is too narrow in potentials used for molecular dynamics on DNA, artifacts may be introduced as a consequence of charged groups that tend to cluster at the cutoff limit. This problem is remedied by extending the range of the switching function from 1 to 4 E, which frees the molecular dynamics from this artifact (14, 111). Cheatham et al (24) recently reported results from a nanosecond dynamics of DNA by using both the spherical cutoff and Ewald summation. These authors showed that more stable structures were observed by applying the Ewald summation. A proper protocol for the treatment of counterions and electrostatic interactions in the molecular dynamics simulations on DNA appears to be well within reach (85, 144, 147).

A sequence of ordered solvent peaks in the electron density map of the minor groove region of AT-rich tracts of the double helix is a characteristic of B-form DNA well established from crystallography (31). This feature, termed the ''spine of hydration,'' has been discussed as a central stabilizing feature of B-DNA, the structure of which is known to be sensitive to environmental effects. Lengthy molecular dynamics simulations on the DNA duplex of sequence d(CGCGAATTCGCG) have been carried out, including explicit consideration of 4000 water molecules and 22 Na^+ counterions (see Figure 2) and based on a new version of the AMBER force field (28a) with the particle mesh Ewald summation used in the treatment of long range interactions. The simulations were carried out for a heretofore unprecedented run length of 1.5 nanoseconds, and support a dynamical model of B-DNA closer to the B form that any previously reported (146b). Analysis of the dynamical structure of the solvent revealed that in over half of the trajectory, a Na^+ ion is found in the minor groove localized at the AT step. This position, the ''ApT pocket,'' was noted previously (72a) to be of uniquely low negative electrostatic potential relative to other positions of the groove, a result supported by the crystal structure of dApU (119a) and by calculations based on continuum electrostatics. The Na^+ ion in the ApT pocket interacts favorably with thymine O2 atom on opposite strands of the duplex, and is well articulated with the water molecules which constitute the remainder of the minor groove spine. This result indicates that counterions may intrude on the minor groove spine of hydration on B form DNA, and subsequently influence the environmental structure and thermodynamics in a sequence dependent manner. The observed narrowing of the minor groove in the AATT region of the d(CGCGAATTCGCG) structure may be due to direct binding effects and also to indirect modulation of the electrostatic repulsions that occur when a counterion resides in the minor groove ApT pocket.

Brownian Dynamics Simulations

Internal motions of DNA, as well as the distribution and dynamics of ions around DNA in the nanosecond regime and beyond, can be studied profitably with the use of Brownian dynamics simulations (84). Few such studies, however, have been reported. Barkley & Zimm (6) reported one of the earliest such calculations. They treated DNA as a semiflexible chain and developed a theoretical account of fluorescence depolarization. In an interesting extension of the Brownian dynamics formalism, which includes the accessible surface area, Kottalam & Case (69) computed Langevin modes of DNA hexamers. More recently, Briki et al (20) reported a Brownian dynamics simulation of a B-DNA $(dA)_5$ $(dT)_5$ oligomer. The lifetime of the base pair and the activation energy for the opening process were calculated to be 15 ms and 20 kcal/mol, respectively, which compared favorably with the corresponding experimental measurements obtained by hydrogen exchange studies.

Reddy et al (115) used stochastic dynamics simulations to probe counterion spin relaxation of the quadrupolar nuclei in the vicinity of DNA. The calculated relaxation behavior was in qualitative accord with experiments. Guldbrand et al (46) recently reported the distribution and dynamics of counterions around B-DNA, with a continuum solvent and an all-atom model of the solute via Brownian dynamics simulations. The continuum solvent model successfully reproduced the results of the fully explicit molecular dynamics simulations for the distribution of those counterions that maintained their hydration shells. The distributions of the counterions far away from the polyelectrolyte in their simulations agreed with the results of the PB approach. The Langevin equation has found applications in theoretical explanations of gel electrophoresis of DNA (76, 150) and in DNA supercoiling (26). Brownian dynamics simulations have been successfully extended to the study of concentrated DNA solutions (31a).

STUDIES ON PROTEIN-DNA AND DRUG-DNA SYSTEMS

The essential contributions to the free energy of association of a protein or a drug molecule with DNA come from (*a*) direct interactions between the protein or drug molecule and the DNA, both intra- and intermolecular; (*b*) release of some of the water molecules bound to both protein or drug and DNA; and (*c*) release of some of the counterions associated with DNA. The last contribution is known as the polyelectrolyte effect.

Polyelectrolyte Effects

Release of condensed counterions is considered to provide an entropic driving force for the DNA-ligand complex formation. Manning (78) and Record et al (113) developed a theoretical framework to understand the role of the ion atmosphere in DNA–ligand association and to quantify the contribution of electrostatic interactions to this process. A plot of log K_{obs} vs log [MX], where K_{obs} is the observed association constant for the DNA–ligand complex and [MX] is the salt concentration, gives a slope of $-Z$ (75). Here, Z is the thermodynamic equivalent of the number of counterions released during the DNA-ligand complex formation, and the phosphate charge is reduced by a value of 0.88, which is a sum of 0.76, to account for the condensed counterions, and 0.12, to approximate the screening effects caused by the remaining uncondensed Debye-Huckel–type diffuse ionic cloud in the system (118). The value of Z in such a log–log plot is interpreted commonly as indicating the number of ionic interactions in the complex. A small value for Z or a positive slope is taken to mean that the electrostatic contribution is not the driving force, whereas a large negative slope is taken to imply that it is the dominant mechanism. Theoretical methods based on molecular simulation can describe the complex in terms of intermolecular interactions and provide a basis for developing more detailed molecular models of the process together with estimates of the corresponding energetics.

The application of continuum electrostatics to the study of salt effects on the binding of minor groove antibiotics to DNA has been described by Honig and coworkers (91). A comparison of Poisson Boltzmann and limiting law counterion binding models provides evidence that the electrostatic free energy in PB contains a significant additional entropy contribution from water reorientation which contributes to the salt dependence of ligand binding (122b). The relationships between limiting law, PB and full MD simulation descriptions of these processes will be fully elaborated as supercomputer power makes molecular simulations on these problems tractable for a wide range of examples. A recent review of the area of salt effects on nucleic acids is due to Sharp and Honig (123a).

Recent free energy simulations on the thermodynamics of λ repressor-operator association have shown that polyelectrolyte effects, in the region that favors protein-DNA association, are short ranged (57). The ion atmosphere contribution to the free energy of association is favorable and is at its maximum when the protein approaches the DNA from a distance of separation of approximately 7 Å, which is typically the radius of the counterion condensate around B-DNA. Displacement of the condensed counterions contributes favorably to the free energy of DNA–ligand complexation. The exact magnitude of this free energy is expected to depend on the nature of the ligand, as well as on the manner in which electrostatic interactions are treated and the direction of approach of the ligand toward the DNA.

The FDPB method was used to analyze salt effects on drug-DNA and protein-DNA complexation (90, 91, 148). Interestingly, the binding free energies turn out to be positive, whereas the slopes ($-Z$) of log K_{calc} vs log (salt) plots are in good accord with experiment. A combined conformational search and PB study to flexible docking in the λ repressor-operator protein-DNA complex has been described (148). At this juncture, there is sufficient indication that theoretical calculations based on the FDPB methodology can account for ion atmosphere effects on protein-DNA and drug-DNA association quantitatively.

GCMC calculations have been reported on the salt dependence of oligocation binding incorporating structural detail of the DNA (98a).

In a recent study on the binding of the Tet repressor to nonspecific and specific DNA monitored by stopped flow techniques, Kleinschmidt et al (67) concluded that nonspecific binding was almost completely driven by the entropy change resulting from the release of three to four Na^+ ions from the double helix on protein binding. Formation of the specific complex was driven by a higher entropy term that resulted from the release of seven to eight Na^+ ions, along with a favorable free energy term that came from nonelectrostatic interactions attributable to specific contacts. Senear & Batey (121) reported a thorough experimental investigation of salt effects on the binding of λ cI repressor to the right operators (OR1, OR2 and OR3) and nonspecific DNA. The thermodynamic equivalent of the number of K^+ ions released varied as 5.9 ± 0.7 for OR1, 6.0 ± 0.6 for OR2, 3.8 ± 0.7 for OR3, and 4.8 ± 0.7 for nonspecific DNA. The observed differences suggested a role for ion binding in site specific recognition, which reflected different deformations of the repressor and/or different conformations of the DNA so that a different number of ions and, presumably, water molecules were removed from the interface on binding.

Other Related Studies

Few simulation studies have yet been conducted on protein-DNA and drug-DNA systems including explicit consideration of water molecules and small ions, largely because of heavy computational requirements. Molecular dynamics simulations with simpler treatment of solvent and counterions, however, have become an integral part of the X-ray and NMR structural refinement procedures for the protein-DNA and drug-DNA complexes (21). Orozco et al (101) analyzed the role of explicit solvent in modeling drug-DNA structures and advocated the use of a distance-dependent dielectric constant. Gago & Richards (40) performed free energy simulations on netropsin binding to poly[d(IC)] poly[d(IC)] and poly[d(GC)] poly[d(GC)] with a consideration of explicit water and counterions. The calculated free energy change ($\Delta\Delta G$) between the ATAT(ICIC) and GCGC sequences was 4.35 kcal/mol, in close correspondence with the experimental value of 4 kcal/mol. The width of the minor groove was proposed to be a determinant of specificity in these sytems. Results of free energy simulations on base specificity of daunomycin and acridine intercalation into DNA with the AMBER force field and implicit incorporation of water and solvated counterions showed qualitative agreement with experiment (27). Swaminathan et al (130) investigated the static and dynamic aspects of the intermolecular hydrogen-bonding network and the organization of water via molecular dynamics simulations on dCpG/proflavin complex. Cardozo & Hopfinger (22) used molecular dynamics simulations to investigate reaction pathways of the complex of dynemicin-A and DNA. A 100-ps molecular dynamics study of a triple helix explicit water and counterions showed that phosphate groups constitute primary centers of Na^+ attachment (92). Sekharudu et al (120) modeled a third poly(dT) strand Hoogsteen base-paired to the major groove of the poly-(dA) poly(dT) Watson-Crick (WC) base-paired duplex in the canonical B-DNA form and performed molecular dynamics simulations for a total period of 400 ps with explicit water and counterions. The geometry of the WC portion of the duplex turned out to be unique, differing from both the A and B forms of DNA. Jin et al (61) characterized the structural aspects of nucleic acid complexes with $2',5''$ and $3',5''$–phosphodiester linkages with the use of a combination of spectroscopic and calorimetric techniques and a computer-assisted search for macromolecular structures that exhibit features consistent with the experimental data. The latter were found to form either a duplex or triplex, depending on the sodium ion concentration, whereas the former formed either a triplex or no complex at all. Singh and coworkers (48) studied the dynamic properties of methyl phosphonate double-stranded DNA, a potential therapeutic agent. Molecular dynamics simulations on DNA decamers with scaled phosphate charges and with disulfide crosslinks attached to exocyclic amines of the adenines in the central 5′-AT-3′ base pairs were performed (34), and the results compared with those of crystal and NMR studies. Brownian dynamics studies (56) that addressed some mechanistic issues of relevance to the kinetics of complexation suggested a nonspecific association of the protein to the DNA followed by a facilitated diffusion (in reduced dimensionality, such as sliding) to the target site, in conformity with the earlier interpretations of experiment by von Hippel and coworkers (146).

More experimental and theoretical studies are needed, and some are in progress in diverse laboratories, to develop a comprehensive view of the molecular aspects of protein-DNA and drug-DNA recognition, DNA fine structure, and the role of solvent and ion atmosphere around DNA.

CONCLUSION

DNA problems are unique in that the molecule per se and its surrounding water and counterions must be treated together as a molecular ecosystem in order to properly describe structure and function. This requires theoretical studies of either well crafted approximations such as continuum electrostatics or of exceedingly high dimensionality as in molecular simulation. Computational research currently has been concerned with the development and assessment of the protocols and force fields that underlie static and dynamic models of DNA in solution (14, 111). There is sufficient justified optimism that the theoretical studies summarized here would evolve very soon into predictive tools for investigations on sequence-dependent fine structure of DNA and protein-DNA recognition.

ACKNOWLEDGMENTS

Funding from the National Institutes of Health, grant GM37909 to DLB is gratefully acknowledged. BJ acknowledges the support received from the Department of Science and Technology, India, and the administration of IIT, Delhi, India. BJ also expresses his gratitude to DLB for funding his visits to the United States over the years and for providing facilities to carry out research at Wesleyan.

Literature Cited

1. Allison SA. 1994. End effects in electrostatic potentials of cylinders: models for DNA fragments. *J. Phys. Chem.* 98:12091
2. Anderson CF, Record MT Jr. 1982. Polyelectrolyte theories and their application to DNA. *Annu. Rev. Phys. Chem.* 33:191
3. Anderson CF, Record MT Jr. 1990. Ionic distributions around DNA and other cylindrical polyions: theoretical descriptions and physical implications. *Annu. Rev. Biophys. Biophys. Chem.* 19:423

3a. Anderson C, Record MT. 1993. Salt dependence of oligoion-polyion binding: a thermodynamic description based on preferential interaction coefficients. *J. Phys. Chem.* 97:7116

4. Bacquet R, Rossky PJ. 1984. Ionic atmosphere of rod-like polyelectrolytes. A hypernetted chain study. *J. Phys. Chem.* 88:2660
5. Bacquet RJ, Rossky PJ. 1988. Ionic distributions and competitive association in DNA/mixed salt solutions. *J. Phys. Chem.* 92:3604
6. Barkley MD, Zimm BH. 1979. Theory of twisting and bending of chain macromolecules: analysis of the fluorescence depolarization of DNA. *J. Chem. Phys.* 70:2991
7. Bashford D. 1991. Electrostatic effects in biological molecules. *Curr. Opin. Struct. Biol.* 1:175
8. Berman HM. 1991. Hydration of DNA. *Curr. Opin. Struct. Biol.* 1:423
9. Berman HM. 1994. Hydration of DNA: take 2. *Curr. Opin. Struct. Biol.* 4:345
10. Berman HM, Olson WK, Beveridge DL, Westbrook J, Gelbin A, et al. 1992. The Nucleic Acid Database: a comprehensive relational database of three dimensional structures of nucleic acids. *Biophys. J.* 63:751
11. Bernstein FC, Koetzle TF, Williams GJB, Meyer EF Jr, Brice MD, et al. 1977. The Protein Data Bank: a computer-based archive file for macromolecular structures. *J. Mol. Biol.* 112:535
12. Beveridge DL, Dicapua FM. 1989. Free energy via molecular simulation: applications to chemical and biomolecular systems. *Annu. Rev. Biophys. Biophys. Chem.* 18:431
13. Beveridge DL, Maye PV, Jayaram B, Ravishanker G, Mezei M. 1984. Aqueous hydration of nucleic acid constituents: Monte Carlo computer simulation studies. *J. Biomol. Struct. Dyn.* 2:261
14. Beveridge DL, Ravishanker G. 1994. Molecular dynamics studies of DNA. *Curr. Opin. Struct. Biol.* 4:246
15. Beveridge DL, Subramanian PS, Jayaram B, Swaminathan S. 1990. In *Theoretical Biochemistry and Molecular Biophysics*, ed. DL Beveridge, R Lavery, pp. 17–38. Schenectady, NY: Adenine
16. Beveridge DL, Swaminathan S, Ravishanker G, Whitka JM, Srinivasan J, et al. 1993. Molecular dynamics simulations on the hydration, structure and motions of DNA oligomers. In *Molecular Dynamics Simulations on the Hydration, Structure and Motions of DNA Oligomers in Water and Biological Molecules*, ed. E Westhof, pp. 165–225. London: Macmillan
17. Bleam ML, Anderson CF, Record MT Jr. 1983. Sodium-23 nuclear magnetic resonance studies of cation-deoxyribonucleic acid interactions. *Biochemistry* 22:5418
18. Bloomfield VA. 1991. Condensation of DNA by multivalent cations: considerations on mechanism. *Biopolymers* 31:1471

18a. Bond JP, Anderson CF, Record MT. 1994. Conformational transitions of duplex and triplex nucleic acid helices: thermodynamic analysis of the effects of salt concentration on stability using preferential interaction coefficients. *Biophys. J.* 67:825

19. Braunlin WH, Bloomfield VA. 1991. ^{1}H NMR study of the base-pairing reaction of d(GGAATTCC): salt effects on the equilibria and kinetics of strand association. *Biochemistry* 30: 754
20. Briki F, Ramstein J, Lavery R, Genest D. 1991. Evidence for the stochastic nature of base pair opening in DNA. A Brownian dynamics simulation. *J. Am. Chem. Soc.* 113:2490
21. Bruenger A, Kuriyan J, Karplus M. 1987. Crystallographic R factor refinement by molecular dynamics. *Science* 235:458
22. Cardozo MG, Hopfinger AJ. 1993. A model for the dynemicin-A cleavage of DNA using molecular dynamics simulation. *Biopolymers* 33:377
23. Chang S-L, Chen S-H, Rill RL, Lin JS. 1990. Measurements of monovalent and divalent counterion distribu-

tions around persistence length DNA fragments in solution. *J. Phys. Chem.* 94:8025
24. Cheatham TE III, Miller JL, Fox T, Darden TA, Kollman PA. 1995. Molecular dynamics simulations on solvated biomolecular systems: The particle mesh Ewald method leads to stable trajectories of DNA, RNA and proteins. *J. Am. Chem. Soc.* 117:4193
25. Chen S-W, Rossky PJ. 1993. Influence of solvent and counterion on $^{23}Na^{+}$ spin relaxation in aqueous solution. *J. Phys. Chem.* 97:10803
26. Chirico G, Langowski J. 1994. Kinetics of supercoiling studied by Brownian dynamics simulation. *Biopolymers* 34:415
27. Cieplak P, Rao SN, Grootenhuis PDJ, Kollman PA. 1990. Free energy calculation of base specificity of drug-DNA interactions: application to daunomycin and acridine intercalation into DNA. *Biopolymers* 29: 717
28. Conrad J, Troll M, Zimm BH. 1988. Ions around DNA: Monte Carlo estimates of distribution with improved electrostatic potentials. *Biopolymers* 27:1711
28a. Cornell WD, Cieplak P, Bayly CI, Gould IA, et al. 1995. A second generation force field for the simulation of proteins, nucleic acids and organic molecules. *J. Am. Chem. Soc.* 117: 5179
29. Das T, Bratko D, Bhuiyan LB, Outhwaite CW. 1995. Modified Poisson-Boltzmann theory applied to linear polyelectrolyte solutions. *J. Phys. Chem.* 99:410
30. Davis ME, McCammon JA. 1990. Electrostatics in biomolecular structure and dynamics. *Chem. Rev.* 90: 509
31. Drew HR, Dickerson RE. 1981. Structure of a B-DNA dodecamer. III. Geometry of hydration. *J. Mol. Biol.* 151:535
31a. Dwyer JD, Bloomfield VA. 1993. Brownian dynamics simulations of probe and self-diffusion in concentrated protein and DNA solutions. *Biophys. J.* 65:1810
32. Erie DA, Suri AK, Breslauer KJ, Jones RA, Olson WK. 1993. Theoretical predictions of DNA hairpin-loop conformations. Correlations with thermodynamic and spectroscopic data. *Biochemistry* 32:436
33. Fenley MO, Manning GS, Olson WK. 1990. Approach to the limit of counterion condensation. *Biopolymers* 30:1191
33a. Fenly MO, Olson WK, Tobias I, Manning GS. 1994. Electrostatic effects in short superhelical DNA. *Biophys. Chem.* 50:255–271
34. Ferentz AE, Wiorkiewicz-Kuczera J, Karplus M, Verdine G. 1993. Molecular dynamics simulations of disulfide cross-linked DNA decamers. *J. Am. Chem. Soc.* 115:7569
35. Forester TR, McDonald IR. 1991. Molecular dynamics studies of the behaviour of water molecules and small ions in concentrated solutions of polymeric B-DNA. *Mol. Phys.* 72: 643
36. Forsen S, Drakenberg T, Wennerstrom H. 1987. NMR studies of ion binding in biological systems. *Q. Rev. Biophys.* 19:83
37. Friedman R, Honig B. 1992. The electrostatic contribution to DNA base-stacking interactions. *Biopolymers* 32:145
38. Friedman RAG, Manning GS, Shahin MA. 1990. The polyelectrolyte correction to site exclusion numbers. In *Drug-DNA Binding in Chemistry and Physics of DNA-Ligand Interactions,* ed. NR Kallenbach, pp. 37–64. Schenectady, NY: Adenine
39. Gabb HA, Lavery R, Prevost C. 1995. Efficient conformational space sampling for nucleosides using internal coordinate Monte Carlo simulations and a modified furanose description. *J. Comp. Chem.* 16:667
40. Gago F, Richards WG. 1990. Netropsin binding to poly[d(ic)] poly-[d(IC)] and poly[d(GC)] poly-[d(GC)]: a computer simulation. *Mol. Pharmacol.* 37:341
41. Garcia AE, Soumpasis DM. 1989. Harmonic vibrations and thermodynamic stability of a DNA oligomer in monovalent salt solution. *Proc. Natl. Acad. Sci. USA* 86:3160
42. Gordon HL, Goldman S. 1992. Simulations of the counterion and solvent distribution functions around two simple models of a polyelectrolyte. *J. Phys. Chem.* 96:1921
43. Gross LM, Strauss UP. 1966. Interactions of polyelectrolytes with simple electrolytes. I. Theory for electrostatic potential and Donnan equilibrium for a cylindrical rod model: the effect of site-binding in chemical physics of ionic solutions. In *Chemical Physics of Ionic Solutions* ed. BE Conway, RG Barradas, pp. 361–89. New York: Wiley & Sons

44. Gueron M, Demaret J-P. 1992. A simple explanation of the electrostatics of the B to Z transition of DNA. *Proc. Natl. Acad. Sci. USA* 89:5740
45. Gueron M, Weisbuch G. 1980. Polyelectrolyte theory. I. Counterion accumulation site binding and their insensitivity to polyelectrolyte shape in solutions containing finite salt concentrations. *Biopolymers* 19:353
46. Guldbrand LE, Forester TR, Lynden-Bell RM. 1989. Distribution and dynamics of mobile ions in systems of ordered B-DNA. *Mol. Phys.* 67:473
47. Harvey S. 1989. Treatment of electrostatic effects in macromolecular modeling. *Proteins* 5:78
48. Hausheer FH, Rao BG, Saxe JD, Singh UC. 1992. Physico chemical properties of)- vs (S)-methylphosphonate substitution on antisense DNA hybridization determined by free energy perturbation and molecular dynamics. *J. Am. Chem. Soc.* 114: 3201
49. Hecht JL, Honig B, Shin Y-K, Hubbell WL. 1995. Electrostatic potentials near the surface of DNA: comparing theory and experiment. *J. Phys. Chem.* 99:7782
50. Hingerty BE, Ritchie RH, Ferrell TL, Turner JE. 1985. Dielectric effects in biopolymers: the theory of ionic saturation revisited. *Biopolymers* 24:427
51. Hirata F, Levy RM. 1989. Salt induced conformational changes in DNA: analysis using the polymer RISM theory. *J. Phys. Chem.* 93:479
51a. Holst M, Kozack R, Saied F. 1994. Protein electrostatics—rapid multigrid-based Newton algorithm for solution of the full nonlinear Poisson-Boltzmann equation. *J. Biomol. Struct. Dyn.* 11:1437
51b. Holst M, Saied F. 1993. Multigrid solution of the Poisson-Boltzmann equation. *J. Comp. Chem.* 14:105
52. Honig B, Nicholls A. 1995. Classical electrostatics in biology and chemistry. *Science* 268:1144
53. Jayaram B, Aneja N, Rajasekaran E, Arora V, Das A, et al. 1994. Modelling DNA in aqueous solutions. *J. Sci. Ind. Res. (India)* 53:88
54. Jayaram B, Beveridge DL. 1990. Free energy of an arbitrary charge distribution imbedded in coaxial cylindrical dielectric continua: application to conformational preferences of DNA in aqueous solutions. *J. Phys. Chem.* 94:4666 (and references therein)
55. Jayaram B, Beveridge DL. 1991. Grand canonical Monte Carlo simulations on aqueous solutions of NaCl and NaDNA: excess chemical potentials and sources of nonideality in electrolyte and polyelectrolyte solutions. *J. Phys. Chem.* 95:2506
56. Jayaram B, Das A, Aneja N. 1995. Energetic and kinetic aspects of macromolecular association: a computational study of repressor-operator complexation. *J. Mol. Struct. (THEOCHEM)*. In press
57. Jayaram B, DiCapua FM, Beveridge DL. 1991. A theoretical study of polyelectrolyte effects in protein-DNA interactions: Monte Carlo free energy simulations on the ion atmosphere contribution to the thermodynamics of λ repressor-operator complex formation. *J. Am. Chem. Soc.* 113:5211
58. Jayaram B, Sharp K, Honig B. 1989. The electrostatic potential of B-DNA. *Biopolymers* 28:975
59. Jayaram B, Swaminathan S, Beveridge DL, Sharp K, Honig B. 1990. Monte Carlo simulation studies on the structure of the counterion atmosphere of B-DNA. Variations on the primitive dielectric model. *Macromolecules* 23:3156
60. Jia X, Marzilli LG. 1991. Zinc ion-DNA polymer interactions. *Biopolymers* 31:23
61. Jin R, Chapman WH, Srinivasan R, Olson WK, Breslow RD, Breslauer KJ. 1993. Comparative spectroscopic, calorimetric and computational studies of nucleic acid complexes with 2′,5″- versus 3′,5″-phosphodiester linkages. *Proc. Natl. Acad. Sci. USA* 90:10568
62. Jovin TM, Soumpasis DM, McIntosh LP. 1987. The transition between B-DNA and Z-DNA. *Annu. Rev. Phys. Chem.* 38:521
63. Kennedy SD, Bryant RG. 1986. Manganese-deoxyribonucleic acid binding modes: nuclear magnetic relaxation dispersion results. *Biophys. J.* 50:669
64. Klement R, Soumpasis DM, Jovin TM. 1991. Computation of ionic distribution around charged biomolecular structures: results for right handed and left handed DNA. *Proc. Natl. Acad. Sci. USA* 88:4631
65. Klement R, Soumpasis DM, Kitzing EV, Jovin TM. 1990. Inclusion of ionic interactions in force field calculations of charged biomolecules—DNA structural transitions. *Biopolymers* 29:1089
66. Klein BJ, Pack GR. 1983. Calcula-

tions of the spatial distribution of charge density in the environment of DNA. *Biopolymers* 22:2331
67. Kleinschmidt C, Tovar K, Hillen W, Porschke D. 1988. Dynamics of repressor-operator recognition: the trilo-encoded tetracycline resistance control. *Biochemistry* 27:1094
68. Kollman P. 1993. Free energy calculations: applications to chemical and biochemical phenomena. *Chem. Rev.* 93:2395
69. Kottalam J, Case DA. 1990. Langevin modes of macromolecules: application to crambin and DNA hexamers. *Biopolymers* 29:1409
70. Lamm G, Pack GR. 1990. Acidic domains around nucleic acids. *Proc. Natl. Acad. Sci. USA* 87:9033
71. Lamm G, Wong L, Pack GR. 1994. Monte Carlo and Poisson-Boltzmann calculations of the fraction of counterions bound to DNA. *Biopolymers* 34:227
72. Lavery R, Hartmann B. 1994. Modelling DNA conformational mechanics. *Biophys. Chem.* 50:33
72a. Lavery R, Pullman B. 1985. The dependence of the surface electrostatic potential of B-DNA on environmental factors. *J. Biomol. Struct. Dyn.* 2: 1021
73. Le Bret M, Zimm BH. 1984. Monte Carlo determination of the distribution of ions about a cylindrical polyelectrolyte. *Biopolymers* 23:271
74. Levitt M. 1983. Computer simulation of DNA double helix dynamics. *Cold Spring Harbor Symp. Quant. Biol.* 47:251
75. Lohman TM. 1986. Kinetics of protein-nucleic acid interactions: use of salt effects to prove mechanisms of interaction. *CRC Crit. Rev. Biochem.* 19:191
76. Madden TL, Deutsch JM. 1991. Theoretical studies of DNA during orthogonal field alternating gel electrophoresis. *J. Chem. Phys.* 94:1584
77. Manning GS. 1972. Polyelectrolytes. *Annu. Rev. Phys. Chem.* 23:117
78. Manning GS. 1978. The molecular theory of polyelectrolyte solutions with application to the electrostatic properties of polynucleotides. *Q. Rev. Biophys.* 11:179
79. Manning GS. 1979. Counterion binding in polyelectrolyte theory. *Acc. Chem. Res.* 12:443
80. Marky LA, Curry J, Breslauer KJ. 1985. Netropsin binding to poly d(AT)-poly d(AT) and to poly dA-poly dT: a comparative thermodynamic study. *Prog. Clin. Biol. Res.* 172:155
81. Matthew JB, Richards FM. 1984. Differential electrostatic stabilization of A, B and Z forms of DNA. *Biopolymers* 23:2743
82. Mazur J, Jernigan RL. 1991. Distance dependent dielectric constants and their application to double helical DNA. *Biopolymers* 31:1615
83. McCammon JA. 1991. Free energy from simulations. *Curr. Opin. Struct. Biol.* 1:196
84. McCammon JA, Harvey SC. 1987. *Dynamics of Proteins and Nucleic Acids*. New York: Cambridge Univ. Press
85. McConnell KJ, Nirmala R, Young MA, Ravishanker G, Beveridge DL. 1994. A nanosecond molecular dynamics trajectory for a B DNA double helix: evidence for substates. *J. Am. Chem. Soc.* 116:4461
86. Mills P, Anderson CF, Record MT Jr. 1985. Monte Carlo studies of counterion-DNA interactions. Comparison of the radial distribution of counterions with predictions of other polyelectrolyte theories. *J. Phys. Chem.* 89:3984
87. Mills P, Paulsen MD, Anderson CF, Record MT Jr. 1986. Monte Carlo simulations of counterion accumulation near helical DNA. *Chem. Phys. Lett.* 129:155
88. Mills P, Anderson CF, Record MT Jr. 1986. Grand canonical Monte Carlo calculations of thermodynamic coefficients for a primitive model of DNA-salt solutions. *J. Phys. Chem.* 90:6541
89. Mills PA, Rashid A, James TL. 1992. Monte Carlo calculations of ion distributions surrounding the oligonucleotide d $(ATATATATAT)_2$ in the B, A, and wrinkled D conformations. *Biopolymers* 32:1491
90. Misra V, Hecht J, Sharp K, Friedman R, Honig B. 1994. Salt effects on protein-DNA interactions: the lambda cI repressor and Eco RI endonuclease. *J. Mol. Biol.* 238:264
91. Misra V, Sharp K, Friedman R, Honig B. 1994. Salt effects on ligand-DNA binding: minor groove antibiotics. *J. Mol. Biol.* 238:245
92. Mohan V, Smith PE, Pettitt BM. 1993. Molecular dynamics simulation of ions and water around triplex DNA. *J. Phys. Chem.* 97:12984
93. Murthy CS, Bacquet RJ, Rossky PJ. 1985. Ionic distributions near polyelectrolytes. A comparison of theoreti-

cal approaches. *J. Phys. Chem.* 89: 701

94. Neidle S, Jenkins TC. 1991. Molecular modeling to study DNA intercalation by antitumor drugs in methods in enzymology. *Methods Enzymol.* 203: 433
95. Nilsson L, Karplus M. 1984. Energy functions from energy minimization and dynamics of nucleic acids. *J. Comput. Chem.* 1:591
96. Nishio T. 1994. Monte Carlo simulations on potentiometric titration of cylindrical polyelectrolytes: introduction of a method and its application to model systems without added salt. *Biophys. Chem.* 49:201
97. Nordenskiold L, Chang DK, Anderson CF, Record MT Jr. 1984. ^{23}Na NMR relaxation study of the effects of conformation and base composition on the interactions of counterions with double-helical DNA. *Biochemistry* 23:4309

97a. Oberoi H, Allewell N. 1993. Multigrid solution of the nonlinear Poisson-Boltzmann equation and calculation of titration curves. *Biophys. J.* 65:48

97b. Olmsted MC, Anderson CF, Record MT. 1989. Monte Carlo description of oligoelectrolyte properties of DNA oligomers. *Proc. Natl. Acad. Sci.* 86:7766

98. Olmsted MC, Anderson CF, Record MT Jr. 1991. Importance of oligoelectrolyte end effects for the thermodynamics of conformational transitions of nucleic acid oligomers: a grand canonical Monte Carlo analysis. *Biopolymers* 31:1593

98a. Olmsted MC, Bond J, Anderson CF, Record MT. 1995. Grand Canonical Monte Carlo molecular and thermodynamic predictions of ion effects on binding of an oligocation to the center of DNA oligomers. *Biophys. J.* 68:634

98b. Olmsted MC, Hagerman PJ. 1995. Excess counterion accumulation around branched nucleic acids. *J. Mol. Biol.* 238:455

99. Olson WK, Zhang P. 1991. Computer simulation of DNA supercoiling. *Methods Enzymol.* 203:403
100. Oosawa F. 1957. A simple theory for thermodynamic properties of polyelectrolyte solutions. *J. Polym. Soc. Jpn.* 23:421
101. Orozco M, Laughton CA, Herzyk P, Neidle S. 1990. Molecular mechanics modelling of drug-DNA structures: the effects of differing dielectric treatment on helix parameters and comparison with a fully solvated structural model. *J. Biomol. Struct. Dyn.* 8:359
102. Otwinowsky Z, Schevitz RW, Zhang R-G, Lawson CL, Joachimiak A, et al. 1988. Crystal structure of trp repressor/operator complex at atomic resolution. *Nature* 335:321

102a. Pack GR. 1993. Counterion condensation theory revisited: limits on its application. *Int. J. Quantum Chem.* 20:213–30

103. Pack GR, Garrett GA, Wong L, Lamm G. 1993. The effect of a variable dielectric coefficient and finite ion size on Poisson-Boltzmann calculations of DNA-electrolyte systems. *Biophys. J.* 65:1363
104. Padmanabhan S, Richey B, Anderson CF, Record MT Jr. 1988. Interaction of an N-methylated polyamine analogue, hexamethonium (2+), with NaDNA: quantitative ^{14}N and ^{23}Na NMR relaxation rate studies of the cation-exchange process. *Biochemistry* 27:4367
105. Penford R, Jonsson B, Nordholm S. 1993. Ion-ion correlations in polyelectrolyte solutions: hard sphere counterions. *J. Chem. Phys.* 99:497
106. Pullman A, Pullman B. 1981. Molecular electrostatic potential of the nucleic acids. *Q. Rev. Biophys.* 14:289
107. Pullman B. 1983. Electrostatics of polymorphic DNA. *J. Biomol. Struct. Dyn.* 1:773
108. Rajasekaran E, Jayaram B. 1994. Counterion condensation in DNA systems: the cylindrical Poisson-Boltzmann model revisited. *Biopolymers* 34:443
109. Ramstein J, Lavery R. 1988. Energetic coupling between DNA bending and base-pair opening. *Proc. Natl. Acad. Sci. USA* 85:7231
110. Rashin AA, Malinsky J. 1991. New method for the computation of ionic distribution of round rod-like polyelectrolytes with the helical distribution of charges. I. General approach and a nonlinearized Poisson-Boltzmann equation. *J. Comp. Chem.* 12: 981
111. Ravishanker G, Auffinger P, Langley DR, Jayaram B, Young MA, Beveridge DL. 1996. Treatment of counterions in computer simulations of DNA. *Rev. Comput. Chem.* In press
112. Ravishanker G, Beveridge DL. *Molecular Dynamics Analysis Tool Chest.* Dept. Chem., Wesleyan Univ., Middletown, CT

112a. Ray J, Manning GS. 1992. Theory of delocalized ionic binding to polynucleotides: structural and excluded volume effects. *Biopolymers* 32: 541–549
113. Record MT Jr, Anderson CF, Lohman T. 1978. Thermodynamic analysis of ion effects on the binding and conformational equilibria of proteins and nucleic acids: the role of ion association and release, screening and ion effects on water activity. *Q. Rev. Biophys.* 11:103
114. Record MT Jr, Anderson CF, Mills P, Mossing M, Roe J-H. 1985. Ions as regulators of protein-nucleic acid interactions in vitro and in vivo. *Adv. Biophys.* 20:109
115. Reddy MR, Rossky PJ, Murthy CS. 1987. Counterion spin relaxation in DNA solutions: a stochastic dynamics simulation study. *J. Phys. Chem.* 91:4923
116. Reed CE, Reed WF. 1991. Monte Carlo study of titration of linear polyelectrolyte effects. *J. Chem. Phys.* 96: 1609
117. Rentzeperis D, Marky LA, Kupke DW. 1992. Entropy volume correlation with hydration changes in DNA-ligand interactions. *J. Phys. Chem.* 96:9612
118. Saenger W. 1984. *Principles of Nucleic Acid Structure*. New York: Springer-Verlag
119. Seeman NC, Rosenberg JM, Rich A. 1976. Sequence-specific recognition of double helical nucleic acids by proteins. *Proc. Natl. Acad. Sci. USA* 73:804
119a. Seeman NC, Rosenberg JM, Suddath FL, Parkkim JJ, Rich A. 1976. *J. Mol. Biol.* 104:109
120. Sekharudu CY, Yathindra N, Sundaralingam M. 1993. Molecular dynamics investigations of DNA triple helical models: unique features of the Watson-Crick duplex. *J. Biomol. Struct. Dyn.* 11:225
121. Senear DF, Batey R. 1991. Comparison of operator-specific and nonspecific DNA binding of the λ cI repressor: [KCl] and pH effects. *Biochemistry* 30:6677
122. Sharp K. 1994. Electrostatic interactions in macromolecules. *Curr. Opin. Struct. Biol.* 4:234
122a. Sharp K. 1995. Polyelectrolyte electrostatics: salt dependence, entropic and enthalpic contributions to free energy in the non-linear Poisson-Boltzmann model. *Biopolymers*. In press
122b. Sharp K, Friedman R, Misra V, Hecht J, Honig B. 1995. Salt effects on polyelectrolyte-ligand binding: comparison of Poisson Boltzmann and limiting law counterion binding models. *Biopolymers*. In press
123. Sharp K, Honig B. 1990. Electrostatic interactions in macromolecules: theory and applications. *Annu. Rev. Biophys. Biophys. Chem.* 19:301
123a. Sharp KA, Honig B. 1995. Salt effects on nucleic acids. *Curr. Opin. Struct. Biol.* 5:323
124. Siebel GL, Singh UC, Kollman PA. 1985. A molecular dynamics simulation of double helical B-DNA including counterions and water. *Proc. Natl. Acad. Sci. USA* 82:6537
125. Skerjanc J. 1990. Contributions of the polyion and counterions to the internal and free energies of polyelectrolyte solutions. *J. Chem. Phys.* 93: 6731
126. Sobel ES, Harpst JA. 1991. Effect of Na^+ on the persistence length and excluded volume of T7 bacteriophage DNA. *Biopolymers* 31:1559
127. Soumpasis DM. 1984. Statistical mechanics of the B-Z transition of DNA: contribution of diffuse ionic interactions. *Proc. Natl. Acad. Sci. USA* 81:5116
128. Stigter D, Dill KA. 1993. Theory for second virial coefficients of short DNA. *J. Phys. Chem.* 97:12995
129. Subramanian P, Ravishanker G, Beveridge DL. 1988. Theoretical considerations on the "spine of hydration" in the minor groove of d(CGC GA AT TCG CG) d (GC GC T-TAAGCGC): Monte Carlo computer simulations. *Proc. Natl. Acad. Sci. USA* 85:1836
130. Swaminathan S, Beveridge DL, Berman HM. 1990. Molecular dynamics simulation of a deoxynucleoside-drug intercalation complex: dCpG/proflavin. *J. Phys. Chem.* 94:4660
131. Swaminathan S, Ravishanker G, Beveridge DL. 1991. Molecular dynamics of B-DNA including counterions and water: a 140 psec trajectory for d(CGCGAATTCGCG) based on the GROMOS force field. *J. Am. Chem. Soc.* 113:5027
132. Swamy KN, Clementi E. 1987. Hydration structure and dynamics of B- and Z-DNA in the presence of counterions via molecular dynamics simulations. *Biopolymers* 26:1901
133. Taylor WH, Hagerman PJ. 1990. Application of the method of phage T4 DNA ligase-catalyzed ring-closure to the study of DNA structure. II. NaCl

dependence of DNA flexibility and helical repeat. *J. Mol. Biol.* 212:363
134. Tomasi J, Persico M. 1994. Molecular interactions in solution: an overview of methods based on continuous distributions of the solvent. *Chem. Rev.* 94:2027
135. Troll MT, Roitman D, Conrad J, Zimm BH. 1986. Electrostatic effects on ion distribution around aqueous macromolecules. *Macromolecules* 19:1186
136. Urabe H, Kato M, Tominaga Y, Kajiwara K. 1990. Counterion dependence of water of hydration in DNA gel. *J. Chem. Phys.* 92:768
137. Valleau JP. 1989. Flexible polyelectrolyte in ionic solution: a Monte Carlo study. *Chem. Phys.* 129:163
138. van Gunsteren WF, Berendsen HJC. 1990. Computer simulation of molecular dynamics: methodology, applications and perspectives in chemistry. *Angew. Chem. Int. Ed. Engl.* 29: 992
139. Vlachy V, Haymet ADJ. 1986. A grand canonical Monte Carlo simulation study of polyelectrolyte solutions. *J. Chem. Phys.* 84:5874
140. Vologodskii AV, Frank-Kamenetskii MD. 1992. Modeling supercoiled DNA. *Methods Enzymol.* 211:467
141. von Kitzing E. 1992. Modeling DNA structures: molecular mechanics and molecular dynamics. *Methods Enzymol.* 211:449
142. Wesnel TG, Meares CF, Vlachy V, Matthew JB. 1986. Distribution of ions around DNA probed by energy transfer. *Proc. Natl. Acad. Sci. USA* 83:3267
143. Westhof E, Beveridge DL. 1989. Hydration of nucleic acids. In *Water Science Reviews: The Molecules of Life*, ed. F Franks, 5:24–136. New York: Cambridge Univ. Press
144. Whitka JM, Swaminathan S, Srinivasan J, Beveridge DL, Bolton PH. 1992. Towards a dynamical structure of duplex DNA in solution: comparison of theoretical and experimental NOE intensities of d(CGCGAATTCGCG). *Science* 255:597
145. Williams AP, Longfellow CE, Freier SM, Kierzek R, Turner DH. 1989. Laser temperature-jump, spectroscopic, and thermodynamic study of salt effects on duplex formation by dGCATGC. *Biochemistry* 28:4283
146. Winter RB, Berg OG, von Hippel PH. 1981. Diffusion driven mechanisms of protein translocation on nucleic acids. 3. The *Escherichia coli lac* repressor-operator interactions: kinetic measurements and conclusions. *Biochemistry* 20:6929
146a. Young M, Jayaram B, Beveridge DL. 1996. Counterions may intrude into the spine of hydration in the minor groove of the B DNA double helix: the ''AT pocket.'' *J. Am. Chem. Soc.* Submitted
146b. Young MA, Ravishanker G, Beveridge DL. 1996. Nanosecond molecular dynamics trajectories for a B DNA oligonucleotide and explicit water and counterions based on the AMBER force field. *Biophys. J.* Submitted
147. York DM, Yang W, Lee H, Darden T, Pedersen LG. 1995. Toward the accurate modeling of DNA: the importance of long-range electrostatics. *J. Am. Chem. Soc.* 117:5001
148. Zacharias M, Luty BA, Davis ME, McCammon JA. 1992. Poisson-Boltzmann analysis of the lambda repressor-operator interaction. *Biophys. J.* 63:1280
149. Zhurkin VB, Ulyanov NB, Gorin AA, Jernigan RL. 1991. Static and statistical bending of DNA evaluated by Monte Carlo simulations. *Proc. Natl. Acad. Sci. USA* 88:7046
150. Zimm BH. 1991. Lakes-straits model of field inversion gel electrophoresis of DNA. *J. Chem. Phys.* 94:2187
151. Zimm BH, LeBret M. 1983. Counterion condensation and system dimensionality. *J. Biomol. Struct. Dyn.* 1: 461

Annu. Rev. Biophys. Biomol. Struct. 1996. 25:395–429

VISUALIZING PROTEIN–NUCLEIC ACID INTERACTIONS ON A LARGE SCALE WITH THE SCANNING FORCE MICROSCOPE

Carlos Bustamante and Claudio Rivetti*

Institute of Molecular Biology and *Howard Hughes Medical Institute, University of Oregon, Eugene, Oregon 97403

KEY WORDS: scanning force microscopy, protein-induced DNA bending

ABSTRACT

CONTENTS

PROSPECTIVES AND OVERVIEW

Until recently, the electron microscope (EM) was the only tool available to characterize the study of the large-scale spatial relationships of nu-

1056-8700/96/0610-0395$08.00

cleoprotein assemblies. This situation has changed rapidly in the last 3 years, as several groups have demonstrated the feasibility of using the scanning force microscope (SFM) to investigate these complex biological systems. With capabilities often complementary to those of the EM, the SFM is becoming a useful tool of structural characterization of protein–nucleic acid complexes and their interactions in air and in aqueous solutions.

The last 15 years have witnessed the extraordinary growth of structural studies in biology, and the impact is being felt in almost all areas of biological research. This growth reflects not only the use of improved analytical methods in biochemistry but also the coming of age of the two most powerful techniques of high-resolution structural determination: X-ray crystallography and NMR. In contrast, the tools available to the biochemist at the next higher level of biomolecular organization are significantly more limited. Because this regime comprises objects with dimensions between 10 and 200 nm, it has been termed "mesoscopic"; the structures are often too complex for X-ray or NMR analysis and, yet, too small to be seen with the optical microscope. For decades, the EM has been the only analytical tool for this structural regime. Despite the power and versatility of the EM, scientists have welcomed the development of alternative methods that can analyze mesoscopic structures under conditions more closely resembling their physiologic environment.

The study of protein–nucleic acid complexes is paradigmatic of this situation. Several nucleic acid binding proteins have been solved at atomic resolution, and their number is increasing rapidly. A much smaller number of high-resolution structures of protein–nucleic acid complexes exist, and their number is increasing at a slower rate. This scenario worsens at the next level of complexity, which involves large multimolecular assemblies that participate in the regulation of transcription and replication.

The invention in 1986 of the SFM (6), also called atomic force microscope, has equipped biologists with a powerful new tool of structural characterization of protein–nucleic acid interactions. In this article, we review the use of SFM to investigate nucleoprotein assemblies of a broad range of dimensions and complexity. Although we include structural studies of chromatin, we exclude studies of the higher order organization of metaphase and polytene chromosomes; their unique requirements of sample preparation and deposition set them apart from the nucleoprotein assemblies considered here. Because of the novelty of these applications, we emphasize several technical aspects of sample preparation and deposition.

THE SCANNING FORCE MICROSCOPE

The SFM is a member of a new class of high-resolution microscopes known generically as scanning probe microscopes. These instruments do not use lenses to form an image but, instead, use a sharply pointed sensor, or "tip," to probe the surface of the sample. In the SFM, the tip is mounted on the end of a flexible cantilever (Figure 1). As the sample is scanned beneath the tip, small forces of interaction with the

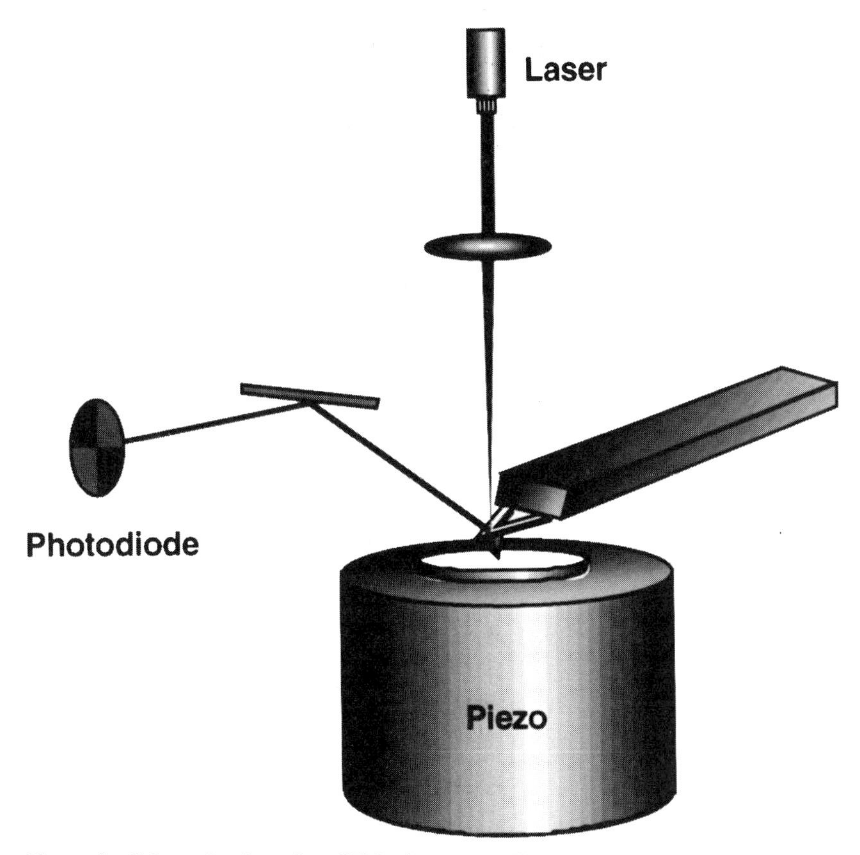

Figure 1 Schematic view of an SFM microscope. The sample is placed on the piezoelectric driver and scanned relative to the tip. In contact mode operation, the surface topography can be obtained directly from the deflection of the cantilever, detected by using the deflection of a laser beam reflected off the back of the cantilever onto a four-quadrant photo diode (optical lever system). Alternatively, the optical lever can be used with a feedback circuit to raise or lower the sample as it is scanned under the tip to cancel the deflection of the cantilever. The image is obtained by converting the voltage applied to the piezo into height information.

sample cause the cantilever to deflect, thereby revealing the sample topography. Deflections as small as 0.01 nm can be detected simply by directing a laser beam on the back of the cantilever and monitoring the amplified deflection of the reflected beam with a four-quadrant photodiode (see 64, 65). The biological applications of this method also have been reviewed recently (9–11, 28).

One of the most attractive features of the SFM is that it can operate at least as well with the cantilever immersed in liquid as in air, which makes it possible to image biological molecules in aqueous buffers. The SFM is, therefore, the first (and so far only) microscope that can achieve nanometer scale resolution on biological samples under physiologic conditions.

Modes of Operation

The SFM can be operated in three different modes: contact, noncontact, and tapping. We consider only contact and tapping applications here, because no biological applications of the noncontact mode have been described to date. In contact mode, the tip is continuously in contact with the surface as it slides over the sample. This operation usually yields stable images, but compression and shear forces generated between the tip and surface may damage the sample. Typical operation forces in contact mode are 1–10 nN. In tapping mode, the cantilever is oscillated up and down at hundreds of kiloHertz as it is scanned relatively to the sample; thus, the tip is allowed to make transient contact with the sample at the bottom of its swing. Because the tip touches the sample, the resolution is usually almost as good as in contact mode. Because the contact is very brief, however, the damage caused by shear forces during scanning is reduced greatly. Imaging of nucleoprotein complexes in air is easier and more reliable in tapping mode than in contact mode.

The tapping mode has been adapted recently for imaging in liquids (29, 60, 61), and its applications to biological molecules in aqueous environments are increasing rapidly. Tapping mode imaging in liquid has several advantages over both the contact mode and the tapping mode in air (9, 29, 60, 61). Tapping in liquid eliminates the liquid–air interface, thereby reducing the capillary forces between the tip and the sample (see below). This feature makes it possible to decrease the amplitude of oscillation of the cantilever without it being captured by the attractive forces. Smaller oscillation amplitudes lead, in turn, to better control of the force acting on the sample and to reduced sample damage (10). Moreover, the smaller shearing forces developed in tapping make stable imaging in liquids possible, as biological samples can

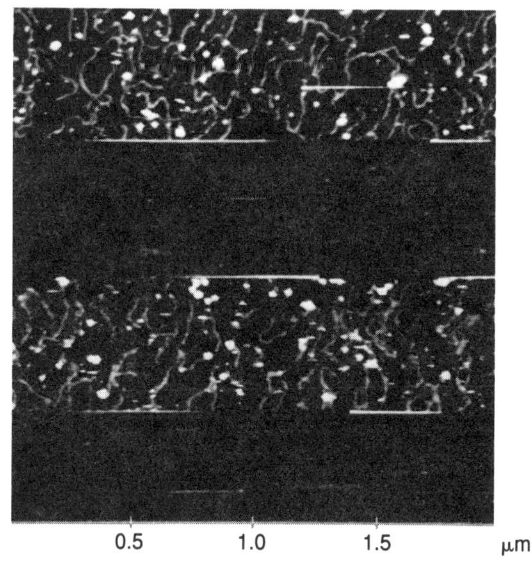

Figure 2 Image of a 1200-bp DNA fragment in buffer (40 mM Tris pH 7.5, 30 mM potassium acetate, 10 mM $MgCl_2$). The first and the third sectors were obtained with tapping mode AFM. In the second and the fourth sectors, the mode of operation was reverted to contact mode. The scan rate was 1.05 Hz, and the tapping frequency was 16.5 kHz. The mean height of the DNA in this image was approximately 1.4 nm. (Image courtesy of G Yang, Institute of Molecular Biology, University of Oregon.)

be detached easily from the surface when they are immersed in an aqueous solution. One example of this effect is illustrated in Figure 2, in which DNA fragments are imaged alternatively in tapping and contact modes. In contact mode, the molecules are swept by the tip, whereas they are imaged reliably in tapping mode.

Imaging of protein–nucleic acid complexes with the use of the tapping mode in liquid is not yet as reliable as in air, probably because of complex hydrodynamic interactions between the sample and the oscillating cantilever. This situation, however, is likely to change in the near future; it is the subject of much attention from several groups interested in visualizing protein–nucleic acid interactions as they occur in nearly physiologic conditions.

Spatial Resolution

The mechanism of image formation in the SFM is very different from those of optical and electron microscopes. The spatial resolution of optical and electron microscopes is a property inherent to the instrument and dependent ultimately on the design and principles of operation of the microscope. In contrast, the spatial resolution of the SFM depends as much on the characteristics of the sample as on the inherent properties of the instrument (10). Thus, although theories of image formation for probe microscopy exist (9, 33, 56), there is no general definition of resolution in force microscopy.

The most important instrument parameter affecting the spatial resolution in SFM is the sharpness of the tip (11), usually described in terms of its end-radius of curvature R_c. Tips with end radii of approximately 10 nm are fabricated easily (11, 32), and resolvable features in an SFM image are typically 5–10 nm apart.

Besides tip dimensions, whether two distinct features of a given object can be resolved depends on the context of each feature in the object (10). This idea is illustrated schematically in Figure 3, where two sharp spikes separated by a distance, *d,* are imaged by a parabolic tip with end-radius R_c. Because the sample is sharper than the tip, the image is the surface defined by the envelope of a pair of inverted tip profiles "hanging" on the spikes (33). The surface and volume of the resulting image is not the "sum" of the images of the individual spikes but rather the union of sets of inverted tip profiles, i.e. the imaging process in SFM is an inherently nonlinear process. In Figure 3*a,* the surface displays a small dimple between the spikes of depth, Δz, which is determined by the tip shape and size and the separation between the spikes. One definition of "resolution" is, then, the minimum separation *d* for which the dimple depth Δz is significantly larger than the instrumental noise. This definition is the closest analogue to the Rayleigh resolution criterion in optical microscopy. The difficulty with this simple idea, however, is shown in Figure 3*b:* As the height difference between the two spikes increases, the depth of the dimple decreases. Two objects that are "resolved" when their heights are nearly equal, therefore, may not be resolved when their heights are unequal. This example shows that resolution in force microscopy is also a function of the height difference Δh between adjacent features and must be determined independently for each feature in the object.

Using the definition of resolution in the preceding paragraph, the minimum separation, *d,* that will result in a dimple of depth Δz for spikes with height difference Δh imaged by a parabolic tip is given by (10):

$$d = \sqrt{2R_c}\left(\sqrt{\Delta z} + \sqrt{\Delta z + \Delta h}\right) \qquad \text{for } d > \sqrt{2R_c\Delta h}. \qquad 1.$$

For features of equal height, and using a parabolic tip with $R_c = 10$ nm, a detectable dimple depth of 0.5 nm yields a minimum resolved

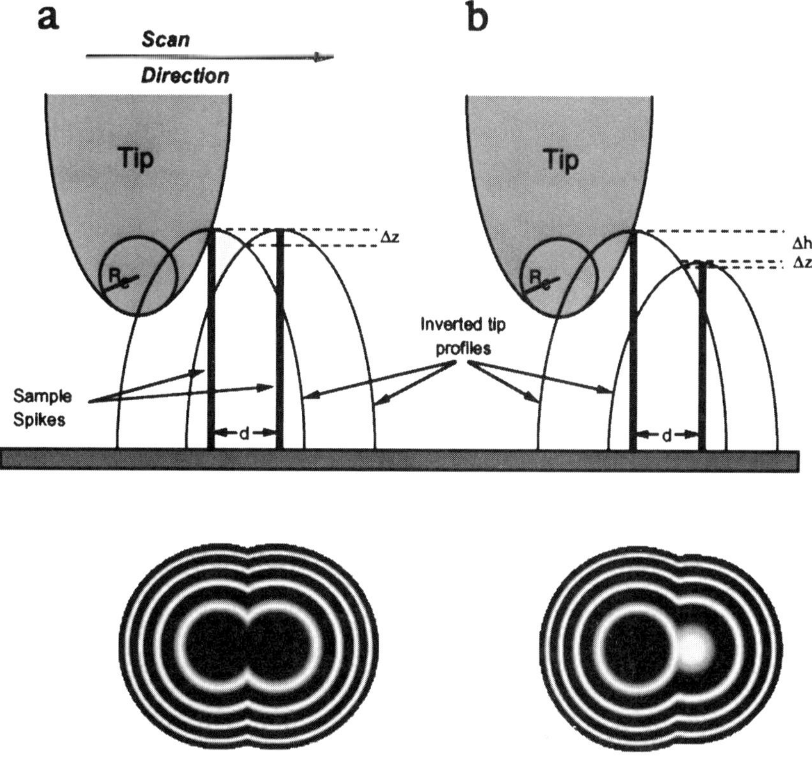

Figure 3 Schematic diagram of the factors that determine the spatial resolution in SFM. Two spikes separated by a distance d are imaged by a parabolic tip with end radius R_c. *(a)* The two spikes have the same height and are well resolved in the image that shows a dimple Δz in between the maxima. *(b)* As the difference in height between the two spikes increases, the resolution decreases (Δz becomes smaller), even though the spikes separation remains constant. Contour plots of the images of the two spikes obtained by the tip in each case are shown below.

separation $d = 6.4$ nm. By comparison, if the height difference is 2.0 nm, the minimum resolved separation is 12.5 nm. This analysis assumes perfectly rigid surfaces for the tip and the sample. In practice, a given feature may be resolved better or worse than this estimate because of sample compliance.

Tip-Sample Interactions

One of the main concerns in the applications of the SFM to study biological structures in general, and protein–nucleic acid assemblies in particular, is the effect of tip-sample forces. Several forces are present between the tip and the sample during imaging (64). Their effect on the sample and on the resulting image depends on their magnitude, the SFM mode of operation, the imaging environment, the nature of the sample, and the sharpness, composition, and shape of the tip (11).

In air, capillary forces and hard-core repulsions between tip and sample are dominant (10). Capillary forces play a major role when imaging in air because, under normal ambient conditions, all samples have a thin layer of water on their surface. Even a few monolayers of adsorbed water can give rise to a water meniscus between the surface and the tip. The resulting attractive force is often strong enough to pin the tip to the surface (11, 13, 64, 77). This force, known as Laplace's force (31), can be as large as a few hundred nN but can be reduced to 1–10 nN by controlling the ambient humidity. Typically, protein–nucleic acid complexes can be imaged reliably in air with a relative humidity of less than 35%, and the samples can be scanned several times without noticeable damage (11, 13, 77, 80). At greater humidity, imaging in contact mode is not stable, because the complexes are swept easily by the tip, unless special care has been taken to strengthen the sample–surface interaction. In contact mode, this component of the total capillary force puts a limit on the lowest imaging tip–sample force and, hence, on the minimum damage to the sample and to the tip. In tapping mode, the molecules are not swept by the tip even at high levels of humidity, but the resulting capillary forces put a lower limit on the amplitude of oscillation required to prevent capture of the oscillating tip by the liquid meniscus and, thus, to determine the minimum operational tip-sample force (9).

Imaging in liquids eliminates the liquid–air interface, thereby reducing the attractive capillary forces greatly. In this case, tip–sample interactions are dominated by van der Waals and electrostatic forces in the range of 0.1 to 1 nN. As illustrated in Figure 2, however, even at these reduced forces, when imaging in contact mode the shear component of the tip–sample interaction may be sufficient to sweep the molecules off the surface.

Attractive forces are balanced by hard-core repulsions between the atoms of the sample and the atoms of the tip (31). If tip and sample are hard, this repulsion defines the sample surface effectively. If the sample is soft and the tip is sharp, however, the pressure caused by the attractive forces can deform or damage the sample. When imaging biological samples, a special effort should be made, therefore, to operate the microscope at the lowest level of possible forces.

SAMPLE PREPARATION AND DEPOSITION

Sample Preparation

Protocols for preparing protein–nucleic acid complexes that are suitable for SFM studies are similar to those used in bulk biochemical studies, but better images are often obtained when the complexes are deposited on the substrate in low salt buffer (<100 mM). Samples that require high salt conditions for the binding reaction (100–500 mM), should be diluted before deposition on the substrate to minimize the formation of salt microcrystals during the drying step. Deposition buffers commonly used are Tris, Hepes, and triethanolamine, with concentrations up to 100 mM of monovalent ions and 1–10 mM of divalent cations, such as Mg^{++}, Mn^{++}, Co^{++}, and Ca^{++}.

Sample purity is essential to obtain reliable imaging with the SFM. Double distilled water and, if possible, nanopure water should be used in the preparation steps. In particular, DNA samples must be purified of contaminating proteins carefully; highly purified proteins are desirable, and protein stabilizers, such as bovine serum albumen (BSA) or glycerol, should be avoided to prevent contamination of the background.

Deposition Protocols

Development of reliable protocols of DNA deposition for SFM have received much attention in the last 4 years. Most of these protocols have not been used with protein–nucleic acid complexes. If the adhesion to the substrate is dominated by the nucleic acid, however, these protocols, with minor modifications, probably could be adapted to image protein–nucleic acid complexes.

The most commonly used substrate has been freshly cleaved mica. The presence of a divalent cation, such as Mg^{++}, in the deposition buffer increases the adhesion of DNA and protein–nucleic acid complexes to the surface greatly (13, 76, 80, 89). Freshly cleaved mica is negatively charged (53); Mg^{++} appears to promote DNA deposition by binding to the mica surface and inverting its charge (11). Increased DNA binding can be obtained also by glow discharging the mica surface before the deposition of the DNA molecules. Other methods of deposition require treating the mica with aminopropyltrimethoxy silane (42, 43, 45); carbon coatings (87); nonionic detergents, such as 2, 4, 6-tris-(dimethylaminomethyl) phenol; cationic detergents, like cetylpyridinium chloride (67); or benzyldimethylalkylamonium chloride (66). Replica methods also have been used recently (14). DNA molecules deposited on mica were overcoated with carbon, and the underside of the carbon layer was imaged in propanol after it was pealed off the mica.

A deposition protocol for imaging DNA molecules and protein–nucleic acid complexes in air involves the following steps: (*a*) placing the sample drop on the freshly cleaved mica; (*b*) allowing the molecules to bind for a deposition time of 1–2 min; (*c*) blotting the excess liquid; (*d*) rinsing with nanopure water; and (*e*) drying the sample under dry N_2 flow.

DEPOSITION KINETICS A recent study has shown that, in good deposition conditions (low salt buffer on mica), the transfer of DNA molecules from solution to the substrate is governed solely by diffusion (C Rivetti, M Guthold, and C Bustamante, in preparation). These authors showed that the density of DNA molecules deposited on mica can be described by the expression:

$$\frac{N^{\circ}\text{ DNA molecules}}{\text{Area}} = \frac{2}{\sqrt{\pi}}[\text{DNA}]\sqrt{Dt}, \qquad 2.$$

where D is the diffusion coefficient of the DNA molecules, [DNA] is the number of DNA molecules/cm^3 in the deposition drop, and t is the time of deposition. Optimal deposition densities of DNA molecules on freshly cleaved mica are obtained at low salt conditions, with DNA concentrations of 0.5–2 nM, and deposition times of 1–2 min. For example, a 0.5-nM solution of DNA in 10 mM NaCl, 2 mM $MgCl_2$, and 4 mM Hepes buffer pH = 7.4 yields an average deposition density of approximately 5300 base pairs (bp)/μm^2 in a deposition time of 30 s and approximately 10,800 bp/μm^2 in 2 min (C Rivetti, M Guthold, and C Bustamante, in preparation). These deposition densities are also ideal for protein–nucleic acid complexes, although, as discussed below, the conditions of deposition must consider the stability and life time of the complexes to prevent complex dissociation.

MOLECULAR MECHANISMS OF DEPOSITION: EQUILIBRATION VS KINETIC TRAPPING Sample deposition and transfer of structures from three-dimensional space onto a two-dimensional surface can potentially alter the spatial relationships between the DNA fragment and its associated proteins, thereby preventing any quantitative characterization of the images. Little is known about the molecular mechanisms of this transfer process and the binding of the complexes to the surface. We present a first approximation here after we discuss the approach of two recent SFM studies conducted on DNA fragments (C Rivetti, M Guthold, and C Bustamante, in preparation) and various protein–nucleic acid complexes (C Walker, WA Rees, DA Erie, and C Bustamante, in preparation).

The outcome of the deposition of nucleic acids or protein–nucleic acid complexes depends on the nature and strength of the molecule–surface interactions. Equilibration occurs if the molecules can approach the surface and search among their accessible states in two dimensions, before the blotting and drying steps. In this case, their appearance and dimensions are indistinguishable from an ensemble of molecules at equilibrium in two dimensions. Alternatively, kinetic trapping results if, during deposition, any segment of the molecules simply sticks to the surface and, thus, remain trapped at the contact sites. In this case, the molecular configurations reflect the history of their approach to the surface. Clearly, these two cases represent extreme alternatives. In practice, a particular deposition protocol may fall somewhere between these two extremes. Whether equilibration is attained may depend on the characteristic time required by the polymers to equilibrate on the surface, relative to the deposition time before blotting and drying. In turn, this characteristic time will depend on such factors as the polymer size, the strength of the molecule–surface interaction, and the number and distribution of the surface binding sites.

The time required by a polymer molecule to access its configurations when it is free to diffuse in two dimensions can be written as $t_0 \sim \nu^N / N\rho_0$, where N is the number of monomers in the polymer, ν is the degrees of freedom per monomer, and ρ_0 is the rate of transition of a monomer among its degrees of freedom. On the other hand, the equilibration time of a polymer molecule with a (steady-state) fraction f_b of its monomers interacting with binding sites on the surface can be written as:

$$t \sim \frac{t_0}{\left\{1 - f_b\left[1 - e\left(\frac{\Delta G_{\text{bind}}}{k_b T}\right)^{-1}\right]\right\}}, \qquad 3.$$

where T is the absolute temperature, k_B is the Boltzman's constant, and ΔG_{bind} is the binding energy of the molecule to each surface site. As indicated by Equation 3, the equilibration time of a polymer molecule that is partly bound to a surface increases with the size of the polymer, the binding energy to the surface, and the density of binding sites on the surface. In general, longer polymers require longer equilibration times before the blotting and drying steps of deposition.

Recently, Rivetti and colleagues addressed these issues (C Rivetti, M Guthold, and C Bustamante, in preparation). These authors compared several deposition procedures of DNA on mica to determine whether the deposition process is dominated by equilibration or by kinetic trapping effects. To this end, DNA fragments of various lengths were deposited on freshly cleaved mica and allowed to equilibrate for 2 min at various ionic strength conditions and various Mg^{++} concentrations. The mean quadratic end-to-end distances of the molecules, $\langle R^2 \rangle$, were measured and compared with the equilibrium values predicted by the wormlike chain model (23) for molecules of the same contour length in two dimensions. According to this model,

$$\langle R^2 \rangle = 4PL\left[1 - \frac{P}{L}\left(1 - e^{-L/2P}\right)\right] \rightarrow 4PL \qquad \text{(for very large } L\text{)}, \qquad 4.$$

where P and L are the persistence length and the contour length of the DNA, respectively. Figure 4 compares $\langle R^2 \rangle$ values determined experi-

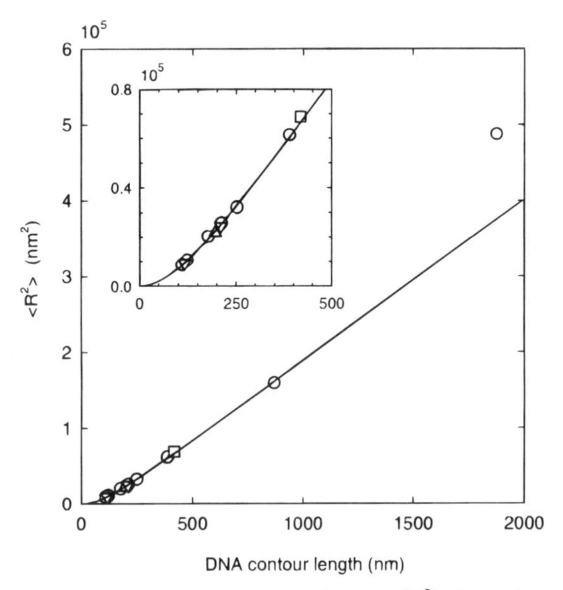

Figure 4 Plot of the mean square end-to-end distance, $\langle R^2 \rangle$, for various DNA sizes as a function of their contour length. The different symbols represent different buffer conditions: *circles,* 4 mM Hepes pH 7.4, 10 mM NaCl, 2 mM $MgCl_2$; *squares,* 4 mM Hepes pH 7.4, 10 mM NaCl, 100 mM $MgCl_2$; *triangles,* 10 mM Hepes pH 8.0, 80 mM NaCl, 5 mM $MgCl_2$; *inverted triangles,* 10 mM Hepes pH 6.8 and 8.0, 5 mM NaCl, 5 mM $MgCl_2$. The value corresponding to the longest DNA fragment (1877 nm) probably reflects the contribution of excluded volume effects. This value is an average of three different depositions in which different incubation times, ranging from 1 to 10 min, were used. The *solid line* is the $\langle R^2 \rangle$ predicted by the wormlike chain model (Equation 4), using a persistence length of 53 nm (12).

mentally for molecules in 4 mM Hepes pH 7.4, 10 mM NaCl, 2 mM $MgCl_2$, with those predicted by Equation 4. Within the range of molecular dimensions studied, the depositions on freshly cleaved mica carried out in low salt buffer lead to adequate equilibration of the molecules on the surface. Moreover, the variance of the square of the end-to-end distance, $\sigma^2_{\vec{R}^2} = 2/3\langle \vec{R}^2 \rangle^2$ (23), also agree quantitatively with the corresponding theoretical values predicted for molecules at equilibrium in two dimensions. In contrast, pretreatment of the mica surface by either glow discharge or monovalent salts yields end-to-end distance values less than those predicted by the equilibrium model. That different mechanisms of deposition operate in these two cases is confirmed by the different appearance of the DNA fragments on the mica observed in the two cases (Figure 5).

Another test of surface equilibrium of DNA molecules on mica can be obtained from the equilibrium distribution of bend angles between two DNA segments separated by a distance, ℓ, along the molecule. This distribution is predicted to be Gaussian by the wormlike and the "hinge" models of polymer statistics (39, 68). According to these models, the bending energy of the molecule in two dimensions is simply (39, 68)

$$E_{\text{bend}} = \frac{k_B T P \theta^2}{2\ell}, \qquad 5.$$

where k_B is the Boltzman constant, T is the absolute temperature, P is the persistence length of DNA, and θ is the bend angle spanning a length of arc ℓ. Although the standard deviation of the angle distribution in two dimensions can be written simply as

$$\langle \theta^2 \rangle = \frac{\ell}{P}. \qquad 6.$$

Figure 6 shows a plot of the $\langle \theta^2 \rangle$ values obtained from the SFM images of DNA fragments of different length deposited in low salt buffer on mica (C Rivetti, M Guthold, and C Bustamante, in preparation). For DNA molecules up to 1500 bp, the persistence length calculated from the slope of the plot is approximately 53 nm, in excellent agreement with values in the literature (12, 25). For longer fragments, the measured persistence length appears larger, probably because of excluded volume effects.

Bezanilla et al (5) recently presented evidence that DNA molecules partially bound to the mica by electrostatic interactions can equilibrate on the surface. These authors imaged DNA molecules under buffer

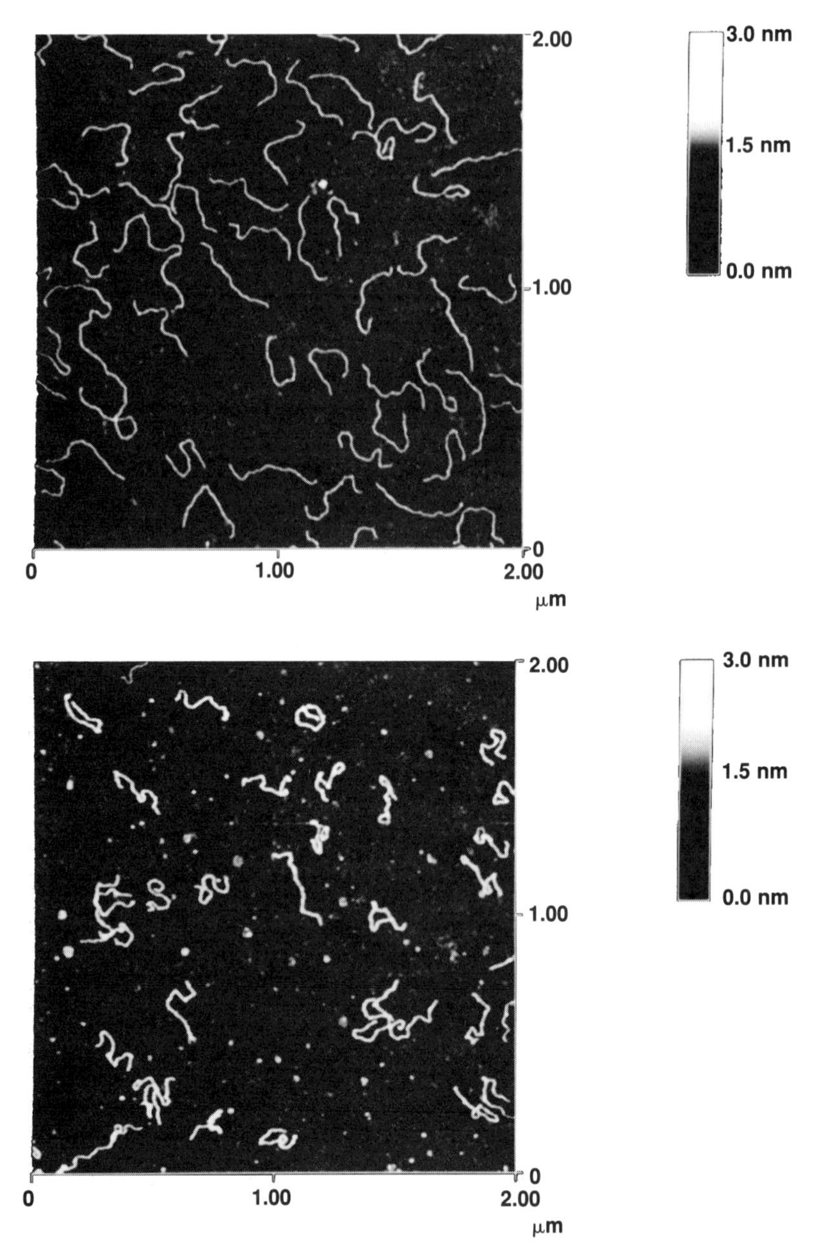

Figure 5 Atomic force microscopic images of a 1258-bp DNA fragment. *(top)* Twenty microliters of a 1 nM DNA solution in low salt buffer (10 mM Hepes pH 7.4, 10 mM NaCl, 2 mM $MgCl_2$) were deposited on freshly cleaved untreated mica and incubated for 2 min before the extra liquid was blotted and the surface was dried. *(bottom)* Twenty microliters of a 0.5 nM DNA solution in the same buffer were deposited on glow discharged mica and incubated for approximately 1 min. The mica disk was glow discharged in vacuum for 10 s.

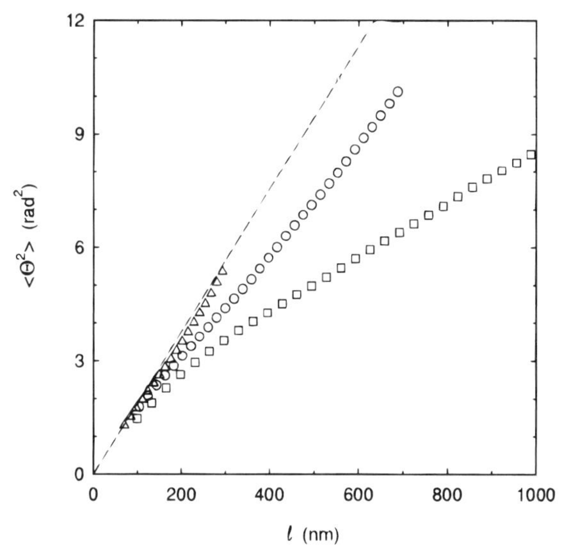

Figure 6 Plot of the mean square angle between two DNA segments as a function of their separation ℓ, for DNA fragments of three different lengths. The persistence length values obtained from the inverse of the slope of the regressions are: *triangles,* 1258 bp, $P = 55.1$ nm; *circles,* 2712 bp, $P = 71.1$ nm; *squares,* 5994 bp, $P = 144.5$ nm. DNA fragments shorter than 1000 bp behave like the 1258-bp fragments (data not shown). The apparent persistence length of the two longer fragments is higher probably caused by excluded volume effects. The DNA molecules were deposited and imaged as described in Figure 5.

that were deposited on mica surfaces treated with Ni^{++}. Under these conditions, they could observe the movement of some of the molecules on the surface. Similar results have been obtained with Mg^{++}-treated mica.

The results described above are likely to be applicable to protein–nucleic acid complexes, as long as the deposition process is dominated by the nucleic acid. Conversely, if the protein–mica interaction is stronger than the DNA–mica interaction, the mean square end-to-end distance of the protein-bound DNA should be smaller than the corresponding value for free DNA. An experiment to determine whether proteins bound to a DNA fragment can affect its equilibration on the mica surface has been performed recently (C Rivetti, M Guthold, and C Bustamante, in preparation). DNA fragments labeled at both bends with streptavidin were deposited in low salt buffer on mica, and their average end-to-end distance was determined by SFM and compared with that of free DNA deposited under identical conditions. The $\langle R^2 \rangle$

value of the streptavidin-labeled molecules was comparable to that of the free DNA. Some nucleic acid–binding proteins may interact more strongly with the mica surface than does streptavidin. Nonetheless, given the electrostatic nature of the molecule–surface interactions on mica, it is unlikely that the protein will dominate the deposition process or determine the global conformation of the complex.

This analysis shows that the strength of the molecule–surface interaction plays an essential role in the nature of the deposition process and its outcome. In general, as predicted by Equation 3, in the limit of strong molecule–surface interactions the deposition is "kinetic," i.e. any given part of the molecule that encounters the surface will remain bound to that location. If the energy of the surface interaction is any times greater than the energies holding the molecular assembly together, the process of deposition can easily alter the spatial relationships between the protein and the nucleic acid, promote complex dissociation, or induce the loss of native structures. In the limit of very weak surface interactions, however, equilibration in two dimensions is very likely, but the weak binding forces may lead to low deposition yield. Between these two extremes, there exists an intermediate range of interactions that result in a good yield of surface-equilibrated molecular depositions. Avoiding kinetic trapping effects is essential for the study of the global conformations of protein–nucleic acid complexes, particularly for the analysis of protein-induced DNA bending, as we discuss later. For these reasons, glow discharged mica or other surface pretreatments, such as monovalent salts, that lead to strong surface interactions should be avoided.

Thermodynamic and Kinetic Factors Affecting the Deposition of Protein–Nucleic Acid Complexes

Quantitative SFM analysis requires an adequate density of molecules on the surface. Optimal DNA surface coverage can be obtained by adjusting the concentration of the DNA in the deposition drop and the deposition time properly (Equation 2). As shown above, optimal deposition densities on mica result when DNA molecules are deposited at nanomolar concentrations with deposition times of 1–2 min. Obtaining optimal deposition densities of protein–nucleic acid complexes is more involved, however, because their concentration and deposition time cannot be varied arbitrarily. Instead, concentrations and time of deposition are constrained by the stability of the complex and by its lifetime.

The simplest equilibrium between a protein P and its DNA binding site D can be written as:

$$PD \Leftrightarrow P + D,$$

and its dissociation constant is given by:

$$K_{diss} = \frac{(R - r)(1 - r)[P_T]}{Rr},$$

where $R \equiv [P_T]/[D_T]$, P_T and D_T are the total concentrations of protein and specific DNA sites in the deposition drop, respectively, and r is the fraction of DNA specific sites occupied by the protein.

Under typical reaction conditions, $R \cup 1$, and half saturation of the DNA sites is obtained when the protein and DNA concentrations $[P_T] = [D_T] = 2K_{diss}$. Moderate to very stable protein–nucleic acid complexes (i.e. $K_{diss} = 10^{-9}$ M $- 10^{-13}$ M), can be deposited with high yield on mica at nanomolar concentrations with deposition times of 1–2 min. If the complexes are relatively unstable, however, (i.e. $K_{diss} = 10^{-6}$ M $- 10^{-7}$ M), similar reaction conditions will yield only 0.1 to 1% of the DNA sites occupied by the protein at equilibrium. Unfortunately, deposition at micromolar concentrations cannot be used, as they will result in excessive surface coverage even for very short deposition times. Likewise, ways to stabilize the complexes by the addition of BSA or of glycerol can seriously compromise the quality of the images and ultimately prevent the binding of the complexes to the surface. One way to overcome this problem is to carry out the binding reaction at concentrations near the dissociation constant of the complex and quickly dilute the sample to nanomolar concentrations before the deposition step. Clearly, the success of this approach will depend ultimately on the lifetime of the complex, τ. Long-lived complexes ($\tau > 5$ min) could be captured on the surface with this method, thereby improving the deposition yield. On the other hand, this approach usually fails for complexes with lifetimes of a few seconds or less, as these complexes tend to dissociate extensively during the 30 s or so minimally required for deposition, blotting, and rinsing. In the latter case, depositions could be carried out at lower temperatures to increase the lifetime of the complex, or the complex could be stabilized by cross-linking. Cross-linking should be considered only as a last resort, however, because it could affect the conformation of the complexes in an unpredictable manner.

PROBING PROTEIN–NUCLEIC ACID INTERACTIONS

During the last 3 years, the SFM has been used to characterize the geometry and the spatial relationships of several protein–nucleic acid complexes. These studies have been directed to characterize the structure of specific and nonspecific protein-nucleic acid complexes (18, 26, 62, 90), determine the stoichiometry of the complexes (18, 85), probe local molecular conformations of DNA by use of specific antibodies (55), and follow protein–nucleic acid interactions in buffer (24).

The nucleoprotein complexes studied by SFM encompass a broad range of protein sizes, from large, multisubunit molecules, such as RNA polymerase ($M_r \sim 450$ kDa) (26, 62, 90), to small transcription regulatory proteins like Cro, which binds to DNA as a dimer ($M_r = 2 \times 7.6$ kDa) (18). This enormous range is mainly due to the extreme sensitivity of the SFM to variations in height and the ability of SFM to image nucleoprotein assemblies without external means of contrast.

Protein-Induced DNA Bending

Many cell processes are regulated by DNA-binding proteins that can bind to target sequences on the DNA with high affinity and specificity. Examples of these processes include transcriptional regulation, hormone receptor–mediated activation, and certain types of site-specific recombination. In the last 10 years, it has become clear that these strong, highly specific interactions use both direct and indirect readout mechanisms (17, 51, 78). Direct readout involves recognition of the nucleic acid sequence by an appropriate protein domain. The protein uses indirect readout, however, to probe the local conformation of the DNA molecule at or in the immediate vicinity of the cognate site. This mechanism exploits the sequence-dependent local conformation and mechanical properties of the nucleic acid molecule (16, 52, 72).

A particular DNA sequence may be considered capable of both adopting a preferred configuration (e.g. localized A conformations, localized B conformations, and bends) and having access to alternative conformations within a given energy range. In other words, the preferred conformation of the site, as well as its flexibility, is likely to play an important role in indirect readout mechanisms (78). Because the binding free energies associated with specific protein–nucleic acid interactions can be 15–30 times greater than the thermal energy, $k_B T$, both the protein and the nucleic acid can modify their conformation and attain a compromise of high affinity and specificity. Protein-induced DNA deformations, such as bending (15), overwinding (8, 37), and underwinding (7, 71), are common occurrence in protein–nucleic acid complexes and appear to be essential in a variety of gene expression processes and their regulation.

An increasing number of prokaryotic and eukaryotic transcription factors have been found to bend the DNA on binding to their specific site (79). DNA bending can regulate transcription in several ways. It can bring distally bound transcription factors together, for example, by facilitating DNA looping. It could facilitate the proper orientation of protein factors relative to one another and relative to the promoter, thereby acting as a transcriptional switch (54). In addition, bending could be required for specific site recognition (18), or the energy stored in the protein-induced bend could facilitate initiation or maturation from initiation to elongation complexes (79).

DNA bending can be characterized by circular permutation analysis (84), phasing analysis (91), cyclization analysis (44, 70), rotational relaxation (57), and high-resolution microscopy (21, 22). One of the main advantages of high-resolution microscopy is that it makes it possible to determine not only mean bend angles but also the bend angle distributions. The shape of these distributions can give information about the flexibility of the protein–DNA complex, the existence of single or multiple populations in solution, and the conformational changes of the protein, the DNA, or both.

The bend angle is defined as the deviation of the DNA from linearity and can be measured as the supplement of the angle determined by the tangents of the DNA molecule at the entry and exit points from the protein. Alternatively, if the DNA fragment is not too long, the bend angle can be determined indirectly from its effect on the end-to-end distance distribution of the DNA fragment (27, 69).

ANALYSIS OF PROTEIN-INDUCED DNA BEND ANGLES: SURFACE EFFECTS

Structural characterization of protein–nucleic acid complexes by SFM is reliable only if the deposition process itself does not affect the conformation of the complexes. As discussed above, kinetic trapping effects can lead to surface-induced configurations easily. In particular, trapping of the DNA on the surface can result in modified bend-angle distributions, with altered means and standard deviations. Meaningful angle distributions, therefore, can be obtained only if the complexes are allowed to equilibrate on the surface during deposition.

The configuration of a complex in which the protein induces a simple bend in the nucleic acid is inherently two dimensional. In this simple case, it is possible to analyze the effect of the binding energy on the mean and the standard deviation of the angle distributions. The explicit angle-dependent energy of a complex in solution that has a mean angle $\langle \theta \rangle$ can be written as (C Walker, WA Rees, DA Erie, and C Bustamante, in preparation):

$$E^{soln}(\theta) = E^{soln}(\langle\theta\rangle) + \frac{1}{2}\frac{f^2E^{soln}(\theta)}{f\theta^2}\bigg|_{\theta=\langle\theta\rangle}(\theta - \langle\theta\rangle)^2 + \ldots$$

$$\Delta E^{soln}(\theta) = \frac{1}{2}\,\mathrm{k}(\theta - \langle\theta\rangle)^2 \qquad 7.$$

$$\mathrm{k} \equiv \frac{f^2E^{soln}(\theta)}{f\theta^2}\bigg|_{\theta=\langle\theta\rangle},$$

where k is the constant of bending rigidity of the complex. The probability distribution of angles θ adopted by the complex in solution is then:

$$P(\theta) = \sqrt{\frac{\mathrm{k}}{2\pi k_B T}}\, e^{-\,\mathrm{k}(\theta - \langle\theta\rangle)^2/2\mathrm{k_B T}} \qquad 8.$$

and its standard deviation is $s = \sqrt{\frac{\mathrm{k_B}T}{\mathrm{k}}}$.

For complexes in contact with a surface, an energy term that describes the interaction of the complex with the surface must also be included:

$$E^{surf} = E^{soln} + E^{surf\,.int}, \qquad 9.$$

and the value of the bending constant, k, must be multiplied by a factor of two to account for the loss of one degree of freedom. If the second term in Expression 9 is independent of the angle θ, its inclusion amounts only to a multiplicative factor in the angular distribution (Equation 8). In this case, the surface energy interaction will not affect the angular distribution. This situation is, in fact, the most likely one given the nonspecific, electrostatic nature of the complex–surface interaction. It is much less likely for the surface binding energy to depend on the bending angle. This analysis shows that, under the most likely scenario, equilibration should give rise to distributions that are unaffected in both their mean and their standard deviation.

The standard deviation of the distributions can provide information about the stiffness of the protein–nucleic acid complex. Because of the finite spatial resolution of the microscope, the tangents to the DNA at the entry and exit points in the complex must be obtained over finite DNA lengths. These tangents are subjected to fluctuations determined by the persistence length of the DNA and the DNA lengths (Equation 6). As a result, the variance of the bend angle distributions contains the contributions of the bending rigidity of the complex, σ_c, and that of the bending rigidity of the DNA arms, σ_b, i.e.:

$$\sigma^2 = \sigma_c^2 + \sigma_b^2 = \sigma_c^2 + \frac{\ell}{P}. \qquad 10.$$

Using Equation 1 for a parabolic tip with a radius of curvature, $R_c = 8$ nm, a limit of detectability, $\Delta z = 0.1$ nm, and a height difference between the protein and the DNA, $\Delta h = 0.4$ nm, gives a DNA arm length, $\ell/2 = 4$ nm. Assuming a value of 53 nm for the persistence length of the DNA molecule, the second term in the above expression predicts a standard deviation, $\sigma = \sigma_b \sim 22^0$. Equation 11 shows that it is possible to obtain information about the rigidity of the protein–nucleic acid complex from the standard deviations of the population, if the contribution caused by the flexibility of the DNA can be estimated.

The analysis presented in this section assumes that the bent complexes are inherently two dimensional. If a protein can bend the DNA in two different locations, however, the bends generally will not be coplanar. Similarly, if the molecule induces torsion in at least one bending site, the result will again be a three-dimensional configuration determined by the individual bend angles and by the dihedral angle between the planes that contain the individual bends. In these cases, the deposition will always involve a transformation of a three-dimensional structure onto a two-dimensional surface. The interpretation of the data in this event can be extremely difficult.

DNA BENDING IN PROKARYOTIC TRANSCRIPTION COMPLEXES Several groups have reported transcription complexes of *Escherichia coli* RNA polymerase with various prokaryotic promoters, imaged in air by SFM (13, 62, 90). Images of open λ-P_L promoter complexes and stalled elongation complexes that harbor a 15-nucleotide long transcript revealed the large-scale morphology of the complexes and the spatial relationships between the polymerase and the DNA. Significantly, the DNA appears bent in both types of complexes (Figure 7*A* and *B*), but the elongation complexes are more severely bent (mean angle, 92°) than the open promoter complexes (mean angle, 54°). This difference appears to be related to the conformational changes associated with the maturation from open promoter complexes to elongation complexes (38, 50). An independent determination of the DNA bending in transcription complexes has been obtained with the use of electric birefringence (48). These studies reported a mean bend angle of 45° for open promoter complexes. In addition, complexes of *E. coli* RNA polymerase with supercoiled plasmids that contain the promoter of the ampicillin-resistance gene have been imaged with the use of SFM (90). The polymerase molecules often appear bound at the terminal loops of the plectonemes, a result consistent with polymerase-induced bending of the DNA.

DNA BENDING IN NONSPECIFIC COMPLEXES Direct visualization of protein–nucleic acid complexes makes it possible to characterize not only the conformation of the specifically bound proteins but also the conformations of the much less understood nonspecific interactions (18). Specific and nonspecific complexes can be distinguished and independently analyzed if the location of the specific site with respect to one end of the molecule is known. Nonspecific interactions are likely to play an important role in the processes of facilitated target location of their cognate site (4, 81). Moreover, practically all DNA binding proteins can also bind DNA nonspecifically; comparison between specific and nonspecific complexes, therefore, can provide important insight into the molecular basis of specificity. Recently, the conformation of specific and nonspecific Cro protein–DNA complexes has been characterized with the use of SFM. A DNA fragment containing the O_{R1}, O_{R2}, and O_{R3} sites of the O_R region of bacteriophage λ located four ninths from one end was used as template. The study revealed that Cro bends the DNA in specific (average bend angle, 69° ± 11°), as well as in nonspecific (average bend angle, 62° ± 23°), complexes. Figure 8 shows the distribution of DNA bend angles determined for nonspecific Cro–DNA complexes. This distribution suggests that bending the DNA at nonspecific sites may be an important component in the mechanism of specific site recognition by Cro. In other words, the protein may sample contacts needed for recognition of its target sequence via DNA bending. If Cro diffuses along the DNA in search of its operator (4, 81), the observed bending of nonspecific DNA would result in the propagation of a "bending wave" along the DNA with Cro riding at

Figure 7 (*A* and *B*) Surface plots of transcription complexes of *E. coli* RNA polymerase deposited on mica and imaged in air by using SFM in contact mode. *(A)* Open promoter complex (OPC); *(B)* elongation complex (C15). The DNA fragment (681 bp) contained the λ P_L promoter located four ninths from one end (62). *(C)* Surface plots of a three Cro dimers (*arrows*) bound to each of the operator sites in a 1-kbp DNA fragment containing the O_R region of bacteriophage λ. The complexes were deposited on mica and imaged in air by using SFM in tapping mode (18). *(D)* Line plot image of DNA-HSF2 complexes deposited on mica and imaged in air by using the tapping mode. The complex on the left shows a single HSF2 trimer (not resolved) bound to an HSE site positioned near the middle of the fragment. The complex on the right shows that two HSF2 trimers (each bound to a corresponding HSE site) are required to form the HSF2-mediated DNA loops associated with maximum transcription activation (85). (e) Line plot image of an NTRC protein (in front) interacting with *E. coli* RNA polymerase (in the back) via DNA looping to form an active transcription complex. The complexes were deposited on freshly cleaved mica and imaged in tapping mode in air. (Image courtesy of M Guthold and K Rippe, Institute of Molecular Biology, University of Oregon.)

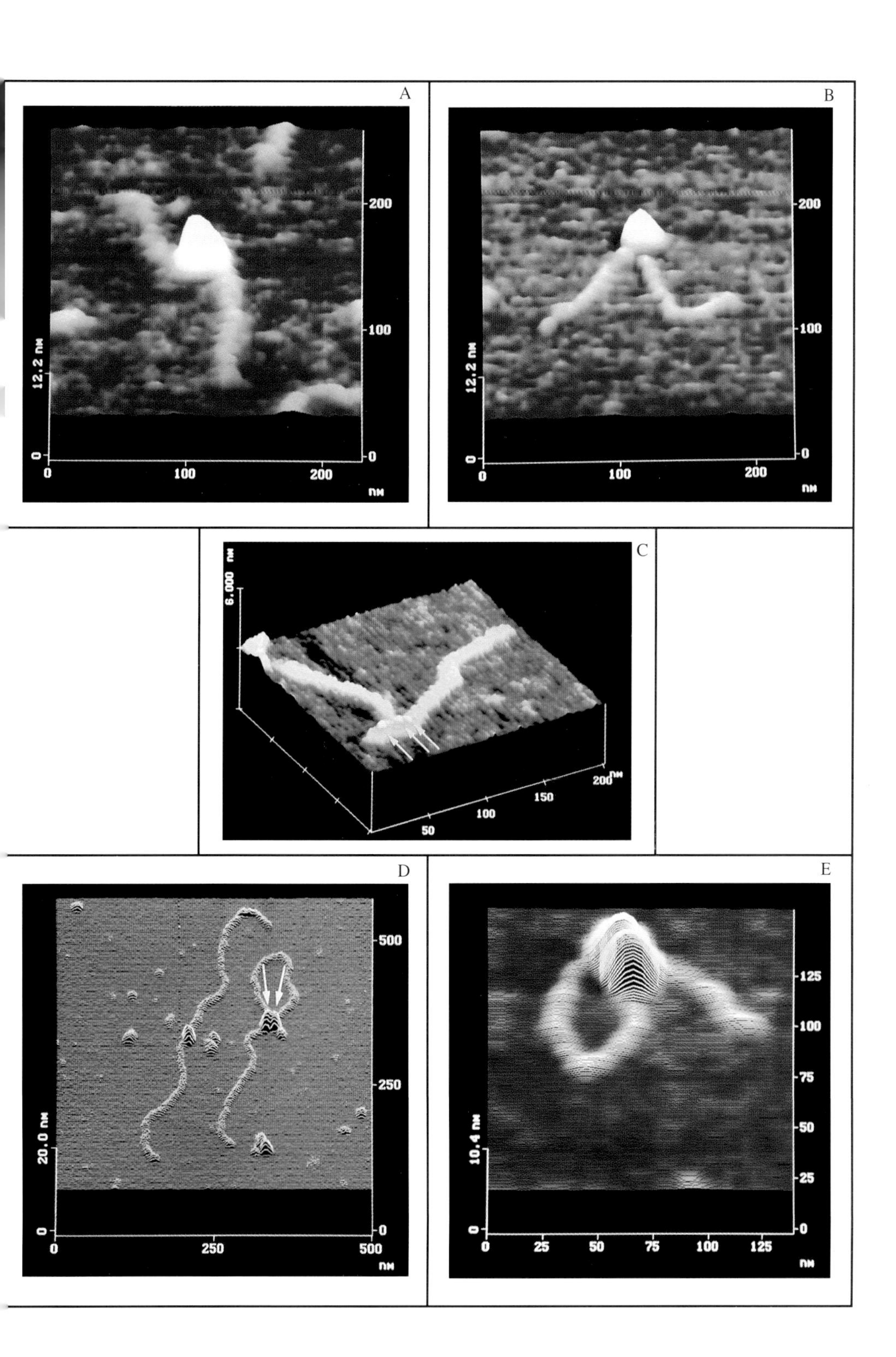

A
B
C
D
E
12.2 nm
200
100
0
100
200
nm
6.000 nm
50
100
150
200 nm
20.0 nm
500
250
0
250
500
nm
10.4 nm
125
100
75
50
25
0
25
50
75
100
125
nm

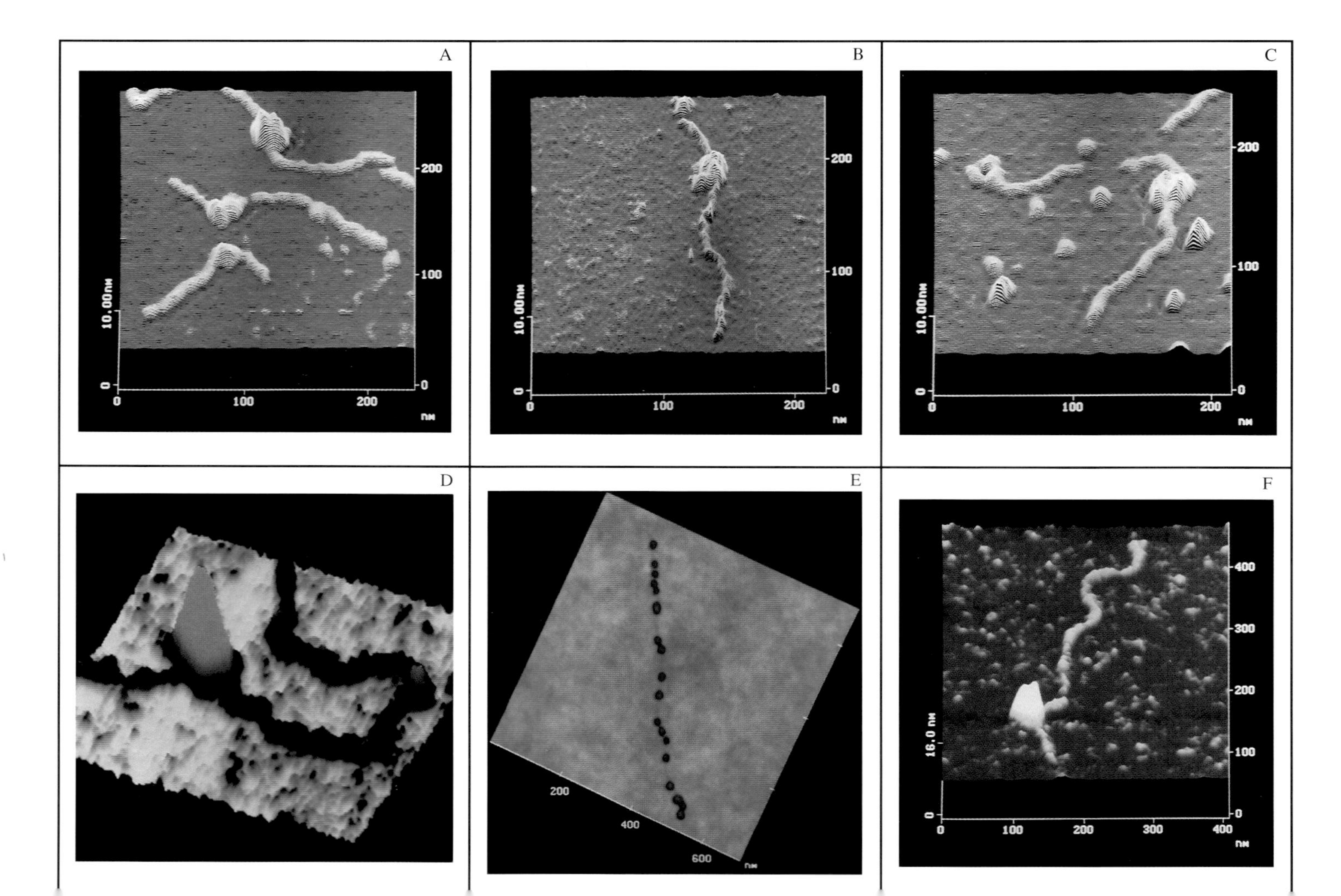
A
10.00nm
0
100
200
nm
B
10.00nm
0
100
200
nm
C
10.00nm
0
100
200
nm
D
E
200
400
600
nm
F
16.0 nm
0
100
200
300
400
nm

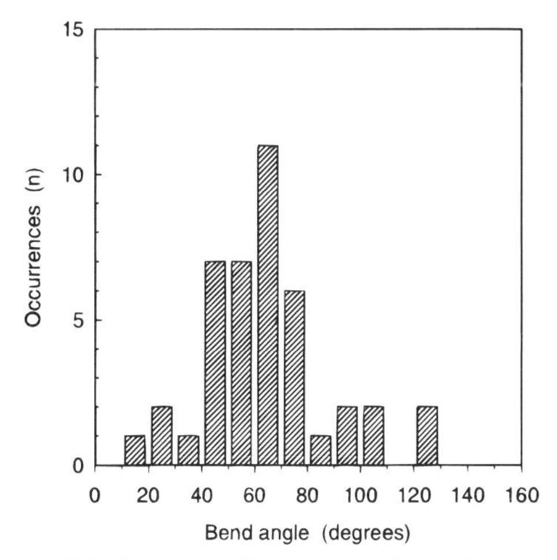

Figure 8 Histogram of the frequency of occurrence of DNA bend angles of nonspecific Cro–DNA complexes measured from SFM images. The mean value was 62° ± 23° (18).

its apex. The diffusion of the bending wave along the DNA can be shown to be energetically feasible (18).

COMPARISON TO OTHER METHODS Table 1 lists the values of the bend angles of various protein–nucleic acid complexes measured by SFM

◄———

Figure 10 *(A)* Line plot image of three different protein–nucleic acid complexes deposited on mica and imaged in air by using the tapping mode. (*bottom*) A complex of a single NTRC dimer (MW = 110 kDa) bound to a single enhancer element located in a 388-bp DNA fragment; (*middle*) a complex of two NTRC dimers (MW = 220 kDa) bound to a strong enhancer element located in a 610-bp DNA fragment; (*top*) an RNA polymerase (MW = 450 kDa) bound to a 1.8-kbp DNA fragment. *(B)* Line plot image of a complex of phosphorylated NTRC and a strong enhancer element. NTRC was phosphorylated in vitro by NTRB and ATP. *(C)* Line plot image of an unphosphorylated constitutive NTRC mutant (S160F NTRC) bound at a strong enhancer site. (From C Wyman, C Bustamante, and S Kustu, in preparation). *(D)* Surface plot of an IgG/Z-DNA complex obtained in air by using the tapping mode (55). (Image courtesy of L Pietrasanta and TM Jovin, Department of Molecular Biology, Max Planck Institute for Biophysical Chemistry.) *(E)* SFM image of linear DNA reconstituted with 18 core histones. The DNA contained a repeated DNA sequence that preferentially phases nucleosomes. The contour length of the chromatin fiber is 693 nm (1). (Image courtesy of MJ Allen, Department of Biological Chemistry, School of Medicine and Microbiology, University of California.) *(F)* SFM image of a stalled elongation complex of *E. coli* RNA-polymerase deposited on freshly cleaved mica and obtained under aqueous buffer in tapping mode with a carbon beam tip. The DNA fragment is 1258 bp long, and the stalling site is located one sixth from one end. (Image courtesy of XZ, Institute of Molecular Biology, University of Oregon.)

Table 1 Bend angle values of several DNA binding proteins obtained by different methods

Protein	Method			
	SFM[1]	Gel bend shift[2]	X-ray[3]	Others[4]
EcoRV[a]	43° ± 29	44°	50°	—
EcoRI endo[b]	10° ± 52	55°	12°	—
EcoRI MTase[c]	51° ± 17	50°	—	—
HhaI MTase[d]	2° ± 28	0°	0°	—
Cro[e]	69° ± 11	30°	40°	45°
RNAP[f]	54° ± 31	—	—	45°

[a1] Scanning force microscopy (SFM) on mica in air (C Walker, WA Rees, DA Erie, and C Bustamante, in preparation); [a2] circular permutation (73); [a3] X-ray (82). [b1] SFM on mica in air (C Walker, WA Rees, DA Erie, and C Bustamante, in preparation); [b2] circular permutation (75); [b3] X-ray (35, 47). [c1] SFM on mica in air (RA Garcia, C Bustamante, and NO Reich, in preparation); [c2] circular permutation (RA García, C Bustamante, and NO Reich, in preparation). [d1] SFM on mica in air (RA García, C Bustamante, and NO Reich, in preparation); [d2] circular permutation (RA García, C Bustamante, and NO Reich, in preparation); [d3] X-ray (36). [e1] SFM on mica in air (18); [e2] circular permutation (34); [e3] X-ray (8); [e4] ligase catalyzed cyclization (42). [f1] SFM on mica in air (62); [f4] electric birefringence (48).

and compares them with the corresponding values obtained by other methods (C Walker, WA Rees, DA Erie, and C Bustamante, in preparation). With the exception of EcoRI endonuclease, the values of the bend angles are in good agreement among the different methods of determination. The discrepancy concerning this protein likely reflects a complex interaction between the protein and the nucleic acid that is not described easily by a simple bend. In fact, the crystal structure indicates that EcoRI induces two phase bends on the DNA (35, 47). As noted above, how the measured angle would depend on the relative orientation of the three-dimensional structure, as it is projected on the surface, is not known.

Protein–Nucleic Acid Stoichiometry

Imaging of biological samples by SFM does not require any external means of contrast, and the spatial resolution of the SFM is limited only by the finite dimensions of the tip and the geometry of the sample (10). This fact makes the SFM particularly suited to visualize multiprotein–nucleic acid assemblies, determine their stoichiometry, and characterize their spatial relationships. Only a few systems have been studied to date; we review three cases here as illustration.

Figure 7*C* depicts a tapping mode image obtained in air of Cro protein molecules bound to the three operator sites in the O_R region of bacteriophage λ (18). Cro binds to each operator site as a dimer to repress transcription from the divergent λ promoters P_R and P_{RM} (59). With a molecular weight of only 14.7 kDa, these dimers (the arrows point to each of three Cro dimers) are among the smallest proteins imaged by any kind of microscopy. The distance between each peak is 7.1 nm. The dimple depth between adjacent peaks is only 0.3 nm, but well within the height resolution of the instrument.

Another application of the SFM to determine the stoichiometry of protein–nucleic acid complexes is illustrated in Figure 7*D*. This image shows the loop structure induced in a DNA fragment by the human transcription activator heat shock factor 2 (HSF2). This factor activates transcription by binding to specific sites on the DNA, called heat shock elements (HSEs). These elements occur close and far from the promoter. At least one close HSE of three nGAAn repeats are necessary for transcription, whereas distally located HSEs function as transcriptional enhancers. This process often results in looped DNA structures bringing together one proximal and one distal HSE to contact the promoter, as shown in Figure 7*D*. Because each HSF2 binds to the HSE as a trimer (not resolved by the SFM), it was not known whether the loops required two HSF2 trimers (one bound to a distal HSE and the other bound to the proximal HSE), or if just one single trimer could effect the looping, as is found for AraC (41). Attempts to resolve this question by EM gave ambiguous results. Scanning force microscopic images obtained with the use of the tapping mode in air closely show the presence of two distinct and symmetrically bound peaks at the closure of the looped activation complex (85). Volume analysis of complexes at the loop junction relative to the HSF2 monomers confirmed that each peak corresponds to an unresolved HSF2 trimer (85).

Transcription initiation by σ^{54}RNA polymerase from *E. coli,* in response to conditions of nitrogen starvation, is controlled by protein factors bound to enhancer sequences far from the promoter, via DNA looping. This system might be considered a primitive analogue of transcription regulation at a distance, as seen often in eukaryotic systems (49). Transcription is activated in an ATP-dependent manner by nitrogen regulatory protein C (NTRC). This protein binds as a dimer (M_r = 110 kDa) to each of the two sites of an enhancer element that are located at -108 and -140 from the start site of transcription. Activation involves the isomerization reaction from close to open promoter complex and requires NTRC phosphorylation, a process carried out in vivo by the autokinase nitrogen regulatory protein B (NTRB) (58). At least two dimers in the complex are necessary for activation, and the function of the enhancer appears to be that of increasing the concentration of NTRC in the vicinity of the glnA promoter. The hypothesis that phosphorylation may promote the formation of multiprotein complexes was postulated on the basis of electron micrographs of phosphorylated NTRC/enhancer complexes (74) and has been demonstrated recently with the use of SFM (C Wyman, C Bustamante, and S Kustu, in preparation). Figure 7*E* shows the looped structure of NTRC and σ^{54}RNA polymerase bound to a 726-bp DNA fragment that contains the glnA promoter and two tandemly arranged enhancer sites.

Determination of molecular weight of multiprotein–nucleic acid assemblies with the use of SFM is difficult because the heights of proteins and nucleic acids often appear smaller than their true values. There could be multiple causes for this situation, including denaturation of the molecules on binding to the surface, mechanical compliance of the molecules under the force exerted by the tip, and differential interaction of the tip with the molecule and the background. The underestimation of the molecular volume resulting from the flattening of the images is compensated, in part, by the overestimation of the lateral dimensions of the protein owing to the finite size of the tip. These two opposite effects may be responsible for the linear proportionality between molecular weight and the volume observed in the range of 110 to 500 kDa (Figure 9).

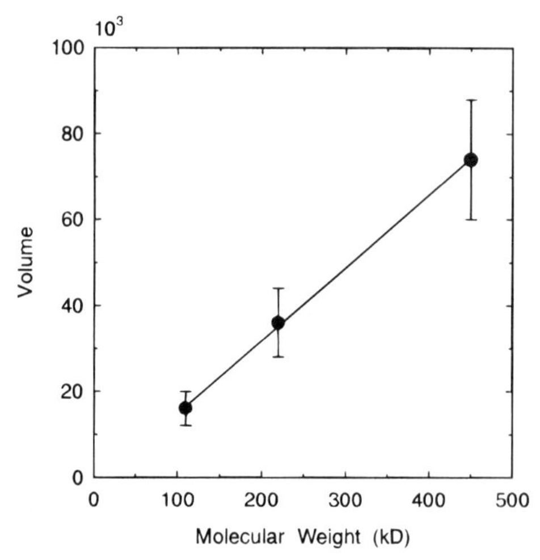

Figure 9 Molecular weight vs volume of three DNA-protein complexes imaged by SFM in air. The data were obtained by averaging 171 single NTRC dimers (MW = 110 kDa), 81 RNA polymerase molecules (MW = 450 kDa), and 123 double NTRC dimers (MW = 220 kDa) bound to specific DNA sequences. The volumes are expressed in arbitrary units, and the error bar represents the standard deviation from the mean for each complex. (From C Wyman, C Bustamante, and S Kustu, in preparation.)

Figure 10*A* shows a single unphosphorylated NTRC dimer bound to a 388-bp DNA fragment that contains a single enhancer element (*bot tom*), two unphosphorylated dimers bound to a 610-bp fragment that contains two tandem enhancer elements (strong enhancer) (*middle*), and a molecule of RNA polymerase bound to a 1.8-kbp fragment that contains the λ P_R promoter (*top*). Figure 10*B*, shows the effect of phosphorylation on the binding of NTRC to the strong enhancer. Under these conditions, NTRC binds as a multimer of dimers. In fact, analysis of many complexes (C Wyman, C Bustamante, and S Kustu, in preparation) showed that 43% of the NTRC-enhancer complexes contain more than two NTRC dimers. For comparison, under identical conditions, only 2% of the unphosphorylated NTRC-enhancer complexes contained more than two NTRC dimers. Significantly, 34% of the complexes formed with a constitutive NTRC mutant (S160F) capable of transcription activation in the absence of phosphorylation contained more than two NTRC dimers (Figure 10*C*).

Conformation-Dependent Molecular Recognition: Immuno-SFM

Recently, an anti-Z DNA antibody was used to probe for the left-handed Z-DNA conformation of a $d(CG)^{11}$ insert in a negatively supercoiled DNA plasmid (55). The SFM images revealed that the antibody remained bound to the DNA even after the plasmids were cleaved with a restriction endonuclease (Figure 10*D*). The bound antibodies had apparent lateral dimensions of 35 nm and a height of 2 nm from the mica surface. The antibodies were often found in a lateral orientation relative to the DNA, an observation consistent with the previous data that indicated a bipedal mode of binding for an anti-Z-DNA IgG. The SFM images suggest that the DNA bends to accommodate the two Fab-combining regions of the antibody.

High-Order Protein–Nucleic Acid Assemblies: Chromatin Studies

Several groups have used the SFM to image whole sperm in buffer (2, 3, 30, 88). These authors have examined the contents of lysed sperm cells in air and in buffer and have shown the toroidal packing organization of the DNA in the nucleus (88). Recently, cell preparation methods from optical and electron microscopy were adapted and used successfully to image lysed cells with SFM in air and in buffer (19, 20).

Fibers of purified chicken erythrocyte chromatin imaged in air by SFM were shown to exhibit an irregular, three-dimensional arrangement of nucleosomes even at low ionic strength (9, 40, 93). The SFM images closely resemble images recently obtained by cryo-electron microscopy of thin nuclear sections (83). Computer simulations carried out by these authors suggested that these three-dimensional irregular structures could result from the natural variability of the DNA linker lengths in chromatin fibers. Computer simulations can also reproduce most of the topographic features of the SFM images and suggest several of the factors that may be responsible for maintaining the integrity of these fibers in solution (86). Among these factors are the presence of the linker histones H1 and H5. Fibers stripped of these histones lack three-dimensional structures that display a "beads-on-a-string" appearance (86). Partially trypsinized chromatin fibers have been also imaged in air by SFM (92). Digestion of the N- and C-terminal domains of the linker histones did not affect the entry and exit angles of the DNA around the nucleosome. This result indicates that the presence of the globular domain of the linker histones is sufficient to maintain this angle in the fiber.

Studies with 21,000-bp *Tetrahymena thermophila* rDNA minichromosomes chromatin fibers recently have shown no less than a sixfold compaction of B-DNA into an irregular fiber, which appears to contain some zig zag regions (46).

Allen and coworkers (1) have studied linear and circular DNA reconstituted with the core histones but without the linker histones H1 and H5. Under these conditions, the reconstituted fibers depict a beads-on-a-string appearance (Figure 10*E*). Using a repeated DNA sequence that preferentially phases nucleosomes, these authors could titrate the eighteen positioning sites on the DNA fragment and characterize the resulting DNA compaction.

Following Protein–Nucleic Acid Interactions in Buffer

Perhaps the most exciting and promising application of the SFM is its ability to operate under liquids in close to physiologic conditions. Although only a few protein–nucleic acid complexes have been imaged in buffer, the following examples illustrate the potential of this capability, unique to the SFM.

Guthold et al (24) used the SFM to follow the assembly of *E. coli* RNA polymerase and DNA. These authors deposited molecules of DNA onto freshly cleaved mica and imaged them in contact mode under low salt aqueous buffer solutions before injecting a solution of RNA polymerase. They showed that it is possible to image the same area of the mica before and just 1 min after the injection of the polymerase. These authors recognized several molecules of RNA polymerase that had bound to the DNA molecules within this time. The results demonstrate that the deposition conditions used could preserve the binding activity of the polymerase and that the presence of the surface does not interfere significantly with the binding process.

Figure 10*F* shows an image of an elongation complex of *E. coli* RNA polymerase and the λ P_R promoter. The molecules were deposited on mica and imaged under a low salt aqueous buffer in tapping mode. Successive images of these complexes showed various parts of the complexes moving somewhat on the mica surface. This observation suggests that it is possible to deposit the complexes with minimal attachment to the mica and to image them with SFM using the tapping mode in liquid. With this capability, it may be possible to use the SFM to visualize and follow in real time the processes of transcription and replication.

CONCLUDING REMARKS

Ten years have passed since the invention of the SFM (6). During this time, much progress has occurred in the development of the instrument and in the design and testing of protocols and methods to image biological samples. With a reliable performance and ease of operation, the ability to image macromolecules without external means of contrast, and the capacity to operate under aqueous solutions with nanometer resolution, the SFM is rapidly becoming a useful tool for structural studies of mesoscopic systems.

In addition to the current capabilities of the SFM reviewed here, a number of important challenges in the area of protein–nucleic acid interactions will soon be able to be addressed for the first time with this technique. For example, biochemists may soon learn how to take advantage of the SFM operation under liquids to follow in real time the assembly of increasingly larger nucleoprotein complexes, characterize their spatial relationships in physiologically relevant conditions, investigate their conformational changes on variations in solution conditions, and follow in real time their biochemical processes.

These efforts, aided by concurrent instrumental and technical developments, may play an important role in helping biochemists integrate the high-resolution and large-scale pictures into a comprehensive view of the mechanisms of control and regulation of transcription and replication.

ACKNOWLEDGMENTS

This work was supported by National Science Foundation grants MBC 9118482 and BIR 9318945 and National Institutes of Health grant GM-

32543. This work was supported in part by a grant from the Lucille P. Markey Foundation to the Institute of Molecular Biology. Claudio Rivetti was supported by a Human Frontier Science Program long-term fellowship.

Literature Cited

1. Allen MJ, Dong XF, O'Neill TE, Yau P, Kowalczykowski SC, et al. 1993. Atomic force microscope measurements of nucleosome cores assembled along defined DNA sequences. *Biochemistry* 32:8390–96
2. Allen MJ, Hud NV, Balooch M, Tench RJ, Siekhaus WJ, Balhorn R. 1992. Tip-radius-induced artifacts in AFM images of protamine-complexed DNA fibers. *Ultramicroscopy* 42–44: 1095–100
3. Allen MJ, Lee C, Lee JD IV, Pogany GC, Balooch M, et al. 1993. Atomic force microscopy of mammalian sperm chromatin. *Chromosoma* 102: 623–30
4. Berg OG, von Hippel PH. 1985. Diffusion-controlled macromolecular interactions. *Annu. Rev. Biophys. Biophys. Chem.* 14:131–60
5. Bezanilla M, Drake B, Nudler E, Kashlev M, Hansma PK, Hansma HG. 1994. Motion and enzymatic degradation of DNA in the atomic force microscope. *Biophys. J.* 67:2454–59
6. Binnig G, Quate CF, Gerber CH. 1986. Atomic force microscope. *Phys. Rev. Lett.* 56:930–33
7. Bramhill D, Kornberg A. 1988. A model for initiation at origins of DNA replication. *Cell* 52:743–55
8. Brennan RG, Roderick SL, Takeda Y, Matthews BW. 1990. Protein-DNA conformational changes in the crystal structure of a lambda Cro-operator complex. *Proc. Natl. Acad. Sci. USA* 87:8165–69
9. Bustamante C, Erie DA, Keller D. 1994. Biochemical and structural applications of scanning force microscopy. *Curr. Opin. Struct. Biol.* 4: 750–60
10. Bustamante C, Keller D. 1995. Scanning force microscopy in biology. *Phys. Today* 48:32–38
11. Bustamante C, Keller D, Yang G. 1993. Scanning force microscopy of nucleic acids and nucleoprotein assemblies. *Curr. Opin. Struct. Biol.* 3: 363–72
12. Bustamante C, Marko JF, Siggia ED, Smith S. 1994. Entropic elasticity of l-phage DNA. *Science* 265:1599–600
13. Bustamante C, Vesenka J, Tang CL, Rees W, Guthold M, Keller R. 1992. Circular DNA molecules imaged in air by scanning force microscopy. *Biochemistry* 3:22–26
14. Butt HJ, Muller T, Gross H. 1993. Immobilizing biomolecules for scanning force microscopy by embedding in carbon. *J. Struct. Biol.* 110:127–32
15. Crothers DM, Gartenberg MR, Shrader TE. 1991. DNA bending in protein-DNA complexes. *Methods Enzymol.* 208:118–46
16. Drew HR, McCall MJ, Calladine CR. 1988. Recent studies of DNA in the crystal. *Annu. Rev. Cell Biol.* 4:1–20
17. Drew HR, Travers AA. 1985. Structural junctions in DNA: the influence of flanking sequence on nuclease digestion specificities. *Nucleic Acids Res.* 13:4445–67
18. Erie DA, Yang G, Schultz HC, Bustamante C. 1994. DNA bending by Cro protein in specific and nonspecific complexes: implications for protein site recognition and specificity. *Science* 266:1562-66
19. Fritzsche W, Schaper A, Jovin TM. 1994. Probing chromatin with the scanning force microscope. *Chromosoma* 103:231–36
20. Fritzsche W, Schaper A, Jovin TM. 1995. Scanning force microscopy of chromatin fibers in air and in liquid. *Scanning* 17:148–55
21. Griffith JD, Makhov A, Zawel L, Reinberg D. 1995. Visualization of TBP oligomers binding and bending the HIV-1 and adeno promoters. *J. Mol. Biol.* 246:576–84

22. Gronenborn AM, Nermut MV, Eason P, Clore GM. 1984. Visualization of cAMP receptor protein-induced DNA kinking by electron microscopy. *J. Mol. Biol.* 179:751–57
23. Grosberg AY, Khokhlov AR. 1994. *Statistical Physics of Macromolecules.* Woodbury, NY: API. 350 pp.
24. Guthold M, Bezanilla M, Erie DA, Jenkins B, Hansma HG, Bustamante C. 1994. Following the assembly of RNA polymerase-DNA complexes in aqueous solutions with the scanning force microscope. *Proc. Natl. Acad. Sci. USA* 91:12927–31
25. Hagerman PJ. 1988. Flexibility of DNA. *Annu. Rev. Biophys. Biophys. Chem.* 17:265–86
26. Hansma HG, Bezanilla M, Zenhausern F, Adrian M, Sinsheimer RL. 1993. Atomic force microscopy of DNA in aqueous solutions. *Nucleic Acids Res.* 21:505–12
27. Hansma HG, Browne KA, Bezanilla M, Bruice TC. 1994. Bending and straightening of DNA induced by the same ligand: characterization with the atomic force microscope. *Biochemistry* 33:8436–41
28. Hansma HG, Hoh J. 1994. Biomolecular imaging with atomic force microscope. *Annu. Rev. Biophys. Biomol. Struct.* 23:115–39
29. Hansma PK, Cleveland JP, Radmacher M, Walters DA, Hillner P, et al. 1994. Tapping mode atomic force microscopy in liquids. *Appl. Phys. Lett.* 64:1738–40
30. Hud NV, Allen MJ, Downing KH, Lee J, Balhorn R. 1993. Identification of the elemental packing unit of DNA in mammalian sperm cell by atomic force microscopy. *Biochem. Biophys. Res. Commun.* 193:1347–54
31. Isrealachvili J. 1992. *Intermolecular and Surface Forces.* New York: Academic. 450 pp.
32. Keller D, Chou CC. 1992. Imaging steep high structures by scanning force microscopy with electron beam deposited tips. *Surf. Sci.* 268:333–39
33. Keller DJ, Franke FS. 1993. Envelope reconstruction of scanning probe microscope images. *Surf. Sci.* 294: 409–19
34. Kim J, Zweib C, Wu C, Adhyn S. 1989. Bending of DNA by gene-regulatory proteins: construction and use of a DNA bending vector. *Gene* 85: 15–23
35. Kim YC, Grable JC, Love R, Greene PJ, Rosenburg JM. 1990. Refinement of EcoRI endonuclease crystal structure: a revised protein chain tracing. *Science* 249:1307–9
36. Klimasauskas S, Kumar S, Roberts RJ, Cheng X. 1994. HhaI methyltransferase flips its target base out of the DNA helix. *Cell* 76:357–69
37. Koudelka GB, Harbury P, Harrison SC, Ptashne M. 1988. DNA twisting and the affinity of bacteriophage 434 operator for bacteriophage 434 repressor. *Proc. Natl. Acad. Sci. USA* 85: 4633–37
38. Krummel B, Chamberlin MJ. 1992. Structural analysis of ternary complexes of *Escherichia coli* RNA polymerase. Deoxyribonuclease I footprinting of defined complexes. *J. Mol. Biol.* 225:239–50
39. Landau LD, Lifshitz EM. 1986. *Theory of Elasticity.* Oxford, NY: Pergamon. 187 pp.
40. Leuba SH, Yang G, Robert C, Samorí B, van Holde K, et al. 1994. Three-dimensional structure of extended chromatin fibers as revealed by tapping-mode scanning force microscopy. *Proc. Natl. Acad. Sci. USA* 91: 11621–25
41. Lobell RB, Schleif RF. 1990. DNA looping and unlooping by AraC protein. *Science* 250:528–33
42. Lyubchenko YL, Jacobs BL, Lindsay SM. 1992. Atomic force microscopy of reovirus dsRNA: a routine technique for length measurements. *Nucleic Acids Res.* 20:3983–86
43. Lyubchenko YL, Oden PI, Lampner D, Lindsay SM, Dunker KA. 1993. Atomic force microscopy of DNA and bacteriophage in air water and propanol: the role of adhesion forces. *Nucleic Acids Res.* 21:1117–23
44. Lyubchenko YL, Shlyakhtenko L, Chernov B, Harrington RE. 1991. DNA bending induced by Cro protein binding as demonstrated by gel electrophoresis. *Proc. Natl. Acad. Sci. USA* 88:5331–34
45. Lyubchenko YL, Shlyakhtenko LS, Harrington RE, Oden PI, Lindsay SM. 1993. Atomic force microscopy of long DNA: imaging in air and under water. *Proc. Natl. Acad. Sci. USA* 90: 2137–40
46. Martin LD, Vesenka JP, Henderson E, Dobbs DL. 1995. Visualization of nucleosomal substructure in native chromatin by atomic force microscopy. *Biochemistry* 34:4610–16
47. McClarin JA, Frederick CA, Wang BC, Greene P, Boyer HW, et al. 1986. Structure of the DNA-EcoRI endonuclease recognition complex at 3

angstrom resolution. *Science* 234: 1526–41

48. Meyer-Alme FJ, Heumann H, Porschke D. 1994. The structure of the RNA polymerase-promoter complex: DNA bending by quantitative electrooptics. *J. Mol. Biol.* 236:1–6
49. North AK, Klose KE, Stedman KM, Kustu S. 1993. Prokaryotic enhancer-binding proteins reflect eukaryote-like modularity: the puzzle of nitrogen regulatory protein C. *J. Bacteriol.* 175: 4267–73
50. Nudler E, Goldfarb A, Kashlev M. 1994. Discontinuous mechanism of transcription elongation. *Science* 265: 793–96
51. Otwinowski Z, Schevitz RW, Zhang RG, Lawson CL, Joachimiak A, et al. 1988. Crystal structure of trp repressor/operator complex at atomic resolution. *Nature* 355:321–29
52. Pabo CO, Sauer RT. 1992. Transcription factors: structural families and principles of DNA recognition. *Annu. Rev. Biochem.* 61:1053–95
53. Pashley RM. 1981. Hydration forces between mica surfaces in aqueous electrolyte solutions. *J. Colloid Interface Sci.* 80:153–62
54. Pérez Martín J, Espinosa M. 1993. Protein-induced bending as a transcriptional switch. *Science* 260:805–7
55. Pietrasanta LI, Schaper A, Jovin TM. 1994. Probing specific molecular conformations with the scanning force microscope. Complexes of plasmid DNA and anti-Z-DNA antibodies. *Nucleic Acids Res.* 22:3288–92
56. Pingali GS, Jain R. 1992. Restoration of scanning force microscope images. In *Proc. IEEE Workshop on Appl. of Comput. Vision*, pp. 282–89
57. Porschke D, Hillen W, Tahahashi M. 1984. The change of DNA structure by specific binding of the cAMP receptor protein from rotation diffusion and dichroism measurements. *EMBO J.* 3: 2873–77
58. Porter S, North AK, Wedel AB, Kustu S. 1993. Oligomerization of NTRC at the glnA enhancer is required for transcriptional activation. *Genes Dev.* 7: 2258–73
59. Ptashne M. 1992. *A Genetic Switch: Phage 1 and Higher Organism.* Cambridge, MA: Blackwell Scientific. 192 pp.
60. Putman AJ. 1994. *In development of an atomic force microscope for biological applications.* PhD thesis. Univ. of Twente, Enschede
61. Putman AJ, Van der Werf KO, De Grooth BG, Van Hulst NF, Greve J. 1994. Tapping mode atomic force microscopy in liquid. *Appl. Phys. Lett.* 64:2454–56
62. Rees WA, Keller RW, Vesenka JP, Yang G, Bustamante C. 1993. Evidence of DNA bending in transcription complexes imaged by scanning force microscopy. *Science* 260:1646–49
63. Richmond TJ, Searles MA, Simpson RT. 1988. Crystals of a nucleosome core particle containing defined sequence DNA. *J. Mol. Biol.* 199: 161–70
64. Rugar D, Hansma PK. 1990. Atomic force microscopy. *Phys. Today* 43: 23–30
65. Sarid D. 1991. *Scanning Force Microscopy with Applications to Electric, Magnetic, and Atomic Forces.* New York: Oxford Univ. Press. 253 pp.
66. Schaper A, Pietrasanta LI, Jovin TM. 1993. Scanning force microscopy of circular and linear plasmid DNA spread on mica with a quaternary ammonium salt. *Nucleic Acids Res.* 21: 6004–9
67. Schaper A, Starink JP, Jovin TM. 1994. The scanning force microscopy of DNA in air and in n-propanol using new spreading agents. *FEBS Lett.* 355: 91–95
68. Schellman JA. 1974. Flexibility of DNA. *Biopolymers* 13:217–26
69. Schutz H, Reinert KE. 1991. DNA-bending on ligand binding: change of DNA persistence length. *J. Biomol. Struct. Dyn.* 9:315–29
70. Shore D, Langowski J, Baldwin RL. 1981. DNA flexibility studied by covalent closure of short fragments into circles. *Proc. Natl. Acad. Sci. USA* 78: 4833–37
71. Spassky A, Kirkegaard K, Buc H. 1985. Changes in the DNA structure of the lac UV5 promoter during formation of an open complex with *Escherichia coli* RNA polymerase. *Biochemistry* 24:2723–31
72. Steitz TA. 1990. Structural studies of protein-nucleic acid interaction: the sources of sequence-specific binding. *Q. Rev. Biophys.* 23:205–80
73. Stöver T, Köhler E, Fagin U, Wende W, Wolfes H, Pingoud A. 1993. Determination of the DNA bend angle induced by the restriction endonuclease EcoRV in the presence of Mg^{2+}. *J. Biol. Chem.* 268:8645–50
74. Su W, Porter S, Kustu S, Echols H. 1990. DNA-looping and enhancer activity: association between DNA-

bound NtrC activator and RNA polymerase at the bacterial glnA promoter. *Proc Natl Acad Sci USA* 87:5504–8
75. Thompson JF, Landy A. 1988. Empirical estimation of protein-induced DNA bending angles: applications to lambda site-specific recombination complexes. *Nucleic Acids Res.* 16: 9687–9705
76. Thundat T, Allison DP, Warmack RJ, Brown GM, Jacobson KB, et al. 1992. Atomic force microscopy of DNA on mica and chemically modified mica. *Scanning Microsc.* 6:911–18
77. Thundat T, Allison DP, Warmack RJ, Ferrel TL. 1992. Imaging isolated strands of DNA molecules by atomic force microscopy. *Ultramicroscopy* 42–44:1101–6
78. Travers AA. 1989. DNA conformation and protein binding. *Annu. Rev. Biochem.* 58:427–52
79. van der Vliet PC, Verrijzer CP. 1993. Bending of DNA by transcription factors. *BioEssays* 15:25–32
80. Vesenka J, Guthold M, Tang CL, Keller D, Delaine E, Bustamante C. 1992. Substrate preparation for reliable imaging of DNA molecules with the scanning force microscope. *Ultramicroscopy* 42-44:1243–49
81. von Hippel PH, Berg OG. 1989. Facilitated target location in biological systems. *J. Biol. Chem.* 264:675–78
82. Winkler FK, Banner DW, Oefner C, Tsernogl D, Brown RS, et al. 1993. The crystal structure of EcoRV endonuclease and of its complexes with cognate and non-cognate DNA fragments. *EMBO J.* 12:1781–95
83. Woodcock CL. 1994. Chromatin fibers observed in situ in frozen hydrated sections. Native fiber diameter is not correlated with nucleosome repeat length. *J. Cell Biol.* 125:11–19
84. Wu HM, Crothers DM. 1984. The locus of sequence-directed and protein-induced DNA bending. *Nature* 308:509–13
85. Wyman C, Grotkopp E, Bustamante C, Nelson HCM. 1995. Determination of heat-shock transcription factor 2 stoichiometry at looped DNA complexes using scanning force microscopy. *EMBO J.* 14:117–23
86. Yang G, Leuba SH, Bustamante C, Zlatanova J, van Holde K. 1994. Role of linker histones in extended chromatin fibre structure. *Nat. Struct. Biol.* 1: 761–63
87. Yang J, Takeyasu K, Shao Z. 1992. Atomic force microscopy of DNA molecules. *FEBS Lett.* 301:173–76
88. Zalensky AO, Allen MJ, Kobayashi A, Zalenskaya IA, Balhorn R, Bradbury EM. 1995. Well-defined genome architecture in the human sperm nucleus. *Chromosoma* 103:577–90
89. Zenhausern F, Adrian M, ten Heggeler-Bordier B, Emch R, Jobin M, et al. 1992. Imaging of DNA by scanning force microscopy. *J. Struct. Biol.* 108: 69–73
90. Zenhausern F, Adrian M, ten Heggeler-Bordier B, Eng LM, Descouts P. 1992. DNA and RNA Polymerase/DNA complex imaged by scanning force microscopy: influence of molecular-scale friction. *Scanning* 14: 212–17
91. Zinkel SS, Crothers DM. 1987. DNA bend direction by phase sensitive detection. *Nature* 328:178–81
92. Zlatanova J, Leuba SH, Bustamante C, van Holde KE. 1995. Role of the structural domains of the linker histones and histone H3 in the chromatin fiber structure at low-ionic strength: scanning force microscopy (SFM) studies on partially trypsinized chromatin. *SPIE Proc.* 2384:22–32
93. Zlatanova J, Leuba SH, Yang G, Bustamante C, van Holde K. 1994. Linker DNA accessibility in chromatin fibers of different conformations: a reevaluation. *Proc. Natl. Acad. Sci. USA* 91: 5277–80

Annu. Rev. Biophys. Biomol. Struct. 1996. 25:431–59

LIPOXYGENASES: Structural Principles and Spectroscopy

Betty J. Gaffney

Chemistry Department, Johns Hopkins University,
Baltimore, Maryland 21218–2685

KEY WORDS: lipoxygenase, X-ray structure, π-helix, non-heme iron, metalloprotein, c-terminus, EMR, EPR, XAS, NIR CD, MCD

ABSTRACT

Lipoxygenases catalyze the formation of fatty acid hydroperoxides, products used in further biochemical reactions leading to normal and pathological cell functions. X-ray structure analysis and spectroscopy have been applied to elucidate the mechanism of lipoxygenases. Two X-ray structures of soybean lipoxygenase-1 reveal the side chains of three histidines and the COO^- of the carboxy terminus as ligands to the catalytically important iron atom. The enzyme contains a novel three-turn π-helix near the iron center. Spectroscopic studies, including electron magnetic resonance, X-ray absorption spectroscopy, infrared circular dichroism, and magnetic circular dichroism, have been applied to compare lipoxygenases from varied sources and with different substrate positional specificity.

CONTENTS

1056–8700/96/0610–0439$08.00

INTRODUCTION: THE LIPOXYGENASE FAMILY OF ENZYMES

The first X-ray structure of a lipoxygenase (5) and the structure of the catalytic site (45) of the same lipoxygenase were reported in 1993. Solutions of other lipoxygenase structures are in progress with the use of X-ray analysis (76, 79, 81) and homology modeling (62). The major domain of lipoxygenase consists largely of α-helices surrounding a central, long (43 amino acids) helix. Cavities lead to the long helix from two sides and pass close to the catalytic, non-heme iron atom. The fold of the major domain of lipoxygenase is not similar to that of known protein structures and has novel features, including three turns of an apparent π-helix and the carboxy terminus as a ligand to the iron atom (5). Delineation of the protein ligands to iron culminates a long search by spectroscopic techniques to determine the nature of this metal center. Ongoing spectroscopic studies are being used to define the iron center in more detail in terms of exchangeable ligands and catalytic intermediates. These results provide the motivation for reviewing the implications of the lipoxygenase structure. Before the principles exhibited by the lipoxygenase structure are discussed, the history and rationale for considering known lipoxygenases as a family of closely related enzymes are presented.

Crystalline lipoxygenase, in the form of colorless plates, was reported as early as 1947, when formation of crystals from ammonium sulfate solutions was used to improve purification of the enzyme from soybeans (83). In that study, Theorell et al found the molecular weight by sedimentation to be 102 kDa. Although we now know that all lipoxygenases contain one iron atom per molecule, these investigators found only 0.28. The enzyme isolated from soybeans was thought to be similar to the ones responsible for similar activity in other plant preparations. The activity, as it was understood at that time (83), involved oxidation of unsaturated fats, accompanied by bleaching of carotenoids. It was not until 1973–1975 that the presence of one iron per 100 kDa was clearly demonstrated (9, 15, 58, 59) and that changes in iron oxidation state, associated with catalysis, were shown by electron magnetic resonance [(EMR), designated EPR in earlier references for electron paramagnetic resonance] spectroscopy (15, 59). The iron center showed no optical spectrum characteristic of heme, and chemical studies gave no evidence of an iron–sulfur center (58). The long-standing goal (26) of chemical studies has been to determine the nature of the iron center in lipoxygen ases and the involvement of oxidation and reduction of the metal in the catalytic mechanism.

Today, interest in lipoxygenases is fueled by the discoveries, during 1967–1981, of animal lipoxygenases and, more recently, of the involvement of products derived from plant and animal lipoxygenase action in normal and pathologic cell functions (67, 72). Research on formation of prostaglandins from arachidonic acid led to discoveries of animal lipoxygenases from, among others, rabbit polymorphonuclear leukocytes (46), human neutrophils (29a), human platelets (32, 51), and rabbit reticulocytes (84). Many other lipoxygenases have been found since 1981. The first sequence of a lipoxygenase was reported in 1987 (71). Now, more than 30 sequences are known for these enzymes (summarized in 62 and 80), and several genomic sequences have also been reported (13 and references cited therein).

Evidence That Lipoxygenases Are a Family of Enzymes

The evidence is summarized as follows (see also 77):

1. Antarafacial removal of hydrogen and addition of oxygen is exhibited by the enzymes in conversion of (1(Z),4(Z))-double bonds in unsaturated fatty acids to a (1,3 (E,Z))-dienyl-5-hydroperoxy system (39).
2. Pairwise sequence identity is 21–27% between plant and animal lipoxygenases, 43–86% among plant sequences, and 39–93% among animal sequences (62).
3. Residues that provide ligands to the catalytic iron are conserved (13, 67, 80).
4. Genomic structures of several animal lipoxygenases are similar (13 and references therein).

In spite of the similarities indicated above, members of the lipoxygenase family of enzymes differ in the regiospecificity of the reaction they catalyze. In forming lipoxygenase products with arachidonic acid as substrate, there are six possible placements of the hydroperoxy group in the product (carbons 5-, 8-, 9-, 11-, 12-, and 15-). Two stereochemical courses could be taken at each position, thereby making a total of 12 hydroperoxide products possible from arachidonic acid. In humans, enzymes that specifically carry out oxidation at one of three positions are well characterized. The products of these enzymes are 5-D_{S-}, 12-L_{S-}, and 15-L_{S-} hydroperoxides. The stereochemical nomenclature for substituents on fatty acids (39) is such that the 5*S*-product actually

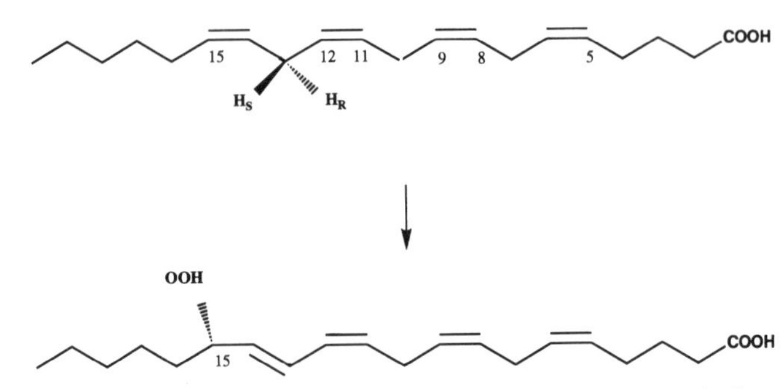

Figure 1 The stereochemical course of the reaction of SBL-1 or reticulocyte 15-LO with arachidonic acid.

has the opposite absolute stereochemistry of that for the 12*S*- and 15*S*-products, so a combination of the D/L Fischer convention with the *R*/*S* nomenclature is used to avoid these complications. The initial removal of hydrogen is antarafacial to the addition of oxygen. Throughout this review, animal lipoxygenases are referred to as 5-LO, 12-LO, and 15-LO. In plants, the substrates are more often 18-carbon chains, so the numbering is different. For simplicity, the major soybean enzyme is referred to as SBL-1, to distinguish it from the other isoforms in soybeans that are designated SBL-2 and SBL-3, for example. (1). Figure 1 shows the stereochemical course of 15-lipoxygenase reactions.

Lipoxygenases also differ in molecular weight. The plant enzymes (90–100 kDa) (71) have an *N*-terminal sequence that is missing in animal lipoxygenases (65–75 kDa) (see references in 13, 62, 80). This region of the plant enzymes is omitted in analysis of sequence identity (62).

BRIEF OVERVIEW OF THE BIOLOGICAL ROLES OF LIPOXYGENASES

Both plants and animals have 5- and 15-lipoxygenases. In each case, there is a cascade of products that results from further enzymatic conversions of the hydroperoxides. In animals, lipoxygenases carry out the first step in the so-called arachidonic acid cascade (67). The branch of this cascade, beginning with 15-LO, leads to lipoxins, whereas the 5-LO branch leads to 5,6-epoxy-leukotrienes. 5-LO activity has been the most studied from the standpoint of pharmacology, because the leukotriene products are involved in a variety of inflammatory responses, including neutrophil chemotaxis, vascular permeability, and smooth muscle contraction, in humans (67). Two reports of mice in which the gene for 5-LO has been inactivated have appeared recently, and the responses of the mice to challenged states were found to be somewhat abnormal (20, 31). In contrast, Nassar et al (47) suggested that animal 15-LO products act as anti-inflammatory agents. This suggests that the contrasting effects of the 5- and 15-LO pathways may be to regulate the extent and magnitude of inflammatory reactions in humans. It is particularly important, therefore, to develop drugs that are specific for one class of the enzyme and not another (44). Ford-Hutchinson et al (19) recently reviewed the pharmacology and cell biochemistry of 5-LO. Novel inhibitors discussed in that article include those that bind not to 5-LO but to the 5-LO activating protein (FLAP). In addition, articles on advances in studies of arachidonic acid metabolism have appeared in recent monographs (29, 67).

The level of current interest in plant lipoxygenases is also high, because the enzymes start some of the pathways for plant self-defense and signaling mechanisms; these subjects are the basis of papers from a 1994 colloquium (66). Seed industries have an interest in inhibiting lipoxygenase products that contribute to rancidification of oils, but some lipoxygenase activity is thought beneficial in bread flour. One strain of commercial soybean has been genetically altered so that the binding site for the catalytic iron is destroyed (87). Although the mutant protein can be expressed in *Escherichia coli*, the inactive protein does not appear at all in the mature soybeans. It will be of interest in the future to learn whether the defense mechanisms are altered in plants or seeds that contain this defective lipoxygenase gene (82).

LIPOXYGENASE STRUCTURAL PRINCIPLES

Overall Structure

Among single-chain protein structures, the lipoxygenase structure is quite large [839 amino acids (71, 80)]. The overall fold has two domains: an N-terminal β-barrel associated with one helix and a major domain composed of 22 helices and 8 β-strands (5). One problem that was solved by obtaining a lipoxygenase structure concerned the difference in molecular weights between the plant (~90 kDa) and the animal enzymes (~70 kDa). The first 146 amino-terminal amino acids of soybean L-1 form an eight-stranded β-barrel. Most of this region is missing from animal lipoxygenase sequences (62, 71, 80). This domain in L-1 may extend farther than the first approximately 200 amino acids, because one α-helix and a large unstructured region is more closely associated with the β-barrel than with the rest of the structure.

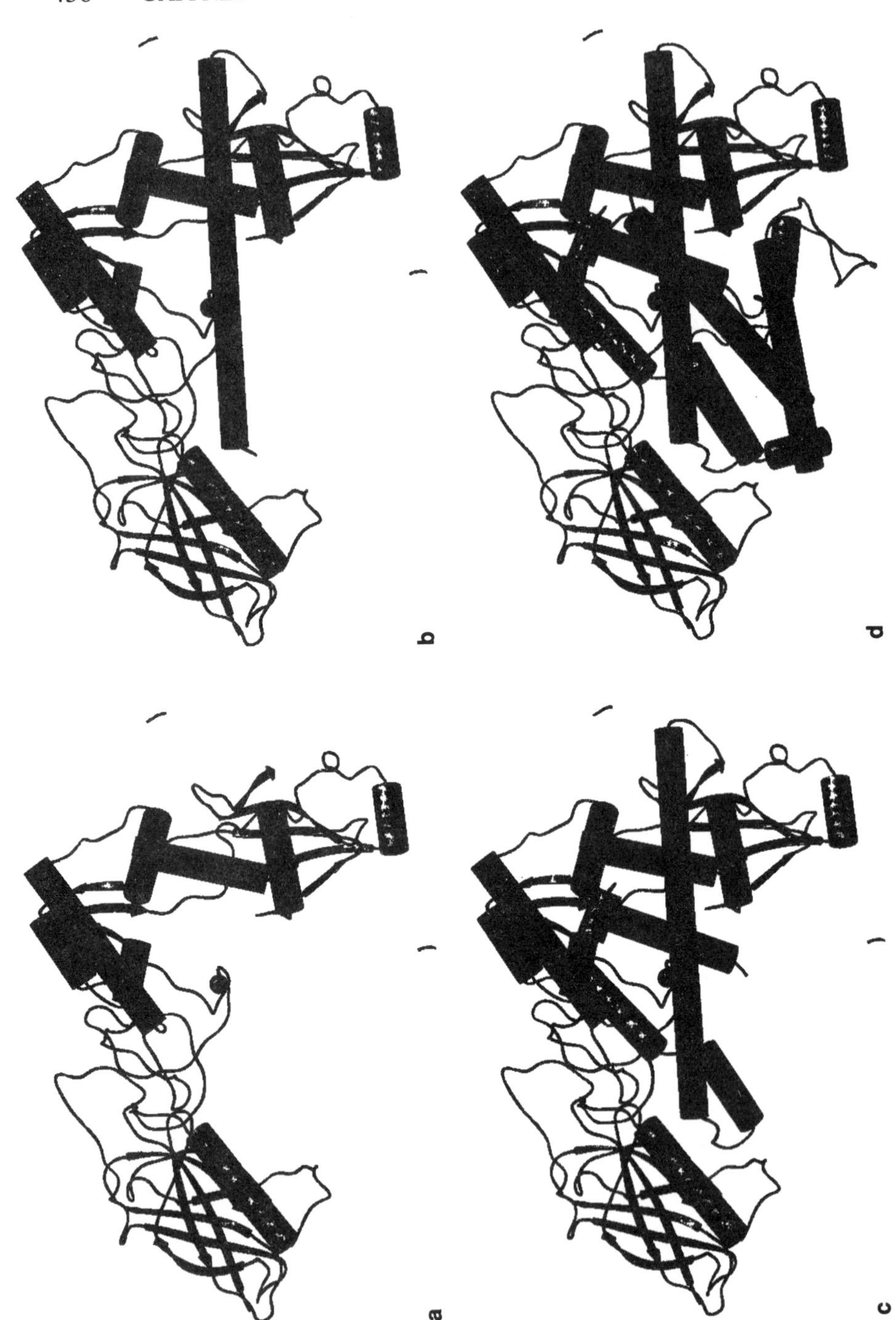
a
b
c
d

Because of the complexity of the lipoxygenase structure, the buildup of the structure, starting from the N-terminus and proceeding to the nearly completed structure, is shown in four sequential steps in Figure 2 (6). The first panel shows the β-barrel at the left, together with structure in the large domain through helix 7. For reference, the iron atom is shown as a sphere in the center. [Most of the structure shown in the first panel (through the fourth helix) can be removed from the intact protein by proteolysis to leave a catalytically active fragment (63).] The second and third panels extend the structure. Panel four extends the structure as far as the second longest helix, helix 18. Helix 18 crosses helix 9 at the iron-binding site.

The complete structure, including the remaining region of helices -19 to -23, together with a diagram giving the numbering of the helices is shown in Figure 3 (23). This figure includes space-filling representations to highlight significant regions of the structure. A funnel-shaped cavity (cavity I) (5), which extends from the surface at the bottom of the molecule toward the buried iron atom, is shown with light gray side chains. A second cavity (cavity II) (5), which extends from the surface on the right, midpart of Figure 3, past the iron site, and up toward helices -11 and -21, is depicted in medium gray. Labeled residues L480 and M341 (light gray) appear to block the entrance to this cavity. The sites of the four sulfhydryl side chains, some of which have been chemically modified, are also given (dark gray). Sulfhydryl derivatives used in the X-ray structurc analysis included C492 labeled with mercury dicyanide, C127 labeled with the same reagent or with mersalyl, and C357 labeled with one of the above-mentioned two reagents or with potassium cyanoaurate (5). C679 in crystalline SBL-1 was not reactive with these heavy atom reagents. Reaction of C492 with methylmercury hydroxide was used in the structure determination by Minor et al (45). Chemical studies have shown that two of the sulfhydryls of SBL-1 can be modified without loss of activity (32b).

The Non-Heme Iron Site

The amino acid side chains that form ligands to iron have been identified from the X-ray structure analyses as three histidines, with $N\epsilon$-

Figure 2 Regions of the lipoxygenase structure are shown with arrows as β-sheets and tubes as α-helices. The sequence of figures shows how the structure progresses from N-terminus through helix 18. (*Upper left*) N-terminus through residue 468; (*upper right*) structure is extended to residue 519 and includes helix 9; (*lower left*) structure is extended to residue 576; (*lower right*) structure is extended to residue 703, thereby giving the structure through helix 18. See Figure 4 for helix numbering. Redrawn from Reference 6.

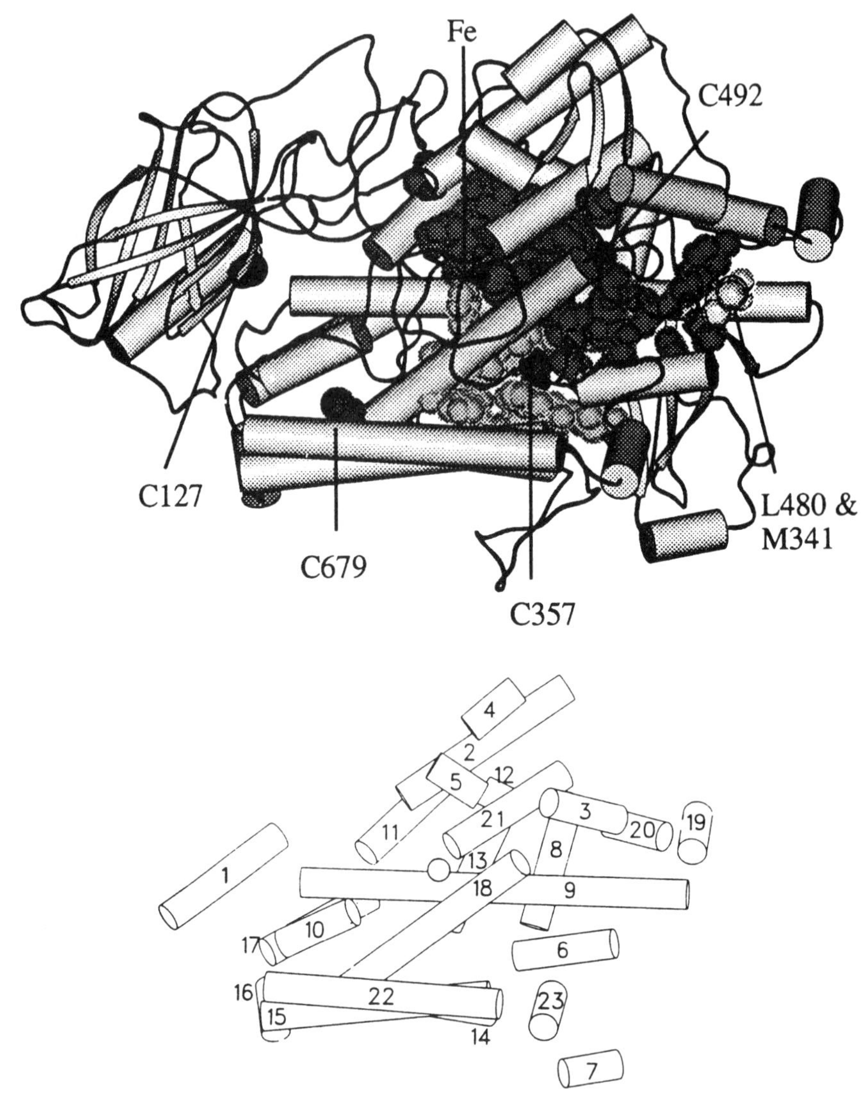

Figure 3 The complete lipoxygenase structure is shown in the upper figure with space-filling representations of side chains that line the cavities that lead to the non-heme iron atom. Light gray = cavity I; medium gray = cavity II. The lower figure gives the numbering of the α-helices. From Reference 23.

ligation, and one oxygen of a carboxyl (5, 45). This carboxyl group is not a side chain but is the COO^- of the carboxy terminus. The terminal COO^- has not been identified previously as a ligand to metal in proteins. The carboxy terminal amino acid is isoleucine in all sequenced lipoxygenases. Of the mutations at the corresponding residue in murine 12-LO, only I663V gave near native activity; I663S and I663N gave 15 and 8% of native activity (13). The group of iron ligands in SBL-1 is chemically the same as the ones in iron superoxide dismutase (4, 81a), although the carboxyl is not the c-terminus in superoxide dismutase and the details of the ligand symmetry appear somewhat different. There are also differences between the two structures of SBL-1 (5, 45). A fifth side chain, asparagine 694, is close enough to be considered an iron ligand in the structure from crystals prepared in polyethylene glycol (pH 5) (45), although it is at least 3 Å from the metal in the structure from crystals in a mixture of salts at high concentration (pH 7) (5). The residue corresponding to this position in some 12- and 15-LOs is histidine. Spectra of these 15-LOs show some differences from those of SBL-1 (see section on *Spectroscopic Studies of Lipoxygenase Structure*). In homology modeling studies based on the SBL-1 structure, two possible conformations, which differ by approximately 90° in χ-1, for this histidine side chain in the modeled human 15-LO (62) were evaluated. Both conformations can be accommodated in the modeled structure, but the δ-nitrogen in the model is predicted to be the iron ligand in the structure with histidine oriented toward iron (62).

Figure 4 gives a view of the iron atom and the side chains of ligands to it from the same perspective as that for Figures 2 and 3. Histidines 499 and 504 lie behind the iron atom on helix 9, whereas the COO^-

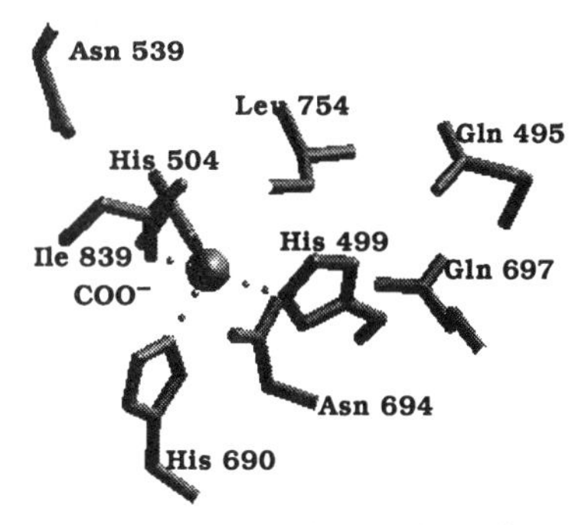

Figure 4 A view of the side chains near the non-heme iron atom (sphere) in SBL-1. The bonds from the iron atom to the terminal COO^- and the three histidine ligands are also shown. The orientation is approximately the same as in Figure 3. Drawing constructed using the program SETOR (18a), with permission.

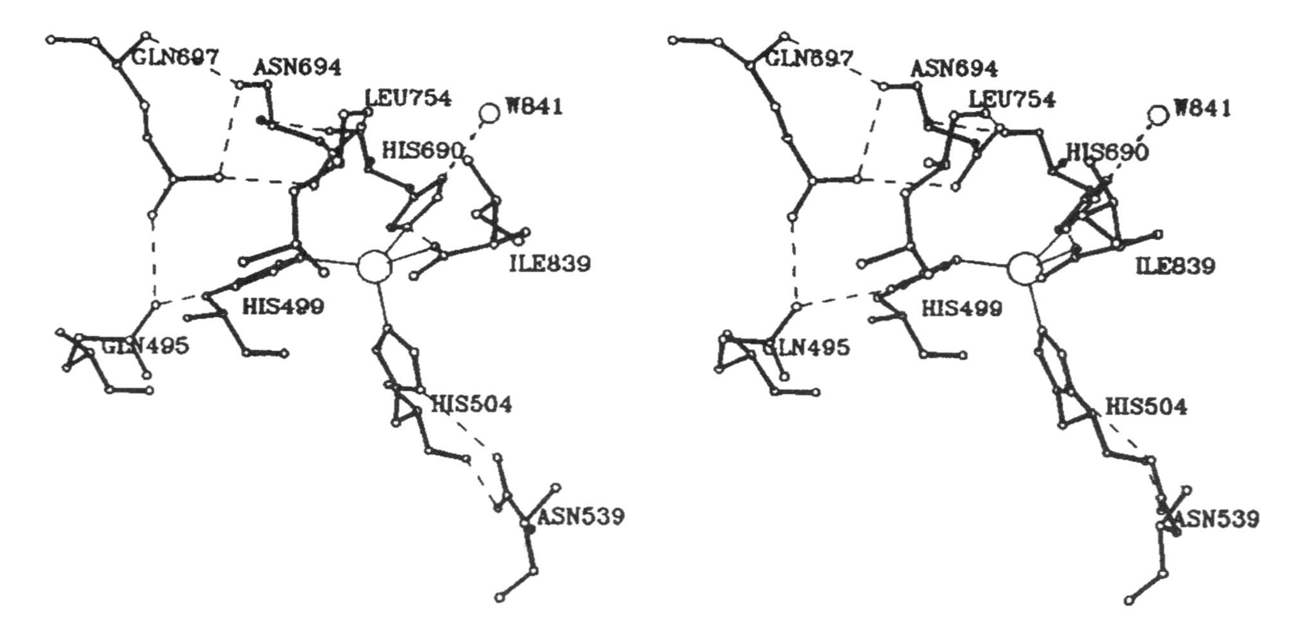

Figure 5 A stereo view of the vicinity near iron from a different orientation than that in Figure 3. Hydrogen bonding near iron (*dashed lines*) is shown. This view also shows the tilt of the His499 ring and the distance between the Asn694 side chain and iron. From Reference 5.

terminus of isoleucine 839 and histidine 690 are in front of the iron atom. Other side chains pictured are involved in hydrogen bonds to the metal ligands. Another view of this region is presented in Figure 5 (5), which displays the hydrogen bonding pattern and has an orientation that shows the space between iron and asparagine 694.

Although the carboxy terminus of lipoxygenase as a metal ligand appears to be a novel structural feature, there are other examples, for instance in hydrogenase (30, 65) and photosystem II (3), in which post-translational modification near the carboxy terminus is essential for assembly and stability of metal binding sites. It is not known whether iron binding to lipoxygenases is a reversible equilibrium; however, the metal in SBL-1 is very tightly bound (58). 5-Lipoxygenase prepared in an expression system incorporates varied levels of metal, and there is evidence that oxidative inactivation of this protein is accompanied by loss of iron (55). These experiments may indicate that iron is bound more weakly in 5-LO than in SBL-1. Loss of iron from reticulocyte 15-LO has also been reported.

A recent example of the *N*-terminus of a protein as a metal ligand was found in cytochrome *f* (42). This ligation imposes a requirement for cleavage by signal peptidase before folding can be complete. Further, this requirement may be a biochemical strategy for maintaining an unfolded state during biosynthesis (42). *C*-terminal ligation to metal can also be a strategy for folding the protein after biosynthesis is complete. Because the carboxy-terminus probably has the only charge associated with the metal ligands in resting lipoxygenase, charge compensation may be an important aspect of assembly of this metal site. It is not yet known whether the carboxy terminus is a metal ligand in those proteins mentioned above in which *C*-terminal processing is important for assembly of the metal site (30, 65).

A Three-Turn π-Helix

Five conserved histidines are found in a short sequence in all but one of the lipoxygenases sequenced (13, 80).When Shibata et al (71) reported the first lipoxygenase sequence, they suggested that some of these histidines might be iron ligands. Results of mutagenesis show that only two of these residues, H499 and H504, are required for enzyme activity in the SBL-1 sequence (80). (Comparisons with other sequences are also given in this reference.) The i and i + 5 spacing of these residues is inconsistent with an α-helical arrangement if both histidines are iron ligands. In fact, these residues are located on the long helix 9, and this helix expands at the metal binding site, as shown

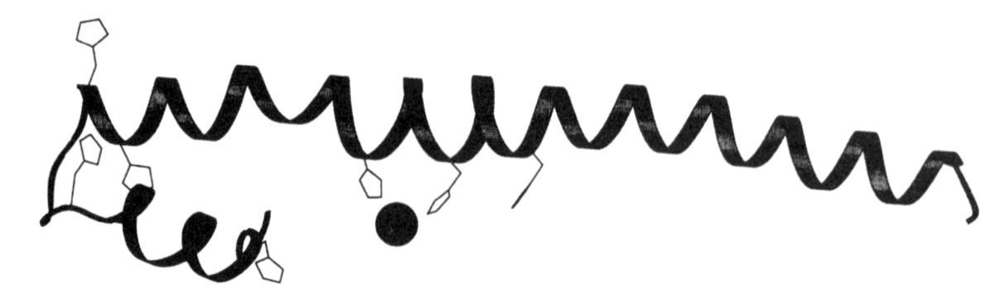

Figure 6 An expanded view of long helix 9, turn, and short helix 10 in SBL-1. Note the expanded, π-helical region. The side chains of the His residues in this region are shown, and the iron atom is represented as a sphere. The left-right orientation is the same as in Figures 2 and 3 (N-terminal end of helix 9 on the right). However, the figure has been rotated about a horizontal axis, relative to the orientation in Figures 2 and 3, to give a better view of the His side chains. Redrawn from Reference 6.

in Figure 6 (5). Most of the carbonyls of the peptide backbone in this region have the same orientation as in an α-helix, and there are eight hydrogen bonds with i to i + 5, instead of i to i + 4 spacing. This is the arrangement of hydrogen bonding in a π-helix. Lipoxygenase has a 14 amino acid stretch, between residues 493 and 506, of π-helix in this region and a shorter, one-turn stretch at H690, where a third histidine-iron bond is located (5). In the 1950s, Low & Baybutt (40) predicted the possible existence of π-helices, and several investigators debated this issue (16, 52, 53). An example of a π-helix with more than one turn has not been reported before in a protein, but a left-handed π-helix with multiple turns has been reported in the synthetic polymer poly(β-phenethyl) L-aspartate (68).

From 1950 to 1952, there was active discussion of possible stable helical arrangements of polypeptides that did not have an integral number of residues per turn. This discussion had a background of 15 years of papers on the principles of protein structure by Huggins, Bragg, Kendrew, Perutz, Pauling, and others. During this period, Pauling and coworkers proposed the α- and γ-helices and evaluated many others (16, 52, 53). Low & Baybutt (40) noted that another helix also fit the criteria for helices laid out by Pauling et al (53) and named it the π-helix. This postulated helix has 4.4 residues per turn and a pitch of 5 Å. Low & Baybutt (40) cited a personal communication from Pauling in which he criticized the π-helix on the basis that it would have a hole down the center. The calculated N-N van der Waals contacts across the middle of the helices are 3.0 Å for a π-helix and 3.2 Å for the γ-helix, compared with 2.9 Å for an α-helix (16). Donohue (16) suggested that water molecules could fill the void in γ-helices but not in π-helices.

Donohue (16) calculated the energetics of π-helices and other helices

Table 1 Helical polypeptide characteristics[a]

Helix name	Atoms in H-bonded ring	H-bond spacing	Residues per turn	Axial translation per residue (Å)	Calculated instability per mole per residue (kcal)
α_{II} ribbon	7		2.2	2.75	0.5
—	8	—	not possible	—	>15.0
3_{10}	10	i → i + 3	3.0	2.00	1.0
—	11	—	not possible	—	>5.0
α	13	i → i + 4	3.6	1.50	0.0
δ	14	i → i − 4	4.3	1.20	2.4
π	16	i → i + 5	4.4	1.15	0.5
γ	17	i → i + 6	5.1	0.98	2.0

[a] From Reference 16.

(Table 1). The calculated stability of the π-helix is only 0.5 kcal/mol less than the α-helix, whereas the 3_{10} helix and the γ-helix are 1.0 and 2.0 kcal/mol less stable. Factors included in the calculation were nonplanarity of the peptide groups, deformation of other single bonds, and nonlinearity of hydrogen bonds. These estimates do not include van der Waals forces or the energies of hydrogen bonds, but ranges of these were discussed by Donohue.

Some irregularities in the lipoxygenase π-helix may be adjustments to maintain a normal protein packing density. Figure 7 shows a quadrant of the Ramachandran plot for the π-helical region compared with that for all of SBL-1. Predicted angles for a π-helix are ϕ=-57, ψ=-70; those for an α-helix are ϕ=-57, ψ=-47; and those for a 3_{10} helix are ϕ=-49, ψ=-26. The rise per residue is 1.15, 1.50 and 2.0 Å for π-, α- and 3_{10}-helices, respectively. Thus, 14 residues of π-helix are shorter along the helix axis than 11 residues of α-helix. The three most abnormal sets of ϕ and ψ in helix 9 of SBL-1 are labeled on the figure. Two of the residues with abnormal angles, M497 and T503, are involved in bifurcated H-bonds from carbonyl oxygens. In addition, residues S498 and N502 have the backbone carbonyl bent out and not participating in a normal H-bond along the helix (S498: ϕ=-63, ψ= -37; N502: ϕ=-65, ψ=-19). This raises the question of whether residues 493–506 in the SBL-1 structure should be regarded as a π-helix or instead form an a-helix with two insertions (32a, 35a).

In the context of this discussion of the π-helical region of the lipoxygenase structure, a mechanistic proposal (62a) based on mutational studies of the first histidine ligand to iron is interesting. Histidine 368

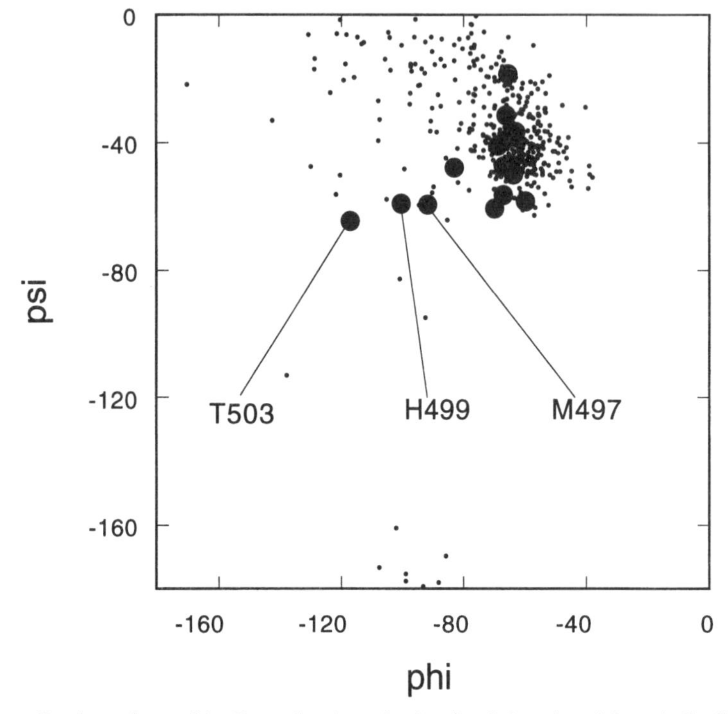

Figure 7 A quadrant of the Ramachandran plot for ϕ and ψ angles of the π-helix (*large dots*) is compared with these angles for all of SBL-1 (*dots*).

in human 5-LO corresponds to H499 in SBL-1. When the mutations H368N, H368Q, or H368S were made in 5-LO, the proteins still could incorporate iron, although the mutated proteins were inactive in catalysis. These results were interpreted to mean that H368 is a replaceable ligand that can be substituted by something else in catalytic intermediates (62a). H499 is uniquely positioned in the SBL-1 structure between helices -9 and -18 and is part of a hydrogen bonding network that involves both helices. It is also tilted from the normal arrangement in which the Fe-N bond is in the plane of the histidine ring. Were the H499 side chain to move substantially, the two cavities in the structure could become connected.

Cavities and Positional Specificity

Before structures were obtained for lipoxygenase, an analysis of sequence differences between 12- and 15-LOs led to a proposal that the 12-/15-positional specificity was governed by residues among the four

Table 2 Residues lining two cavities in soybean lipoxygenase-1[a]

Cavity I:
C357, V358, I359, R360, D408, Y409, I412, Y493, M497, S498, **H499,** L501, N502, T503, V570, N573, W574, V575, D578, Q579, L581, D584, K587, **R588,** Y610, W684, L689, **H690** and V693
Cavity II:
T259, **W340, F346,** E349, M350, G353, V354, **N355,** V358, L407, Y409, L480, **K483,** I487, **D490,** S491, H494, Q495, L496, **H499,** W500, **H504, I538,** L541, A542, L546, I547, I553, T556, F557, **Q697,** G701, I704, M705, **N706,** R707, P708, T709, Y734, S747, L748, V750, I751, I753, **L754** and **I839.**

[a] Residues conserved in known lipoxygenase sequences, regardless of positional specificity, are given in boldface in the preceding lists.

conserved differences between 12- and 15-LOs (77). Mutagenesis was used to test the proposal, and the triple mutant Q416K/I417A/M418V of human 15-LO was found to give 12-LO and 15-LO products in the ratio of 15:1, compared with the wild-type protein, which gave a 1:9 ratio. These residues correspond to T555, T556, and F558 in SBL-1 (21, 80), and the latter two have been identified as among the residues that line one of the cavities in the structure of SBL-1 (5) (see Table 2). The reverse mutation made in human platelet 12-LO (12) (K416Q/A417I/V418M) converted the 12-LO, with negligible 15-LO activity, to an enzyme 15-LO: 12-LO products in the ratio 1:9 to 1:4. To achieve a ratio of 2:1, all residues between 398 and 429 of platelet 12-LO were changed to those of human 15-LO (12). Other investigators have obtained variable results in mutations in the same region of other 12- and 15-LOs (82). Shen et al (70a) suggested that 12-/15-LO activity may not be encoded in the protein sequence but may be a function of protein folding or posttranslational modification. They examined the specificity of the enzyme resulting from expression of a 15-lipoxygenase gene in macrophages of transgenic rabbits and found that the ratio of 12- to 15-LO products varied from 0.3 to 0.4 in different isolates of these macrophages.

The two cavities in the major domain of SBL-1 have been designated cavity I and cavity II, as noted in Figure 3 (23). The residues lining cavity I are given in Table 2. From this list, we find several other candidates that can modulate the positional specificity of lipoxygenases through mutagenesis. For instance, cavity residues T556 and F557 in SBL-1, the sites of significant mutations in other 15- and 12-LOs (12, 77), are located at the end of helix 12, but helix 21 (residues 741–755) is also close to this region (62), as can be seen in Figure 3. Although mutations that succeed in interconverting 15-LO and 5-LO specificity have not been reported yet (39, 56), the platelet 12-LO sequence has been replaced with 5-LO sequences in the region between residues 399 and 418 (corresponding to residues 538–557 in SBL-1) (39). The mutated proteins had good immunoreactivity but minimal 12-LO activity and no 5-LO activity. These studies were designed to address whether the mechanisms of these two enzymes are similar with respect to substrate binding in cavity II. The opposite stereochemistry (5-D and 15-L) of the products of 5- and 15-LO suggests that the fatty acids may bind in the same manner in both enzymes, except with the carboxyl and methyl ends of the fatty acids in reverse positions in the binding site (39).

An interesting feature of cavity II in SBL-1 is a restriction caused by a salt bridge before the cavity reaches from the surface to the iron center. The salt bridge is formed by D490 and R707 (5), but R707 is not a conserved residue in lipoxygenase sequences. Prigge et al (62) have inspected computer models of other lipoxygenases and have suggested that arginine side chains from neighboring structural elements can occupy the same region as R707 in SBL-1, which implies that the restriction in cavity II is conserved by other arginine to aspartate-490 salt bridges.

SPECTROSCOPIC STUDIES OF LIPOXYGENASE STRUCTURE

Three reviews of non-heme iron proteins that provide background to this section have been published (25, 33, 78).

Characterization

Beginning with demonstrations that the native enzyme gives no significant visible absorption and no EMR signal (15, 59), applications of spectroscopy have been used to examine the SBL-1 structure and function. Addition of one equivalent of 13-L_S-hydroperoxy linoleic acid (HPOD, 13-L_S-hydroperoxy-[*E,Z*]-9,11-octadecadienoic acid) converted the enzyme to a yellow form that gives an EMR signal characteristic of high-spin iron. Magnetic susceptibility studies have since shown that both the ferric and ferrous forms of soybean lipoxygenases are high spin (ferrous form, $S = 4/2$; ferric form, $S = 5/2$) (11, 57). In addition, the level of fluorescence intensity of enzyme preparations is lower for the ferric form than for the ferrous form (18). The ferric enzyme spectroscopic signatures revert to those of the ferrous enzyme with the addition of reducing agents or the anaerobic addition of substrate. A recent addition to spectroscopy of the lipoxygenases is characterization of 5-LO (10). 5-Lipoxygenase exhibited the same spectroscopic properties as SBL-1, with one significant exception: 13-HPOD could oxidize the iron (10) but did not react further to form the purple intermediate (49, 59, 75) that is observed with SBL-1. The purple intermediate is discussed further below. The paper on 5-LO spectroscopy (10) contains references to many of the papers on spectroscopy of lipoxygenase that have been published since the 1970s.

The caveat in most of the spectroscopic studies of lipoxygenases is that frozen samples that have a high concentration of protein are required. There are numerous examples of metal centers in proteins that yield different low-temperature spectra, depending on the solutes present. Two examples in non-heme systems are the manganese cluster in photosystem II (3) and the ferric center in phenylalanine hydroxylase (2, 27, 86). Figure 8 gives an example similar effects in lipoxygenase and illustrates the range of changes possible in the EMR spectra of ferric SBL-1 (24, 25, 73). The well-studied case of the addition of ethanol to SBL-1 samples gives a line shape similar to, but slightly sharper than, the spectrum shown in Figure 8 for imidazole addition (24, 73). Because the SBL-1 spectra are also influenced by the concentration of buffer components, ionic interactions at the surface of the

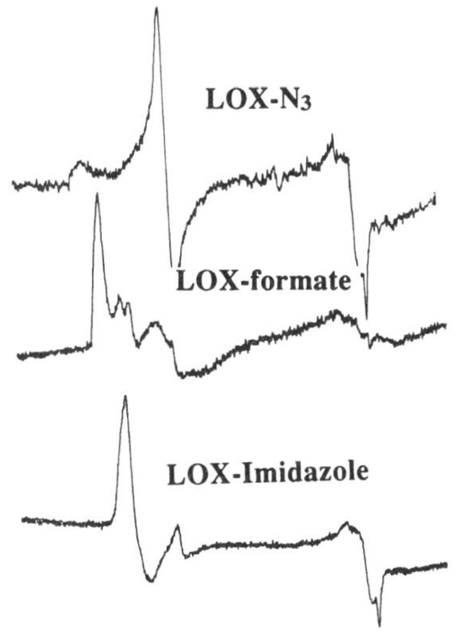

Figure 8 EMR spectra at 4 K of SBL-1 in (*a*) 0.2 M phosphate, 0.5 M sodium azide, pH 7.0; (*b*) 0.2 M sodium formate, pH 7.4; and (*c*) 0.5 M imidazole added to the buffer for *b*. Background signals were not subtracted and contribute to the $g = 2$ region on the right of the spectra. Lipoxygenase was 60 μM. The horizontal axis extends from 0 to 400 mT. From Reference 24.

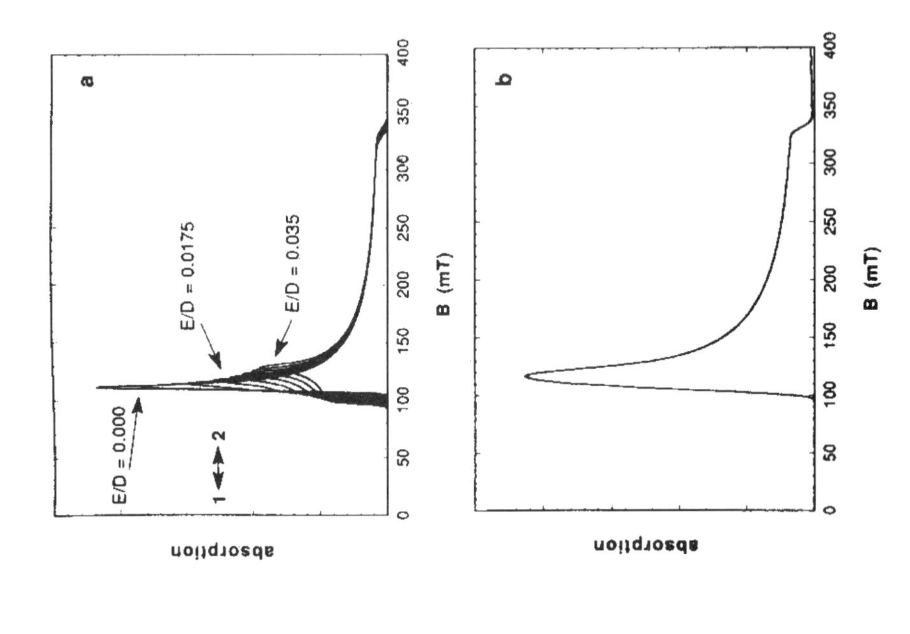
a
b
E/D = 0.000
E/D = 0.0175
E/D = 0.035
1 ↔ 2
absorption
B (mT)

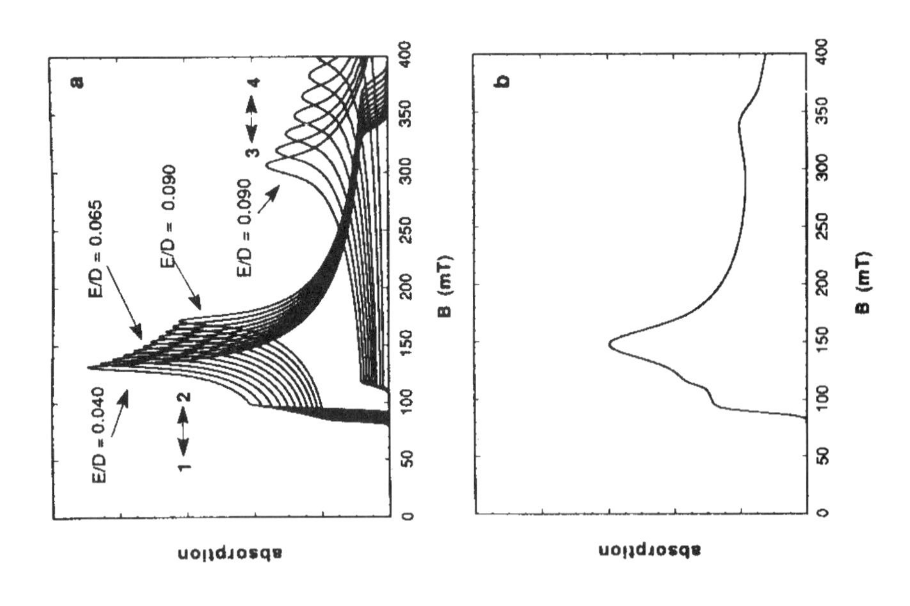
a
b
E/D = 0.040
E/D = 0.065
E/D = 0.090
E/D = 0.090
1 ↔ 2
3 ↔ 4
absorption
B (mT)

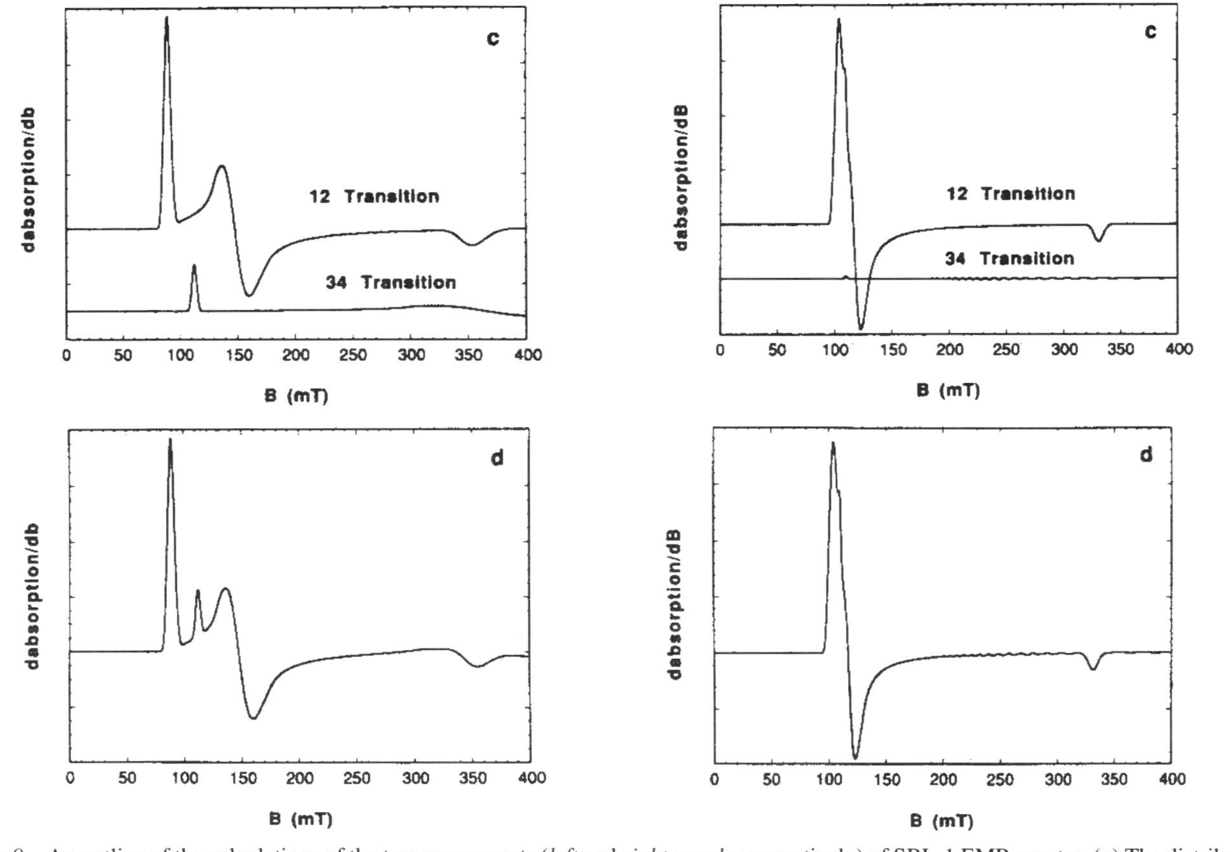

Figure 9 An outline of the calculations of the two components (*left* and *right panels*, respectively) of SBL-1 EMR spectra. (*a*) The distribution of absorption spectra used in the calculation, including absorption between energy levels 1 and 2 and between levels 3 and 4, is shown. (*b*) The signal amplitudes in *a* are multiplied by a Gaussian amplitude factor and added to give the simulated absorption spectrum shown. (*c*) The derivative of the absorption for transitions between each of the pairs of levels is shown separately. (*d*) The sum of the spectra in *c* is shown. From Reference 24.

protein, when it is frozen, may result in altered structural details in the vicinity of the metal center [see Yang & Brill (89) for a discussion of the effects of freezing on hemeproteins]. These effects in SBL-1 studies are not limited to frozen solutions, however, because the effects of glassing agents on solution circular dichroism (CD) and Raman spectra of lipoxygenases have also been reported (61).

Electron magnetic resonance spectra of both SBL-1 (24, 74) and 5-LO (10) have been analyzed quantitatively (90). One study of SBL-1 (24) included the determination of conditions under which the ferric enzyme gave one or the other of two signals but not a mixture of both, e.g. the spectra in Figure 8 for SBL-1 with formate or imidazole (24). Spectra that had been a minor component in other reports were obtained by oxidation of dilute solutions of resting SBL-1 in phosphate or formate buffer (pH 6–8) and rapid removal of fatty acid by-products of the oxidation. Figure 9 (24) shows steps in simulation of the two types of EMR spectra, whereas Figure 10 (24) shows the overlay of the calculated and experimental spectra for the component favored by phosphate or formate buffers. Calculations of the EMR spectra of ferric 5-LO (10) were used to determine the relative proportions of signals at g_{eff} = 4.3 (similar to the SBL-1 spectra with azide in Figure 8) and at g_{eff} values greater than 6. In this context, g_{eff} is an effective g-value that satisfies $h\nu = g_{eff}\beta B$, where h is the Planck constant; ν is the spectrometer frequency; β is the Bohr magneton, and B is the magnetic field at which EMR absorption occurs. Both zero-field and Zeeman terms contribute to g_{eff} (25). The conclusion (10) was that the g_{eff} = 4.3 signal, although of significant intensity, represented less than 3%

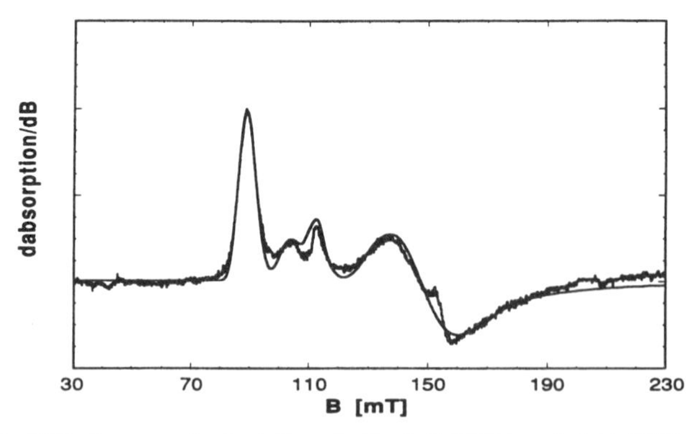

Figure 10 With the use of calculations in the left (93%) and right (7%) columns of Figure 9, a simulated spectrum is overlaid on the low-field portion of the experimental spectrum of SBL-1 in 0.2 M sodium phosphate, pH 7.0. From Reference 23.

of the iron in the sample. The signals at higher g_{eff} values had a shape very similar to the signal from SBL-1. This study, and the ones of His-substituted lipoxygenases (6, 8, 43, 54) that are discussed below, indicate that lipoxygenase EMR signals are related more to the immediate iron ligand environment but do not distinguish between lipoxygenases of different positional specificity (see 38, however, for an exception).

Changes in Iron Ligation

Differences between oxygen and nitrogen ligation to iron are difficult to determine by spectroscopic techniques and usually uncertainties are ± 1 N or O. Nevertheless, the combined techniques of X-ray absorption spectroscopy (XAS) (85), magnetic circular dichroism (MCD) (88, 91), Mössbauer (17, 22), and resonance Raman (of a catecholate complex with SBL-1) (14) together with mutagenesis (mentioned above) were sufficient to identify the nature and location of the potential protein ligands to iron before the lipoxygenase structure was completed. Most spectroscopic studies also suggest the presence of one or two additional oxygen ligands to iron, at least one of which was not apparent in the X-ray structures at 2.6 Å resolution. The earlier results of measurements by XAS, MCD, and resonance Raman have been summarized in several recent papers (54, 69). The discussion here is limited to those experiments that address changes in the coordination environment: an Asn to His substitution in some 12- and 15-LOs and the differences between native ferrous and activated ferric SBL-1.

Changes in Iron Ligation Caused by Amino Acid Substitutions

Two residues in Figures 4 and 5 are not conserved throughout lipoxygenase sequences. In the SBL-1 numbering, these residues are Q495 and N694; in other lipoxygenases, however, these positions may be occupied by E and H, respectively. (In SBL-1, N694 was found at different distances from iron in the two X-ray structures, as noted earlier.) Human and rabbit 15-LOs and some 12-LOs have the histidine for asparagine substitution at the second of these positions (see references in 12, 62, 80) and have been compared with SBL-1 by several forms of spectroscopy (6, 8, 54). Related mutagenesis experiments have shown that when asparagine in this position in 5-LO is changed to glutamine or aspartate (67) or, in another study of SBL-3 (37), to histidine, alanine, or serine, the mutant proteins incorporate iron, but only the histidine mutant has activity similar to the native.

Studies (54) by near-infrared CD (NIR CD) and MCD spectroscopy are the most detailed to date to address differences in iron center coordination for lipoxygenases in which asparagine, at the position corresponding to SBL-1 694, is replaced by histidine. The comparison was made with samples in the resting, ferrous state. Sucrose was used as a glassing agent for the study, because addition of it to buffered SBL-1 or 15-LOs did not alter the solution (3°C) NIR CD spectra. The spectroscopic results were consistent with six-coordination for the 15-LOs and a mixture of five- and six-coordination for SBL-1. The values of the zero-field splitting parameter, D, are of different sign and magnitude in comparing the His-substituted (D negative) and Asn-substituted (D positive) enzymes, in either five- or six-coordinate species. This finding is consistent with substitution of the stronger His-ligand for an O-ligand or a vacant ligand position. In the same work, the results of XAS were average bond lengths, assuming a hexacoordinate species, of 2.16 ± 0.03 Å for both the soybean and the reticulocyte native enzymes. Results of another XAS study of native SBL-1 (69) in a similar buffer environment led to the same conclusion. For comparison, the average of the iron-ligand bond distances deduced from the 2.6 Å X-ray analysis is 2.2 Å (5). Calculations from XAS data based on fewer than six ligands led to shorter average bond lengths (e.g. 2.12 ± 0.01 Å for five coordinate and 2.0 Å for four-coordinate) (54).

Electron magnetic resonance spectroscopy of lipoxygenases in the ferric form has also been used to examine whether there are spectroscopic differences between proteins with Asn or His in SBL-1 position 694. The EMR spectra of rabbit reticulocyte 15-LO (His) have been reported by two groups (6, 8, 43). This lipoxygenase had little EMR intensity in the resting form and EMR spectra characteristic of high-spin ferric iron after oxidative activation with one equivalent of 13-HPOD, similar to the results obtained with SBL-1 (24, 59, 74) and 5-LO (10) (see above). Electron magnetic resonance spectra of reticulocyte 15-LO were somewhat varied, however, and suggested a broader distribution of spectral parameters characterizing the spectra of the reticulocyte enzyme compared with those for SBL-1. In contrast, EMR spectra of a SBL-3 Asn to His mutant, in the ferric form, resembled those of wild-type ferric SBL-3 more than they did the reticulocyte proteins (37).

Changes Between Ferrous and Ferric States

The second question addressed by spectroscopic studies is whether there are changes in the iron ligation of activated ferric lipoxygenases compared with the resting ferrous ones. Although an X-ray structure of a ferric lipoxygenase has not been reported yet, XAS spectroscopy has been applied to the question (69, 85), and SBL-1 samples have been used in the studies. One study interpreted differences in data from the two forms of lipoxygenase in terms of replacement of one histidine N-ligand with an O-ligand (85). The results of the more recent study (69) were interpreted in terms of six-coordinate ferric iron with one short bond (1.88 Å), which was not present in the ferrous enzyme sample, and five others with an average bond length of 2.11 Å. The observation of a short bond for ferric SBL-1, but not for the ferrous form, is consistent with a water ligand in ferrous SBL-1 and ionization of this ligand in the ferric enzyme, thereby giving a hydroxyl ligand with the short bond length. The existence of water coordination to iron in ferric SBL-1 was suggested earlier by broadening of EMR signals by ^{17}O-water (48).

RECENT HIGHLIGHTS OF MECHANISTIC STUDIES

A future goal is to relate lipoxygenase structure to mechanism. Some of the insights and unresolved questions regarding the mechanism of lipoxygenases are summarized briefly here.

The current working model of the mechanism of lipoxygenase has the features given in Figure 11. The outer cycle, in which a substrate radical dissociates, has important consequences. Lipoxygenases are inhibited by reducing agents that consume the hydroperoxide, P, and thus trap the enzyme in the ferrous form, E (10, 19, 36). Recently, unusually large kinetic isotope effects have been observed in the SBL-1 reaction (28, 28a, 34), and Figure 11 would have to be modified to be consistent with these effects. The kinetic constants have been found to be a function of temperature, pH, viscosity, substitution of deuterium for hydrogen in the substrate, and substitution of solvent water by deuterium oxide (28a).

Other tests of the mechanism summarized in Figure 11 have focused on two questions: Does lipoxygenation proceed by a free radical mechanism? And, is there involvement of the Δ^9-double bond as well as involvement of the Δ^{12}-bond?

Many investigators have addressed the question of whether lipoxygenation proceeds by a free radical mechanism. Because free radicals are so easily formed by auto-oxidation of polyunsaturated fatty acids (60), it has been challenging to demonstrate that they are true catalytic intermediates of lipoxygenase. In reactions with lipoxygenase, low concentrations of oxygen favor observation of radicals derived from linoleic acid or other substrates. With respect to Figure 11, therefore, either

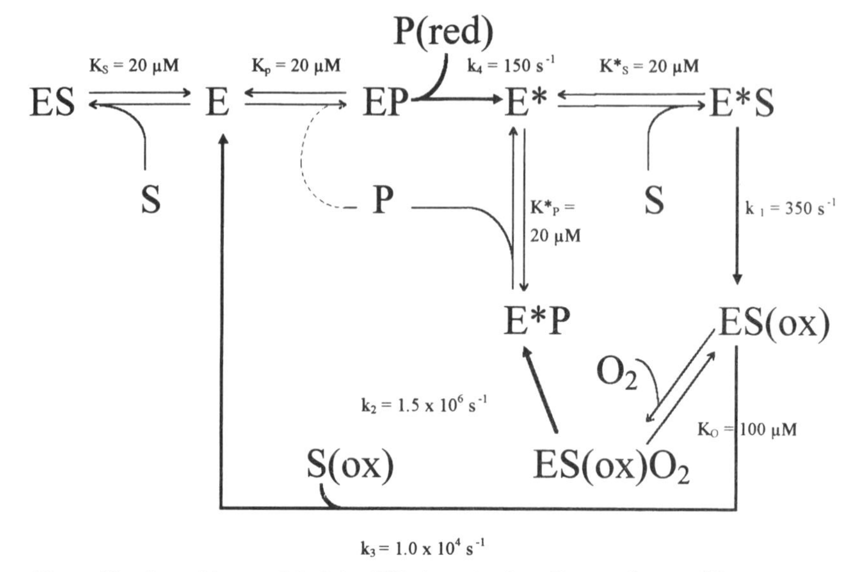

Figure 11 A working model of the SBL-1 mechanism. Ferrous forms of lipoxygenase are designated E and ferric forms are E*. P = product hydroperoxide, and S = substrate. Numerical values shown are for SBL-1 and are taken, largely, from Schilstra et al (70). Kinetic constants are considerably different for rabbit reticulocyte 15-LO (41). Adapted from References 41 and 70.

ES(ox) or S(ox) radicals would build up if conversion to ES(ox)O_2, and then to E*P, is diminished by low oxygen concentrations.

Recent spectroscopic studies of 5-LO provide important new insight into the mechanism regarding formation of a purple intermediate in the lipoxygenase mechanism (10). The oxidation of iron in native SBL-1 is specific for a fatty acid hydroperoxide and cannot be achieved using hydrogen peroxide (59). In a report on 5-LO, Chasteen et al (10) found that this reaction is not specific with respect to the position of the peroxyl group on the fatty acid, as 13-HPOD oxidizes both 15- and 5-LOs. Addition of one equivalent of 13-HPOD to ferric SBL-1, however, gives a purple color (49, 59), whereas the same addition to ferric 5-LO gives no further color over the yellow of the ferric enzyme (10). Assuming that this observation does not simply reflect kinetic differences in the two enzymes, it lends support to the idea that the purple intermediate is a catalytic intermediate in SBL-1 reactions. It also suggests that the structure of the ferric enzyme is different from the ferrous form in ways that affect hydroperoxide binding. Several studies in which free radicals have been detected in purple lipoxygenase are interpreted in terms of conversion of an E*P complex to EP(ox) and subsequent free radical reactions (35, 50). This reasoning might add intermediates to the scheme in Figure 11 that include ternary complexes of lipoxygenase, product, and substrate.

The possibility that the lipoxygenase reaction proceeds by a radical pair reaction has been examined by studies of the D/H kinetic isotope effect for reactions in magnetic fields to 0.2 Tesla (34). No effect was found, thereby ruling out a radical pair lifetime in the range of approximately 10^{-10} to 10^{-6} s. Hwang & Grissom (34) note that relaxation by nearby iron might put the radical pair lifetime outside of this range in lipoxygenase. Typical relaxation times at room temperature for ferric and ferrous iron range from 10^{-12} to 10^{-10} s (1a).

Several indications of involvement of two double bonds in the lipoxygenase mechanism have been noted. Although oxygen is added by SBL-1 to C-13 of linoleic acid, a linoleic acid isomer in which the Δ^9-double bond originally had the Z-geometry [(9Z, 12E)-9,12-octadecadienoic acid] was converted to the thermodynamically less favored E-geometry after enzymatic reaction (21). Also, there is some evidence of 9-peroxyl radicals in purple lipoxygenase (50). Secondary steps in converting 15-hydroperoxy linoleic acid to 14,15-leukotriene A_4 also involve hydrogen abstraction three carbons removed from the site involved in the first step in lipoxygenase catalysis (7). Radicals that contain a 12,13-epoxy linoleate, which is a structure related to leukotriene A_4 (LTA_4), have been identified in SBL-1 reactions by combined EMR, gas chromatography (GC), and mass spectroscopic analysis (35).

ACKNOWLEDGMENTS

I thank JO Wrabl for preparing Figure 11. Support by the National Institutes of Health grant GM36232 is gratefully acknowledged.

Literature Cited

1. Axelrod B, Cheesbrough TM, Laakso S. 1981. Lipoxygenase from soybeans. *Methods Enzymol.* 71:441–51
1a. Banci L. 1993. *Biological Magnetic Resonance*, Vol. 12: *EMR of Paramagnetic Molecules*, ed. LJ Berliner, J Reuben, p. 91. New York: Plenum
2. Bloom LM, Benkovic SJ, Gaffney BJ. 1986. Characterization of phenylalanine hydroxylase. *Biochemistry* 25: 4204–10
3. Boerner RJ, Nguyen AP, Barry BA, Debus RJ. 1992. Evidence from directed mutagenesis that aspartate 170 of the D1 polypeptide influences the assembly and/or stability of the manganese cluster in the photosynthetic water-splitting complex. *Biochemistry* 31:6660–72
4. Borgstahl GEO, Page HE, Hickey MJ, Beyer WF Jr, Hallewell RA, Tainer JA. 1992. The structure of human mitochondrial manganese superoxide

dismutase reveals a novel tetrameric interface of two 4-helix bundles. *Cell* 71:107–18
5. Boyington JC, Gaffney BJ, Amzel LM. 1993. The three-dimensional structure of an arachidonic acid 15-lipoxygenase. *Science* 260:1482–6
6. Boyington JC, Gaffney BJ, Amzel LM, Doctor KS, Mavrophilipos DV, et al. 1995. The X-ray structure and biophysical studies of a 15-lipoxygenase. *Ann. NY Acad. Sci.* 744:310–3
7. Bryant RW, Schewe T, Rapoport SM, Bailey JM. 1985. Leukotriene formation by a purified reticulocyte lipoxygenase enzyme. *J. Biol. Chem.* 260: 3548–55
8. Carroll RT, Muller J, Grimm J, Dunham WR, Sands RH, Funk MO Jr. 1993. Rapid purification of rabbit reticulocyte lipoxygenase for electron paramagnetic spectroscopy characterization of the non-heme iron. *Lipids* 28: 241–4
9. Chan HW-S. 1973. Soya-bean lipoxygenase: an iron-containing dioxygenase. *Biochim. Biophys. Acta* 327:32–5
10. Chasteen ND, Grady JK, Skorey KI, Neden KJ, Riendeau D, Percival MD. 1993. Characterization of the non-heme iron center of human 5-lipoxygenase by electron paramagnetic resonance, fluorescence, and ultraviolet-visible spectroscopy: redox cycling between ferrous and ferric states. *Biochemistry* 32:9763–71
11. Cheesbrough TM, Axelrod B. 1983. Determination of the spin state of iron in native and activated soybean lipoxygenase-1 by paramagnetic susceptibility. *Biochemistry* 22:3837–40
12. Chen X-S, Funk CD. 1993. Structure-function properties of human platelet 12-lipoxygenase: chimeric enzyme and in vitro mutagenesis studies. *FASEB J.* 7:694–701
13. Chen X-S, Kurre U, Jenkins NA, Copeland NG, Funk CD. 1994. cDNA cloning, expression, mutagenesis of c-terminal isoleucine, genomic structure, and chromosomal localizations of murine 12-lipoxygenase. *J. Biol. Chem.* 269:13979–87
14. Cox DD, Benkovic SJ, Bloom LM, Bradley FC, Nelson MJ, et al. 1988. Chatecholate LMCT bands as probes for the active sites of nonheme iron oxygenases. *Biochemistry* 110: 2026–32
15. de Groot JJMC, Veldink GA, Vliegenthart JFG, Boldingh J, Wever R, van Gelder BF. 1975. Demonstration by EPR spectroscopy of the functional role of iron in soybean lipoxygenase-1 *Biochim. Biophys. Acta* 377:71–9
16. Donohue J. 1953. Hydrogen bonded helical configurations of the polypeptide chain. *Proc. Natl. Acad. Sci. USA* 39:470–8
17. Draheim JE, Carroll RT, McNemar TB, Dunham WR, Sands RH, Funk MO Jr. 1989. Lipoxygenase isoenzymes: a spectroscopic and structural characterization of soybean seed enzymes. *Arch. Biochem. Biophys. Res. Commun.* 269:208–18
18. Egmond MR, Finazzi-Agro A, Fasella PM, Veldink GA, Vliegenthart JFG. 1975. Changes in the fluorescence and absorbance of lipoxygenase-1 induced by 13-L_S-hydroperoxylinoleic acid and linoleic acid. *Biochim. Biophys. Acta* 397:43–9
18a. Evans SV. 1993. Hardware-lighted 3-dimensional solid model representations of macromolecules. *J. Mol. Graphics* 11:134–8
19. Ford-Hutchinson AW, Gresser M, Young RN. 1994. 5-Lipoxygenase. *Annu. Rev. Biochem.* 63:383–417
20. Funk CD, Chen X-S, Kurre U, Griffis G. 1995. Leukotriene-deficient mice generated disruption of the 5-lipoxygenase gene. See Ref. 67, pp. 145–50
21. Funk MO Jr, Andre JC, Otsuki T. 1987. Oxygenation of trans polyunsaturated fatty acids by lipoxygenase reveals steric features of the catalytic mechanism. *Biochemistry* 26:6880–4
22. Funk MO Jr., Carroll RT, Thompson JF, Sands RH, Dunham WR. 1990. Role of iron in lipoxygenase catalysis. *J. Am. Chem. Soc.* 112:5375
23. Gaffney BJ, Boyington JC, Amzel LM, Doctor KS, Prigge ST, Yuan SM. 1995. Lipoxygenase structure and mechanism. See Reference 67, pp.11–16
24. Gaffney BJ, Mavrophilipos DV, Doctor KS. 1993. Access of ligands to the ferric center in lipoxygenase-1. *Biophys. J.* 64:773–83
25. Gaffney BJ, Silverstone HJ. 1993. In *Biological Magnetic Resonance*, Vol. 13: *EMR of Paramagnetic Molecules*, ed. LJ Berliner, J Reuben, pp. 1–57. New York: Plenum
26. Gibian MJ, Galaway RA. 1977. Chemical aspects of lipoxygenase reactions. In *Bioinorganic Chemistry*, ed. EE Van Teamelen, pp. 117–35. New York: Academic
27. Glasfeld E, Xia YM, Debrunner P, Caradonna JP. 1993. *Spectroscopic characterization of the active site in phenylalanine hydroxylase*. Presented

at Am. Chem. Soc. Meet., 205th, Denver, CO
28. Glickman MH, Wiseman JS, Klinman JP. 1994. Extremely large isotope effects in the soybean lipoxygenase-linoleic acid reaction. *J. Am. Chem. Soc.* 116:793–4
28a. Glickman MH, Klinman JP. 1995. Nature and rate-limiting steps in the soybean lipoxygenase-1 reaction. *Biochemistry* 34:14077–92
29. Goetzl EJ, Lewis RA, Rola-Pleszczynski, eds. 1994. *Cellular Generation, Transport. and Effects of Eicosanoids. Biological Roles and Pharmacological Intervention. Ann. NY Acad. Sci.* 744:1–340
29a. Goetzl EJ, Sun F. 1979. Generation of unique mono-hydroxy-icosatetraenoic acids from arachidonic acid by human neutrophils. *J. Exp. Med.* 150: 406–11
30. Gollin DJ, Mortenson LE, Robson RL. 1992. Carboxyl-terminal processing may be essential for production of active NiFe hydrogenase in *Azotobacter vinelandii. FEBS Lett.* 309:371–5
31. Goulet JL, Snouwaert JN, Latour AM, Coffman TM, Koller BH. 1994. Altered inflammatory responses in leukotriene-deficient mice. *Proc. Natl. Acad. Sci. USA* 91:12852–6
32. Hamberg M, Samuelsson B. 1974. Prostaglandin endoperoxides. Novel transformations of arachidonic acid in human platelets. *Proc. Natl. Acad. Sci. USA* 71:3400–4
32a. Heinz DW, Baase WA, Zhang X-J, Blaber M, Dahlquist FW, Matthews BW. 1994. Accomodation of amino acid insertions in an alpha-helix of T4 lysozyme: structural and thermodynamic analysis. *J. Mol. Biol.* 236: 869–86
32b. Höhne WE, Kojima N, Thiele B, Rapoport SM. 1991. Lipoxygenases from soybeans and rabbit reticulocytes: inactivation and iron release. *Biomed. Biochim. Acta* 50:125–38
33. Howard JB, Rees DC. 1991. Perspectives on non-heme iron protein chemistry. *Adv. Protein Chem.* 42:199–280
34. Hwang CC, Grissom CB. 1994.J. Unusually large deuterium isotope effects in soybean lipoxygenase is not caused by a magnetic isotope effect. *J. Am. Chem. Soc.* 116:795–6
35. Iwahashi H, Parker CE, Mason RP, Tomer KB. 1991. Radical adducts of nitrosobenzene and 2-methyl-2-nitrosopropane with 12,13-epoxylinoleic acid radical, 12,13-epoxylinoleic acid radical and 14,15-epoxyarachidonic acid radical *Biochem. J.* 276:447–53
35a. Keefe LJ, Sondek J, Shortle D, Lattman EE. 1993. The alpha aneurism: a new structural motif in an insertion mutant of staphylococcal nuclease. *Proc. Natl. Acad. Sci. USA* 90:3275–9
36. Kemal C. 1987. Reductive inactivation of soybean lipoxygenase 1 by catechols: a possible mechanism for regulation of lipoxygenase activity. *Biochemistry* 26:7064–72
37. Kramer JA, Johnson KR, Dunham WR, Sands RH, Funk MO Jr. 1994. Position 713 is critical for catalysis but not iron binding in soybean lipoxygenase 3. *Biochemistry* 33:15017–22
38. Kroneck PMH, Cucurou C, Ullrich V, Ueda N, Suzuki H, et al. 1991. Porcine leukocyte 5- and 12-lipoxygenases are iron enzymes. *FEBS Lett.* 287:105–7
39. Kühn H, Schewe T, Rapoport SM. 1986. The stereochemistry of the reactions of lipoxygenases and their metabolites. Proposed nomenclature of lipoxygenases and related enzymes. *Adv. Enzymol.* 88:273–311
40. Low BW, Baybutt RB. 1952. The π-helix—a hydrogen bonded configuration of the polypeptide chain. *J. Am. Chem. Soc.* 74:5806–7
41. Ludwig P, Holzhütter HG, Colosimo A, Silvestrini MC, Schewe T, Rapoport SM. 1987. A kinetic model for lipoxygenases based on experimental data with the lipoxygenase of reticulocytes. *Eur. J. Biochem.* 168:325–7
42. Martinez SE, Huang D, Szczepaniak A, Cramer WA, Smith J. 1994. Crystal structure of chloroplast cytochrome *f* reveals a novel cytochrome fold and unexpected heme ligation. *Structure* 2: 95–105
43. Mavrophilipos DV. 1986. *Characterization of the iron environment of lipoxygenases.* PhD thesis. Johns Hopkins Univ., Baltimore, MD. 139 pp.
44. McMillan RM, Walker ERH. 1992. Designing therapeutically effective 5-lipoxygenase inhibitors. *Trends Pharmacol. Sci.* 13:323–30
45. Minor W, Steczko J, Bolin JT, Otwinowski Z, Axelrod B. 1993. Crystallographic determination of the active site iron and its ligands in soybean lipoxygenase-1. *Biochemistry* 32:6320–3
46. Narumiya S, Salmon JA, Cottee FH, Weatherley BC, Flower RJ. 1981. Arachidonic acid 15-lipoxygenase from rabbit peritoneal polymorphonuclear leukocytes. *J. Biol. Chem.* 256: 9583–92
47. Nassar GM, Morrow JD, Roberts LJ

II, Lakkis FG, Badr KF. 1994. Induction of 15-lipoxygenase by interleukin-13 in human blood monocytes. *J. Biol. Chem.* 269:27631–4
48. Nelson MJ. 1988. Evidence for water coordinated to the active site iron in soybean lipoxygenase-1. *J. Am. Chem. Soc.* 110:2985–6
49. Nelson MJ, Cowling RA. 1990. Observation of a peroxyl radical in samples of "purple" lipoxygenase. *J. Am. Chem. Soc.* 112:2820–1
50. Nelson MJ, Cowling RA, Seitz SP. 1994. Structural characterization of alkyl and peroxyl radicals in solutions of purple lipoxygenase. *Biochemistry* 33:4966–73
51. Nugteren H. 1975. Arachidonic lipoxygenase in blood platelets. *Biochim. Biophys. Acta* 380:299–307
52. Pauling L, Corey RB. 1952. Configuration of polypeptide chains with equivalent *cis* amide groups. *Proc. Natl. Acad. Sci. USA* 38:86–93
53. Pauling L, Corey RB, Branson HR. 1951. The structure of proteins: two hydrogen-bonded helical configurations of the polypeptide chain. *Proc. Natl. Acad. Sci. USA* 37:205–11
54. Pavlosky MA, Zhang Y, Westre TE, Gan Q-F, Pavel EG, et al. 1995. Near-infrared circular dichroism, magnetic circular dichroism, and X-ray absorption spectral comparison of the non-heme ferrous active sites of plant and mammalian 15-lipoxygenases. *J. Am. Chem. Soc.* 117:4316–27
55. Percival MD. 1992. Human 5-lipoxygenase contains an essential iron. *J. Biol. Chem.* 266:10058–61
56. Percival MD, Ouellet M. 1992. The characterization of 5-histidine-serine mutants of human 5-lipoxygenase. *Biochem. Biophys. Res. Commun.* 186: 1265–70
57. Petersson L, Slappendel S, Vliegenthart JFG. 1985. The magnetic susceptibility of native soybean lipoxygnase-1. Implications for the symmetry of the iron environment and possible coordination of dioxygen to Fe (II). *Biochim. Biophys. Acta* 828:81–5
58. Pistorius EK, Axelrod B. 1974. Iron, an essential component of lipoxygenase. *J. Biol. Chem.* 249:3183–6
59. Pistorius EK, Axelrod B, Palmer G. 1976. Evidence for participation of iron in lipoxygenase reaction from optical and electron spin resonance studies. *J. Biol. Chem.* 251:7144–8
60. Porter NA, Weber BA, Weenen H, Khan JA. 1980. Autooxidation of polyunsaturated lipids. Factors controlling the stereochemistry of product hydroperoxides. *J. Am. Chem. Soc.* 102:5597–601
61. Pourplanche C, Lambert C, Berjot M, Marx J, Chopard C, et al. 1994. Conformational changes of lipoxygenase (LOX) in modified environments. *J. Biol. Chem.* 269:31585–91
62. Prigge ST, Boyington JC, Gaffney BJ, Amzel LM. 1996. Structure conservation in lipoxygenases: structural analysis of soybean lipoxygenase-1 and modeling of human lipoxygenases. *Proteins: Struc., Funct. Genet.* In press
62a. Rådmark O, Zhang Y-Y, Hammarberg, Lind B, Hamberg M, et al. 1995. 5-Lipoxygenase: structure and stability of recombinant enzyme, regulation in Mono Mac 6 cells. See Ref. 67, pp. 1–10
63. Ramachandran S, Carrol RT, Dunham WR, Funk MO Jr. 1992. Limited proteolysis and active-site labeling studies of soybean lipoxygenase 1. *Biochemistry* 31:7700–6
65. Rossman R, Sauter M, Lottspeich F, Bock A. 1994. Maturation of the large subunit (HYCE) of *Escherichia coli* hydrogenase 3 requires nickel incorporation followed by C-terminal processing at Arg537. *Eur. J. Biochem.* 220:377–84
66. Ryan CA, Lamb CJ, Jagendorf AT, Kolattukudy PE, eds. 1995. *Proc. Natl. Acad. Sci. USA* 92:4075–205
67. Samuelsson B, Ramwell P, Paoletti R, Folco G, Granström E, et al, eds. 1995. *Advances in Prostaglandin, Thromboxane, and Leukotriene Research,* Vol. 23. New York: Raven. 573 pp.
68. Sasaki S, Yasumoto Y, Uematsu I. 1981. π-Helical conformation of poly (β-phenethyl) L-aspartate. *Macromolecules* 14:1797–801
69. Scarrow RC, Trimitsis MG, Buck CP, Grove GN, Cowling RA, Nelson MJ. 1994. X-ray spectroscopy of the iron site in soybean lipoxygenase-1: changes in coordination upon oxidation or addition of methanol. *Biochemistry* 33:15023–35
70. Schilstra MJ, Veldink GA, Vliegenthart JFG. 1994. Kinetic analysis of the induction period in lipoxygenase catalysis. *Biochemistry* 33:3974–9
70a. Shen J, Kühn H, Petho-Schramm A, Chan L. 1995. Transgenic rabbits with the integrated human 15-lipoxygenase gene driven by a lysozome promoter: macrophage-specific expressison and variable positional specificity of the

transgenic enzyme. *FASEB J.* 9: 1623–37
71. Shibata D, Steczko J, Dixon JE, Hermodson M, Yazdanparast R, Axelrod B. 1987. Primary structure of soybean lipoxygenase-1. *J. Biol. Chem.* 262: 10080–5
72. Sigal E. 1991. The molecular biology of mammalian arachidonic acid metabolism. *Am. J. Physiol.* 260: L13–L28
73. Slappendel S, Aasa R, Malmström BG, Verhagen J, Veldink GA, Vliegenthart JFG. 1982. Factors affecting the line-shape of the EPR signal of high-spin Fe(III) in soybean lipoxygenase-1. *Biochim. Biophys. Acta* 708: 259–65
74. Slappendel S, Veldink GA, Vliegenthart JFG, Aasa R, Malmström BG. 1981. EPR Spectroscopy of soybean lipoxygenase-1: description and quantification of the high-spin Fe(III) signals. *Biochim. Biophys. Acta* 667: 77–86
75. Slappendel S, Veldink GA, Vliegenthart JFG, Aasa R, Malmström BG. 1983. A quantitative optical and EPR study on the interaction between soybean lipoxygenase-1 and 13-L-hydroperoxylinoleic acid. *Biochim. Biophys. Acta* 747:32–6
76. Sloane DL, Browner MF, Dauter Z, Wilson K, Fletterick RJ, Sigal E. 1990. Purification and crystallization of 15-lipoxygenase from rabbit reticulocytes. *Biochem. Biophys. Res. Commun.* 173:507–13
77. Sloane DL, Leung R, Craik CS, Sigal E. 1991. A primary determinant for lipoxygenase positional specificity. *Nature* 354:149–52
78. Solomon EI, Zhang Y. 1992. The electronic structures of active sites in nonheme iron enzymes. *Acc. Chem. Res.* 25:343–52
79. Stallings WC, Kroa BA, Carroll RT, Metzger AL, Funk MO Jr. 1990. Crystallization and preliminary X-ray characterization of a soybean seed lipoxygenase. *J. Mol. Biol.* 211:685–7
80. Steczko J, Donoho GP, Clemens JC, Dixon JE, Axelrod B. 1992. Conserved histidine residues in soybean lipoxygenase: functional consequences of their replacement. *Biochemistry* 31: 4053–7
81. Steczko J, Minor W, Stojanoff V, Axelrod B. 1995. Crystallization and preliminary X-ray investigation of lipoxygenase-3 from soybeans. *Protein Sci.* 4:1233–5
81a. Stoddard BL, Howell PL, Ringe D, Petsko GA. 1990. The 2.1 Å resolution structure of iron superoxide dismutase from *Pseudomonas ovalis. Biochemistry* 29:8885–93
82. Suzuki H, Kishimoto K, Yoshimoto T, Yamamoto S, Kanai F, et al. 1994. Site-directed mutagenesis studies on the iron-binding domain and the determinant for the substrate oxygenation site of porcine leukocyte arachidonate 12-lipoxygenase. *Biochim. Biophys. Acta* 1210:308–16
83. Theorell H, Holman RT, Åkeson Å. 1946. Crystalline lipoxydase. *Acta Chem. Scand.* 1:571–6
84. Thiele BJ, Belkner J, Andree H, Rapoport TA, Rapoport SM. 1979. Synthesis of non-globin proteins in rabbit-erythroid cells. *Eur. J. Biochem.* 96: 563–9
85. van der Heijdt LM, Feiters MC, Navaratnam S, Nolting HF, Hermes C, et al. 1992. X-ray Absorption spectroscopy of soybean lipoxygenase-1. *Eur. J. Biochem.* 207:793–802
86. Wallick DE, Bloom LM, Gaffney BG, Benkovic SJ. 1984. The reductive activation of phenylalanine hydroxylase and its effect on the redox state of the nonheme iron. *Biochemistry* 23: 1295–302
87. Wang WH, Takano T, Shibata D, Kitamura K, Takeda G. 1994. Molecular basis of a null mutation in soybean lipoxygenase-2: substitution of glutamine for an iron-ligand histidine. *Proc. Natl. Acad. Sci. USA* 91: 5828–32
88. Whitaker JW, Solomon EI. 1988. Spectroscopic studies on ferrous nonheme iron active sites: magnetic circular dichroism of mononuclear Fe sites in superoxide dismutase and lipoxygenase. *J. Am. Chem. Soc.* 110: 5329–39
89. Yang A-S, Brill AS. 1991. Influence of the freezing process upon fluoride binding to hemeproteins. *Biophys. J.* 59:1050–63
90. Yang A-S, Gaffney BJ. 1989. Determination of relative spin concentration in some high-spin ferric proteins using E/D-distribution in electron paramagnetic resonance simulations. *Biophys. J.* 51:55–67
91. Zhang Y, Gebbhard MS, Solomon EI. 1991. Spectroscopic studies of the non-heme ferric active site in soybean lipoxygenase: magnetic circular dichroism as a probe of electronic and geometric structure. Ligand-field origin of zero-field splitting. *J. Am. Chem. Soc.* 113:5162–75

SUBJECT INDEX

S

CUMULATIVE INDEXES

CONTRIBUTING AUTHORS, VOLUMES 21–25

CHAPTER TITLES, VOLUMES 20–24

INDEXED BY KEYWORD